HETEROGENEOUS CATALYSIS RESEARCH PROGRESS

HETEROGENEOUS CATALYSIS RESEARCH PROGRESS

MATHIAS B. GUNTHER
EDITOR

Nova Science Publishers, Inc.
New York

For permission to use material from this book please contact us:
Telephone 631-231-7269; Fax 631-231-8175
Web Site: http://www.novapublishers.com

LIBRARY OF CONGRESS CATALOGING-IN-PUBLICATION DATA

Heterogeneous catalysis research progress / Mathias B. Gunther (editor).
p. cm.
Includes bibliographical references.
ISBN 978-1-60456-979-7 (hardcover)
1. Heterogeneous catalysis. I. Gunther, Mathias B.
QD505.H468 2008
541'.395--dc22 2008030512

Published by Nova Science Publishers, Inc. ✦ New York

CONTENTS

PREFACE

Heterogeneous catalysis is a chemistry term which describes catalysis where the catalyst is in a different phase (i.e., solid, liquid and gas, but also oil and water) to the reactants. Heterogeneous catalysts provide a surface for the chemical reaction to take place on. In order for the reaction to occur, one or more of the reactants must diffuse to the catalyst surface and adsorb onto it. After reaction, the products must desorb from the surface and diffuse away from the solid surface. Frequently, this transport of reactants and products from one phase to another plays a dominant role in limiting the reaction rate. Understanding these transport phenomena and surface chemistry such as dispersion is an important area of heterogeneous catalyst research. Catalyst surface area may also be considered. Mesoporous silicates, for example, have found utility as catalysts because their surface areas may be in excess of 1000 m2/g, which increases the probability that a reactant molecule in solution will come in contact with the catalyst surface and adsorb. If diffusion rates are not taken into account, the reaction rates for various reactions on surfaces depend solely on the rate constants and reactant concentrations. Asymmetric heterogeneous catalysis can be used to synthesize enantiomerically pure compounds using chiral heterogeneous catalysts. The field is of great industrial and environmental importance

Chapter 1 - The purification of automobile exhaust gases such as nitrogen oxides (NO_x), CO, and hydrocarbons (HCs), which cause photochemical smog and acid rain respectively leading to respiratory diseases and ecological damages, has become one of the main objectives of environmental catalysis. Perovskite-typed mixed oxides are regarded as promising materials for catalytic removal of those exhaust pollutants due to their excellent redox properties and nonstoichiometric structure and are thought as ideal but yet cheap substitutes for the commercial noble metal autocatalysts. The traditional ceramic method involving high-temperature calcination allows only prepare perovskites having low specific surface areas which limit their practical applications as catalysts. By contrast, a novel technique designated as reactive grinding avoids this thermal treatment and the crystallization proceeds at nearly ambient temperature thus yielding extraordinary high surface areas.

The oxygen vacancies, overstoichiometric, atomic and lattice oxygen of series of Co-Mn-, Fe-based perovskites were studies by means of XPS, O_2-isotope exchange, O_2-TPD and H_2-TPR techniques. Moreover, their catalytic behaviors during NO reduction using C_3H_6 and CO reductants were respectively discussed. NO reduction by propene in the presence of oxygen over La(Co, Mn, $Fe)_{1-x}$(Cu, $Pd)_xO_3$ is believed to involve an organo nitrogen mechanism with the appearance of R-NO_2 and an isocyanate intermediate, while a mechanism for CO + NO reaction was proposed which involves dissociation of chemisorbed NO forming N_2 and/or N_2O, and oxidized perovskite surface which was continuously reduced by CO producing CO_2.

A moderate H_2O deactivation occurred over perovskite due to the occupation of available active sites by water vapor, which was verified to be strictly reversible when removing steam from the feed. Two distinct SO_2 poisoning mechanisms depending on SO_2 feed concentration and poisoning time were also reported. The reversible poisoning was due to the competitive adsorption of SO_2 and reactants as well as the coverage of activate sites by surface sulfite and sulfate species. By contrast, the high concentration SO_2 (80 ppm) leads to a severe sulfatation with destruction of the perovskite structure and the generation of $La_2(SO_4)_3$ and Fe_2O_3 phases resulting in an irreversible loss of activity in NO reduction by C_3H_6.

The NO reduction in the presence of an excess oxygen was performed by using the perovskite-alumina combined catalysts. Finally, a recent advance on NO reduction by using perovskites, supported metals and ion-exchanged zeolites was reviewed.

Chapter 2 - The introduction of organic groups into nanoporous materials results in functionalized nanostructured heterogeneous catalysts. By various self-assembly routes including a new solvent-assisted grafting and co-assembly of multiple types of organic groups, the authors and others have recently demonstrated the synthesis of a broad range of nanoporous catalysts containing high surface areas, site-isolated catalytic sites, nanometer pores and efficient and selective catalytic properties for various types of reactions. The grafting of site-isolated organic groups with a polar solvent such as ethanol or isopropanol is among the newest synthetic strategies that allow the synthesis of novel heterogeneous catalysts with enhanced catalytic efficiency. While the distributions and site-isolation of the catalytic sites in these materials can be controlled simply by changing the solvents used for grafting the catalytic functional groups, the degree of site-isolation of the catalytic sites can be probed by a simple new colorimetric method involving measuring of the d-d electronic spectra of metal complexes with the catalytic sites. By introducing appropriate secondary functional groups into the nanoporous catalyst structure, selective catalysts for specific reactants can also be produced. Furthermore, by a judicious choice of the active catalytic sites, selectivity for a specific product by these catalysts has been demonstrated recently.

Chapter 3 - Mesoporous silica containing $-(CH_2)_2CN$ groups on the surface has been prepared from tetraethylorthosilicate (TEOS) and (2-cyanoethyl)triethoxysilane (CTES) in aqueous ethanol in the presence of n-dodecylamine as the templating agent. Unlike in the cases of zeolitic materials or other porous variants of silica such as MCM-41, no high temperature calcinations are required for removing the templating molecules because it is possible to obtain the functionalized silica material of high porosity by either washing at an elevated temperature or soxhlet extraction. The cyanopropyl groups on the chemically modified silica (CMS) are converted to $-(CH_2)_2CO_2H$ groups via acid hydrolysis. The resultant materials, CMS-$(CH_2)_2CN$ and CMS-$(CH_2)_2CO_2H$, are found to be highly porous displaying very high BET surface areas of 1400 and 1071 m^2/g respectively. The tetranuclear cobalt(III) oxo clusters of the type $Co_4(\mu_3\text{-}O)_4(\mu\text{-}O_2CR)_4L_4$ with R = CH_3 or Ph and L = pyridine or a substituted pyridine have been chemically immobilized on CMS-$(CH_2)_2CO_2H$. N_2 adsorption data on CMS-$(CH_2)_2CO_2H$ and the supported reagent obtained after immobilization of the cubane-like clusters of cobalt(III) indicate retention of high surface area and the porous nature of the organosilica support. Most pores in the supported reagent are found to have diameters in the 3-6 nm range. More importantly, both the silica support and also the immobilized cobalt(III) complexes are hydrolytically stable. The supported materials, which have been characterized by various methods including spectroscopy and electron microscopy, are suitable as heterogeneous oxidation catalysts for substrates such as alkylaromatic hydrocarbons, α-pinene and alcohols using air, oxygen and TBHP as oxidants. The high reaction yields, improved selectivities, and possibilities for reusing the catalysts suggest that the cobalt(III) based catalysts developed by us may be seen as a promising class of supported reagents.

Chapter 4 - The purpose of this chapter is to review the status of spinels for selective alkylation of aromatic compounds with functional groups such as –OH, $-NH_2$. Alkylation of aromatic compounds is an important industrial reaction due to the usage of alkylated products in a wide variety of applications. Traditional Friedel-Crafts alkylation suffers from many disadvantages, and solid acid catalysts are preferred for several known advantages. Most of the known catalysts systems, such as zeolites, MgO, various metal oxides, lead to non-selective alkylation or involves high operating temperature ($\geq 400^0$C). Among the metal oxide catalysts used for aromatic alkylation, spinel (AB_2O_4) compounds have been effective and emerging as one of the potential systems for selective alkylation of aromatic compounds. Main advantages associated with spinels are the selective alkylation of aromatic compounds with good to very good yield and stable activity, with molecules such as phenol, aniline, pyridine in fixed-bed reactors around 300^0C. Methanol was employed as the methylating agent in most of the cases, and higher alcohols for higher alkylation. Few detailed characterization studies reported in the literature reveals the heterogeneous distribution of metal ions on the surface, and hence the associated synergism might be responsible for high catalytic activity. In situ infrared spectroscopic studies reveal the reaction mechanism aspects of selective alkylation of phenol and aniline. Medium range acid-base property of spinels is another important factor that makes them indispensable for different reactions with different cations. Indeed inverse spinel systems $B(AB)O_4$ are the most interesting due to the versatile nature of cation distribution in tetrahedral (T_d) and octahedral (O_h) sites, which contributes to the structural integrity and enhanced activity apart from the ability to tune cation site distribution with different metal ions. Although there are many positive factors with spinels, such as selective alkylation, more needs to be explored to exploit the spinel system for commercial applications.

Chapter 5 - The main experimental drawback raised in a homogeneous Fenton catalyst is analyzed in this chapter. In order to overcome its deficiency a heterogeneous catalyst is proposed.

In the homogeneous catalysis approach it was observed that hydrogen peroxide can be activated either by ions: Fe^{2+} or Fe^{3+}. The iron on its two oxidation states shows similar efficiencies on the degradation of a wide range of highly toxic organic molecules under mildly acid conditions. Amongst the organic compounds that can be mentioned are: synthetic textile dyes (amaranth, reactive black 5, acid green 25, basic blue 9, direct red 23 and acid orange 7), phenol, catechol, hydroquinone, p-benzoquinone, aniline, cresol and small open chain organic acids such as oxalic. Hence, strong oxidants such as $HO^•$, $HO_2^•$, FeO^{2+} among others are responsible of such degradation. Although homogeneous catalysts in the Fenton and Fenton-type reactions are working well, the catalytic activity ceases after a critical time because the iron ions undergo a chemical speciation during the reaction. It is not possible to restart it again either with the addition of more iron ions or with hydrogen peroxide. In order to maintain the catalytic activity after certain reaction time, a heterogeneous catalyst is proposed. This will improve the control over iron ions during the oxidation reaction. Iron (III) supported on fly ash and iron oxide supported on cerium oxide were used as heterogeneous Fenton catalysts and their preparations were performed by standardized thermal procedures. Chemical oxidation of two azo dyes, reactive black 5 (RB5) and basic orange 2 (BO2), was conducted by using the heterogeneous Fenton catalysis. Experimental results show that it is possible to oxidize 0.061 mM RB5 (in a media 50 mM Na_2SO_4, ≈pH 2.8 adjusted with H_2SO_4) using a stoichiometric amount of H_2O_2. After 2 hours of treatment, reactive solutions were effectively colorless and the 80% of the original chemical oxygen demand was removed. A complete decoloration and 100% COD abatement was achieved for the degradation of 0.1 mM BO2 in 5 hours in the presence of 4 mM H_2O_2, 0.1 g/l catalyst (50 % iron oxide) at initial solution pH of 3.0. Low Fe(III) leaching in heterogeneous supported catalyst were observed.

Chapter 6 - In this chapter, the author's research on the decomposition of Freon-12 (CCl_2F_2) on solid acids, the oxidation of CO using functionalized gold nanocatalysts, the methodological pitfalls in the reaction testing of phenol hydroxylation, and the liquid-solid interfaces of catalysis relevance is delineated from the perspectives of applied catalysis and fundamental model studies. One strategy of the author's applied catalysis investigation is to identify key catalytic functionalities to be achieved, and then rationally design or functionalize catalytic materials according to the desired functionalities. On the other hand, one strategy of the author's fundamental model studies is to distill unanswered questions from applied catalysis research or the relevant literature, and then design simplified or idealized model systems in order to answer these questions and obtain general trends. Herein, these two strategies are reflected by selected examples from the author's research, and a few review articles are introduced for further reference.

Chapter 7 - The design of active, selective, stable and easily recyclable heterogeneous catalysts is a challenging goal of liquid phase oxidation catalysis. Early transition-metal oxygen-anion clusters or polyoxometalates (POMs) have received increasing attention as oxidation catalysts because of their unique ensemble of properties, such as inorganic nature, metal oxide-like structure, thermodynamic stability to oxidation, thermal and hydrolytic stability, tunable acidities, redox potentials and solubility, etc. The apparent structural analogy of POMs and metal oxide surfaces allows considering POMs as discrete, soluble fragments of extended metal oxide lattices, which can be comprehensively investigated at the atomic level. In the structure of transition-metal-monosubstituted POMs (M-POMs), the active center M is isolated in and strongly bound to an inert metal oxide matrix and is thus prevented from hydrolysis and oligomerization, which makes M-POMs perspective building blocks for designing single site heterogeneous catalysts. The development of strategies for converting homogeneous POMs to solid, true heterogeneous catalysts is of primary importance. A simple wet impregnation usually did not allow obtaining catalysts which would be stable to leaching in polar reaction media. In the past decade, a great deal of work was directed to immobilize POMs on specially modified supports *via* the formation of a chemical bond (ionic, covalent or dative). Embedding POM into an inert matrix using sol-gel method may also lead to stable heterogeneous catalysts. Importantly, the molecular structure of POM can be maintained after immobilization, which ensures uniformity of spatially separated active centers on the catalyst surface or inside pores. In this chapter, the author surveys recent achievements in the synthesis of POM-based heterogeneous catalysts using different immobilization techniques, their characterization, and application for liquid phase selective oxidation of organic compounds. Special attention is paid to analyze the frequently success-limiting issues of catalyst activity and selectivity after immobilization and, especially, after recycling, catalyst stability to leaching and nature of catalysis.

Chapter 8 - The heterogenization of chiral homogeneous catalysts, which endows homogeneous systems with attractive features such as easy product separation and catalyst recovery by simple filtration, constitutes a rapidly expanding research area in asymmetric catalysis. Among various organic transformations, chiral transition-metal complexes catalyzed asymmetric ring opening reaction is one of the most fascinating areas because the enantiopure end-products have wider application in pharmaceuticals, fine chemicals and as chiral auxillaries. In this direction attempts were made to develop highly active and enantioselective catalysts based on chiral ligands viz. BINOLs, SALENs, etc., with various transition metals. As chiral catalysts are expensive, focuses on low catalyst loading and/ or their recovery and re-use are important aspects. Since various homogeneous catalysts have been found to be efficient in asymmetric catalysis their immobilization on solid supports is of great interest. This led research to make these systems recyclable by way of supporting the catalyst on organic and inorganic polymeric materials or making use of ionic liquids or manipulating solubility of the catalyst by increasing the molecular weight of catalyst with simultaneous increase in active sites so that the catalyst is easily recovered by simple precipitation method in a post catalytic workup process. Therefore, this chapter will give an in-sight of epoxide asymmetric ring opening reaction based on chiral recyclable catalysts and would bring about the latest trends in this area of research.

Chapter 9 - Development of heterogeneous metal oxide catalytic system with selective redox transition and highly recyclic feature is a challenging research in synthetic organic chemistry. Yet, very limited attention has been paid in consideration with stability and reusability of the catalytic system. This article reviews the authors' recent development on lead-ruthenate pyrochlore ($Ru_2Pb_2O_7$, Pyc) modified Nafion® membrane catalyst (designated as |NPyc|) for selective organic functional group transformation including oxidation of alcohol to aldehyde and ketone, organic sulfide to sulfoxide under ambient conditions. Note that Nafion® is a rigid perfluropolymer backbone polymer bearing ion-exchangeable sulfonic acid terminal group ($-SO_3^-H^+$) and is extensively used as a solid-state protonic conductor in fuel cell application. It is seldom used for chemical modification in synthetic organic chemistry. This chapter covers preparation of |NPyc|, characterization by XRD, SEM, SECM, AFM and TGA, and catalytic organic synthesis in triphasic medium with co-oxidants such as H_2O_2, NaOCl and O_2. Under optimal working conditions, the membrane catalyst showed very good selectivity and good turnover for wide range of organic compounds. The |NPyc| catalyst can be recycled over > 25 times without any catalytic degradation. As to the mechanism, a high valent-oxo ruthenium redox species, perruthenate/ruthenate redox couple (Pyc-RuO_4^- /Pyc-RuO_4^{2-}) exists in the |NPyc|, is believed to participate in the oxidation reaction with the co-oxidants.

Chapter 10 - Solids with rough surfaces are extremely common in nature and they appear in many industrial processes. The effect of surface heterogeneity in physical, chemical and biological processes is consequently of great interest for many practical purposes. Due to its importance in both basic research and practical applications, the study of geometric heterogeneity have attracted extensive attention and propelled intensive research activities in this field. In heterogeneous catalysis, reaction performance depends strongly on the structure and geometry of the environment in which the catalytic process takes place. The structure of a real catalyst is never a perfect crystalline one. More information about the complex structure of surfaces and materials could be obtained by adopting some new approaches or theories. The role of surface heterogeneity in heterogeneous catalysis will be reviewed in this chapter with special attention to the surface heterogeneity effect on Eley-Rideal reaction mechanism, which is an elementary step in many complex interfacial processes. More detailed characterization of heterogeneous structures can be obtained by applying multifractal scaling analysis (MSA). Some discussion about its development and its application to the heterogeneous reaction is also given in this chapter. The applicability of the multifractal scaling analysis to the characterization of the heterogeneous process will be given with a focus on the heterogeneous catalytic reaction. The time dependence of the heterogeneous reaction over heterogeneous surface will be discussed with the help of dynamic scaling theory.

Chapter 11 - In the last decade increasing numbers of articles have been published focusing on the combination of heterogeneous catalysis and continuous flow methods. The publication of such an overview of the field is particularly justified by the recent practical applications involving a large number of novel instruments designed to perform continuous flow reactions in the presence of heterogeneous or immobilized homogeneous catalysts. In a typical experiment, a solution of the substrate/reactant is pumped through the heterogeneous catalyst optionally under elevated temperature and/or pressure. This application of flow methodology offers many advantages in organic chemistry. Reactions are typically performed in a continuous flow mode on a small scale, thus, reaction temperatures can easily be controlled. Products are eluted from the reactors after minutes, so information about the outcome of the reaction can be quickly obtained. Parameters, such as flow rate, temperature or pressure may be adjusted during the reaction allowing for fast reaction optimization and the results are highly reproducible. Scale up of reactions can be realized by either running the reaction over a long period of time or by performing many reactions in parallel. Automation of reactions can also be easily achieved through automated liquid handlers which inject solutions of the substrate before contacting the catalyst and collect the reaction mixture at the reactor output. In this chapter the authors give a comprehensive overview of the field of continuous flow techniques using heterogeneous catalyst to perform reactions. The topics covered will begin with reactions under ambient conditions, continue towards reactions under supercritical conditions and conclude with the description of microwave assisted continuous flow systems.

In: Heterogeneous Catalysis Research Progress
Editor: Mathias B. Gunther, pp. 1-79
ISBN 978-1-60456-979-7

Chapter 1

APPLICATION OF PEROVSKITES FOR AUTOMOBILE EXHAUST PURIFICATION

Runduo Zhang[1] and Serge Kaliaguine[2]

[1] Institute of Modern Catalysis, State Key Laboratory of Chemical Resource Engineering, Beijing University of Chemical Technology, Beijing, 100029 China

[2] Department of Chemical Engineering, Faculty of Science and Engineering, Laval University, Quebec, G1K 7P4 Canada

ABSTRACT

The purification of automobile exhaust gases such as nitrogen oxides (NO_x), CO, and hydrocarbons (HCs), which cause photochemical smog and acid rain respectively leading to respiratory diseases and ecological damages, has become one of the main objectives of environmental catalysis. Perovskite-typed mixed oxides are regarded as promising materials for catalytic removal of those exhaust pollutants due to their excellent redox properties and nonstoichiometric structure and are thought as ideal but yet cheap substitutes for the commercial noble metal autocatalysts. The traditional ceramic method involving high-temperature calcination allows only prepare perovskites having low specific surface areas which limit their practical applications as catalysts. By contrast, a novel technique designated as reactive grinding avoids this thermal treatment and the crystallization proceeds at nearly ambient temperature thus yielding extraordinary high surface areas.

The oxygen vacancies, overstoichiometric, atomic and lattice oxygen of series of Co-Mn-, Fe-based perovskites were studies by means of XPS, O_2-isotope exchange, O_2-TPD and H_2-TPR techniques. Moreover, their catalytic behaviors during NO reduction using C_3H_6 and CO reductants were respectively discussed. NO reduction by propene in the presence of oxygen over La(Co, Mn, Fe)$_{1-x}$(Cu, Pd)$_x$O$_3$ is believed to involve an organo nitrogen mechanism with the appearance of R-NO_2 and an isocyanate intermediate, while a mechanism for CO + NO reaction was proposed which involves dissociation of chemisorbed NO forming N_2 and/or N_2O, and oxidized perovskite surface which was continuously reduced by CO producing CO_2.

A moderate H_2O deactivation occurred over perovskite due to the occupation of available active sites by water vapor, which was verified to be strictly reversible when removing steam from the feed. Two distinct SO_2 poisoning mechanisms depending on SO_2 feed concentration and poisoning time were also reported. The reversible poisoning was due to the competitive adsorption of SO_2 and reactants as well as the coverage of activate sites by surface sulfite and sulfate species. By contrast, the high concentration SO_2 (80 ppm) leads to a severe sulfatation with destruction of the perovskite structure and the generation of $La_2(SO_4)_3$ and Fe_2O_3 phases resulting in an irreversible loss of activity in NO reduction by C_3H_6.

The NO reduction in the presence of an excess oxygen was performed by using the perovskite-alumina combined catalysts. Finally, a recent advance on NO reduction by using perovskites, supported metals and ion-exchanged zeolites was reviewed.

1. Basic Concept of Perovskites

The special electric, magnetic, optical, superconductive and catalytic properties of perovskite-typed oxides make this group of materials attracting and widely used. Perovskites were named according to the similarity of their structure with the $CaTiO_3$ compound. The ideal perovskite structure is cubic with a space group of $Pm3m\text{-}O_h^1$ and its unit can be formulated as ABO_3, where A is a larger cation in a twelvefold coordination to oxygen anions and B is a smaller cation surrounded by six oxygens, in the octahedral coordination. It was established by Raveau[1] that the ReO_3 framework played an important role as a host structure in the construction of numerous structures of complex oxides. Similarly, the perovskite structure can also be thought as a superstructure with a ReO_3-type framework built up through incorporating A cations into the BO_6 octahedra (Figure 1).

A great diversity of perovskites can be obtained by a proper adjustment of their compositions with different matching ions in A or B sites by satisfying certain limits of the tolerance factor (t) defined by Goldschmidt[2]: $t = (r_A + r_O) / \sqrt{2}\,(r_B + r_O)$, where r_A (> 0.09 nm), r_B (> 0.051 nm) and r_O are the ionic radii for A, B and O ions. Up to now, about 90% of metallic natural elements of the periodic table are known to be stable in a perovskite-type oxide structure. Rare-earth metals, alkali and alkali-earth metals can occupy the A sites, while transition metals such as cobalt, manganese, iron, nickel, vanadium, chromium, gallium, copper etc are the most commonly found as the B-site cations.

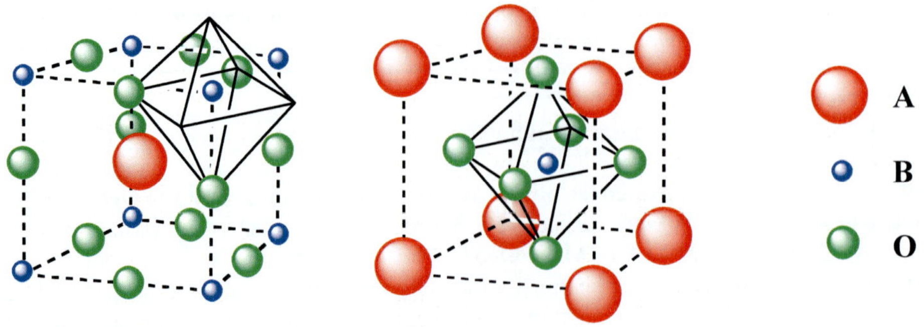

Figure 1. Schematic structure of ideal ABO_3 perovskite.

It is commonly accepted that the B-site ions play a crucial role for various redox reactions and the role of A-site cations is to stabilize the B-O octahedra as well as modify the chemical environment of the B-site cations.

The ideal cubic perovskite structure appears just in a few cases where *t*-values are close to 1 and at high temperature. In fact, the cubic symmetry was seldom achieved for a practically used perovskite. For example, the $CaTiO_3$ compound was originally thought to be cubic, but its true symmetry was later verified to be orthorhombic[3]. Lower *t*-values ($0.75 < t < 1.0$) yielding a slight deviation from the ideal situation would take place in most cases. Deviations from the cubic structure may be realized via a simple distortion or an enlargement of the unit cell, or a combination of both, resulting in the distorted structures with orthorhombic, rhombohedral, tetragonal, monoclinic, and triclinic symmetries as so far known. Because the latter three structures are very rare and poorly characterized, only the orthorhombic and rhombohedral were described in the literature. A cooperative buckling of corner-shared octahedra occurs for *t* values within the range of $0.75 < t < 0.90$, leading to the orthorhombic distortion which has a space group of *Pbnm*. When there is no octahedral buckling, a small deformation from cubic to rhombohedral symmetry may take place as the tolerance factor ranges at $0.9 < t < 1.0$.

The simple perovskite structure may be appropriately modified by incorporating two types of A or/and B ions with suitably different sizes and charges having the general formula of the perovskite as $A_{1-y}A'_yBO_3$ or $AB_{1-x}B'_xO_3$ or $A_{1-y}A'_yB_{1-x}B'_xO_3$. If the charge between A and A' or B and B' is different, in the ordered structure the oxygens are slightly shifted toward the more charged cation although the octahedral symmetry of B and B' cations is preserved. Considering the electroneutrality fulfilled by perovskites, the partial substitution of A and B ions by corresponding cations with different valencies would also give rise to the unbalance of the total perovskite charge, subsequently, yielding anion or cation deficiency for compensation.

2. Perovskite Synthesis

2.1. Methods

The various preparation methods of perovskite-type oxides can be roughly classified as equimolecular oxide mixing synthesis (I) and decomposition synthesis (II). The choice of a particular synthesis basically depends on the expected application of these oxides. In the field of heterogeneous catalysis, the emphasis is placed on the specific surface area and porosity of powdered perovskites. The traditional ceramic method requires a high calcination temperature (usually >1000 °C) for the solid state reaction to occur and often leads to the formation of coarse aggregation[4]. As a result, the grain size of the powder obtained by this method is relatively large and the specific surface area is very small (several m^2/g), making it unsuitable for catalytic applications. In addition, the purity of perovskite phase of product generated by this method is not satisfactory.

Several approaches have been proposed toward gaining a better degree of mixing and lowering the temperature necessary for complete crystallization. These encompass the methods of spray-drying[5], freeze-drying[6], coprecipitation[7], microemulsion[8], citrate complexation[9], flame-pyrolysis[10]. Their objective was a high specific surface area, a good crystallinity, and a relatively narrow particle-size distribution. The spray-drying, freeze-drying, and coprecipitation methods for preparing $La_{1-x}M_xMnO_3$ (M=Pb, K, Ce, Co, Ni, Mg, Li) perovskites were compared in reference[11]. The relative order of catalytic activity per unit surface area was freeze-drying > spray-drying > coprecipitation, which is related to difference in the calcination temperature. Moreover, Tascón et al.[12] have carried out a comparative study of methods involving the coprecipitation either as a single cyanide compound, or as an oxalate mixture, or by citric complexation for the preparation of $LaMO_3$ (M=V, Cr, Mn, Fe, Co, Ni) catalysts. The heating temperature necessary to obtain a single perovskite phase was higher for oxalate decomposition (900-1000 °C) than for cyanide or citrate decomposition, exhibiting surface areas varying in the ranges of 2-6 m^2/g (oxalates), 10-40 m^2/g (cyanides), and 4-55 m^2/g (citrates).

Up to recently reactive grinding (or mechanical alloying) had mostly been used as a method of preparation of nanocrystalline metal alloys[13]. This method was also applied to the alloying of such brittle materials as Si and Ge[14], to the synthesis of composite materials incorporating preformed oxide or carbide particles into a metallic matrix[13] or to the nitrogenation of iron or steel under nitrogen or ammonia[15]. Reactive grinding also allows preparation at rather low temperatures of crystalline oxide phases normally produced at very high temperatures. This is exemplified by high-energy-ball milling of alumina which showed that at essentially room temperature any transition alumina can be converted to α-alumina by reactive milling[16], while normally this conversion requires a temperature of 1200 °C. In addition, this work showed that if grinding is performed in the presence of a reactive gas such as oxygen, the α-alumina could be produced with an unusually high surface area on the order of 100 m^2/g. Series of ABO_3 perovskites were synthesized by Kaliaguine's group via this method using the individual simple oxides of A and B cations as the reactants[17]. The crystallization process for perovskite usually takes 8 h followed by a 10 h refining process with various leachable grinding additives to further develop the surface area. Perovskites investigated in the present chapter were prepared by this novel method.

2.2. Physical Properties

The chemical composition of perovskites prepared was estabilished by atomic absorption spectroscopy (AAS) using a Perkin-Elmer 1100B spectrometer for Fe, Co, Mn, Cu contents and an inductively coupled plasma (ICP) spectrometer (Perkin-Elmer, Optima 4300DV) for La and Pd contents, usually showing values close to the nominal formulation (Table 1). The BET specific surface areas and pore diameters of these materials were determined from nitrogen adsorption isotherms measured at -196 °C using an automated gas sorption system (NOVA 2000, Quantachrome Corporation) operating in continuous mode and were also reported in Table 1. Specific surface areas of 20-30 m^2/g were obtained for cobalt lanthanates and approximately 40 m^2/g for manganese lanthanates and 30-50 m^2/g for iron lanthanates even after calcination at 500 °C for 5 h. The crystallite phase identification of solid solutions was accomplished by means of X-ray diffraction (XRD) using a Siemens D 5000 diffractometer and Cu Kα radiation (λ=1.5406Å) with the results being given in Table 1. The comparison of experimental spectra with JCPDS charts indicates that all Co- and Mn-containing samples are essentially rhombohedral perovskite-type mixed oxides and with the formula $LaMnO_{3.15}$ for the unsubstituted lanthanum manganite. The crystallite phase analysis reveals a main orthorhombic $LaFeO_3$ perovskite-type structure (JCPDS card 74-2203) for all prepared Fe-containing catalysts. Diffraction lines corresponding to PdO or CuO are not observed, suggesting that Pd and Cu metals are fully incorporated into the perovskite structure or the formed particles of PdO and CuO are too tiny to be detected by XRD. The crystallite sizes were calculated using Scherrer equation with values lower than 20 nm. Based on the result that both pore diameters and crystallite sizes are around 10-20 nm, the porous structure of present perovskites was believed to be realized via clustering of their individual nanoscale primary particles in agreement with reference[18].

3. OXYGEN NONSTOICHIOMETRY

The defect perovskites usually associate with oxygen vacancies or cationic vacancies (exhibiting an oxygen excess). Oxygen vacancies are more common than cationic vacancies. Many oxygen-deficient perovskites can be described on the basis of complex perovskite-related super-structures of general formula $ABO_{3-\lambda}$, in which the stacking depends on the sizes, electronic configurations and coordination numbers of A and B cations. Although most of the 3d cations have led to the formation of these $ABO_{3-\lambda}$ phases, the cobalt family has been better studied so far. Oxygen excess nonstoichiometry in perovskite oxides is not as common as anion-deficient nonstoichiometry probably because introduction of interstitial oxygen in perovskite structure is thermodynamically unfavorable. The best-characterized perovskite showing oxygen overstoichiometry is $LaMnO_{3+\lambda}$.

The oxygen nonstoichiometry as well as the redox properties of Co-, Mn- and Fe-based perovskites was investigated by means of O_2-TPD and H_2-TPR and XPS techniques. The amounts of O_2 released during O_2-TPD were calculated after deconvolution of the O_2 desorption curves using Lorentzian peak shape and they are reported in Table 2.

Table 1. Properties of Co-, Mn-, and Fe-based catalysts after calcination at 500 °C for 5h*[19-21]

Sample	Chemical composition	Specific surface area (m^2/g)	Crystallite size (nm)	Pore volume (cm^3/g)	Pore diameter (nm)	symmetry
$LaCoO_3$	$La_{1.05}Co_{1.0}O_{3\pm\delta}$	29.2	11.8	0.16	18.1	rhombohedral
$LaCo_{0.9}Cu_{0.1}O_3$	$La_{1.0}Co_{0.86}Cu_{0.14}O_{3\pm\delta}$	23.8	10.8	0.17	15.6	rhombohedral
$LaCo_{0.8}Cu_{0.2}O_3$	$La_{0.97}Co_{0.76}Cu_{0.24}O_{3\pm\delta}$	22.4	9.2	0.13	14.7	rhombohedral
$LaMnO_3$	$La_{0.98}Mn_{1.0}O_{3\pm\delta}$	40.6	11.3	0.15	10.8	rhombohedral
$LaMn_{0.9}Cu_{0.1}O_3$	$La_{0.97}Mn_{0.91}Cu_{0.09}O_{3\pm\delta}$	40.7	13.0	0.20	12.4	rhombohedral
$LaMn_{0.8}Cu_{0.2}O_3$	$La_{1.1}Mn_{0.78}Cu_{0.22}O_{3\pm\delta}$	42.6	14.4	0.20	10.8	rhombohedral
$LaFeO_3$	$La_{0.99}Fe_{1.0}O_{3\pm\delta}$	30.5	18.7	0.13	16.2	orthorhombic
$LaFe_{0.8}Cu_{0.2}O_3$	$La_{0.97}Fe_{0.80}Cu_{0.20}O_{3\pm\delta}$	41.6	18.9	0.16	11.8	orthorhombic
$LaFe_{0.97}Pd_{0.03}O_3$	$La_{0.96}Fe_{0.97}Pd_{0.03}O_{3\pm\delta}$	48.1	16.8	0.15	14.6	orthorhombic

* Iron can be detected as contaminant with a weight percent less than 2% for all samples.

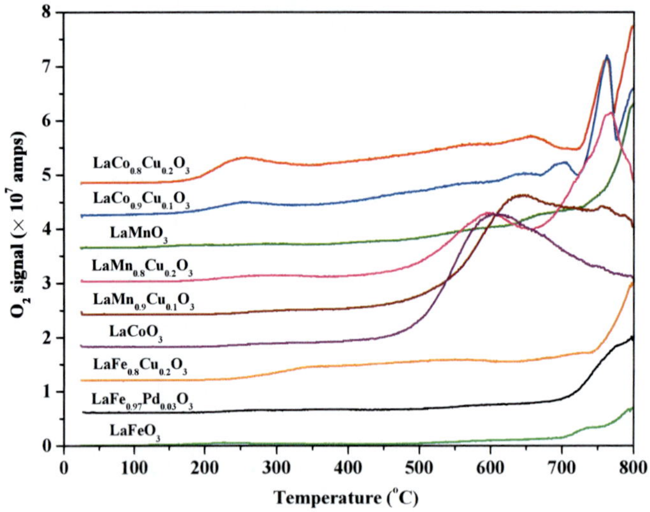

Figure 2. O_2-TPD profiles over Co-, Mn-, and Fe-based perovskites [19-21].

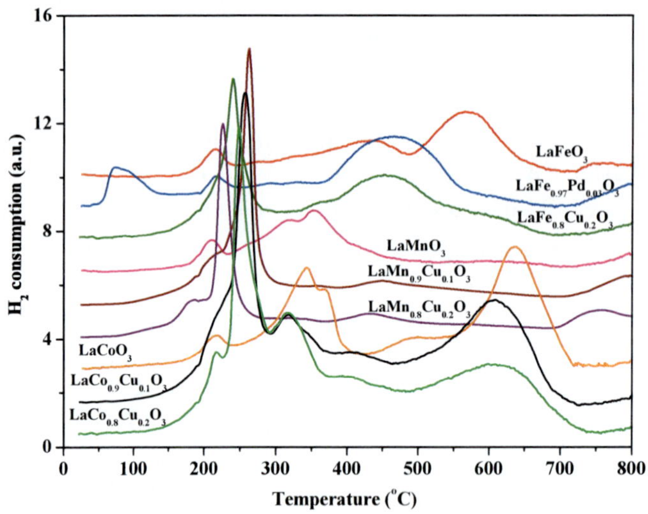

Figure 3. H_2-TPR profiles of Co-, Mn-, and Fe-based perovskites [19-21].

3.1. O_2-TPD Study

Two kinds of desorption peaks are present in the O_2-TPD traces (Figure 2) of $LaCoO_3$: the broad one appearing in the range of 150-700 °C was ascribed to α-O_2 desorption and related to molecular O_2 adsorbed on oxygen vacancies[22]. The latter peak with a maximum at 797 °C was designated as β-O_2 desorption and attributed to the liberation of O_2 from the lattice[22]. With the partial substitution of Co by Cu, the desorption of α-O_2 is significantly enhanced proportionally to the degree of substitution. This result indicates that more oxygen vacancies can be generated over Co-based perovskites after Cu incorporation into B-site likely due to a positive charge compensation in agreement with the previous report[23]. Additionally, the β-O_2 desorption of Cu-substituted lanthanum cobaltites can take place at relatively lower temperature (with the former ones at about 765 °C and the latter ones at about 795 °C) with respect to $LaCoO_3$ suggesting the mobility of lattice O_2 was also improved via Cu substitution.

A minor α-O_2 peak at T<450 °C was found in the O_2-TPD profile for $LaMnO_3$ associated with a slight enhancement in its intensity upon increasing Cu content. However, these α-O_2 desorptions are severely suppressed compared to $LaCo_{1-x}Cu_xO_3$ which are associated with the presence of overstoichiometric oxygen. A following intense desorption with a maximum at 608 °C was observed over lanthanum manganite, showing a drop in its intensity after Cu substitution. The introduction of Cu ions in $LaMnO_3$ perovskite lattice was reported to suppress its overstoichiometric (excess) oxygen[23]. This peak is therefore associated with the desorption of overstoichiometric oxygen of lanthanum manganite. This overstoichiometric oxygen desorbed from the parent $LaMnO_3$ was quantified and found to be consistent with the formula of $LaMnO_{3.15}$ determined by XRD.

Only a small amount of α-O_2 desorbed from $LaFeO_3$ at T<700 °C, while more β-O_2 desorbed at T>700 °C. Compared to $LaFeO_3$, a similar α-O_2 desorption but a more intense β-O_2 desorption was found over $LaFe_{0.97}Pd_{0.03}O_3$ during TPD of O_2 experiments.

Table 2. Amount of O_2 desorbed from perovskites shown in Table 1 during O_2-TPD experiments [19-21]

Sample	Amount of oxygen desorbed[a]			Number of monolayers desorbed[b]		
	α-O_2 (μmol·g^{-1})	excess O_2 (μmol·g^{-1})	β-O_2 (μmol·g^{-1})	α-O_2	excess O_2	β-O_2
$LaCoO_3$	<700 °C 133.2	-	700-800 °C 112.4	1.14	-	0.96
$LaCo_{0.9}Cu_{0.1}O_3$	<700 °C 252.2	-	700-800 °C 163.6	2.65	-	1.72
$LaCo_{0.8}Cu_{0.2}O_3$	<700 °C 285.6	-	700-800 °C 171.6	3.19	-	1.91
$LaMnO_3$	324 °C 16.8	609 °C 316.7	- -	0.10	1.95	-
$LaMn_{0.9}Cu_{0.1}O_3$	323 °C 22.6	646 °C 189.3	760 °C 155.7	0.14	1.16	0.96
$LaMn_{0.8}Cu_{0.2}O_3$	301 °C 40.1	599 °C 106.8	763 °C 196.1[c]	0.23	0.62	1.15
$LaFeO_3$	<700 °C 38.2		700-800 °C 76.2	0.31		0.63
$LaFe_{0.8}Cu_{0.2}O_3$	<700 °C 196.5		700-800 °C 202.2	1.18		1.22
$LaFe_{0.97}Pd_{0.03}O_3$	<700 °C 62.8		700-800 °C 173.1	0.33		0.90

[a] Calculated by deconvolution of the O_2 desorption curves

[b] Calculated with 4 μmol·m^{-2} of oxygen per monolayer

A broad plateau-like α-O_2 desorption peak at 200-700 °C together with a sharp β-O_2 peak maximum at 796 °C was observed in the O_2-TPD profile for $LaFe_{0.8}Cu_{0.2}O_3$ perovskite. This significant enhancement of adsorbed α-O_2 is likely related to the surface oxygen vacancies generated upon Cu substitution.

3.2. H_2-TPR Study

The TPR profiles of five Co-based perovskites prepared by reactive grinding with ZnO as additive for different milling times are shown in Figure 4(a-e), Figure 4f represents a $LaCoO_3$ perovskite prepared by the citrate method (reference sample).

All TPR profiles show two steps of reduction generally observed in $LaCoO_3$ perovskites. Their positions on the temperature scale are strongly dependant on the morphology of the samples. A minor peak around 200°C was ascribed to the reduction of adsorbed oxygen and is accordingly increasing with BET surface area as listed in Table 3. The hydrogen consumption was corresponding to a complete reduction of the Co^{3+} to the metal after each TPR measurement.

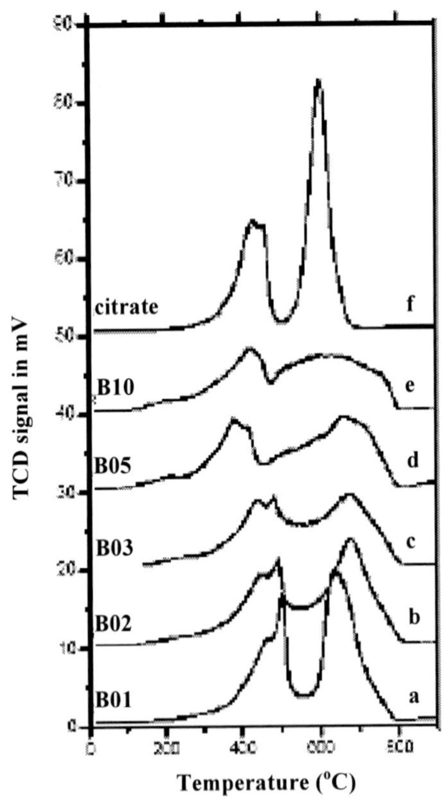

Figure 4. TPR profiles of perovskites [24]: a) $LaCoO_3$ washed after miling 1 hour with ZnO. b) $LaCoO_3$ washed after miling 2 hour with ZnO, c) $LaCoO_3$ washed after miling 3 hour with ZnO, d) $LaCoO_3$ washed after miling 5 hour with ZnO, e) $LaCoO_3$ washed after miling 10 hour with ZnO, f) $LaCoO_3$ synthesized by the citrate method.

Additionally, the formations of Co^0 and La_2O_3 were also confirmed by XRD patterns. Although an agreement for completely reducing $LaCoO_3$ into metallic cobalt was achieved, a debate for the detail TPR behavior of $LaCoO_3$ is still going on. Some authors report that Co^0 is formed only during the second step of reduction and attribute the first peak only to the reduction of Co^{3+} to Co^{2+}[25, 26]. Bedel et al.[27] however found that 8.6 wt % of Co^0 can be extracted by reduction from $LaCo_{0.4}Fe_{0.6}O_3$ with orthorhombic symmetry during the first step of reduction (410°C) but no evidence was found of Co^0 being extracted during the first step from $LaCoO_3$ and $LaCo_{0.75}Fe_{0.25}O_3$ perovskites having rhombohedral symmetry. The reduction profile of our sample allows us to suppose a more complex reduction process than the simple two steps developed by several authors[25-27].

In our case the hydrogen consumption of the former deconvolution peak is always indicating a reduction exceeding the mere $Co^{3+} \rightarrow Co^{2+}$. At one hour of grinding time sample B01 shows a 17.5% conversion to the metal, which first increases to 31% after two hours (sample B02) and then decreases regularity to 13.7% after 10 hours of milling (sample B10). The citrate sample, with the lowest surface area (see Table 1), yields only 2.3% conversion to the metal. Obviously some Co^0 must be already formed at this intermediate temperature, well below the second peak corresponding to the complete $Co^{2+} \rightarrow Co^0$ reduction of the bulk cobalt.

In several contributions from our group we have shown that the perovskites prepared by reactive grinding had a particular morphology, which is constituted of agglomerates of nanoparticles[18, 28]. The interface between the nanoparticles or grain boundaries have been shown to be especially important in gas-solid reactions exemplified by the sulfatation of $LaCoO_3$ by SO_2 traces leading to catalysis deactivation by sulphur[18]. These grain boundaries have also been shown to have high oxygen mobility leading to a particular behavior in oxygen isotopic exchange[28]. Increasing specific surface area by secondary grinding in the presence of a grinding additive is due to the partial separation of these nanocrystalline domains which also leads to the decrease in volume of the grain boundaries. The decrease in Co^0 content observed after TPR to intermediate temperature in sample B02 to B10 is therefore associated with the decrease in volume of the grain boundaries corresponding to the increased surface area.

Table 3. Properties of the $LaCoO_3$ Perovskite sample

Sample designation	Milling time (h)	XRD analysis[a]	Average crystallite size D (nm)[b]	BET surface area (m^2/g)	Chemical composition $La_xCo_yFe_zO_{3-\delta}$		
					x	y	z
BO1	1	Perovskite structure	17.5	16	1	0.99	0.01
BO2	2	..	15.7	30	1	1.04	0.01
BO3	3	..	16.3	43	1	1.02	0.01
BO5	5	..	14.0	58	1	1.05	0.01
BO10	10	..	10.2	61	1	1.03	0.01
Citrate	-	..	> 100	5.6	1	0.89	< 0.01

[a] XRD spectra were compared to JCPDS file

[b] Calculated using Scherrer equation (Klug and Alexander, 1974).

This easier reducibility of the Co^{2+} in the grain boundaries would also be the result of a higher oxygen mobility in these highly defective zones.

The reducibility of perovskites was investigated by H_2-TPR to check the state of the catalyst surface and bulk, and get information needed for mechanistic studies. Since La^{3+} is nonreducible under the conditions of H_2-TPR, the observed H_2 consumption peaks in the TPR profile of $LaCoO_3$ (Figure 3) are due to the reduction of Co^{n+} cations. As seen from Figure 3, the H_2 consumption provides evidence of the complete reduction of Co^{3+} to Co^0 occurring in two steps, from Co^{3+} to Co^{2+} with a peak maximum at 344 °C and from Co^{2+} to Co^0 centered at 635 °C, in agreement with the results of previous studies[17, 29]. In the case of Co-based samples, the small H_2 consuming peak at 217 °C was ascribed to the reduction of α-oxygen, which is enhanced by the Cu substitution. The tall peak at approximately 250 °C observed over Cu-substituted samples can be assigned to the reduction of Cu cations in lattice from Cu^{2+} to Cu^+, while the complete reduction from Cu^+ to Cu^0 was related to the shoulder at 505 °C. These values are very similar to those observed in H_2-TPR of Cu/MCM-41[30].

The existence of overstoichiometric oxygen in lanthanum manganite generated by reactive grinding was already confirmed by XRD as discussed above. This kind of overstoichiometric oxygen is believed to be associated with Mn^{4+}. It is especially reactive and could easily be reduced due to its low coordination number with cation in perovskite lattice, so that the peak centered at 212 °C in the profile of $LaMnO_3$ was attributed to the reduction of overstoichiometric oxygen. The subsequent peak at 350 °C and the shoulder at its rising part could be ascribed to the reduction of $Mn^{3+} \rightarrow Mn^{2+}$ in bulk and over surface. The third peak observed above 700 °C was associated with the partial reduction of the $Mn^{2+} \rightarrow Mn^0$. After Cu-substitution, the sharp reduction peak related to $Cu^{2+} \rightarrow Cu^+$ was observed at 250 and 225 °C for x=0.1 and x=0.2, respectively, simultaneously with copper reduction.

Small amounts of highly reducible Fe^{4+} were reported present in $LaFeO_3$[31-33] with its reduction to Fe^{3+} occurring at T<300 °C during H_2-TPR experiments[32]. The minor reduction peak at 215 °C in the H_2-TPR profile of $LaFeO_3$ is thus attributed to the reduction of Fe^{4+} to Fe^{3+}. The quantitative analysis of the successive Fe^{3+} reduction peaks with maxima at 440 and 573 °C correlated well with the H_2 consumption necessary for the reduction of Fe^{3+} to Fe^{2+}. Those two peaks were therefore ascribed to $Fe^{3+} \rightarrow Fe^{2+}$ reduction occurring over the surface and in the bulk of parent $LaFeO_3$, respectively. Finally, the minor reduction peak above 700 °C was ascribed to the partial reduction of Fe^{2+} to metallic iron. The low temperature (78 °C) H_2 consumption which occurs over $LaFe_{0.97}Pd_{0.03}O_3$, corresponds to the reduction of Pd^{2+} to Pd^0 according to the literature[34]. This result reveals the excellent redox properties of Pd substituted perovskite, possibly leading to a good catalytic performance especially at low temperature. The sharp peak at 241 °C observed in the H_2-TPR profile of Cu substituted lanthanum ferrite was again ascribed to the $Cu^{2+} \rightarrow Cu^+$ lattice reduction.

3.3. XPS study

Figure 5 shows the O 1s XPS signals of $LaCoO_3$ samples as prepared and calcined at 200, 400 and 900 °C. At first glance, the spectra show two main photolines at 529.1 eV and 531.2-531.7 eV. According to Tejuca et al.[35] the low binding energy peak must be assigned to lattice oxygen, and the high binding energy one to adsorbed oxygen species normally designated as α-oxygen.

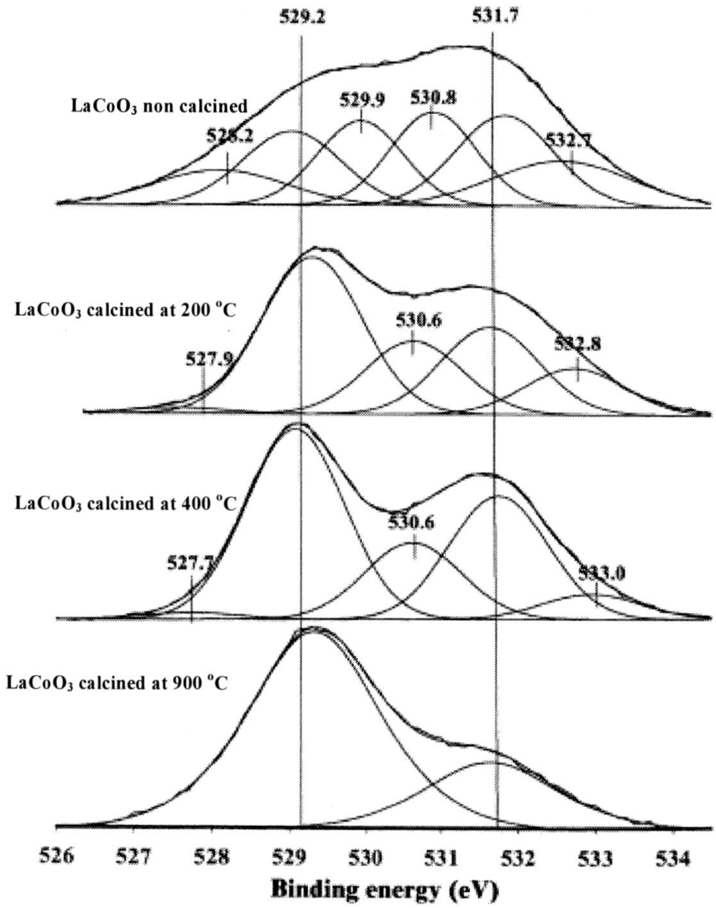

Figure 5: Deconvolution of O 1s photolines of $LaCoO_3$ samples [17] (fresh sample, calcined at 200, 400 and 900 °C).

The non-calcined sample shows six components, two of which have already essentially disappeared upon calcination at 200 °C. The peak at 529.2 ± 0.1 eV is ascribed to regular lattice oxygen O^{2-}. The low binding energy peak at 527.9 ± 0.3 eV is likely to be associated with molecularly physisorbed O_2[36]. This essentially vanished after calcination at 200 °C. A third line at 529.9 eV, which also disappears after calcination at 200 °C was ascribed to adsorbed water found at 532.8 ± 0.2 eV. This peak is likely associated with adsorbed O_2^- oxygen. Its intensity decreases regularly with calcination temperature and is absent after calcination at 900 °C.

3.4. Isotopic Exchange Experiments

Isotopic exchange experiments[37] were introduced for the investigation of oxygen exchange behavior over lanthanum cobaltite perovskites prepared by methods including solid-state (SS), coprecipitation (COP), citrate complexation (CIT), and reactive grinding of single oxides (RG) and of an amorphous precursor (COPRG).

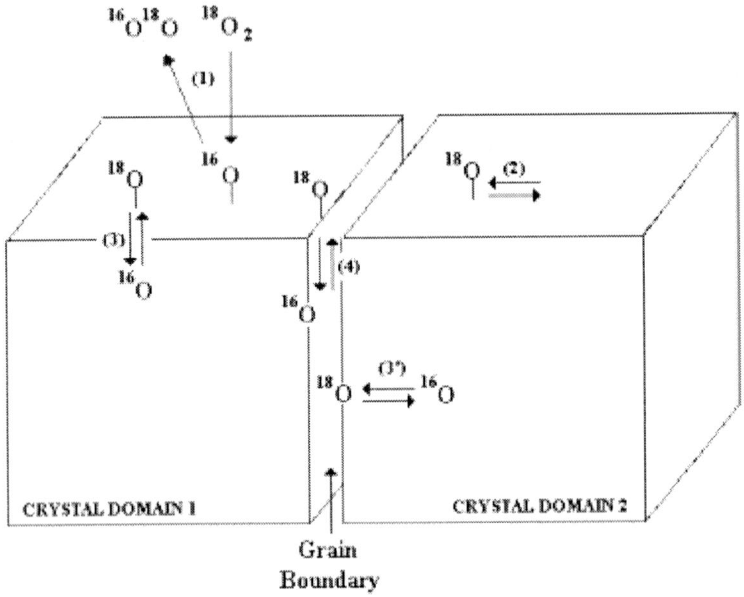

Figure 6. Scheme of oxygen exchange on $LaCoO_3$ [37].

It was found that the surface and the grain boundaries, where diffusion proceeds at a higher rate than bulk diffusion, can logically be considered to be quickly exchanged and present the same concentrations of ^{18}O and ^{16}O as the gas phase. Reactive grinding favors the generation of more grain boundaries yielding a better performance catalyst for oxidative reactions with respect to the other preparation methods tested.

3.5. Redox Properties

Based on the above investigations, the redox properties of perovskites were further discussed. A broad plateau-like α-O_2 desorption was observed over the $LaCoO_3$ perovskite. The intensity of this α-O_2 desorption is enhanced by Cu substitution because of a positive charge deficiency after substitution of Co^{3+} by Cu^{2+} compensated by oxygen vacancies[17]. This α-O_2 desorption was ascribed to oxygen adsorbed on anion vacancies, which can be readily removed at relatively low temperature. A rather low α-O_2 coverage of $LaMnO_3$ after O_2 adsorption was found with respect to $LaCoO_3$ although its coverage can be slightly enhanced by Cu substitution, indicating a low density of anion vacancies over lanthanum manganites.

According to our previous work[17], the following process was proposed to occur during calcination of perovskite:

1) Transient generation of anion vacancy during calcination of fresh sample

$$\underset{M^{n+}}{\overset{OH}{|}} + \underset{M^{n+}}{\overset{OH}{|}} \xrightarrow{-H_2O} M^{(n-1)+}\square + \underset{M^{n+}}{\overset{O^-}{|}} \quad (1)$$

(where M=Co or Mn or Fe or Cu or Pd; n=2 or 3 or 4; □ stands for anion vacancy)

2) Instantaneous adsorption of α-O_2 at the anion vacancy

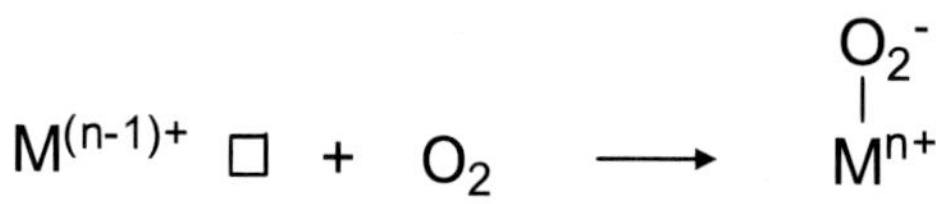

(2)

The two reactions (3) and (4) are proposed for the α-O_2 desorption process:

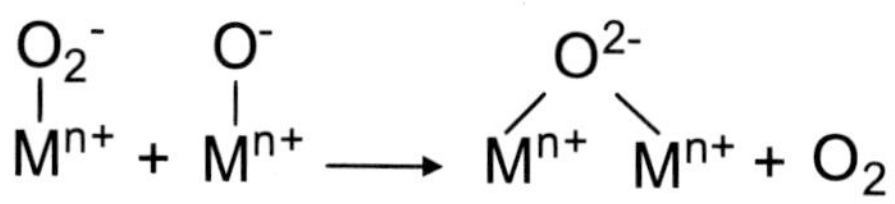

(3)

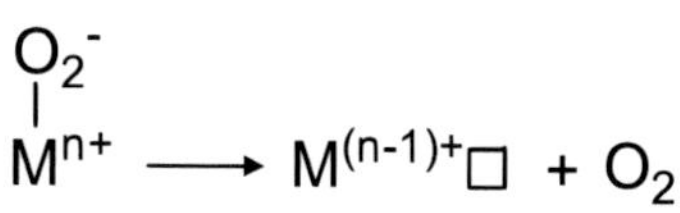

(4)

A significant desorption of overstoichiometric oxygen (excess O_2) at 450-650 °C was observed over $LaMnO_3$ during O_2-TPD experiments, whose intensity was suppressed by increasing Cu substitution. This may be formulated as:

$$4\ Mn^{4+} + 2\ O^{2-}(\square) \longrightarrow 4\ Mn^{3+} + O_2 \qquad (5)$$

In Equation (5) the excess oxygen charge is compensated by cation vacancies (noted here as (□)) following the suggestions of van Roosmalen et al [38].

The β-oxygens in the surface lattice of two series of perovskites can only be desorbed at high temperature (T>700 °C) leading to the B-site ion reduction and, surface anion vacancy generation.

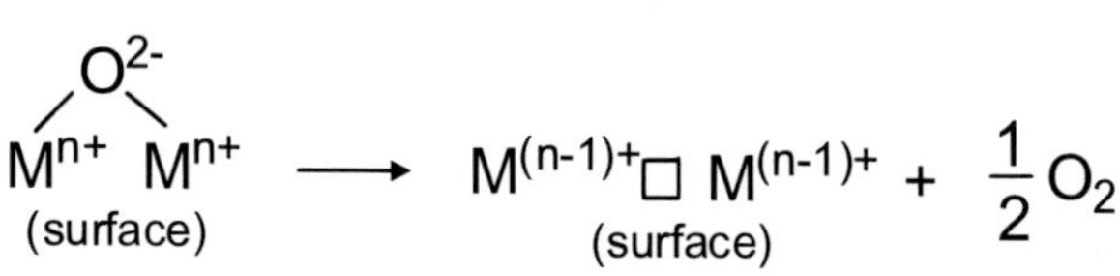

(6)

β-O_2 desorption is maintained via the diffusion of lattice oxygen from the bulk to the surface:

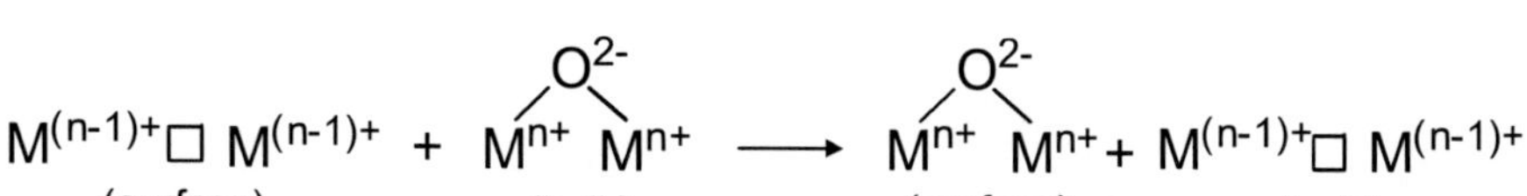

(7)

This β-oxygen desorption is promoted by Cu substitution as indicated by the results of O_2-TPD experiments in accordance with the conclusion that Cu incorporation enhanced the mobility of lattice O_2 according to our studies of the reducibilities of $LaCo_{1-x}Cu_xO_3$[19], $LaMn_{1-x}Cu_xO_3$[20] and $LaFe_{1-x}(Cu, Pd)_xO_3$[21] by H_2-TPR.

4. NO Reduction by HCs in the Presence of Oxygen Over Perovskites

Nitrogen oxides are one class of pollutants from automobile engine mainly due to the thermal reaction between N_2 and O_2 from the air at high temperature caused by the ignition in the combustion chamber. Nitric oxide (NO) is the main component of these oxides with a molar ratio above 90%. Close to one hundred hydrocarbons components present in the engine exhaust and are designated below as total hydrocarbons (THCs). They are generated from the incomplete combustion of fuel. Among them, C_3H_6 is often used as a model reducing hydrocarbon gas for the study of NO catalytic reduction. Attempts to use Co-, Mn-, and Fe-based perovksites as catalysts for the SCR of NO by C_3H_6 in the presence of O_2 have been performed in our group. The research was carried out aiming at establishing the different physicochemical properties of $La(Co, Mn, Fe)_{1-x}(Cu, Pd)_xO_3$, as well as illustrating the correlation between the modification of their structure and their catalytic behavior. Another objective of this work was to propose a reaction mechanism for NO reduction over series of perovskites.

4.1. Adsorption of Reactant Molecules

4.1.1. TPD of NO + O_2

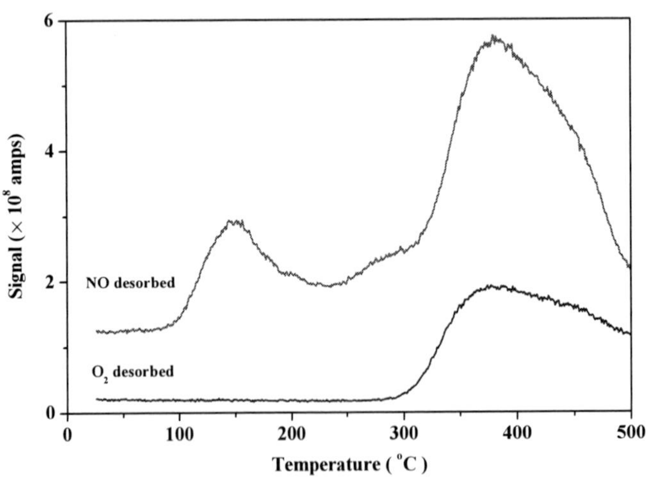

Figure 7. TPD of NO + O_2 profiles over $LaCo_{0.8}Cu_{0.2}O_3$ perovskite [19].

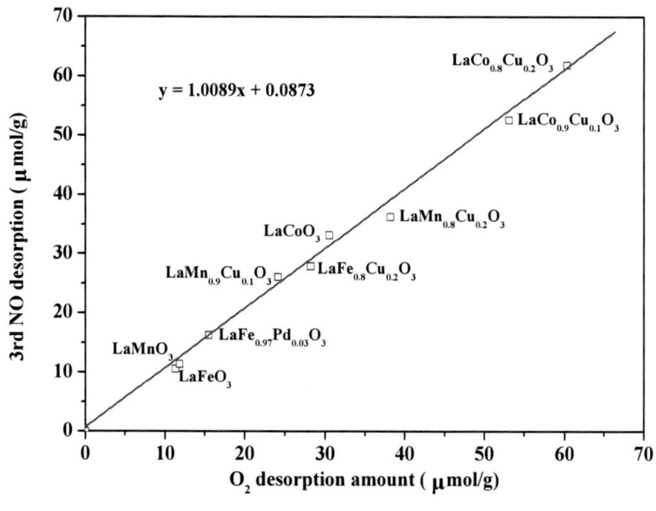

Figure 8. Relationship between O_2 and the corresponding NO desorption at T > 300 °C during NO + O_2-TPD over Co-, Mn-, and Fe-based perovskites [19-21].

Different kinds of species including nitrosyl[39-41], nitrite[40, 42, 43] and nitrate species[40, 42-44] were observed in the literature by means of TPD and FTIR during NO adsorption over ion-exchanged ZSM-5 and MCM-41, metal supported Al_2O_3, lanthanum cobaltites and ferrites perovskites. In the present study, the adsorbed species found during co-adsorption of NO and O_2 over Co-, Mn-, Fe-based perovskites were investigated and compared by means of recording the MS signals of NO (m/e=30), O_2 (m/e=32), N_2O (m/e=44) and N_2 (m/e=14) in the effluent of NO+O_2-TPD tests. The amounts of various species were quantified from the deconvoluted peaks as listed in Table 4. The NO desorption trace (m/e=30) for $LaCo_{0.8}Cu_{0.2}O_3$, reported in Figure 7, shows essentially three features: a small peak centered at 150-170 °C, a broad shoulder at 250-300 °C and a major peak with a maximum at 380-410 °C.

Table 4. Amounts of NO, N_2O, N_2 and O_2 desorbed from Co-, Mn-, and Fe-based perovskites during NO+O_2-TPD experiments [19-21]

sample	NO ($\mu mol \cdot g^{-1}$) 1	2	3	total	O_2 ($\mu mol \cdot g^{-1}$)	3rd NO/O_2 [a]	N_2O (amp·s)	N_2 ($\mu mol \cdot g^{-1}$)
$LaCoO_3$	150 °C	280 °C	380 °C	48.1	30.5	1.09	-	-
	9.7	5.2	33.2					
$LaCo_{0.9}Cu_{0.1}O_3$	146 °C	277 °C	392 °C	59.8	53.1	0.99	-	-
	3.1	4.1	52.6					
$LaCo_{0.8}Cu_{0.2}O_3$	164 °C	280 °C	403 °C	64.5	60.3	1.03	-	-
	1.0	1.6	61.9					
$LaMnO_3$	167 °C	283 °C	394 °C	35.9	11.8	0.95	5.8	3.3
	5.8	18.7	11.4					
$LaMn_{0.9}Cu_{0.1}O_3$	184 °C	268 °C	386 °C	45.0	24.1	1.08	6.7	3.7
	6.6	12.3	26.1					
$LaMn_{0.8}Cu_{0.2}O_3$	177 °C	271 °C	422 °C	58.8	38.2	0.97	8.3	4.6
	2.3	21.1	36.4					

Table 4. (Continued)

sample	NO ($\mu mol \cdot g^{-1}$) 1	2	3	total	O_2 ($\mu mol \cdot g^{-1}$)	3rd NO/O_2 [a]	N_2O (amp·s)	N_2 ($\mu mol \cdot g^{-1}$)
$LaFeO_3$	176 °C	272 °C	360 °C	49.2	11.3	0.94	2.1	3.1
	19.1	19.5	10.6					
$LaFe_{0.8}Cu_{0.2}O_3$	207 °C	293 °C	382 °C	63.5	28.2	0.99	4.5	5.7
	15.3	20.3	27.9					
$LaFe_{0.97}Pd_{0.03}O_3$	191 °C	279 °C	364 °C	51.6	15.5	1.05	2.8	4.6
	15.5	19.8	16.3					

[a] The molar ratio between amount in the third NO desorption peak (T>300 °C) and amount of O_2 desorbed over the same temperature range.

The oxygen trace shows that oxygen desorbs from perovskite simultaneously with the high temperature NO peak in amount that essentially increases with Cu substitution (see Table 4). It is remarkable that the high temperature NO peaks and O_2 peaks are essentially located at the same temperature and even the shapes of the peaks look alike (Figure 7). This is likely related to the report of Coq et al.[45] that the desorption of NO_3^- species would be as a NO desorption peak appearing at high temperature in the range 300-400 °C with a parallel O_2 desorption. Moreover, the numbers of moles of NO desorbed at high temperature are plotted against those of O_2 for all investigated Co-, Mn-, Fe-based perovskites as described in Figure 8. Given the imprecisions associated with curve fitting of NO traces, the data are reasonably fitted with a line of slope of 1 indicating that the desorption ratio of NO/O_2 is essentially 1 over these catalysts. This strongly suggests again that the high temperature NO desorption is associated with the nitrate species having the general formula NO_3^-. This may indicate that the oxidation of NO involves essentially α-oxygen as O_2^- ion radicals formed upon O_2 adsorption during calcination:

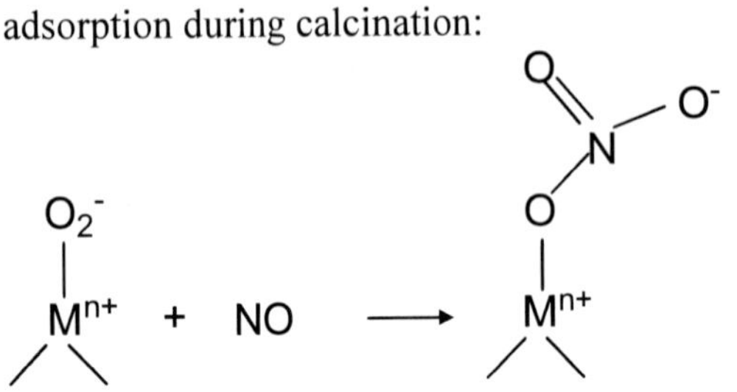

(8)

In Eq.(8), the monodentate nitrate species is written as an example of surface nitrate species which does not exclude possible bridging and bidentate species.

Desorption of nitrate species leads to the formation of NO and O_2 with a molar ratio

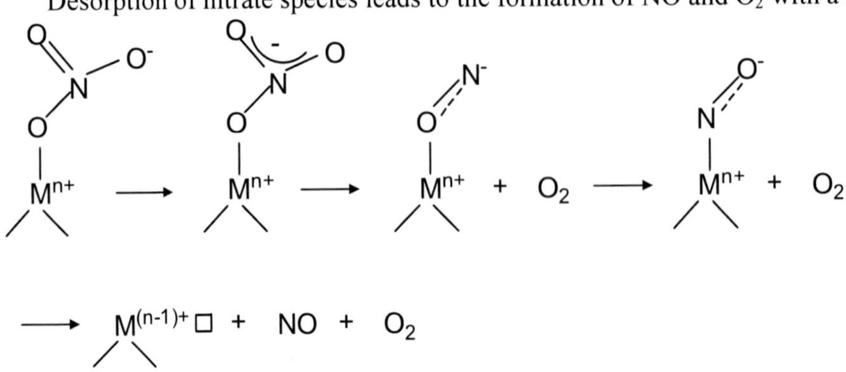

(9)

Experiments over Cu/ZSM-5, Centi and Perathoner[46] concluded that the thermal stability of adsorbed species increased with increasing oxidation state of nitrogen in nitrogen adspecies. Valyon and Hall[41] also found that the nitrosyl species adsorbed on Co-ferrierite is weakly adsorbed compared to nitrate species. Therefore, the second NO desorption peak occurring at medium temperatures likely corresponds to the nitrosyl species, according to their thermal stability.

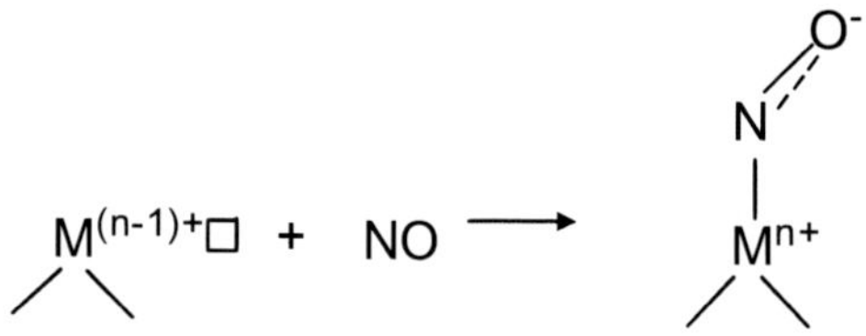

(10)

Finally, the first NO desorption peaks occurring at temperatures lower than 200 °C in NO + O_2-TPD patterns of prepared perovskites were ascribed to physically adsorbed NO.

4.1.2. TPD of C_3H_6

The adsorption and transformation of C_3H_6 over lanthanum cobaltites, manganites, and ferrites were also investigated via C_3H_6-TPD experiments with quantitative analysis of the various gases desorbed from perovskites as listed in Table 5. CO and CO_2 are the main carbonaceous species over tested samples during C_3H_6-TPD. Significant C_3H_6 desorption occurs merely over unsubstituted samples at relatively low temperature range of 50-70 °C. These results indicated that a transformation from C_3H_6 into CO_2 had already taken place over perovskites during TPD process, resulting in a CO_2 desorption at high temperature. This C_3H_6 transformation was promoted by Cu substitution, leading to the disappearance of the C_3H_6 desorption peak replaced by CO and CO_2 desorption peaks at elevated temperature. Taking into account the result that α-oxygen of Co-, Mn-, Fe-based catalysts is increased by Cu substitution, it was thus thought that α-O_2 bonding to oxygen vacancies plays an important role in the C_3H_6 activation. From our C_3H_6-TPD study, the following scheme may be proposed for the adsorption C_3H_6 over various perovskites.

Table 5. Amount of C_3H_6, CO, and CO_2 desorbed from Co-, Mn-, and Fe-based perovskites during C_3H_6-TPD experiments [19-21]

Sample	C_3H_6 (μmol·g^{-1})	CO (μmol·g^{-1})	CO_2 (μmol·g^{-1})
$LaCoO_3$	2.83 (54 °C)	14.2 (> 210 °C)	155.8 (> 270 °C)
$LaCo_{0.9}Cu_{0.1}O_3$	0.84 (50 °C)	148.1 (> 210 °C)	268.3 (> 270 °C)
$LaCo_{0.8}Cu_{0.2}O_3$	0.52 (49 °C)	161.4 (> 210 °C)	301.9 (> 270 °C)
$LaMnO_3$	16.5 (70 °C)	10.2 (> 250 °C)	52.3 (> 250 °C)
$LaMn_{0.9}Cu_{0.1}O_3$	5.3 (60 °C)	21.3 (> 220 °C)	226.9 (> 250 °C)
$LaMn_{0.8}Cu_{0.2}O_3$	3.1 (65 °C)	22.1 (> 220 °C)	283.4 (> 250 °C)
$LaFeO_3$	3.5 (60 °C)	57.2 (> 250 °C)	72.9 (> 200 °C)
$LaFe_{0.8}Cu_{0.2}O_3$	2.1 (49 °C)	34.0 (> 250 °C)	239.4 (> 200 °C)
$LaFe_{0.97}Pd_{0.03}O_3$	2.3 (60 °C)	174.1 (> 250 °C)	342.7 (> 200 °C)

Using the site described by the left hand side of Equation (3), the olefin should adsorb on the electron deficient part of the site, namely:

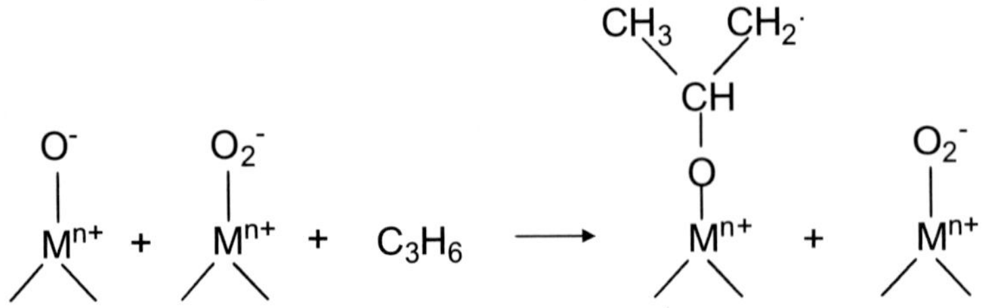

(11)

The -$OC_3H_6^{\cdot}$ radical group (allylic adspecies) can now react with the surrounding α-O_2 to form CO_2 and water as described in ref. [47].

4.1.3. TPSR of NO + O_2 in C_3H_6/He flow

$LaCo_{0.8}Cu_{0.2}O_3$ perovskite was pretreated under an atmosphere of 3000 ppm NO + 1% O_2, and the surface nitrogen-containing species formed were further reacted with 1000 ppm C_3H_6/He while rising temperature up to 500 °C at a heating rate of 5 °C/min. The corresponding TPSR profile is shown in Figure 9. It is clearly observed that the desorption features related to physically and chemically adsorbed NO at T<300 °C in Figure 9, are similar to those obtained in the TPD of NO+O_2 reported in Figure 7. The NO desorption at high temperature (T>300 °C) in Figure 7, which was ascribed to nitrate species, is not present any more in the traces of Figure 9 suggesting that among the adsorbed NO, only the nitrate specie is highly active toward C_3H_6, leading to its complete consumption by this reducing agent. N_2 was also detected during this experiment again suggesting that the reduction of the nitrate surface species by propylene is associated with the SCR process.

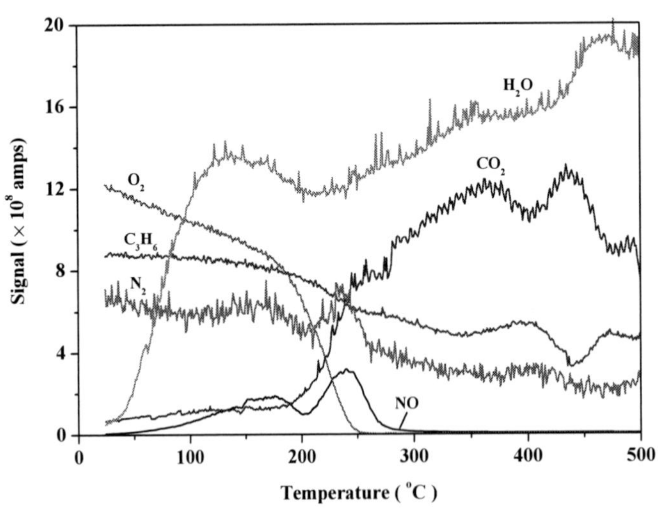

Figure 9. MS signals during TPSR of NO + O_2 in C_3H_6/He flow over $LaCo_{0.8}Cu_{0.2}O_3$ perovskite [19].

4.2. Catalytic Performance

The temperature dependence of NO conversion to N_2 by reaction with propene over La(Co, Mn, Fe)$_x$(Cu, Pd)$_x$O$_3$ is shown in Figure 10a. Among unsubstituted samples the worst NO conversion was obtained over $LaFeO_3$, whereas the best one was achieved over $LaMnO_3$. An enhanced NO conversion was observed over these perovskites after Cu substitution resulting in the NO conversions roughly following the order: $LaFeO_3 < LaCoO_3 < LaMnO_3 \approx LaCo_{0.8}Cu_{0.2}O_3 < LaMn_{0.8}Cu_{0.2}O_3 \approx LaFe_{0.8}Cu_{0.2}O_3$. The best performance for this reaction was achieved in the case of $LaFe_{0.97}Pd_{0.03}O_3$. Organo-nitrogen compounds (ONCs) were detected in the effluent at high temperature as depicted in Figure 10b. These organo-nitrogen compounds are widely mentioned in the literature as intermediates during SCR of NO by hydrocarbons[43, 47, 48]. The catalytic behavior of Co-, Mn- and Fe-based perovskites in C_3H_6 conversion is described in Figure 9c. It was found that C_3H_6 conversion could take place at relatively low temperature compared to NO conversion to N_2, with a sharp increase up to 60-80% in the range of 200-400 °C (except for $LaFeO_3$) and approaching a stable level at further elevated temperature. Cu substitution promotes the C_3H_6 conversion especially in the region of 300-400 °C.

The catalytic performances, including yields of N_2, NO_2, N_2O, and organo-nitrogen compounds, conversions of C_3H_6 and O_2, as functions of O_2 concentration over $LaCo_{0.8}Cu_{0.2}O_3$ at 450 °C are depicted in Figure 11. With an increase in O_2 concentration, N_2 production was gradually inhibited. The formation of N_2O and organo-nitrogen compounds was lower than 6%, and totally suppressed when O_2 concentration is higher than 1%. On the contrary, the NO_2 production was only detected at O_2 concentration above 1% approaching a maximum of about 5% yield at 2% O_2.

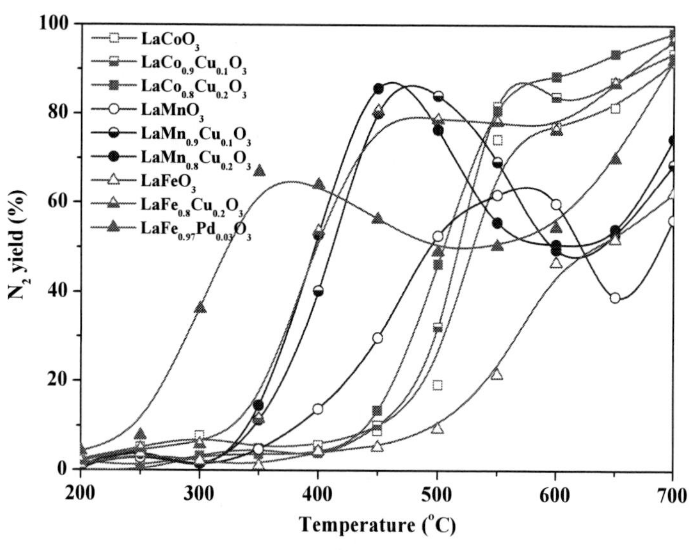

Figure 10a. N2 yield in C3H6 + NO + O2 reaction over perovskites [19-21] Conditions: GHSV=55,000 h-1, 3000 ppm C3H6, 3000 ppm NO, 1% O2.

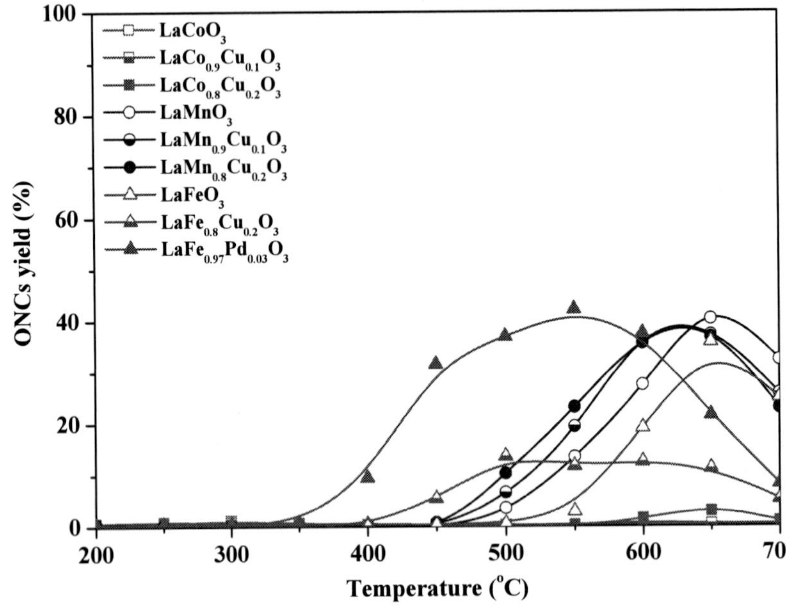

Figure 10b. ONCs yield in C_3H_6 + NO + O_2 reaction over perovskites [19-21] Conditions: GHSV=55,000 h^{-1}, 3000 ppm C_3H_6, 3000 ppm NO, 1% O_2.

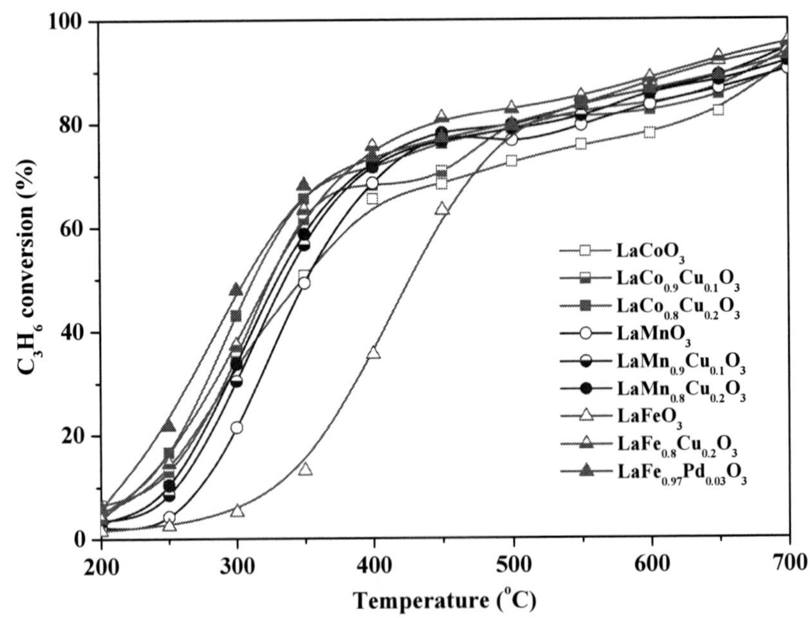

Figure 10c. C_3H_6 conversion in C_3H_6 + NO + O_2 reaction over perovskites [19-21] Conditions: GHSV=55,000 h^{-1}, 3000 ppm C_3H_6, 3000 ppm NO, 1% O_2.

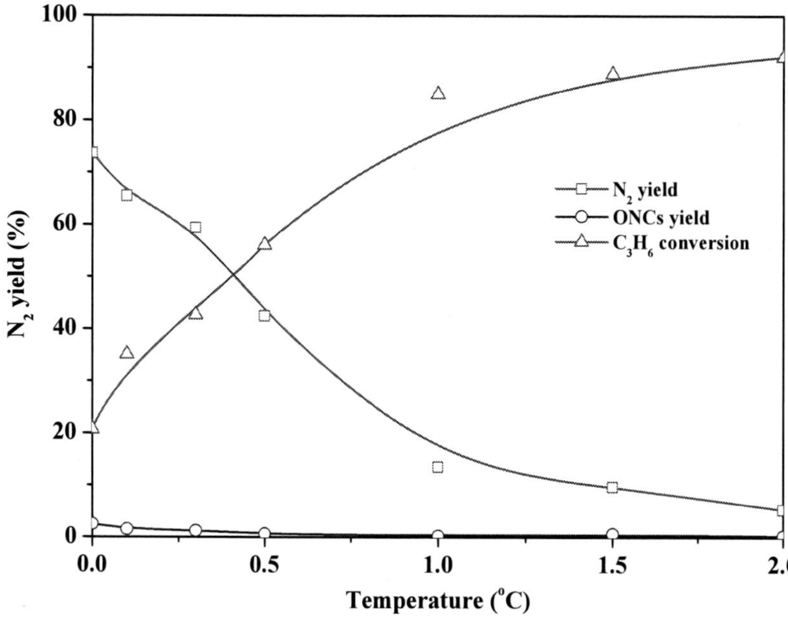

Figure 11. Effect of O_2 feed concentration on NO reduction by C_3H_6 over $LaCo_{0.8}Cu_{0.2}O_3$ [19]. Conditions: T=450 °C, GHSV=55,000 h^{-1}, 3000 ppm C_3H_6, 3000 ppm NO, 0-20,000 ppm O_2.

4.3. Reaction Mechanism

The high activity of nitrate species to react with propene was established via TPSR of $NO+O_2$ in C_3H_6/He tests indicating that the formation of nitrate species is the first important step for NO reduction. This is in agreement with our previous finding in C_3H_6-SCR of NO over Cu-MCM-41[30]. Furthermore, organo nitrogen compounds, identified to be mainly composed of 1-nitropropane (C_3H_7-NO_2) by GC-MS, were detected in the region of 500-700 °C during the NO reduction (Figure 10b), which strongly implies a mechanism with organo-nitrogen compounds as intermediates. The formation of organo nitrogen compounds via reaction of adsorbed hydrocarbons with the surface nitrate was reported as the rate-determining step of the whole SCR of NO process[49-51]. The rate of nitrate consumption was also proven to be close to that of N_2 formation over Ce/ZSM-5[51] and Cu/ZSM-5[42] according to FT-IR kinetic study. Hence, the formation of 1-nitropropane should involve a reaction between the adsorbed species formed in reaction (11) and the nitrate species:

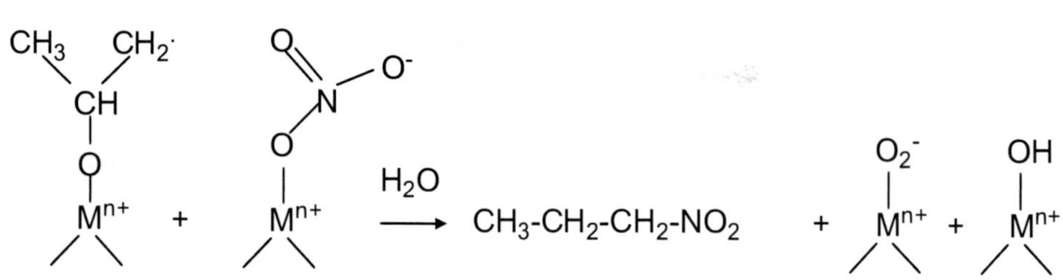

(12)

Haj et al. [52] investigated the reactivity of 1-nitropropane and found it can be decomposed under 10% O_2 or react with NO + O_2 exhibiting an extraordinarily high activity compared to the gas mixture of C_3H_6 + NO, the final products being N_2, CO_2 and H_2O. Although the detailed scheme for 1-nitropropane reaction wasn't well established, this result stresses the fact that such nitro compound species are highly reactive in the presence of O_2. In our case the corresponding surface coverage of nitro compounds was probably too low to allow detection under oxidative atmosphere. As a consequence, 1-nitropropane which would be consumed rapidly, would rather act as an intermediate than as a byproduct under the usual reaction conditions with a mixture of 3000 ppm C_3H_6 + 3000 ppm NO + 1% O_2. It was however also detected in the effluent of NO-SCR by C_3H_6 in the absence of O_2. Based on the suggestion of Blower and Smith[53], a further transformation of R-NO_2 into isocyanate (R-NCO) by a cyclic intermediate could be formulated as:

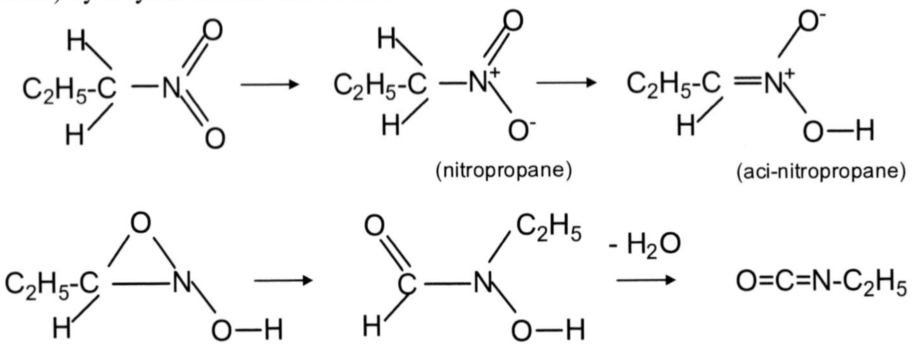

(13)

Finally, the C_2H_5NCO was proposed to react with NO via the coupling of nitrogen atoms to yield the product of N_2 and CO_2 according to the assumption of Witzel et al[54]. At the same time, the participation of O_2 can promote the oxidation of the ethyl-radical into CO_2 and H_2O and accelerate the formation of N_2 via the coupling of nitrogen atoms.

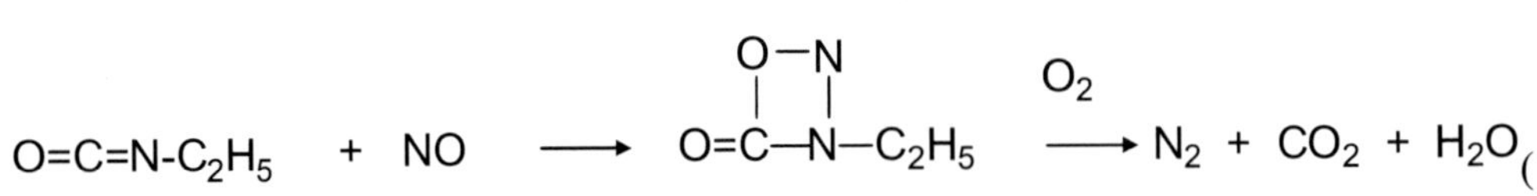

(14)

In brief, the global mechanism, as mentioned above, can be summarized as follows: firstly, adsorbed propene reacts with the nitrate species, which are formed via NO oxidation by α-oxygen over perovskites, resulting in the formation of organo-nitrogen compounds mainly composed of 1-nitropropane. Then, the organo-nitrogen compounds are converted to the isocyanate. Finally, the N_2 is formed by the coupling of nitrogen atoms from isocyanate species and NO. On the other hand, propene was oxidized not only by O_2 but also by the nitrate species.

5. NO Catalytic Reduction by CO over Perovskites

As early as 1952, perovskites were proposed as catalytic materials for CO oxidation[55]. Twenty years later, their potential application as catalysts for automobile exhaust purification was pointed out by Libby[56]. Much attention has been paid in the last three decades to perovskites as candidates for the NO reduction using CO as a reducing gas to simultaneously control the emission of those two pollutants from motor vehicles, with the studies mainly focusing on the following compositions: La(Fe, Mn)O_3[57], $La_{1-x}Sr_xFeO_3$[58], $La_{1-x-y}Sr_xCe_yFeO_3$[59], La_2CuO_4[60], La(Cr, Mn, Co, Ni)O_3[60], $LaCoO_3$[22]. The same three series of La(Co, Mn, Fe)O_3 and their derivatives of B-site substituted by Cu^{2+} and Pd^{2+} were used for the CO + NO reaction. A mechanism has also been proposed.

5.1. Adsorption of Reactant Molecules

5.1.1. TPD of NO

Table 6. Amount of NO desorbed from Co-, Mn-, and Fe-based perovskites during NO-TPD experiments [61, 62]

	Physically desorbed NO (μmol·g^{-1})	Nitrosyl species (μmol·g^{-1})	Nitrate species (μmol·g^{-1})	Total (μmol·g^{-1})	Total (μmol·m^{-2})
$LaCoO_3$	90 °C / 2.4	201 °C / 27.6	388 °C / 9.4	37.0	1.27
$LaCo_{0.9}Cu_{0.1}O_3$	-	255 °C / 39.9	-	39.9	1.68
$LaCo_{0.8}Cu_{0.2}O_3$	-	283 °C / 45.7	-	45.7	2.04
$LaMnO_3$	100 °C / 8.8	202 °C / 15.9	311 °C / 5.8	21.7	0.53
$LaMn_{0.9}Cu_{0.1}O_3$	106 °C / 5.7	210 °C / 23.5	323 °C / 6.0	29.5	0.72
$LaMn_{0.8}Cu_{0.2}O_3$	-	209 °C / 29.5	361 °C / 13.7	43.2	1.01
$LaFeO_3$	84 °C / 4.5	214 °C / 20.2	283 °C / 1.8	22.0	0.72
$LaFe_{0.97}Pd_{0.03}O_3$	80 °C / 4.1	222 °C / 24.2	317 °C / 2.9	27.1	0.65
$LaFe_{0.8}Cu_{0.2}O_3$	86 °C / 2.1	225 °C / 36.3	334 °C / 3.3	39.6	0.82

The MS signals of NO (m/e=30) and O_2 (m/e=32) as functions of desorption temperature were recorded during NO-TPD under He flow for perovskites (Table 6). A broad NO desorption centered at 200 °C with one minor shoulder at 90 °C and another one at 388 °C was observed in the TPD of NO profiles for $LaCoO_3$. Upon Cu substitution the low temperature peak is entirely suppressed, the 200 °C one is substantially increased and shifted to higher temperature so that it interferes with the minor high temperature one. Similar NO desorption features were found for lanthanum manganites and ferrites after NO adsorption, representing three superimposed desorption peaks. In Section 4, TPD of NO + O_2 studies over

Co-, Mn-, and Fe-based perovskites, NO peaks appearing at similar temperatures as those in NO-TPD experiments (apart from a minor desorption at medium temperature and an intense one at high temperature) were also found and ascribed to the desorptions of physically adsorbed NO, nitrosyl species and nitrate species in the order of their thermal stability. The NO peaks obtained at low (80-110 °C), medium (200-210 °C), and high (310-390 °C) temperatures in the present NO-TPD analyses were thus correlated to physically adsorbed NO, nitrosyl and nitrate species.

The intensities of nitrosyl species for unsubstituted $LaCoO_3$, $LaMnO_3$ and $LaFeO_3$ is enhanced upon Cu or Pd substitution. Based on the fact that more oxygen vacancies can be formed by substituting trivalent cations (Co^{3+}, Mn^{3+}, Fe^{3+}) by bivalent ones (Cu^{2+}, Pd^{2+}) (due to a positive charge deficiency confirmed by O_2-TPD experiments, see Table 2), this intense NO desorption peak related to nitrosyl species is likely formed by NO adsorption on anion vacancies as verified by Shin et al. via FT-IR study[63]. The NO chemisorption as a negatively charged form (NO^-) was thought to be the first important step for NO + CO reaction [64] because of the back-donation which occurs through antibonding orbitals, determines a weakening of the N-O bond when net electron transfer is from metal to NO. The formation of nitrosyl species is formulated on the basis of the above discussion (Eq.10).

5.1.2. TPD of CO

The desorption of CO (m/e=28) and CO_2 (m/e=44) was recorded by MS during the CO-TPD experiments for $La(Co, Mn, Fe)_{1-x}(Cu, Pd)_xO_3$ samples. A quantitative analysis of the various carbonaceous gases desorbed from perovskites is summarized in Table 7. Both CO and CO_2 desorptions were observed indicating that the oxidation of CO into CO_2 occurs over those lanthanum perovskites during CO-TPD experiments.

Table 7. Amounts of CO and CO_2 desorbed from Co- and Mn-based perovskites during CO-TPD experiments [61, 62]

Sample	CO		CO_2		Total carbonaceous species
	T (°C)	amount (μmol·g^{-1})	T (°C)	amount (μmol·g^{-1})	amount (μmol·g^{-1})
$LaCoO_3$	>200	26.6	>200	116.4	143.0
$LaCo_{0.9}Cu_{0.1}O_3$	>200	22.5	>150	166.4	188.9
$LaCo_{0.8}Cu_{0.2}O_3$	>200	18.2	>150	195.0	213.2
$LaMnO_3$	>250	40.5	>200	66.7	107.2
$LaMn_{0.9}Cu_{0.1}O_3$	>250	56.9	>200	95.3	152.2
$LaMn_{0.8}Cu_{0.2}O_3$	>250	70.5	>200	104.3	174.8
$LaFeO_3$	>250	53.2	>350	43.9	97.1
$LaFe_{0.97}Pd_{0.03}O_3$	>250	36.9	>200	102.5	139.4
$LaFe_{0.8}Cu_{0.2}O_3$	>225	25.0	>220	136.9	161.9

According to the reports of Tascón et al., the inhibiting effect of NO on subsequent CO adsorption is larger than the inhibiting effect of CO on NO adsorption over both $LaCoO_3$ and $LaMnO_3$[65]. It is thus speculated that CO appears to be more weakly adsorbed than NO at the same site of perovskites. The CO adsorption thus likely involves an adsorption at anion vacancies via a coordinative bond without electron transfer:

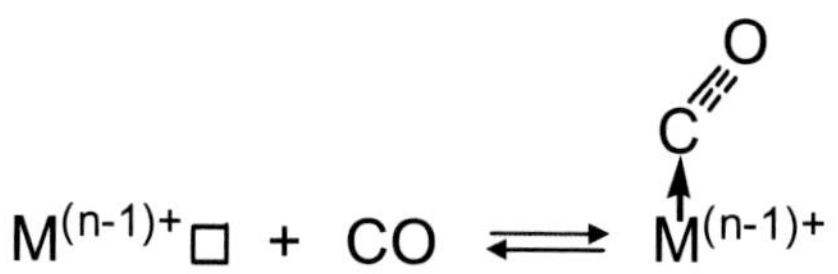

(15)

CO is oxidized into CO_2 over perovskites during CO-TPD experiments (Table 7) and this CO oxidation is promoted by Cu substitution leading to an enhancement in CO_2 desorption. It was reported that CO oxidation happens via a suprafacial catalytic process involving a surface lattice atomic oxygen (O^-)[5]. The CO oxidation therefore was proposed as the following step:

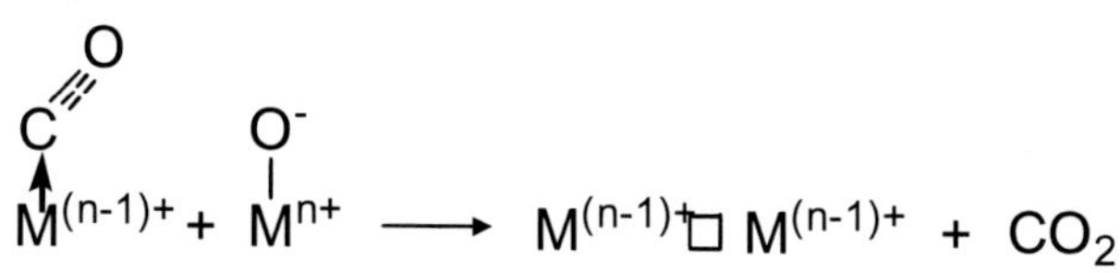

(16)

In the presence of gaseous O_2, this CO oxidation can also be accelerated by α-O_2 over perovskite[66] expressed as the following reaction.

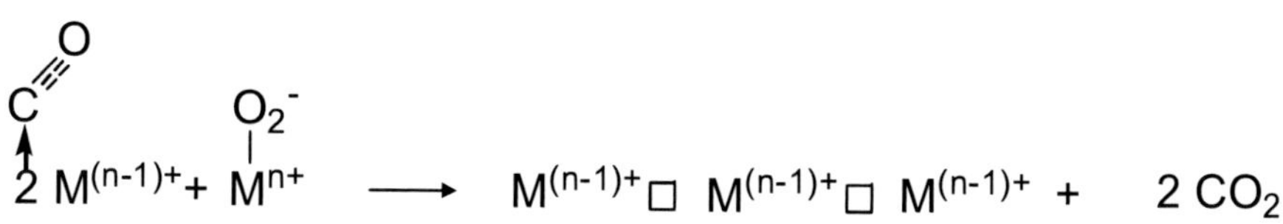

(17)

5.2. Catalytic Performance

Figure 12a shows the variations of N_2 yields with temperature. Among the six tested samples, the lowest conversion to N_2 is obtained in the case of $LaMnO_3$ with a progressive increase up to 76% at 500 °C. A higher N_2 yield with a maximum of 93% for $LaCoO_3$ and 95% for $LaFeO_3$ was observed at the same temperature. An obvious enhancement in N_2 yield is realized by Cu (or especially at Pd) partial substitution for the three series of perovskites resulting in an order of $LaFe_{0.97}Pd_{0.03}O_3 > LaCo_{0.8}Cu_{0.2}O_3 > LaCo_{0.9}Cu_{0.1}O_3 > LaMn_{0.8}Cu_{0.2}O_3 > LaFe_{0.8}Cu_{0.2}O_3 > LaMn_{0.9}Cu_{0.1}O_3 > LaFeO_3 > LaCoO_3 > LaMnO_3$. N_2O is detectable in the effluent with maxima in the range of 270-330 °C depending on the solid composition as illustrated in Figure 12b. More significant N_2O production and CO conversion were found over $LaCoO_3$ compared to those obtained over $LaMnO_3$ (see Figures 12b and 12c). The N_2O production and CO conversion obtained over lanthanum cobaltite and lanthanum manganite can be further improved via Cu substitution. In general, the best performance is achieved over $LaFe_{0.97}Pd_{0.03}O_3$. No obvious NO_2 formation was detected during activity tests of these perovskites under the atmosphere of 3000 ppm NO and 3000 ppm CO.

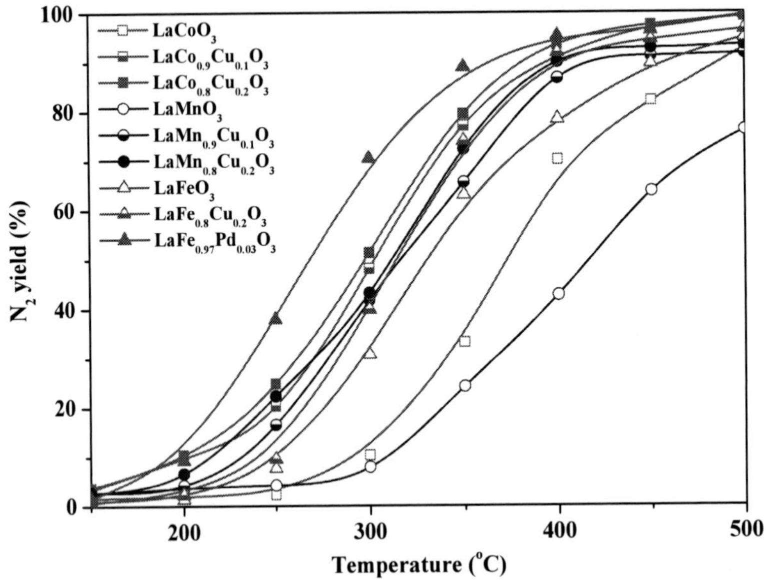

Figure 12a. N_2 yield in CO + NO reaction over perovskites [61, 62]. Conditions: GHSV=50,000 h^{-1}, 3000 ppm NO, 3000 ppm CO, balanced by He.

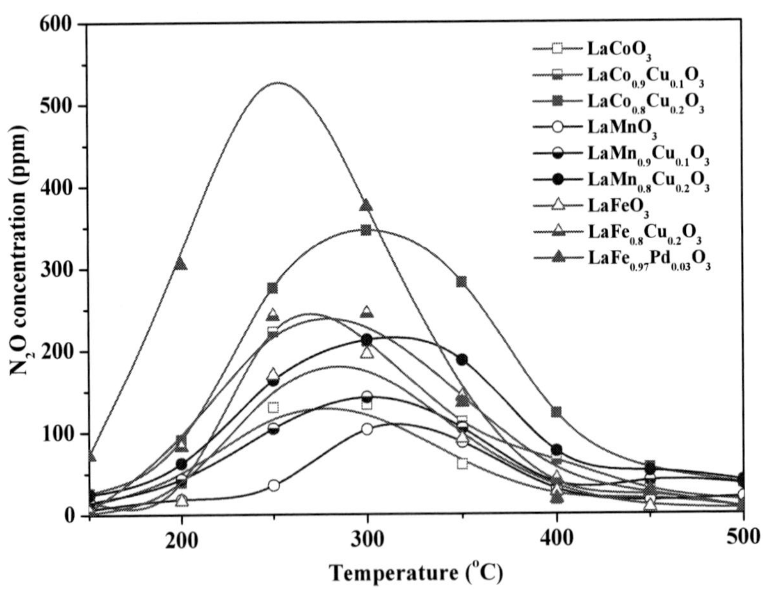

Figure 12b. N_2O concentration in CO + NO reaction over perovskites [61, 62]. Conditions: GHSV=50,000 h^{-1}, 3000 ppm NO, 3000 ppm CO, balanced by He.

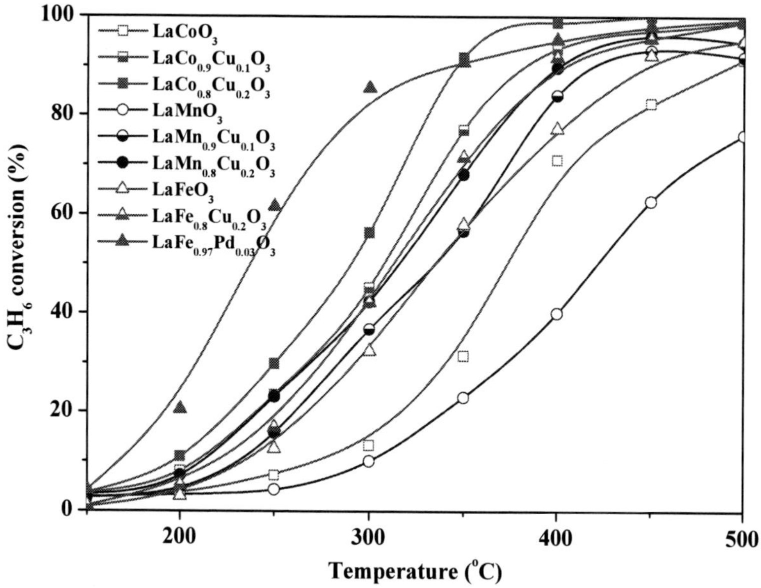

Figure 12c. C3H6 conversion in CO + NO reaction over perovskites [61, 62]. Conditions: GHSV=50,000 h-1, 3000 ppm NO, 3000 ppm CO, balanced by He.

For the $LaCo_{0.8}Cu_{0.2}O_3$, N_2 yields in the reactions of NO + CO, N_2O + CO, NO decomposition and N_2O decomposition were investigated and the results are shown in Figure 13a with the objective of better understanding the NO and CO transformation in this catalytic reduction. In the absence of reducing agent, a little NO decomposition (<5%) occurs at 350-500 °C. N_2O decomposition is much easier than NO decomposition starting at 300 °C and achieving a N_2O conversion of 68% at 500 °C, possibly related to the easy cleavage of N-O bond in N_2O. Nevertheless, it is noticed that N_2O decomposition does not take place at T<300 °C. The transformation of nitrogen oxides (NO and N_2O) into N_2 is significantly improved by the CO reducing agent, resulting in N_2 yields of 51% in NO + CO reaction and 74% in N_2O + CO reaction at 300 °C and almost reaching 100% in both reactions at 500 °C. The better N_2 yield in N_2O reduction by CO is ascribed to the easier molecular activation of N_2O compared to NO.

CO conversions in reaction of NO + CO, N_2O + CO, CO oxidation by perovskite (no oxygen), CO + O_2 over $LaCo_{0.8}Cu_{0.2}O_3$ are compared in Figure 13b. CO oxidation can still carry on with a maximum conversion of 42% at 400 °C in the absence of O_2, which is likely realized by consuming α-oxygen from the perovskite surface. With the consumption of adsorbed O_2 species by CO oxidation, a decline in CO oxidation occurs. Nitrogen oxides (NO and N_2O) promote the CO oxidation up to 56% and 71% at 300 °C, respectively. N_2O is again more active than NO to donate its oxygen atom and oxidize CO. Gaseous oxygen with the same atomic O concentration as 3000 ppm NO or N_2O leads to the highest CO oxidation achieving a value of 87% even at 300 °C. The above results suggest an order of the efficiency in oxygen utilization for CO oxidation as: oxygen molecule (O_2) or α-oxygen > oxygen atom (O^{2-}) in NO or N_2O > lattice oxygen (O^{2-}) or β-oxygen.

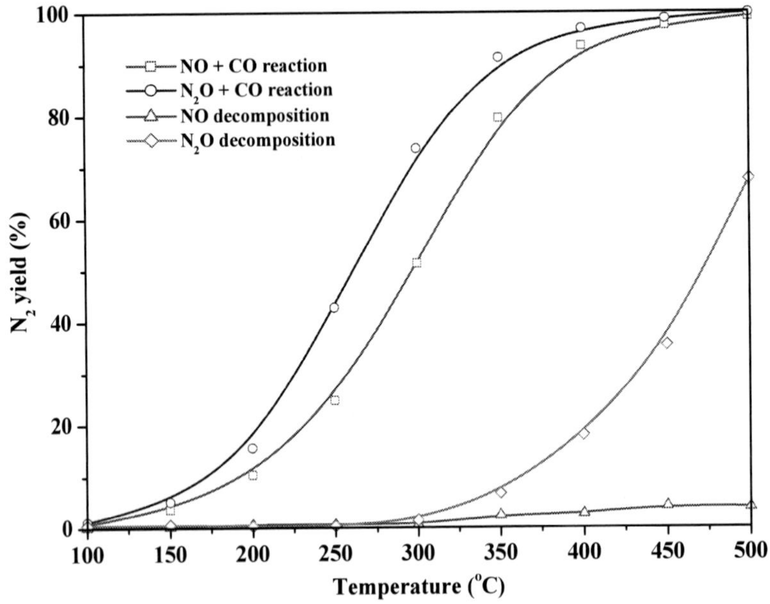

Figure 13a. N_2 yield in various reactions over $LaCo_{0.8}Cu_{0.2}O_3$ [61]. Conditions: GHSV=50,000 h^{-1}, 3000 ppm NO, 3000 ppm N_2O, 3000 ppm CO, balanced by He

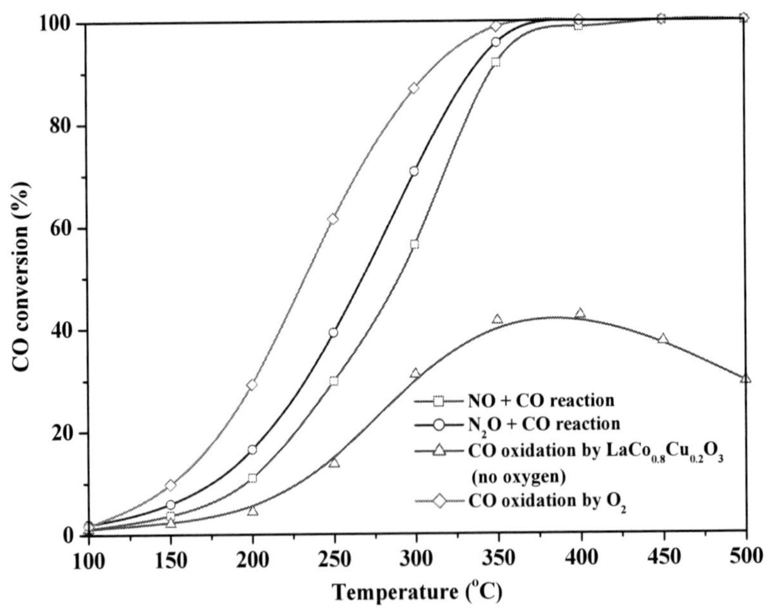

Figure 13b. CO conversion in various reactions over $LaCo_{0.8}Cu_{0.2}O_3$ [61]. Conditions: GHSV=50,000 h^{-1}, 3000 ppm NO, 3000 ppm N_2O, 3000 ppm CO, balanced by He.

5.3. Reaction Mechanism

N_2O decomposition over $LaCo_{0.8}Cu_{0.2}O_3$ showed a better activity in terms of N_2 yield (68% at 500 °C) than that achieved in NO decomposition (4% at 500 °C). N_2O was known to have a linear structure N-N-O in which the N-N bond length is somewhat longer than that in triply bonded N_2 (1.09 Å) and in which the N-O distance is increased by about 0.03 Å relative to free nitric oxide with a bond order of 2.5, resulting in an easy cleavage of its N-O bond and better dissociation behavior. However, enough energy is still necessary for N-O bond breakage according to the result that N_2O was only decomposed as temperatures in excess of 300 °C (Figure 13a).

With the assistance of CO, N_2 yields for both NO and N_2O transformations were significantly improved up to 80% in NO + CO reaction and 91% in N_2O + CO reaction even at 350 °C. Chien[67] et al. observed a higher NO adsorption rate for actived (reduced) than for unactivated $LaCoO_3$ and $La_{0.85}Ba_{0.15}CoO_3$. A reduced perovskite surface with a large amount of anion vacancies was reported to be crucial for the NO dissociation according to FT-IR and EPR spectroscopic analyses[68]. The role of CO seems therefore to maintain a reduced surface, which is necessary for successive NO dissociation. CO oxidation is also promoted by the O atom from NO or N_2O (Figure 13b). From the above results it seems that these tested perovskites act as "oxygen reservoir" transferring oxygen atoms from nitrogen oxides to CO thus achieving NO reduction and CO oxidation simultaneously due to their outstanding redox properties.

At low temperature, the dissociation of adsorbed NO species occurring over reduced perovskite and yielding N_2O and N_2 was recognized as the rate determining step for catalytic reduction of NO by CO. The dimeric species of NO, such as N_2O_2, can be an intermediate, the formation of which involves the N-N bond formation and N-O bond cleavage[40]. Two parallel reactions for chemisorbed NO dissociation occurring over a reduced surface with N_2O and N_2 as the respective products were assumed:

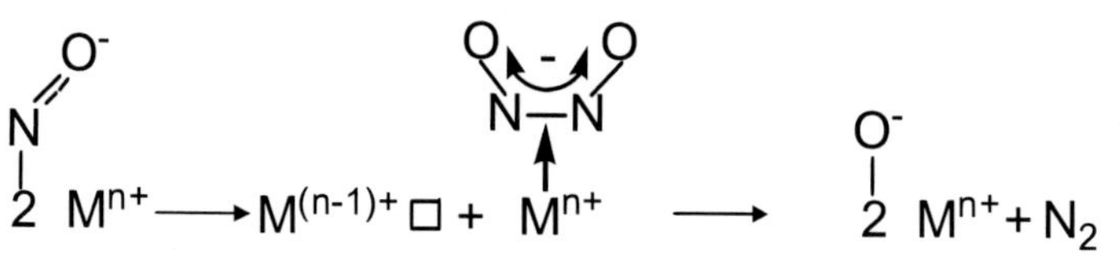

(18)

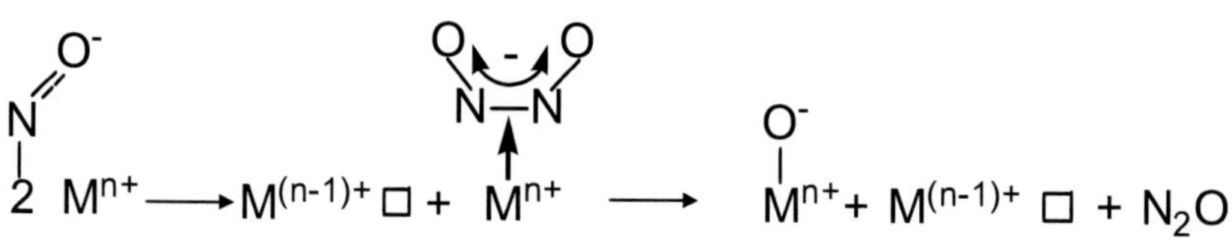

(19)

Reaction (19) involves the breakage of one N-O bond making this reaction much easier than reaction (18) which results in two N-O bond cleavages. This assumption is in agreement with the activity results shown in Figure 13a, indicating that reaction (19) is dominating at low temperature. CO can be oxidized by $M^{n+}O^-$ species as illustrated in Eq.16 together with regeneration of anion vacancies on the surface for a continuous NO dissociation.

In Section 4, we have studied the catalytic reduction of NO by propylene over the same catalysts. In these studies we have found that the main NO reduction process involved surface nitrate species NO_3^-. This is in contrast with the conclusions of the present work which considers the nitrosyl species NO^- as the main adsorbed reactant. The difference between the two systems is due to the higher efficiency of CO (compared to propene) in reducing the perovskite surface.

VOCs Combustion Over Perovskites

Volatile organic compounds (VOCs) are precursory of pollutant smog components[69]. Some VOCs are contained in the automobile exhaust due to the leak, evaporation, and incomplete combustion of fuel. In addition, methane, which can be taken as a reference compound because it belongs to the most difficult to oxidize class of VOCs[70], is abundant in the gaseous exhaust of car driven by natural gas energy. The processes aiming at a control of these compounds in atmospheric emissions are the objects of intense investigations[71-73]. Perovskite-type oxides are of special interest in this context because of their great versatilities and their redox properties since Voorhoeve et al. reported their high activities in catalytic oxidation[74]. The oxidation of hexane and methane over perovskite-type materials was investigated and the corresponding mechanisms and kinetics are discussed in this section.

Two catalysts of $LaCoO_3$ (47.7 m^2/g) and $LaFeO_3$ (22 m^2/g) were prepared by reactive grinding and calcined at 300 °C. Lanthanum cobaltite was shown more active per unit surface area than lanthanum ferrite (see Figure 14). A detailed kinetic study of the catalytic oxidation of n-hexane on those samples was thereafter performed. At low temperature a Langmuir-Rideal mechanism involving non-dissociated adsorbed α-oxygen was proposed, while a Mars-van Krevelen mechanism which uses surface atomic β-oxygen originating from the bulk occurs at relatively higher temperature[75].

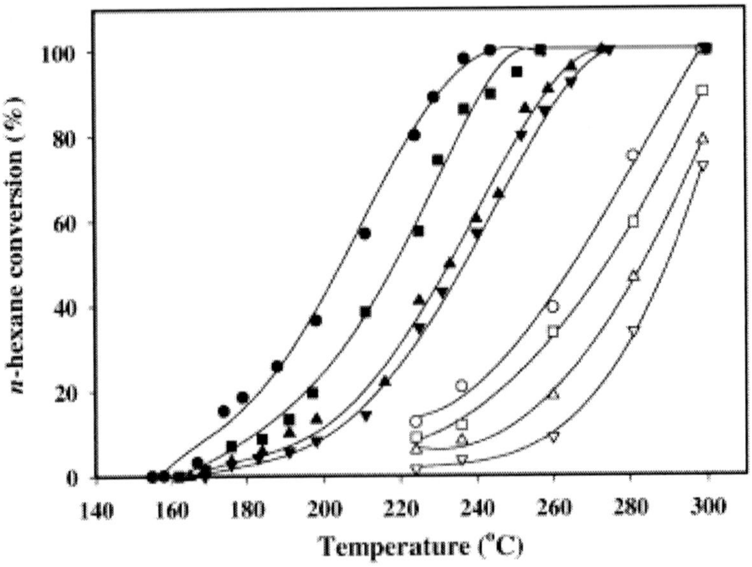

Figure 14. Oxidation of n-hexane over $LaCoO_3$ and $LaFeO_3$. For $LaCoO_3$: (●) 5625 h^{-1},(■) 11250 h^{-1}, (▲) 22500 h^{-1}, (▼) 27750 h^{-1}; for $LaFeO_3$: (○) 5625 h^{-1}, (□) 11250 h^{-1}, (△) 11250 h^{-1}, (▽) 22500 h^{-1} [76].

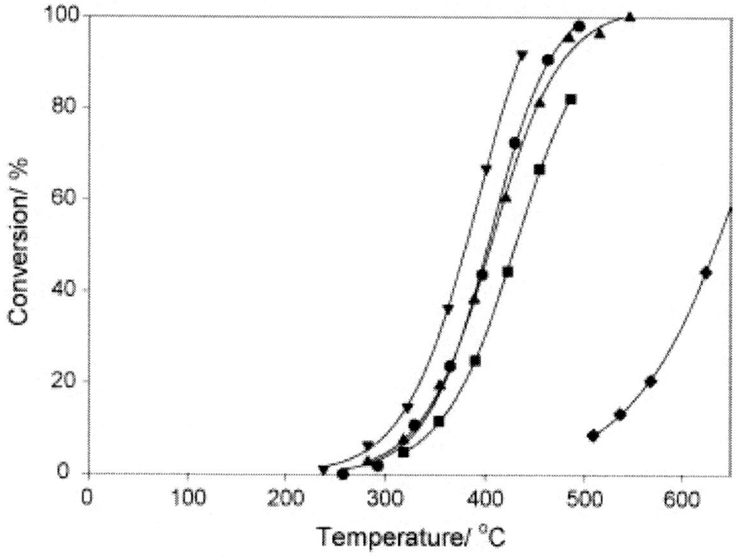

Figure 15. Steady state conversions obtained for the oxidation reaction of 0.25% CH_4 as a function of the reaction temperature for: (▼) COPRG; (●) RG; (▲) CIT; (■) COP; (♦) SS [77].

Five $LaCoO_3$ samples were prepared by different methods: solid state (SS), coprecipitation (COP), citrate complexation (CIT) and reactive grinding of the single oxides (RG) and of an amorphous precursor (COPRG), exhibiting large differences in morphology. Kinetic analysis for the CH_4 oxidation reaction showed that the specific reaction rate is significantly affected by the synthesis procedure as described in Figure 15. Grinding an amorphous precursor results in a perovskite which combines a high specific surface area and a high activity. The RG sample presents a lower activity than COPRG, due to a higher iron contamination from the milling balls.

The mechanism of α-O_2 desorption supposes the formation of some surface anionic vacancies for example on the edge of the crystal dislocations. Because of the higher diffusional rates observed in the grain boundaries, it is supposed that β-oxygen is mainly coming from these intercrystal boundaries. The amounts of oxygen desorbed at these two temperatures were found to be mainly dependant on the specific surface area of the sample, but were not much influenced by the synthesis method. β-oxygen, the amount of which is regarded as characteristic of oxygen mobility in the solid, was found to desorb in amounts strongly dependent on the solid morphology. The results suggest that main β-oxygens do not originate from the bulk of the crystal domains but rather from the grain boundaries formed due to incompatibility of lattice orientations between two neighbor crystal domains. For the CH_4 oxidation reaction, the grain boundary oxygens, which are distinguished from bulk oxygen and are however quickly exchanged with the gaseous oxygen during the isotopic exchange reaction, are proposed to be the source of oxygen active in the oxidation. Then, the oxygen mobility in the bulk of the solid is of special significance for the activity in this case.

7. Deactivation of DeNO$_X$ Perovskite Catalysts by Water Vapor

Steam is invariably present in a real exhaust gas of motor vehicles in relatively high concentration due to the fuel combustion. The influence of water vapor on catalytic performances should not be ignored when dealing with the aim to develop a practical TWCs. Cu/ZSM-5 catalysts once were regarded as suitable substitutes to precious metal catalysts for NO elimination[78], nevertheless, they are susceptible to hydrothermal dealumination leading to a permanent loss of activity[79]. Perovskites have a higher hydrothermal stability than zeolites[35]. Although perovskites were expected to be potential autocatalysts in the presence of water[80], few reports related to the influence of water on the reactants adsorption, the perovskite physicochemical properties, and the catalytic performance in NO-SCR were previously documented. The H_2O deactivation mechanism is also far from well established.

Three series of $LaCo_{1-x}Cu_xO_3$, $LaMn_{1-x}Cu_xO_3$, $LaFe_{1-x}(Cu, Pd)_xO_3$ perovskites prepared by reactive grinding were characterized by X-ray diffraction (XRD), scanning electron microscopy (SEM), temperature programmed desorption (TPD) of O_2, NO + O_2, and C_3H_6 in the absence or presence of H_2O, Fourier transform infrared (FTIR) spectroscopy as well as activity evaluations without or with 10% steam in the feed. This research was carried out with the objective to investigate the water vapor effect on the catalytic behavior of the tested perovskites. An attempt to propose a steam deactivation mechanism and to correlate the water resistance of perovskites with their properties has also been done.

7.1. Effect of Steam on Reactants Adsorption over $LaFe_{0.8}Cu_{0.2}O_3$

The effect of water vapor on the adsorption of reactants (O_2, NO, and C_3H_6) was investigated over $LaFe_{0.8}Cu_{0.2}O_3$ by comparison of TPD of O_2, NO + O_2, C_3H_6 in the absence of H_2O with the same TPD experiments conducted in the presence of H_2O. The amounts of the various gases desorbed from the perovskite were subsequently calculated after deconvolution of the TPD traces and are summarized in Table 8.

Under a dry atmosphere the oxygen desorption from $LaFe_{0.6}Cu_{0.2}O_3$ at T<700 °C during the TPD of O_2 experiment was attributed to α_1-O_2 and α_2-O_2 species formed via molecular oxygen bound to the anion vacancies of perovskites according to the ascription in previous sections. The oxygen desorbed at T>700 °C is thereafter designated as β-oxygen which was described as an oxygen species liberated from the lattice. Both adsorbed C_3H_6 (at 49 °C) and generated CO, CO_2 (>250 °C) were released from $LaFe_{0.8}Cu_{0.2}O_3$ after C_3H_6 adsorption suggesting the partial oxidation of adsorbed C_3H_6 occurs with CO and CO_2 products. Finally, three overlapped NO desorption peaks at 105, 263, 396 °C were observed during NO + O_2-TPD process and respectively ascribed to the desorption of physically adsorbed, nitrosyl, and nitrate species.

The transient generation of anion vacancy of fresh perovskite during calcination as well as the generation of adsorbed oxygen (O_2^-) was believed to involve the process as formulated in Eqs (1) and (2).

Table 8. Amounts of various gases desorbed from $LaFe_{0.8}Cu_{0.2}O_3$ during TPD experiments [81]

TPD of 10% O_2	O_2 desorbed (μmol·g^{-1})	107.5 (258 °C)	87.2 (671 °C)	204.3 (793 °C)
TPD of 10% O_2 + 5% H_2O		11.2 (392 °C)	18.4 (661 °C)	27.9 (799 °C)
TPD of 3000 ppm NO + 1% O_2	NO desorbed (μmol·g^{-1})	21.2 (105 °C)	17.6 (263 °C)	28.1 (396 °C)
TPD of 3000 ppm NO + 1% O_2 + 5% H_2O		13.1 (170 °C)	11.7 (274 °C)	16.4 (385 °C)
TPD of 3000 ppm NO + 1% O_2	O_2 desorbed (μmol·g^{-1})	-	-	27.8 (391 °C)
TPD of 3000 ppm NO + 1% O_2 + 5% H_2O		-	-	16.0 (377 °C)
TPD of 3000 ppm C_3H_6	C_3H_6 desorbed (μmol·g^{-1})	2.4 (49 °C)	-	
TPD of 3000 ppm C_3H_6 + 5% H_2O		0.2 (<300 °C)	-	
TPD of 3000 ppm C_3H_6	CO desorbed (μmol·g^{-1})	-	35.6 (>250 °C)	
TPD of 3000 ppm C_3H_6 + 5% H_2O		-	3.2 (>250 °C)	
TPD of 3000 ppm C_3H_6	CO_2 desorbed (μmol·g^{-1})	-	241.3 (>250 °C)	
TPD of 3000 ppm C_3H_6 + 5% H_2O		-	91.8 (>250 °C)	

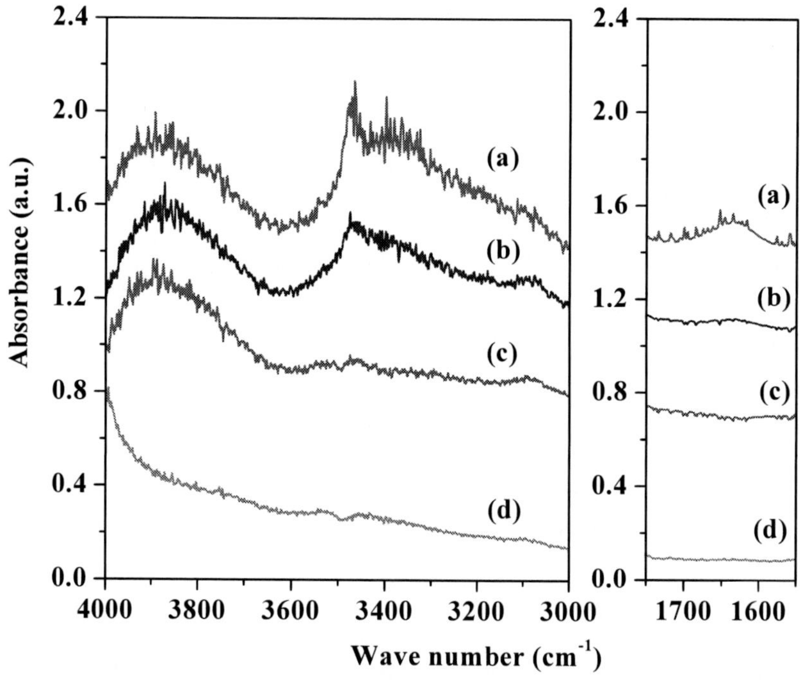

Figure 16. FTIR of $LaFe_{0.8}Cu_{0.2}O_3$ perovskite before and after H_2O deactivation: (a) exposure to water vapor of 70 Torr; (b) (a) after drying at 250 °C; (c) (a) after calcination at 500 °C [81].

The presence of 5% H_2O leads to a clear suppression of not only α-oxygens but also β-oxygens (Table 8), indicating a competitive adsorption between water vapor and O_2 at the same sites. Some authors pointed out that water vapor adsorbs easily at low temperature on all metal oxides including perovskites, but the mode of the adsorption (molecular or dissociative) varies with temperature and the oxide character[82, 83]. The hydration and dehydration of perovskite were further studied by IR spectroscopy via exposure of $LaFe_{0.8}Cu_{0.2}O_3$ to 70 Torr water vapor yielding broad bands at 1600-1750 cm^{-1}, 3200-3600 cm^{-1}, 3700-4000 cm^{-1} as illustrated in Figure 16a. Peri and Hannan reported similar bands related to stretching and bending H_2O molecule, with frequencies at around 1650 cm^{-1} and 3300 cm^{-1}, hydroxyl groups with maxima around 3698, 3737, 3795 cm^{-1} over γ-alumina after water adsorption[84]. Therefore, the broad bands at 1600-1750 cm^{-1} and 3200-3600 cm^{-1} were respectively ascribed to the stretching and bending frequencies of water, while the frequencies at 3700-4000 cm^{-1} were attributed to the hydroxyl (-OH) groups of $LaFe_{0.8}Cu_{0.2}O_3$ (Figure 16a). After drying at 250 °C the bands in the regions of 1600-1750 cm^{-1} and 3200-3600 cm^{-1} almost vanished, while the 3700-4000 cm^{-1} IR bands remained essentially unchanged (Figure 16b) indicating the surface was mainly covered by hydroxyl groups (3700-4000 cm^{-1}) under these conditions. These IR results suggest that water can exist under the forms of adsorbed H_2O and -OH groups over perovskite after hydrothermal treatment.

Taking into account the suppression of adsorbed oxygen species by water vapor through a competitive adsorption as verified by TPD, the adsorption of water was proposed to take place at anion vacancies as illustrated in Eq. 20. The re-adsorption (either dissociative or not) may also occur with simultaneous decrease in the sample concentration of oxygen vacancies:

$$Fe^{2+}\square + H_2O \rightleftharpoons \overset{H_2O^-}{\overset{|}{Fe^{3+}}} \quad (20)$$

$$Fe^{2+}\square + \overset{O^-}{\overset{|}{Fe^{3+}}} + H_2O \longrightarrow \overset{OH^-}{\overset{|}{Fe^{3+}}} + \overset{OH^-}{\overset{|}{Fe^{3+}}} \quad (21)$$

The competitive adsorption of water at anion vacancies (Eqs. 20 and 21) makes the formation of nitrosyl and nitrate species more difficult (Eqs. 10 and 8). The latter involves the adsorption of NO or O_2 molecules at the same sites.

7.2. Catalytic Performance

Variation of the H_2O inhibition with temperature over all perovskites is illustrated in terms of relative activities (conversion in the presence of water/conversion in the absence of water) for NO conversion to illustrate the resistance to deactivation by H_2O of the prepared samples (Figure 17). Among the perovskites tested, the heaviest inhibition in NO conversion due to 10% water addition was achieved by $LaMnO_3$. The resistance to H_2O deactivation of NO conversion for unsubstituted perovskites followed the order of $LaMnO_3 < LaFeO_3 < LaCoO_3$. This can be enhanced via Cu substitution. The least suppression in NO conversion after H_2O deactivation was observed over $LaFe_{0.8}Cu_{0.2}O_3$ among transition metal perovskites (except for $LaFe_{0.97}Pd_{0.03}O_3$). $LaFe_{0.97}Pd_{0.03}O_3$ shows the best resistance to H_2O deactivation resulting in the least deactivation in NO conversion. This good performance in the presence of water is likely ascribed to the excellent characteristics of Pd. Indeed according to a previous report, H_2 generated from the steam reforming of propene adsorbed over palladium can be dissociated into hydrogen atoms by this noble metal and efficiently utilized for NO reduction[85].

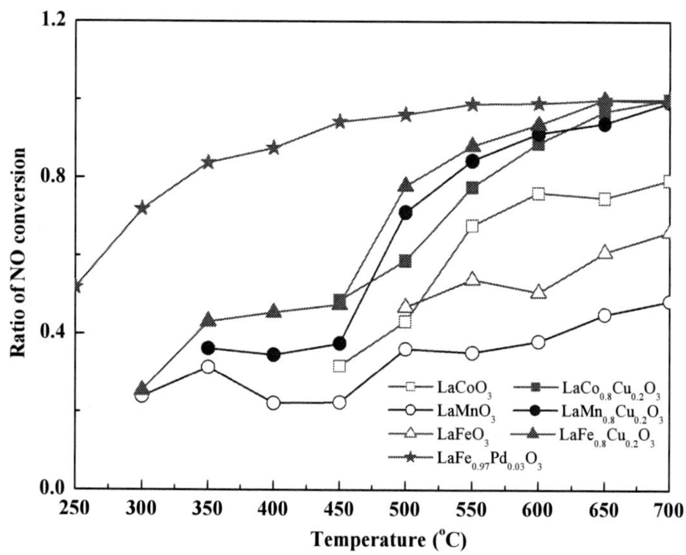

Figure 17. Ratio of NO conversion with and without 10% water [81]. Conditions: 3000 ppm NO, 3000 ppm C_3H_6, 1% O_2, 0 or 10% H_2O, 50,000 h^{-1}.

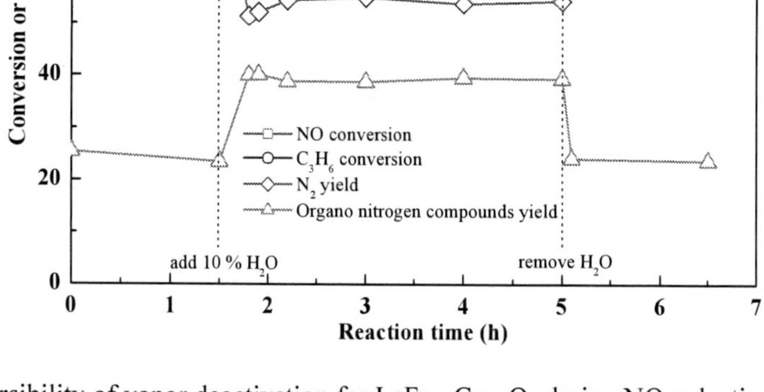

Figure 18. Reversibility of vapor deactivation for $LaFe_{0.8}Cu_{0.2}O_3$ during NO reduction by propene in the presence of oxygen [81]. Conditions: 3000 ppm NO, 3000 ppm C_3H_6, 1% O_2, 0 or 10% H_2O, 600 °C, 50,000 h^{-1}.

The reversibility of H_2O deactivation was studied by adding 10% H_2O vapor and subsequently removing H_2O from the reactant stream at 600 °C as depicted in Figure 18. Although some inhibition of NO and C_3H_6 conversions and N_2 yield takes place after 10% H_2O addition, the H_2O deactivation is fully reversible upon removal of the H_2O from the feed.

8. Sulfur Dioxide Poisoning of Perovskites

8.1. SO_2 Poisoning During C_3H_6 + NO + O_2 Reaction

Since fuels contain residual sulfur, the resistance to SO_2 poisoning has become one of the main issues in the development of technologies in which NO reduction is involved. This problem has become less stringent owing to the recent regulations of several countries which limit the sulfur content of fuels. In Canada for example diesel fuel cannot contain more than 15 ppm sulfur. Unfortunately, perovskites are sensitive to SO_2 poisoning due to the basic properties of lanthanum, iron, manganese, cobalt oxides. A crucial problem for perovskite utilization is therefore the ease of SO_2 poisoning. It is thus necessary to clarify the SO_2 poisoning mechanism during catalytic reduction of NO by C_3H_6 in order to improve our understanding of the key factors related to poisoning and propose the composition of better performing perovskite-based catalysts with relatively higher resistance to SO_2 poisoning. Activity tests towards NO reduction by C_3H_6 in the absence and presence of SO_2 were performed in order to study the influence of SO_2 in gas phase on the reactants adsorption and physicochemical properties of $LaFe_{0.8}Cu_{0.2}O_3$ perovskite. Attempts to propose a SO_2 poisoning mechanism as well as to regenerate the poisoned samples have also been performed in this study.

8.1.1. Physical Properties

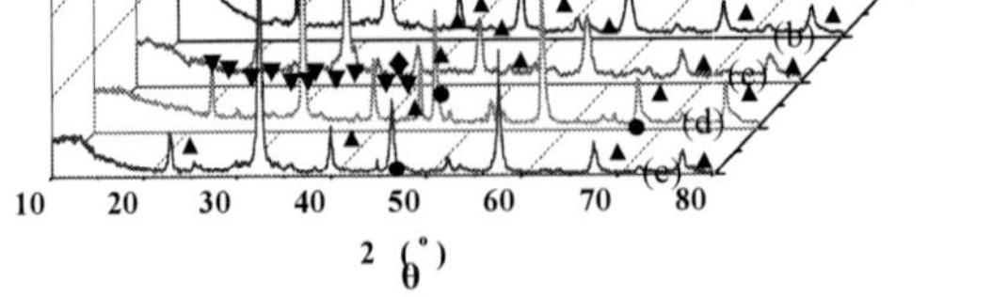

Figure 19. XRD patterns of corresponding $LaFe_{0.8}Cu_{0.2}O_3$ mixed oxides. (a) fresh $LaFe_{0.8}Cu_{0.2}O_3$; (b) $LaFe_{0.8}Cu_{0.2}O_3$ after 80 ppm SO_2 poisoning at 500 °C for 3 h; (c) $LaFe_{0.8}Cu_{0.2}O_3$ after 80 ppm SO_2 poisoning at 500 °C for 36 h; (d) $LaFe_{0.8}Cu_{0.2}O_3$ after 5% H_2 reduction up to 900 °C; (e) sample (d) after oxidation by 5% O_2 at 900 °C [86]

Nearly pure orthorhombic $LaFeO_3$ perovskite-type structure (JCPDS card 74-2203) for fresh $LaFe_{0.8}Cu_{0.2}O_3$ sample (42 m^2/g) was confirmed by XRD bands in pattern (a) in Figure 19. No significant difference between the fresh $LaFe_{0.8}Cu_{0.2}O_3$ (curve (a) in Figure 19) and the poisoned sample which was exposed to 80 ppm SO_2 for 3 h (21 m^2/g) (curve (b) in Figure 19) was found by XRD, suggesting that the SO_2 poisoned phases exist either in amorphous form or in particle sizes below the detection limit. Interestingly, minor new phases including $La_2(SO_4)_3$ (JCPDS card 45-0904) and Fe_2O_3 (JCPDS card 86-0550) appear besides the main perovskite structure after the exposure was prolonged to 36 h (16 m^2/g, curve (c) in Figure 19) in coincidence with XPS analysis reported by Alifanti et al.[87] that it is mainly lanthanum of $La_{1-x}Ce_xMn_{1-y}Co_yO_3$ perovskite oxides which forms surface sulfate during sulfur poisoning. It is clearly seen that the intensity of the diffraction peak at $2\theta \approx 32^o$ for $LaFe_{0.8}Cu_{0.2}O_3$ was significantly suppressed after 36 h poisoning by 80 ppm SO_2 (compare curves (a) and (c) in Figure 19), indicating that the perovskite structure had been partially destroyed simultaneously with the appearance of the new phases ($La_2(SO_4)_3$ and Fe_2O_3) although the ABO_3 perovskite is still the major phase. The stability of $LaFe_{0.8}Cu_{0.2}O_3$ perovskite synthesized by reactive grinding was also investigated as described by XRD traces (d) and (e) in Figure 19. The results show that the prepared $LaFe_{0.8}Cu_{0.2}O_3$ solid is considerably stable and could maintain its perovskite structure even under a reducing atmosphere of 5% H_2/Ar up to 900 °C (curve (d) in Figure 19), accompanied with the production of some metallic iron (JCPDS card 06-0696) due to the partial reduction of iron cations in B-sites. The reconstruction of perovskite structure for the reduced sample was realized by a subsequent oxidation by 5% O_2/He at 900 °C for 3 h (curve (e) in Figure 19). This process is believed to proceed via a solid phase reaction of Fe, Fe_2O_3 and La_2O_3 around grain boundaries under an oxidative atmosphere[37].

8.1.2. Temperature Programmed Analysis

Table 9. Amounts of the various gases desorbed ($\mu mol \cdot g^{-1}$) from $LaFe_{0.8}Cu_{0.2}O_3$ perovskite during TPD experiments [86]

Gas composition during adsorption	Desorbed molecule	Amount desorbed ($\mu mol \cdot g^{-1}$) (T/°C)		
10% O_2	O_2	107.5 (258 °C)	87.2 (671 °C)	204.3 (793 °C)
10% O_2 + 20 ppm SO_2	O_2	47.0 (253 °C)	50.8 (663 °C)	149.8 (800 °C)
Ratio*		0.44	0.58	0.73
3000 ppm NO + 1% O_2	NO	21.2 (105 °C)	17.6 (265 °C)	28.1 (395 °C)
3000 ppm NO + 1% O_2 + 20 ppm SO_2	NO	17.1 (109 °C)	15.8 (262 °C)	16.5 (385 °C)
Ratio*		0.81	0.90	0.59
3000 ppm NO + 1% O_2	O_2			27.8 (391 °C)
3000 ppm NO + 1% O_2 + 20 ppm SO_2	O_2			16.7 (383 °C)
Ratio*				0.60
3000 ppm C_3H_6	C_3H_6	2.4 (49 °C)		
3000 ppm C_3H_6 + 20 ppm SO_2	C_3H_6	0.6 (47 °C)		
Ratio*		0.25		
3000 ppm C_3H_6	CO		35.6 (>250 °C)	
3000 ppm C_3H_6 + 20 ppm SO_2	CO		19.2 (>250 °C)	
Ratio*			0.54	
3000 ppm C_3H_6	CO_2		241.3 (>250 °C)	
3000 ppm C_3H_6 + 20 ppm SO_2	CO_2		158.7 (>250 °C)	
Ratio*			0.66	

* molar amount desorbed with SO_2 in adsorption gas/molar amount desorbed without SO_2 in adsorption gas.

The effects of SO_2 in gas phase on the adsorption behavior of reactants (O_2, NO, and C_3H_6) over $LaFe_{0.8}Cu_{0.2}O_3$ were investigated via TPD of O_2, NO + O_2, and C_3H_6, by comparing TPD traces obtained in the absence and in the presence of SO_2 in the gas phase during adsorption (Table 9). Three O_2 desorption peaks at 258, 671, and 793 °C occurring in O_2-TPD process for fresh $LaFe_{0.8}Cu_{0.2}O_3$ were respectively assigned to α_1-O_2, α_2-O_2 and β-O_2. 20 ppm SO_2 inhibited the formation of adsorbed oxygens (especially α_1-O_2) and less affected the β-O_2 desorption from $LaFe_{0.8}Cu_{0.2}O_3$ lattice. Three overlapping NO peaks at 105, 265, and 395 °C appearing during the NO + O_2-TPD were assigned to the respective desorption of mononitrosyl, dinitrosyl, and nitrate species. It seems that the suppression of nitrate due to SO_2 is somewhat stronger than that of the nitrosyl species. A significant inhibition in the amounts of C_3H_6, CO and CO_2 desorbed in the presence of 20 ppm SO_2 was also found during C_3H_6-TPD.

8.1.3. FTIR study

IR spectroscopy was also implemented to study the sulfur species formed over $LaFe_{0.8}Cu_{0.2}O_3$ after SO_2 poisoning at 500 °C, with the results being represented in Figure 20. The reference sample $La_2(SO_4)_3$ showed IR bands in the frequency range of 1000-1300 cm^{-1} (line(f)). Very similar bands were observed in reference[88] for CeO_2 sulfated by progressive reaction with SO_2. No IR bands in the range of 1000-1300 cm^{-1} were observed for fresh $LaFe_{0.8}Cu_{0.2}O_3$ (line (a)).

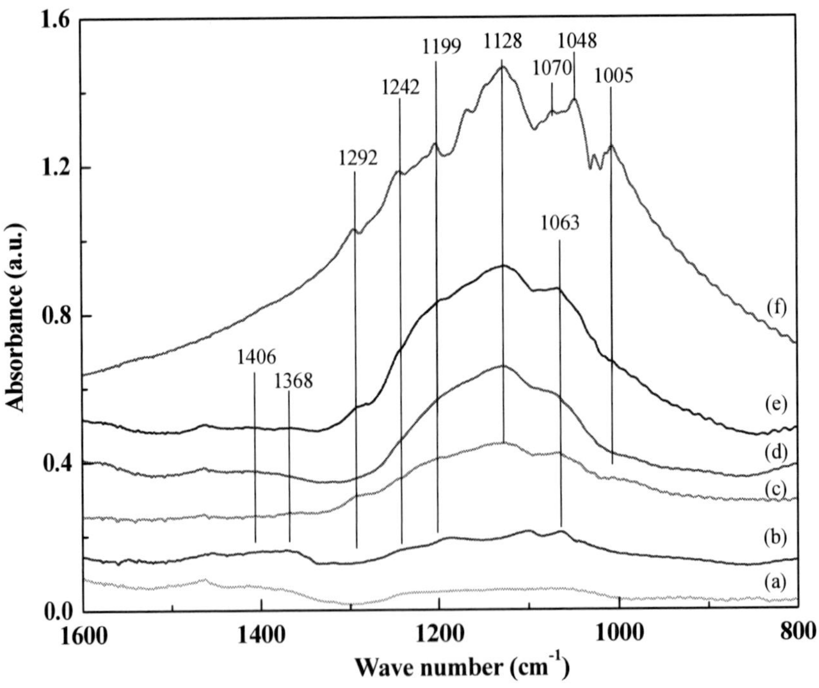

Figure 20. FTIR spectra of fresh and SO_2 poisoned $LaFe_{0.8}Cu_{0.2}O_3$ perovskites [86]. (a) fresh $LaFe_{0.8}Cu_{0.2}O_3$; (b) $LaFe_{0.8}Cu_{0.2}O_3$ after 20 ppm SO_2 poisoning at 500 °C for 3 h; (c) $LaFe_{0.8}Cu_{0.2}O_3$ after 80 ppm SO_2 poisoning at 500 °C for 3 h; (d) Sample (e) after 5% H_2 regeneration at 500 °C for 3 h; (e) $LaFe_{0.8}Cu_{0.2}O_3$ after 80 ppm SO_2 poisoning at 500 °C for 36 h; (f) $La_2(SO_4)_3$

A 3 h exposure of $LaFe_{0.8}Cu_{0.2}O_3$ to 20 ppm SO_2 at 500 °C led to the appearance of very weak bands near 1063, 1199, 1242, 1368 and 1406 cm^{-1} (line (b) in Figure 20). In reference[88] a 1340 cm^{-1} line was ascribed to the υ (S=O) vibration of surface sulfate species with only one S=O bond and a 1400 cm^{-1} band was thought to belong to SO_3 like species. Such ascriptions may also be valid here for the 1368 and 1406 cm^{-1} lines but the very weak signals preclude any definitive assignment. The same reference[88] describes the formation of sulfite species over SO_2 poisoned samples with IR bands near 1063 cm^{-1}. The presence of this band in spectra (20b) to (20e) confirms the formation of such species upon reaction with SO_2. The bands at 1128, 1199 and 1242 cm^{-1} which are present in the $La_2(SO_4)_3$ spectrum shown as line (f) in Figure 20 indicate the formation of sulfate species in lines (b) to (e). Although these sulfate species cannot be detected in XRD patterns (line (b) in Figure 19) possibly due to their limited contents, their IR bands are observable. After 36 h deactivation by 80 ppm SO_2 at 500 °C, the formation of sulfates became significant appearing as IR bands of 1000-1300 cm^{-1} (line (e) in Figure 20) in accordance with the generation of $La_2(SO_4)_3$ phase confirmed by XRD (line (c) in Figure 19). H_2 reduction diminishes the density of sulfate as depicted by line (d) in Figure 20.

The effects of poisoning time on NO conversion were studied via continuously exposing the $LaFe_{0.8}Cu_{0.2}O_3$ sample to a reaction atmosphere containing 0-80 ppm SO_2 for 3 h at 500 °C and on line monitoring the concentrations of the various gases in the effluent during this time, as shown in Figure 21. As the SO_2 feed concentration remains into the range of 0-20 ppm, stable NO conversion which represents about 3-25% diminution of their original conversions can be achieved within 40 min and remain stable at these levels with reaction time.

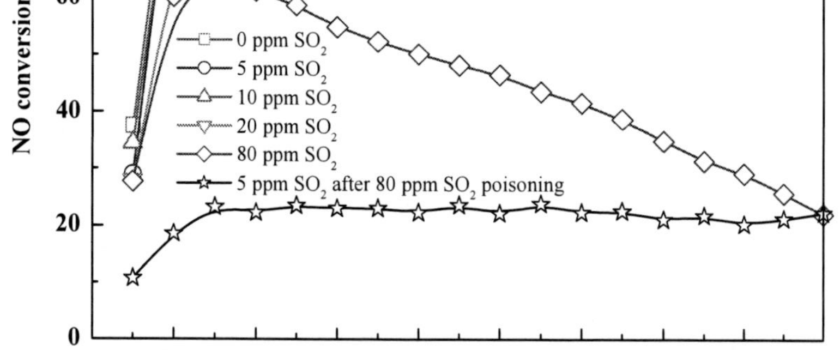

Figure 21. Effect of reaction time on NO conversion in $C_3H_6 + NO + O_2$ reaction over $LaFe_{0.8}Cu_{0.2}O_3$ perovskite under various SO_2 feed concentrations [86]. Conditions: GHSV=50,000 h^{-1}, 3000 ppm C_3H_6, 3000 ppm NO, 1% O_2, 0-80 ppm SO_2, 500 °C.

In the presence of 80 ppm SO_2, the deactivation of $LaFe_{0.8}Cu_{0.2}O_3$ however became severe and the loss in catalytic activity was proportional to the poisoning time. Furthermore, the original activities of spent $LaFe_{0.8}Cu_{0.2}O_3$ cannot be recovered after switching the stream from 80 ppm SO_2 containing feed to a 5 ppm SO_2 one.

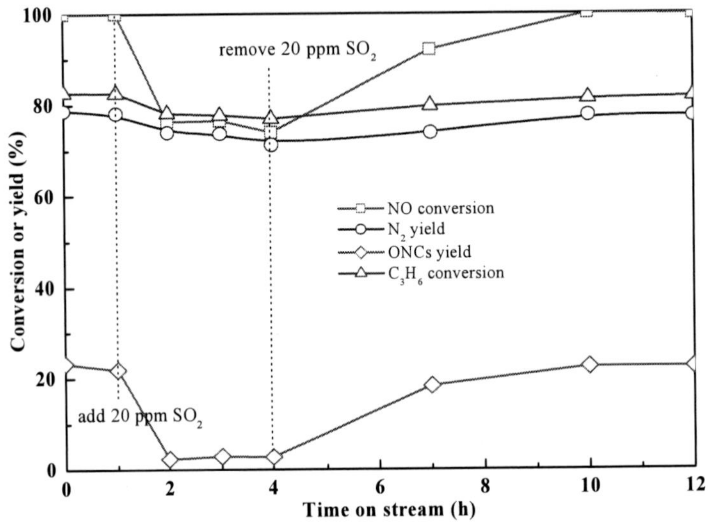

Figure 22a. Regeneration of $LaFe_{0.8}Cu_{0.2}O_3$ perovskite poisoned by 20 ppm SO_2 for 3 h [86]. Conditions: GHSV=50,000 h^{-1}, 3000 ppm C_3H_6, 3000 ppm NO, 1% O_2, 0 or 20 ppm SO_2, 500 °C.

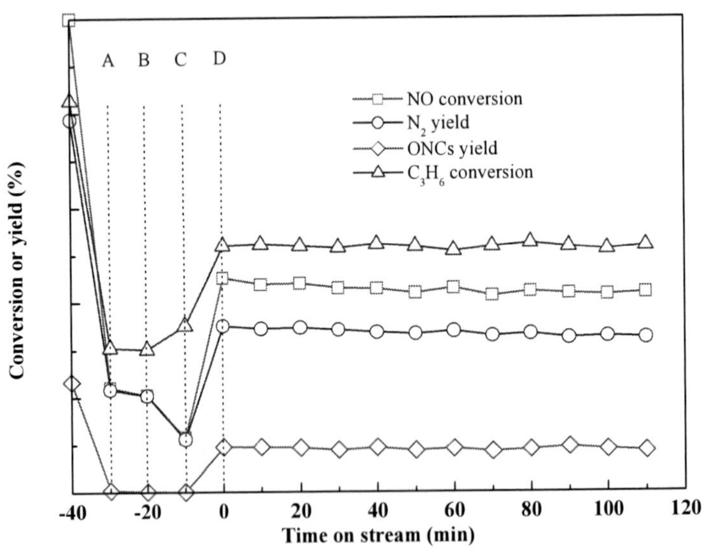

Figure 22b. Regeneration of $LaFe_{0.8}Cu_{0.2}O_3$ perovskite poisoned by 80 ppm SO_2 for 36 h [86]. Conditions: GHSV=50,000 h^{-1}, 3000 ppm C_3H_6, 3000 ppm NO, 1% O_2, 0 or 80 ppm SO_2, 500 °C. (A) poisoned by 80 ppm SO_2 for 3 h; (B) after stage A, treated by the flow of 3000 ppm NO + 3000 ppm C_3H_6 + 1% O_2 for 2 h; (C) after stage A, treated by 5% O_2/He flow for 3 h, and then 3000 ppm NO + 3000 ppm C_3H_6 + 1% O_2 for 2 h; (D) after stage A, treated by 5% H_2/He flow for 3 h, and then 3000 ppm NO + 3000 ppm C_3H_6 + 1% O_2 for 2 h.

The reversibility of SO_2 poisoning was also investigated as illustrated in Figs.22a and b. With the exposure to 20 ppm SO_2, the activity at 500 °C of $LaFe_{0.8}Cu_{0.2}O_3$ was inhibited with a diminution of 23% in NO conversion, 10% in N_2 yield, 89% in ONCs yield, and 7% in C_3H_6 conversion. Interestingly, this deactivation can be reversed by removing SO_2 from the feed stream (Figure 22a). Severely poisoned $LaFe_{0.8}Cu_{0.2}O_3$ (after 80 ppm poisoning for 3 h at 500 °C) could however not be regenerated just by flushing the spent sample with a reactant mixture containing no SO_2. A 5% O_2/He treating at 500 °C led to a further minor decrease in NO conversion while the opposite occurred in C_3H_6 conversion. Surprisingly, the activity can be partially regenerated by treating the spent sample by 5% H_2/He flow for 3 h at 500 °C and then switching to a normal reaction atmosphere (Figure 22b). A significant formation of ONCs can be again observed after H_2 treatment accompanied with a partial recovery of catalytic activity in the $C_3H_6 + NO + O_2$ reaction.

8.1.4. Effect of SO_2 on the Formation of Sulfur Species and Various Poisoning

O_2 species formed over $LaFe_{0.8}Cu_{0.2}O_3$ after O_2 adsorption were investigated via O_2-TPD experiments as described in Table 9, showing α-O_2 peaks at 253 and 671 °C and β-O_2 peak at 793 °C. In the presence of 20 ppm SO_2 in the adsorption gas, a diminution of adsorbed O_2 species (especially α_1-O_2) formed over $LaFe_{0.8}Cu_{0.2}O_3$ was found (see Table 9), indicating a competitive adsorption between gaseous SO_2 and O_2 at the same site. Acidic SO_2 was believed preferentially adsorbed on the surface of perovskite compared to O_2 due to the basicity of this solid. A process similar to α-O_2 adsorption upon anion vacancies is involved:

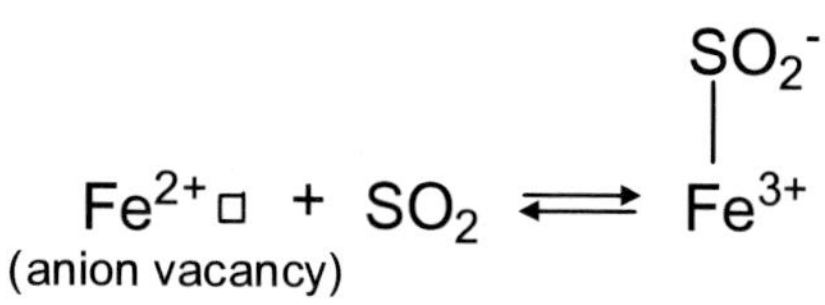

(22)

This SO_2 chemical adsorption consumes the anion vacancies leading to the inhibition of α-O_2 adsorption. Additionally, the formation of sulfite and sulfate species over the surface is possible via a subsequent reaction of chemisorbed SO_2 with $Fe^{3+}O^-$ and $Fe^{3+}O_2^-$. Although those sulfite and sulfate species formed after 3 h poisoning under 80 ppm SO_2 were not detected by XRD (line (b) in Figure 19) possibly due to their limited amounts, their formation was confirmed by FTIR (line (c) in Figure 20):

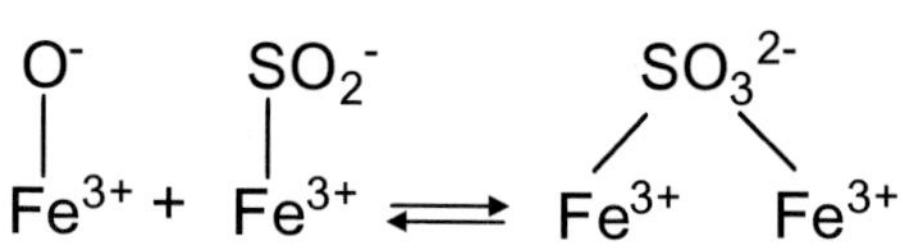

(23)

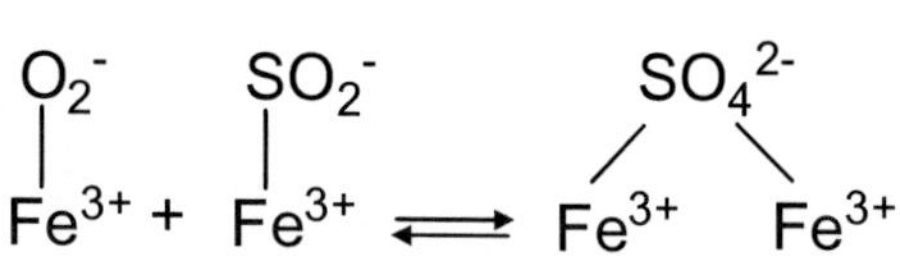

(24)

In the absence of SO_2, adsorption of NO, C_3H_6 and O_2 involved the steps formulated in Eqs. 10, 11 and 2. The competitive adsorption of SO_2 at anion vacancies (Eq.22) reduces the formation of both nitrosyl and nitrate species (Eqs.10 and 8). An inverse relation between the amount of sulfates formed and NO_x adsorption capacity was also pointed out by Hodjati et al.[89] who noticed the competition between SO_2 and nitrogen oxides for the same adsorption sites.

The C_3H_6 desorption was essentially inhibited in the presence of SO_2 because sulfur species can react with $Fe^{3+}O^-$ radical to form a relatively stable $Fe^{3+}SO_3^{2-}Fe^{3+}$ (see Eq. 23), resulting in a significant decline in the density of available adsorption sites for C_3H_6. Simultaneously, the scarcity of α-O_2 surface species ($Fe^{3+}O_2^-$) due to a competitive SO_2 adsorption (Eq. 22) leads to a decrease in both rates of propene oxidation and carbonaceous species (CO and CO_2) formation.

Bulk $La_2(SO_4)_3$ along with some sulfite/sulfate is largely formed over spent $LaFe_{0.8}Cu_{0.2}O_3$ after 36 h poisoning by 80 ppm SO_2 as evidenced by XRD (line (c) in Figure 19) and FTIR (line (e) in Figure 20). The above results suggest that SO_2 initially affects the adsorption of reactants via a competitive adsorption on the anion vacancies at low concentration ($\leq$ 20 ppm) yielding chemisorbed SO_2^-, SO_3^{2-} and SO_4^{2-} over the perovskite surface. The poisoning effect becomes more significant at high SO_2 concentration (80 ppm), where the sulfatation of perovskites is serious enough to provoke a solid phase transformation from surface sulfate associated with B-site ions to lanthanum sulfate due to the more basic properties of A-site La^{3+} ions. This leads to the perovskite structure destruction with appearance of new phases:

$$6\ Fe^{3+}(SO_4^{2-})Fe^{3+} + 4\ La^{3+} + 18\ O^{2-}_{(lattice)} \longrightarrow 2\ La_2(SO_4)_3 + 6\ Fe_2O_3 \quad (25)$$

This destruction is considered as the main reason leading to the permanent deactivation of the perovskite catalyst.

8.2. Resistance of $LaCo_{1-x}Fe_xO_3$ To SO_2 Poisoning During Methane Oxidation

8.2.1. Characterization And Activity Test

Three $LaCoO_3$ samples (I, II, and III) with different specific surface areas were prepared by reactive grinding. In the case of $LaCoO_3$ (I), only one step of grinding was performed. This step allowed us to obtain a crystalline $LaCoO_3$ phase. $LaCoO_3$ (II) and $LaCoO_3$ (III) were prepared in two grinding steps: a first step to obtain perovskite crystallization and a second step with additive to enhance specific surface area. The obtained compounds (perovskite + additive) were washed repeatedly (with water or solvent) to free samples from any trace of additive. The physical properties of the three catalysts are presented in Table 10. $LaCoO_3$ (I) was designed to present a very low specific surface area for comparison purposes. NaCl used as the additive in the case of $LaCoO_3$ (II) led to a lower surface area than ZnO used for $LaCoO_3$ (III), even if the crystallite size calculated with the Sherrer equation led to similar values for the three catalysts. The three catalysts prepared were perovskites having specific surface areas between 4.2, 10.9 and 17.2 m^2/g after calcination at 550 °C. A second milling step was performed in the presence of an additive, yielding an enhanced specific surface area.

Table 10. Properties of perovskite samples $LaCoO_3$ (I, II, and III) after grinding and calcination at 550 °C [(a)] [18]

Catalyst	$LaCoO_3$ (I)	$LaCoO_3$ (II)	$LaCoO_3$ (III)
element composition	$La_{0.94}Co_{0.97}Fe_{0.03}O_{3-\delta}$	$La_{0.93}Co_{0.90}Fe_{0.10}O_{3-\delta}$	$La_{1.00}Co_{0.80}Fe_{0.20}O_{3-\delta}$
SBET (m^2/g)	4.2	10.9	17.2
D1 (nm)	9.37	10.10	9.40
S_{Th} (m^2/g)	87.8	81.5	87.6
D2 (nm)	196	75.5	47.6
S_{BET}/S_{Th}	0.048	0.134	0.196
crystalline phase[b]	P, C	P	P

[(a)] D1, particle size calculated from the Sherrer equation; S_{Th}, specific surface area calculated assuming cubic particles and a density of $LaCoO_3$ equal to 7.29; D2, equivalent cubic particle size calculated from BET surface area. [(b)] P, perovskite; C, cobalt oxide.

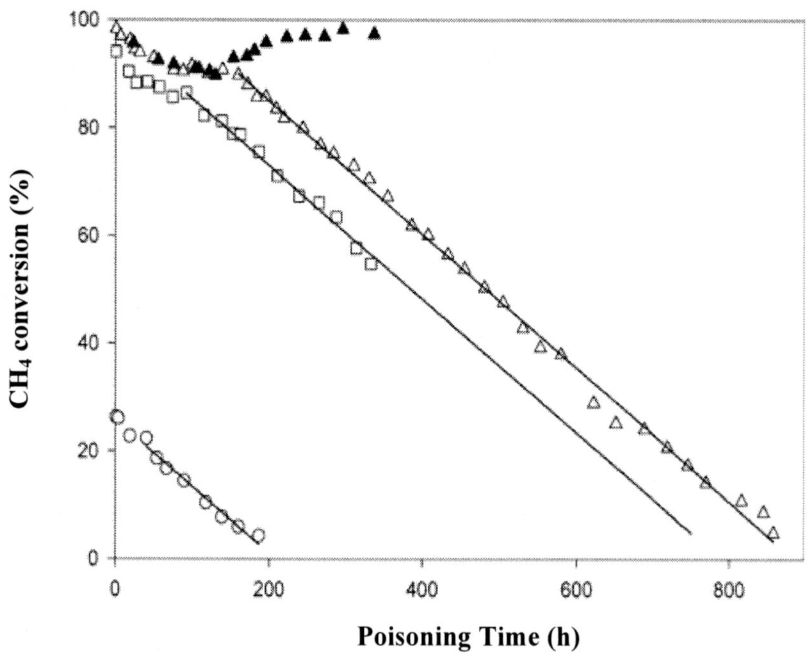

Figure 23. Effect of SO_2 on the activity of the catalysts for the reaction of CH_4 oxidation (500 °C, HVSV=8437 h^{-1}, 12.5 ppm SO_2, 0.25% CH_4, 10% O_2, balanced by He): ○, $LaCoO_3$ (I); □, $LaCoO_3$ (II); △, $LaCoO_3$ (III); ▲, $LaCoO_3$ (III) reaction with SO_2 during 130 h, end of the reaction without SO_2. [18].

Figure 23 shows the effect of injecting 12.5 ppm of SO_2 in the feed on the activity of the catalysts for the CH_4 oxidation reaction. Two regimes of poisoning can be distinguished for the three catalysts. In the first one, the conversion followed an exponential decrease to a plateau, which represents about a 10-15% diminution of the conversion. At this stage, deactivation is reversible. After 130 h of poisoning, $LaCoO_3$ (III) recovered all its activity in 210 h after removal of SO_2 from the feed. At the end of the first poisoning stage, the specific surface area of $LaCoO_3$ (III) drops from 17.2 to 4.8 m^2/g and 2.03 wt% sulfur (determined by

ICP) is accumulated on the catalyst. The elementary particles seem to be detached and developing a surface at the separation. Accompanying the drastic drop in specific surface area, the particle diameter obtained by the Sherrer equation (15.9 nm) showed higher perovskite crystal size than before poisoning (9.4 nm). During the second poisoning regime, the conversion is decreasing linearly with the reaction time until essentially complete deactivation. The second step of deactivation is a linear decrease of the conversion with poisoning time. Suppressing the SO_2 gas from the feed did not lead to any recovery of the activity, and thus deactivation is now irreversible. An important loss of the perovskite structure and clear signals for lanthanum sulfate and cobalt oxide was confirmed by XRD (not shown). Moreover, an important modification of the structure of the catalysts (Figure 24d) was observed by SEM after a long poisoning time. The poisoned catalyst is not constituted by agglomerates of nanometric elementary particles (Figure 24b) any more. The catalyst is now constituted of elementary particles of larger size, which is found to be between 200 to 800 nm.

Figure 24. SEM photographs obtain after poisoning at 500 °C on (a) $LaCoO_3$ (III) fresh, (b) $LaCoO_3$ (III) poisoned to the end of the first step, (c) $LaCoO_3$ (III) poisoned for 130 h in the first step followed by regeneration for 207 h, and (d) $LaCoO_3$ (III) poisoned until total deactivation [18].

Table 11. XPS surface composition (atom %) of sample Co3 at various poisoning stages [18]

$LaCoO_3$ (III) stage	Co	Fe	La	O	S	Fe/ (Co + Fe)	(Co + Fe)/ La
fresh	9.9	7.5	15.8	66.8	0.0	0.43	1.10
first stage	5.8	5.3	13.1	70.6	5.2	0.48	0.85
first stage, regenerated	7.6	5.6	14.6	70.9	1.2	0.43	0.90
totally poisoned	5.2	2.0	6.9	73.1	12.8	0.28	1.05

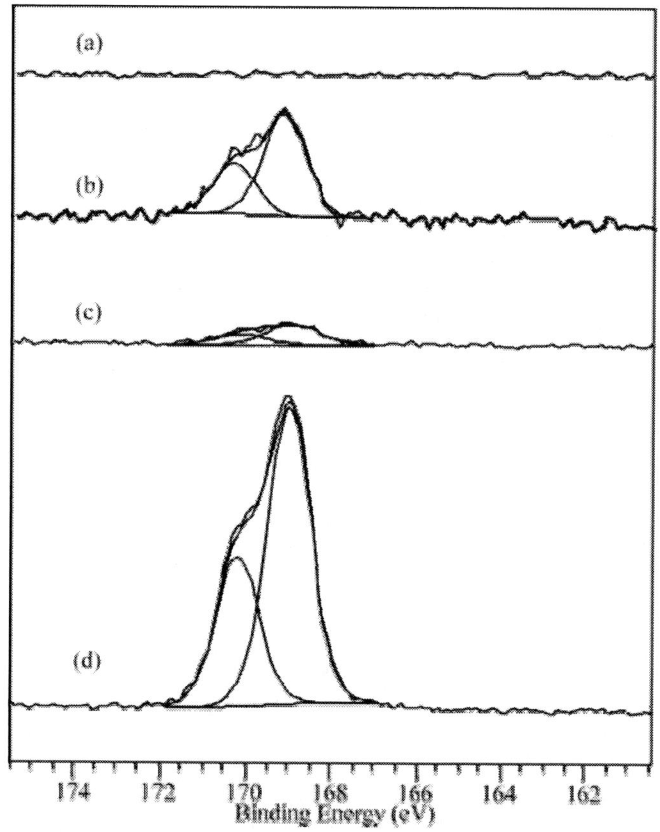

Figure 25a. S2p XPS spectra of $LaCoO_3$ at different poisoning steps: (a) $LaCoO_3$ fresh, (b) $LaCoO_3$ end of first stage, (c) $LaCoO_3$ regenerated in reaction, (d) $LaCoO_3$ totally deactivated. [18].

The solids were analyzed by XPS using a Kratos AXIS Ultra spectrometer. The XPS spectra were obtained using a monochromatized Al Kα source (15 W, 14 kV), for a small area (700 × 300 micron). A high surface concentration of sulfur was observed by XPS (Table 11), clearly higher than in the bulk (26.4% of the lanthanum was sulfated on the surface compared to 15.4% in the bulk).

The S2p lines of sample Co3 are shown in Figure 25a. The plot (a) corresponds to complete absence of sulfur, as expected. At the end of the first poisoning stage, the S2p line characteristic of sulfur in sulfate anions (plot (b)) does not disappear entirely upon regeneration (plot (c)).

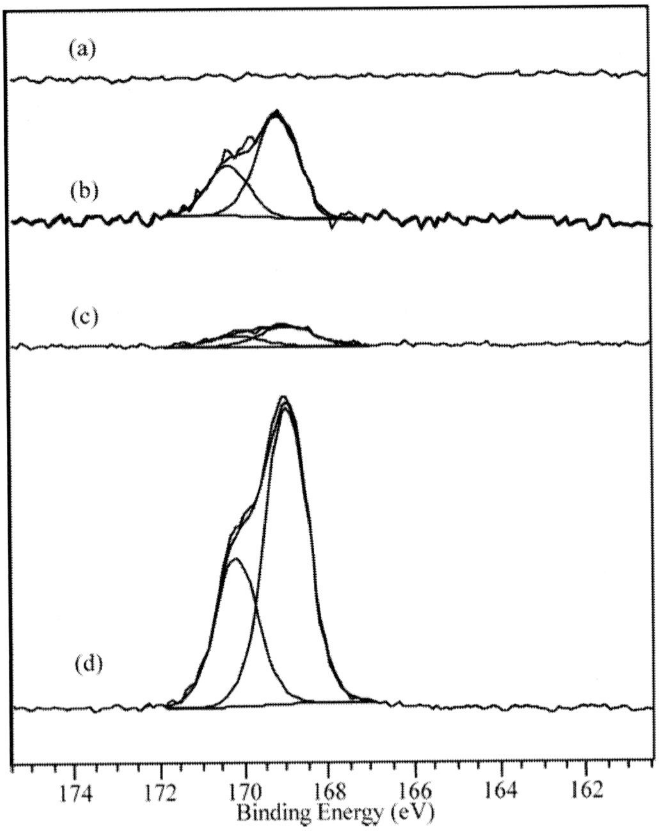

Figure 25b. O1s XPS spectra of $LaCoO_3$ (III) at different poisoning steps: (a) fresh, (b) end of first stage, (c) regenerated in reaction, (d) totally deactivated. [18].

Important modifications of the O1s spectra (Figure 25b) were also noted. Before poisoning (Figure 25ba), two peaks were observed[17, 90]. The first one at 529.3 eV was ascribed to regular perovskite lattice oxygen O^{2-}. The second at 531.9 eV was ascribed to O^- oxygen species. After poisoning to the end of the first step (Figure 25b), the spectrum could be fitted with the two previous peaks, a third peak at 532.2 eV attributed to oxygen in SO_4^{2-} [91] and a fourth one at 530.1 eV, which can be ascribed to oxygen in Co_3O_4[92] or, as previously observed, to oxygen in physisorbed water[17] The fifth peak at 533.2 eV is due to adsorbed O_2^-.

8.2.2. SO_2 Poisoning Mechanism

All above results can be reconciled by considering the findings in our recent work about the poisoning behavior of $LaCo_{1-x}Fe_xO_3$ perovskites by 12.5 ppm SO_2 during methane oxidation[18]. At 500 °C two different poisoning regimes happened successively. In the early stage a reversible minor deactivation was observed corresponding to the formation of adsorbed sulfur species competing with α-oxygen adsorption.

The plateau in the activity curve at the end of the first regime (Figure 23) may then correspond to the formation of a layer of sulfated material on the surface of the elementary particles in equilibrium with the SO_2 in the gas phase (Scheme 1). The bulk of the elementary particles is not sulfated during this step and that is the reason a reversible activity is observed (surface sulfate, which were found to be less stable than bulk sulfate, can decompose under these reaction conditions). $LaCoO_3$ (III) contains 1.5 wt % sulfur after 207 h of regeneration; thus, sulfur desorption is rather slow. The fact that even with 1.5 wt % sulfur, activity can be totally recovered is explained by the fact that only the external surface of the accessible elementary particles participates to the oxidation reaction. The differences observed in the first poisoning regime time for the three catalysts (Figs.23 and 24) are strongly dependent on the BET surface area. This is obviously associated with the differences in morphology of the fresh samples. In $LaCoO_3$ (I), the low S_{BET}/S_{Th} suggests a more compact agglomerate and therefore a stronger attachment of the 10 nm nanoparticles. By contrast, a looser agglomerate of the nanoparticles is observed in $LaCoO_3$ (III). The formation of surface sulfate during the first step involves a change in density, which implies strong constraints on the attachments of the nanoparticles in the agglomerates. The weaker agglomerates in $LaCoO_3$ (III) yield the creation of a new perovskite surface, which contributes to an apparent decrease of the deactivation. The plateau corresponds to this reactivation effect compensating for the progressing deactivation of the exposed perovskite surface in the fresh sample.

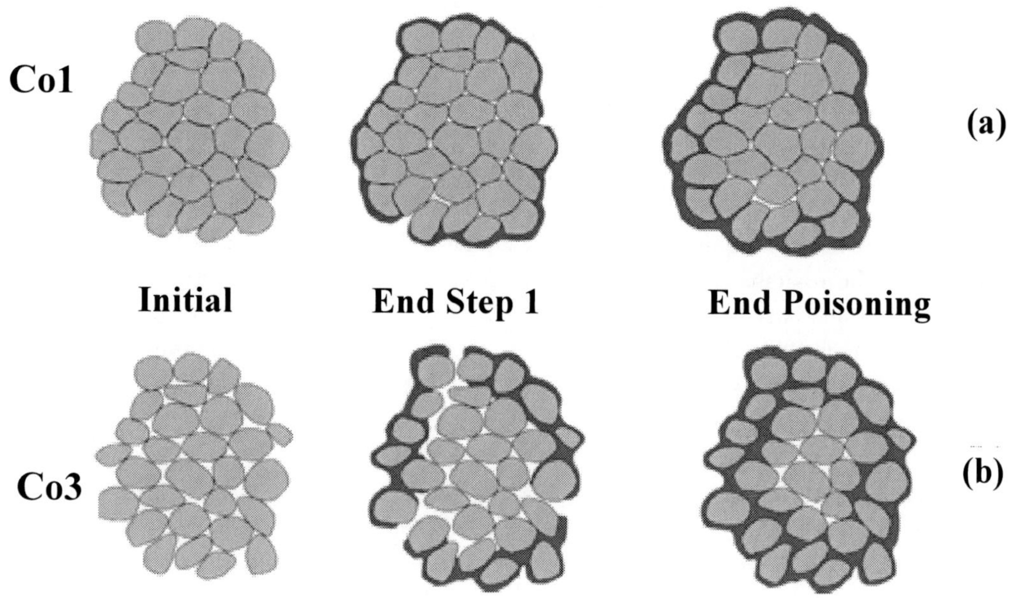

Scheme 1. Deactivation mechanism in the case of (a) dense agglomerates and (b) looser agglomerates [18].

In the second poisoning regime, the significant formation of bulk sulfate involves even more drastic changes in morphology. The surface sulfates are rearranged into bulkier compact particles (Figure 24c) and the forces imposed on the unconverted nanoparticles are significant and also lead to generation of a new accessible perovskite surface. The constant deactivation rate observed for the three samples in Figure 23 would therefore be associated with the same rate of perovskite surface regeneration in the three samples. This is now due to a bulk phenomenon and would therefore not depend on the morphology (and BET surface area) of the fresh sample.

Scheme 1 describes the evolution of samples $LaCoO_3$ (I) and $LaCoO_3$ (III) during the two poisoning regimes ($LaCoO_3$ (II) is intermediate between these two and therefore not represented). It shows that the differences in morphology of the fresh samples yield differences in the conversion to sulfate necessary to reach the end of the first regime as well as differences in this conversion at complete deactivation (end of the second regime).

9. Lean NO Reduction over the Perovskite-Ag/Alumina Combined Catalysts

The internal combustion engines of motor vehicles are commonly classified into gasoline engine and diesel engine[93]. The polluting emission gases (NO, HCs, CO) from the gasoline engine present a reductant/oxidant ratio close to stoichiometric. They can be simultaneously eliminated using ABO_3 perovskite-type oxides as three-way catalysts (TWCs) through a catalytic reduction of NO by unburned hydrocarbons (HCs) and CO in the exhaust effluent [94]. In order to achieve a better fuel economy, the lean-burn gasoline engine and diesel engine are commonly used to realize a more complete combustion of fuel at excess oxygen conditions. However, the deNO_x activities of perovskite-based catalysts are rather poor under a simulated exhaust gas for the diesel engine fed with overstoichiometric oxygen based on our previous investigation[19, 21]. In contrast, alumina supported silver was thought as effective material for lean NO_x reduction as documented in the literature[95-97].

The main drawback of using Ag/Al_2O_3 as auto catalyst was the significant formation of CO, a harmful by-product, due to the partial oxidation of HCs and relatively poor deNO_x activity under a three-way atmosphere with stoichiometric O_2[98]. The addition of a small amount of perovskite into alumina perhaps facilitates the transformation of HCs and CO as suggested in the reviews related to perovskite[66, 99]. Therefore, the blend of perovskite-based and alumina-based silver catalysts for NO catalytic reduction by propene at various oxygen contents is studied with the purpose to obtain a mixture having optimal performance for NO and C_3H_6 removals under rich (for gasoline engine) or/and lean (for lean-burn gasoline and diesel engines) conditions.

The physicochemical characterization (BET, AAS, O_2- and NO + O_2-TPD, and reference catalytic tests (NO reduction by C_3H_6 in the presence of 0-10% oxygen) for $La_{0.88}Ag_{0.12}FeO_3$, Ag/Al_2O_3, Mixture (I): 10% $La_{0.88}Ag_{0.12}FeO_3$ + 90% Al_2O_3, Mixture (II): 3% Ag/(10% $La_{0.88}Ag_{0.12}FeO_3$ + 90% Al_2O_3), and Mixture (III): 10% $La_{0.88}Ag_{0.12}FeO_3$ + 90% (3% Ag/Al_2O_3) were investigated. An attempt to diminish the oxidation ability of $La_{0.88}Ag_{0.12}FeO_3$ by means of mixing with less oxidative alumina for the sake of optimizing the deNO_x efficiency especially under lean-burn conditions has been made.

9.1. Relationship between Redox Properties and Catalytic Behavior of Ag-Containing Perovskite and Alumina

The redox properties of perovskite-based and alumina-based silver catalysts were studied by means of O_2-TPD (Figure 26). Those α-oxygens formed over perovskite (117 μmol/g) and Mixtures (I-III) (3-15 μmol/g) are believed to make a contribution for good oxidative properties and facilitate both the nitrate formation and the propene oxidation[17, 96, 100], while lattice oxygen becomes mobile and merely reactive at high temperature. The results in Figure 26 indicate that the oxidation ability (represented by the content of surface α-O_2) of the investigated samples followed the order of $La_{0.88}Ag_{0.12}FeO_3$ >> Mixture (II) > Mixture (III) > Mixture (I) > Ag/Al_2O_3. The distinct redox properties between $La_{0.88}Ag_{0.12}FeO_3$ and Ag/Al_2O_3 will likely result in their different catalytic behaviors in the NO + C_3H_6 + O_2 reaction.

The reaction mechanism for NO catalytic reduction by propene in the presence of oxygen had already been established over La(Fe, Mn, and Co)$_{1-x}$Cu$_x$O$_3$ perovskites as involving organo nitrogen compounds (ONCs) generated from the interaction between nitrate species and adsorbed C_3H_6. An isocyanate intermediate formed from ONCs can react with NO and/or O_2 to yield the final products[21]. A similar mechanism for C_3H_6-SCR of NO in excess oxygen over Ag/alumina catalysts was also documented in the literature with the formation of nitrate (or NO_2) species as the first important step and R-NCO as a key intermediate[100].

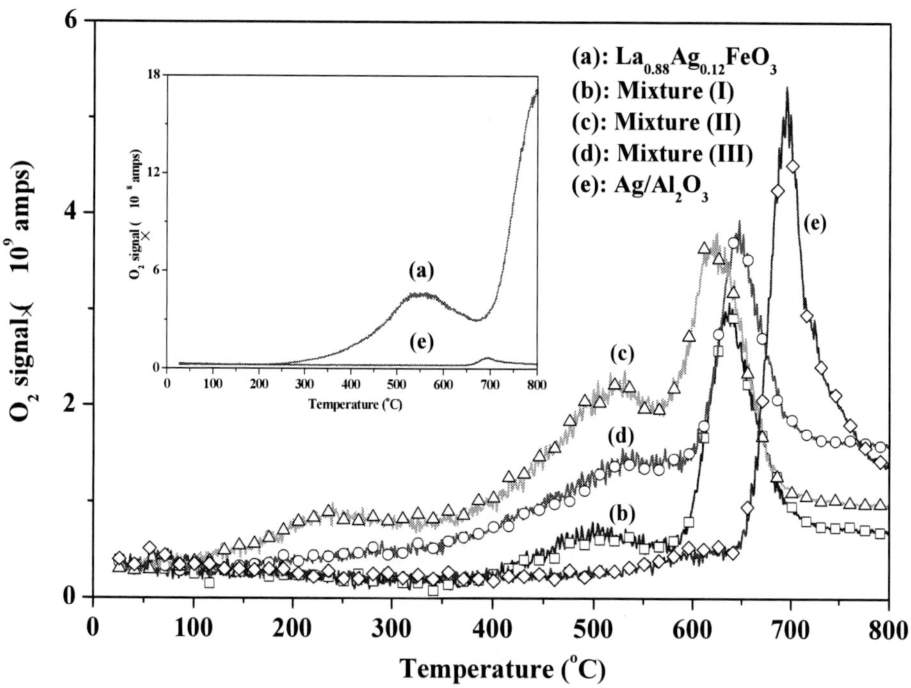

Figure 26. TPD of O2 profiles over prepared catalysts [101].

9.2. Synergistic Effect of Perovskite and Alumina

NO and C_3H_6 conversions at 500 °C as functions of O_2 feed concentration over $La_{0.88}Ag_{0.12}FeO_3$, Ag/Al_2O_3 and their corresponding mixtures are shown in Figs.27a and b. As seen from Figure 27a, a NO conversion higher than 97% was obtained over $La_{0.88}Ag_{0.12}FeO_3$ at O_2 feed concentration less than 1%, which decreased dramatically at 2% O_2. NO conversion at O_2 >4% was mainly due to oxidation of NO into NO_2 as evidenced by the NO_2 signal detected in the gas leaving the reactor (not shown). The NO conversion achieved over Ag/Al_2O_3 in the absence of oxygen is low (19%). This value increased proportionally to the oxygen feed concentration at 0-2% O_2, almost reaching 100% at >3% O_2. Furthermore, Mixtures (I-III) always resulted in a better C_3H_6 conversion compared to Ag/Al_2O_3 alone no matter the O_2 contents in the feed as illustrated in Figure 27b. In brief, perovskite shows good activity for both NO and C_3H_6 removals at O_2 atmosphere close to stoichiometric. High efficiency for NO reduction by C_3H_6 over Ag/Al_2O_3 was only realized under an excess of oxygen. Mixture (I) of 90% alumina with 10% $La_{0.88}Ag_{0.12}FeO_3$ maintained a NO conversion of 75% at 2% O_2 and ~29% at >3% O_2. 3% Ag impregnation of Mixture (I) for the preparation of Mixture (II) can further promote the NO conversion. Interestingly, the blend of $La_{0.88}Ag_{0.12}FeO_3$ and Ag/Al_2O_3 produced a bi-component catalyst (mixture (III)) showing good performance for both NO reduction and C_3H_6 oxidation at 500 °C over the whole oxygen range from stoichiometric (1%) to lean burn conditions (10%).

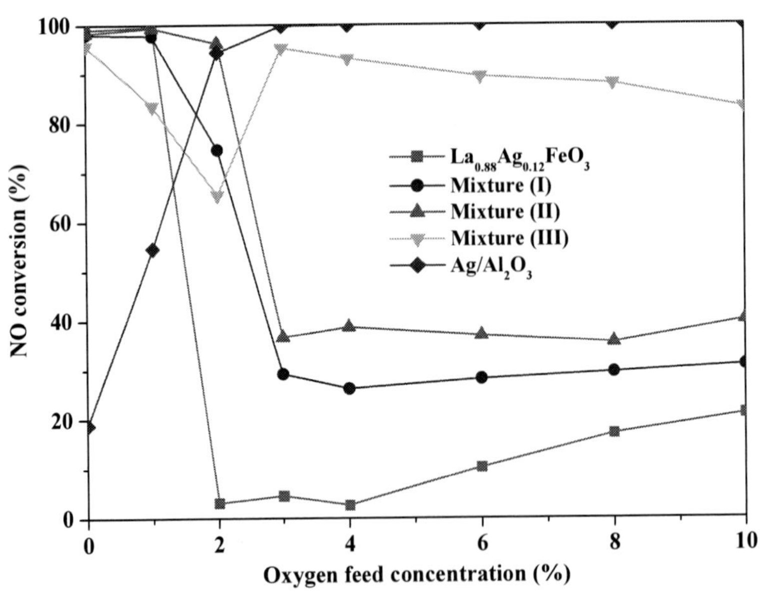

Figure 27a. Effect of O_2 feed concentration on NO conversion in C_3H_6 + NO + O_2 reactions over $La_{0.88}Ag_{0.12}FeO_3$, Ag/Al_2O_3 and Mixture (I, II, III). Conditions: 3000 ppm C_3H_6, 3000 ppm NO, 500 °C, 30,000 h^{-1} [101].

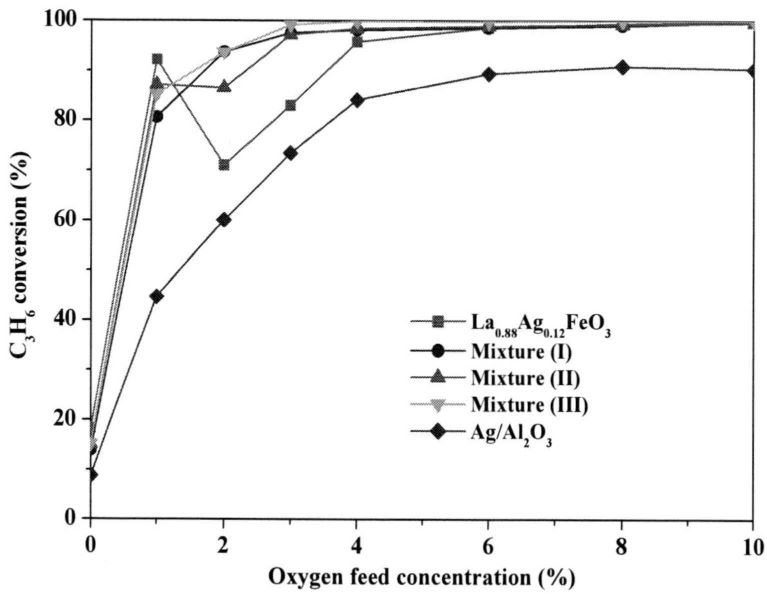

Figure 27b. Effect of O_2 feed concentration on C_3H_6 conversion in C_3H_6 + NO + O_2 reactions over $La_{0.88}Ag_{0.12}FeO_3$, Ag/Al_2O_3 and Mixture (I, II, III). Conditions: 3000 ppm C_3H_6, 3000 ppm NO, 500 °C, 30,000 h^{-1} [101].

10. Recent Advances on NO Reduction Related to Mixed Oxides, Supported Metals and Ion-Exchange Mesoporous Materials

10.1. Ga-Based Mixed Perovskites

Ga-based catalysts are of interest because of the high selectivity towards N_2 of Ga-ZSM-5 and the good activity of Ga_2O_3/Al_2O_3 during NO catalytic reduction by hydrocarbons. However, their applications were limited either because of the possible dealumination under hydrothermal treatment and pore blockage caused by carbonaceous deposit for Ga-ZSM-5, or because of poisoning by water vapor and thermal sintering for Ga_2O_3/Al_2O_3[102-105]. Mixed oxides act as catalytic materials but with the advantages of high thermal and mechanical stability, structure and chemical diversity, excellent redox properties, better resistance against steam and SO_2 deactivation compared to their simple oxides[35]. Among them, $LaGaO_3$ perovskite and $ZnGa_2O_4$ spinel are two kinds of mixed oxides commonly used for NO catalytic abatement[21, 56, 106, 107]. Recently, we prepared $LaFe_{1-x}(Cu, Pd)_xO_3$ perovskite-type oxides achieving satisfactory performances in NO reduction by C_3H_6 under a simulated atmosphere of the gasoline engine exhaust at approximately stoichiometric O_2 content[21]. Facing up the energy crisis, the use of diesel engine which achieves a more complete combustion of fuel at overstoichiometric oxygen (lean burn conditions) became a popular tendency. NO_x abatement under such oxidizing conditions is a great challenge for the conventional TWCs. The spinel-type Ga_2O_3-Al_2O_3-ZnO was claimed to be able to reduce NO under an excess of oxygen[106, 107]. The distinct catalytic behaviors at different oxygen contents of Ga-based oxides with perovskite and spinel structure during NO reduction attracted our attention.

The various O_2 species formed over the catalysts were investigated by O_2-TPD, as illustrated in Figure 28. 31 μmol O_2 desorbed per gram $LaGaO_3$ at 658 °C is designated as α-O_2 and ascribed to oxygen species bound to the surface anion vacancies of perovskite, whereas the β-O_2 appears above 720 °C (18 μmol/g). In contrast, very low α-O_2 (2.2 μmol/g) and rather low β-O_2 (6.7 μmol/g) were formed over $ZnGa_2O_4$ spinel (Figure 28) although its specific surface area (40 m^2/g) is superior to the one of $LaGaO_3$ perovskite (19 m^2/g). This suggests largely different redox characteristics between $ZnGa_2O_4$ and $LaGaO_3$, indicating an oxidation ability of $LaGaO_3$ largely superior to that of $ZnGa_2O_4$.

The different structural environments between $LaGaO_3$ and $ZnGa_2O_4$ are deemed to be responsible for their distinct redox properties. Besides the coordination with Ga^{3+} ions, 12 oxygens are also coordinated with La^{3+} in a large cavity in the perovskite, resulting in a relatively weak La-O bonding. The opposite occurs in a spinel structure displaying 4 oxygens firmly bonded to Zn^{2+} ions in a ZnO_4 tetrahedron. The smaller ion radii of tetrahedral Zn^{2+} (0.74 Å) with respect to 12 coordinated La^{3+} (1.46 Å) also implies a stronger Zn-O bonding. The relatively lower metal-O bond energy of $LaGaO_3$ compared to that of $ZnGaO_4$ also increases the opportunity of creating anion vacancies, which are adsorption sites for α-O_2. These structural factors lead to a higher mobility of both surface oxygen and lattice oxygen in the $LaGaO_3$ perovskite structure compared to that in the $ZnGa_2O_4$ spinel structure in accordance with O_2-TPD results.

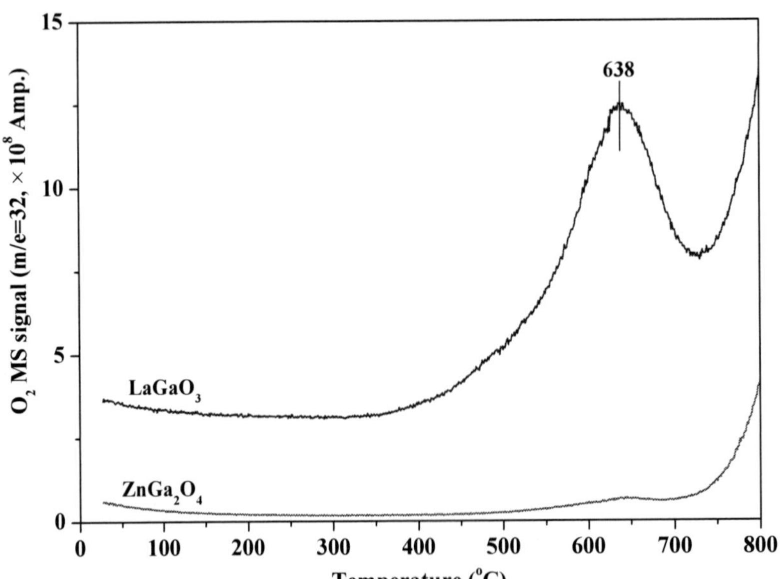

Figure 28. O_2-TPD profile of $ZnGa_2O_4$ and $LaGaO_3$ mixed oxides [108].

Catalytic performance including N_2, ONCs, NO_2, CO, CO_2 yields and NO conversion at 500 °C as functions of O_2 feed concentration over $LaGaO_3$ and $ZnGa_2O_4$ were investigated under an atmosphere of 1000 ppm NO, 3000 ppm C_3H_6, and 0-10% O_2 as shown in Figs.29 and 30, respectively. N_2 is the main N-containing product in addition to minor ONCs and NO_2 byproducts over Ga-based oxide in the NO + C_3H_6 + O_2 reaction. This result is in agreement with previous reports that Ga-based catalysts yield a high selectivity towards N_2 during NO reduction[102]. As seen from Figure 29, the N_2 yield obtained over $ZnGa_2O_4$ in the absence of oxygen is low (15%). This value was slightly increased at higher concentration of oxygen reaching a N_2 yield of 28% at 10% O_2. On the contrary, a satisfactory N_2 yield (72%) was obtained over $LaGaO_3$ with no oxygen in the feed, but the N_2 yield decreased dramatically (down to 16%) at 1% O_2 and was lower than 10% at higher O_2 feed concentrations (Figure 30). A significant ONCs yield was observed over Ga-based perovskites in the absence of oxygen. Nevertheless, the maximum ONCs yield over $ZnGa_2O_4$ appeared at 6% O_2. NO_2 production was only apparent over the perovskite sample in excess oxygen giving a yield of 4% for $LaGaO_3$ at 10% O_2. CO_2 yield over the two tested oxides increased gradually with the increase of oxygen feed concentration, showing values following the trend of $LaGaO_3$ > $ZnGa_2O_4$. On the contrary, the CO due to the partial oxidation of C_3H_6 was found to be more readily formed over $ZnGa_2O_4$ spinel compared to $LaGaO_3$.

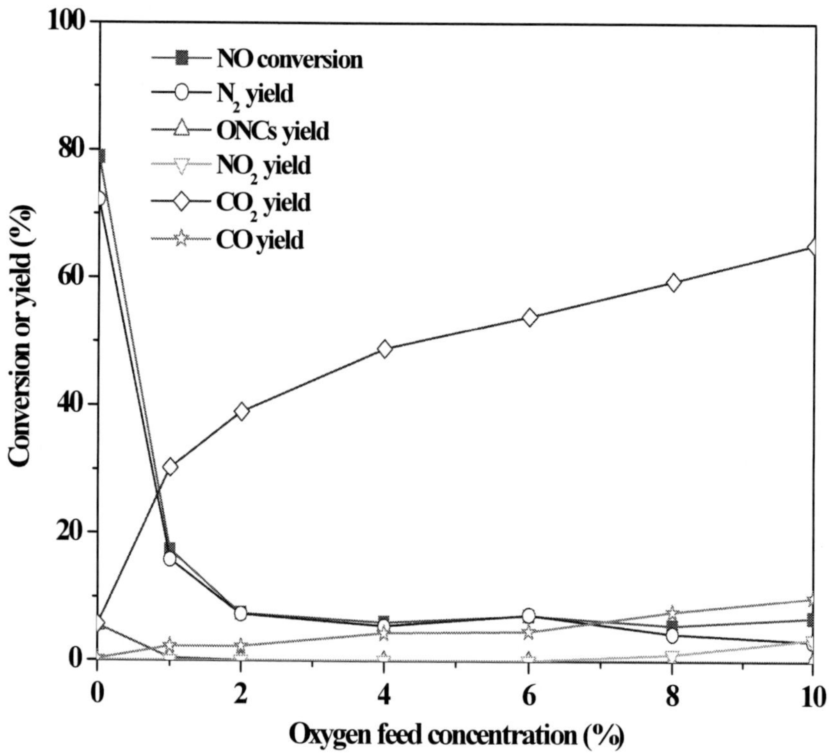

Figure 29. Catalytic performance in NO reduction by C_3H_6 over $LaGaO_3$ [108], Conditions: 1000 ppm NO, 3000 ppm C_3H_6, 0-10% O_2, 500 °C, 30,000 h^{-1}.

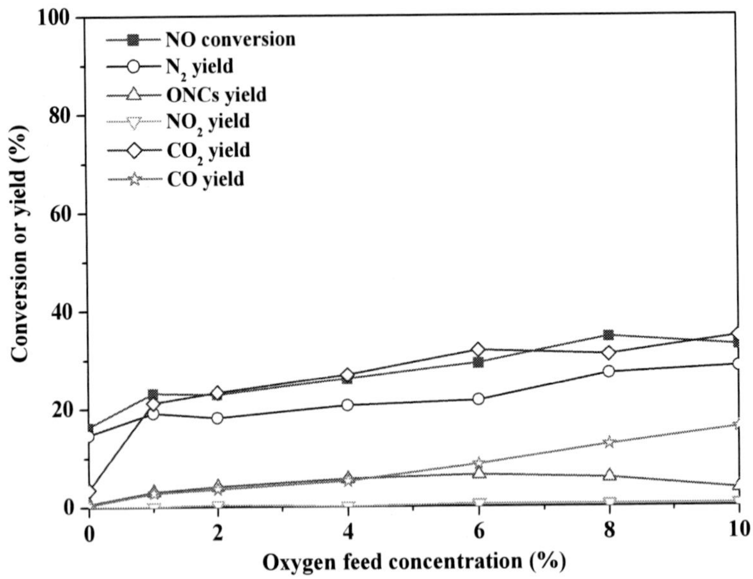

Figure 30. Catalytic performance in NO reduction by C_3H_6 over $ZnGa_2O_4$ [108], Conditions: 1000 ppm NO, 3000 ppm C_3H_6, 0-10% O_2, 500 °C, 30,000 h^{-1}.

Considering the high stability of La^{3+} and Zn^{2+} ions against reduction[109], the GaO_6 octahedra of $LaGaO_3$ and $ZnGa_2O_4$ oxides were deemed to play a major role in the catalytic reduction of NO using propene as the reducing agent. This is in accordance with the reported finding that highly dispersed GaO_x species are responsible for the good SCR activity of Ga_2O_3/Al_2O_3 and Ga/ZSM-5[102, 103]. Nevertheless, Seiyama pointed out that the reactivity of B-site cations in mixed-oxides is strongly dependent on the nature of the neighboring metal (A-site) cations as well as the bonding distance between this metal and oxygen[22]. He also thought that spinels were suitable materials for partial oxidation reactions while perovskites were active for complete combustion[22].

Based on the proposed SCR mechanism for mixed oxides (Section 4), the ease of α-O_2 generation on anion vacancies of $LaGaO_3$ perovskite even under an O_2 poor atmosphere is beneficial to the formation of nitrate species and ONCs, leading to a satisfactory N_2 yield at O_2 concentrations less than 1%. The ONCs generation over $LaGaO_3$ was detectable in this O_2 concentration range. On the contrary, the conversions towards N_2 occurring over Ga-based perovskites were inhibited simultaneously with a diminution of ONCs yield, at oxygen concentrations exceeding 2%. Similarly, a decline in N_2 yield parallel with the increase in O_2 concentration was also found over lanthanum cobaltites and ascribed to a severe C_3H_6 combustion facilitated by α-O_2, which reduces the amount of reductant necessary for NO reduction[19]. Over the poorly oxidative $ZnGa_2O_4$, the presence of O_2 was found to be essential for the reduction of nitrogen oxide. An increase in O_2 feed concentration enhanced the N_2 yield, likely through a promotion in NO_3^- formation by gaseous O_2. A maximum ONCs yield was observed at 6% O_2 for this reason. Enough reducing agent can be maintained even at excess oxygen conditions over this spinel, resulting however in a moderate N_2 yield. The C_3H_6 conversions obtained over $ZnGa_2O_4$ and $LaGaO_3$ coincide with their oxidation abilities as established by O_2-TPD.

10. 2. Alumina Supported Silver Catalysts

Silver catalysts were deemed as the most effective materials for SCR of NO in excess oxygen and investigated intensively since 1990s[95-97]. Most of reports have already suggested that highly isolated Ag cations, strongly bonded to alumina, are the active sites for the NO-SCR process[100]. For the purpose of obtaining the best SCR performing catalysts, a colloidal solution containing AlOOH or $Al(OH)_3$ has been employed as a precursor of alumina support in a study of ours dealing with supported silve catalysts. It is hoped that their -OH groups might contribute to Ag species being evenly dispersed and anchored on the alumina surface. Some attractive interaction is also expected during the activation procedure converting Boehmite or aluminum hydroxide to γ-alumina. Ag/Al_2O_3 catalyst was also prepared using commercial γ-Al_2O_3 as a support for a comparative study. Emphasis will be placed on clarifying the relation between the redox properties of the silver catalysts with their catalytic activities for both NO reduction and C_3H_6 oxidation.

10.2.1. Preparation Method

Alumina-supported silver catalysts were prepared by impregnating the commercial AlOOH (Catapal 18N80), Al_2O_3 (Catabx SBa 150) and $Al(OH)_3$ (Pfaltz and Bauer) supports with an aqueous solution containing desired amounts of silver nitrate (99 %, Aldrich) under an intense stirring at room temperature for 4 h. These mixtures were condensed under vacuum using a rotor evaporator (R110, Brinkmann), thereafter, dried at 110 °C for 4 h and calcined at 500 °C for 5 h under a 10% O_2/He gas flow to obtain the final products designated as Ag/Al_2O_3 (I), Ag/Al_2O_3 (II) and Ag/Al_2O_3 (III), respectively. Further calcination of Ag/Al_2O_3 (III) at 800 and 950 °C for 5 h results in the samples denoted as Ag/Al_2O_3 (IV) and Ag/Al_2O_3 (V). Corresponding aluminas (Al_2O_3 (I), Al_2O_3 (II) and Al_2O_3 (III)) were also prepared by the same procedure except for the addition of silver nitrate.

10.2.2. Physicochemical Properties of Alumina Supported Silver Catalysts and Their Supports

Table 12. Physicochemical properties of alumina-based catalysts [110]

Sample	Specific surface area (m^2/g)	Pore volume (ml/g)	Pore size (nm)	Crystal domain size (nm)	Ag weight percent (%)
Al_2O_3 (I)	143	0.42	11.9	7.7	-
Ag/Al_2O_3 (I)	140	0.40	10.8	6.6	2.9
Al_2O_3 (II)	144	0.47	9.5	6.8	-
Ag/Al_2O_3 (II)	144	0.47	9.5	6.9	3.1
Al_2O_3 (III)	64	0.26	3.8	-	-
Ag/Al_2O_3 (III)	35	0.15	3.8	-	3.2
Ag/Al_2O_3 (IV)	79	0.22	10.8	7.9	3.1
Ag/Al_2O_3 (V)	61	0.22	14.6	6.8	3.1

The crystal phase analysis was performed by XRD (not shown), revealing a main γ-Al_2O_3 phase (JCPDS 29-0063) for Ag/Al_2O_3 (I, II, IV, V), but Ag/Al_2O_3 (III) which was synthesized from the $Al(OH)_3$ source and calcined at 500 °C, exhibited an amorphous structure XRD profile. BET specific surface area, pore volume and size, crystallite size and Ag content of the prepared alumina-based catalysts are summarized in Table 12. The supports originating from AlOOH or γ-Al_2O_3 and denoted as Al_2O_3 (I) or Al_2O_3 (II) display a high surface area of ~140 m^2/g. Ag loading over these two carriers scarcely affects the values of their surface areas. By contrast, a relatively lower surface area (about 64 m^2/g) was found for Al_2O_3 (III) prepared from $Al(OH)_3$. Subsequently, Ag impregnation further decreased its surface area which reached a value of 35 m^2/g. The surface area of Ag/Al_2O_3 (III) was significantly enhanced by means of calcination at 800 and 950 °C reaching values of 79 m^2/g for Ag/Al_2O_3 (IV) and 61 m^2/g for Ag/Al_2O_3 (V). Taking into account the XRD data which show a transformation from the amorphous phase for Ag/Al_2O_3 (III) to γ-Al_2O_3 phase for Ag/Al_2O_3 (IV) and Ag/Al_2O_3 (V), the formation of the gamma alumina phase is deemed as the main reason for the enhancement in surface area.

10.2.3. Temperature Programmed Analysis

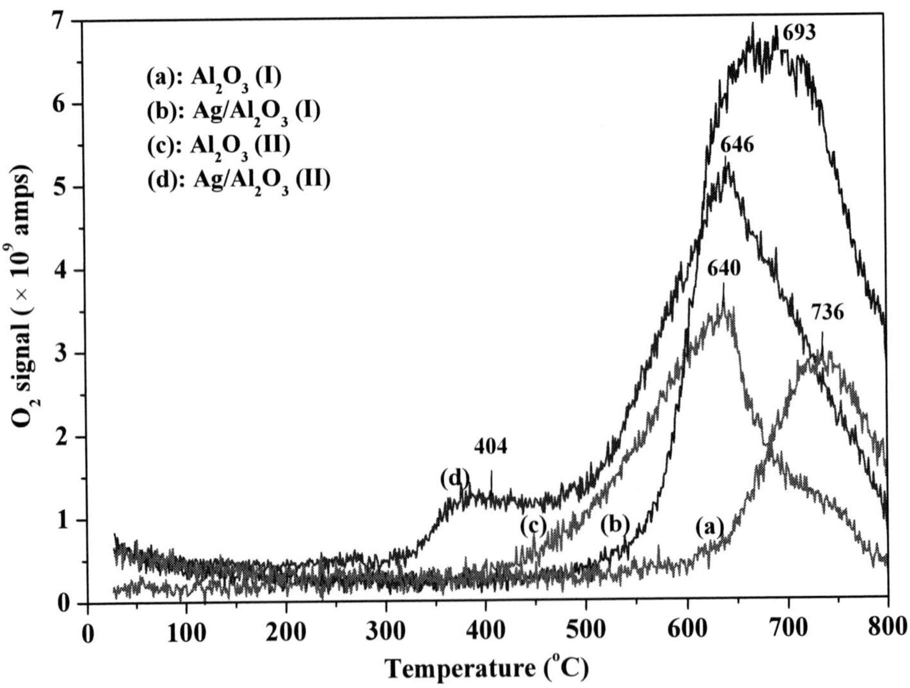

Figure 31a. O_2-TPD profiles of Al_2O_3 (I), Ag/Al_2O_3 (I), Al_2O_3 (II) and Ag/Al_2O_3 (II) samples [110].

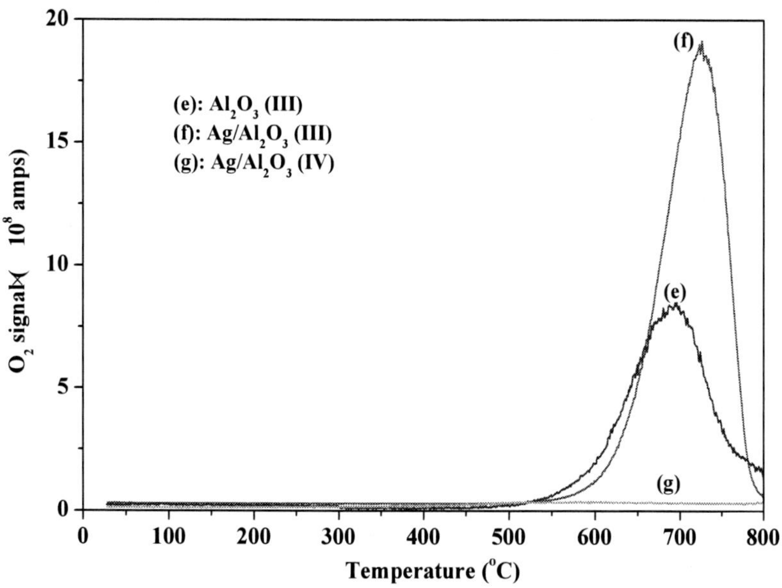

Figure 31b. O_2-TPD profiles of Al_2O_3 (III), Ag/Al_2O_3 (III) and Ag/Al_2O_3 (IV) samples [110].

Table 13. Amount of O_2 desorbed from the alumina-based samples during O_2-TPD experiments [110]

Sample	Amount of oxygen desorbed	
	Low temperature (μmol/g)	High temperature (μmol/g)
Al_2O_3 (I)	-	4.2 (736 °C)
Ag/Al_2O_3 (I)	-	14.2 (693 °C)
Al_2O_3 (II)	-	6.3 (640 °C)
Ag/Al_2O_3 (II)	1.4 (404 °C)	9.6 (646 °C)
Al_2O_3 (III)	-	113.4 (688 °C)
Ag/Al_2O_3 (III)	-	193.8 (717 °C)
Ag/Al_2O_3 (IV)	1.6 (<500 °C)	8.1 (>500 °C)

Various oxygen species on the silver surface were previously investigated by Raman spectroscopy and roughly classified into physisorbed molecular oxygen, chemisorbed molecular oxygen ($Ag-O_2$) and the ionic oxygen incorporated into the substrate[111]. The oxygen species formed over alumina-based catalysts after 10% O_2 adsorption were investigated by O_2-TPD experiments, as illustrated in Figs.31a and b.

Only one O_2 peak for Al_2O_3 (I) occurring at 736 °C was likely related to the liberation of surface lattice oxygen associated with aluminum (Al-O) (line (a) in Figure 31a). A clear enhancement in the intensity of this O_2 peak after Ag loading, as seen from line (b) in Figure 31a for Ag/Al_2O_3 (I), may be explained by the incorporation of silver into lattice to form an Ag-O-Al structure, which is likely more reducible than an Al-O-Al structure. The O_2 desorption occurring at 640 °C over Al_2O_3 (II) was again assigned to the release of lattice oxygen bound to aluminum (line (c) in Figure 31a). In the profile of O_2-TPD for Ag/Al_2O_3 (II), the O_2 desorption appearing at 646 °C was accordingly attributed to the lattice oxygen associated with silver or aluminum, whereas, another O_2 desorption peak at 404 °C was assigned to the molecular oxygen chemisorbed upon silver due to its low desorption temperature falling into the temperature range for α-oxygen desorption as previously described in Ref. [17] (line (d) in Figure 31a). An attempt was made to correlate the distinct O_2-TPD behaviors of Ag/Al_2O_3 (I) and Ag/Al_2O_3 (II) with the nature of their alumina sources (Ag/Al_2O_3 (I) was prepared from a Boehmite, while Ag/Al_2O_3 (II) was prepared from γ-phase Al_2O_3). During the preparation procedure for Ag/Al_2O_3 (I), an exchange between the -OH groups of Boehmite and Ag^+ ions of silver nitrate might take place, which is favorable to the formation of an Ag-O-Al local structure and the high dispersion of Ag over alumina:

$$\rangle Al{-}OH + Ag^+ \longrightarrow \rangle Al{-}O{-}Ag + H^+ \quad (26)$$

As for the preparation of Ag/Al_2O_3 (II), the chance to yield Ag-O-Al structure is reduced due to the scarcity of hydroxyl group over γ-phase alumina compared to Boehmite. The higher lattice oxygen desorption at 646 °C for Ag/Al_2O_3 (II) (9.6 μmol/g) compared to that at 640 °C for parent Al_2O_3 (II) (6.3 μmol/g) could be ascribed to the contribution from the highly dispersed Ag_2O particles over surface, having a strong interaction with support, or Ag^+ ions associated with lattice oxygen. The low-temperature peak at 404 °C in the O_2-TPD profile for Ag/Al_2O_3 (II) reflects its peculiar redox characteristic. It may be due to the formation of some larger Ag_mO (m>2) clusters over Ag/Al_2O_3 (II). The increasing particle size can accelerate the metallic Ag yield due to Ag_2O decomposition. Moreover, the stability of Ag_mO clusters will decrease because of a relatively poor interaction with alumina along with the increase of their sizes. The above reasons seem to explain this O_2 desorption appearing at relatively low temperature. The O_2 desorbed from perovskites at similar temperatures was previously designated as α-O_2, and was thought to be crucial for hydrocarbons oxidation in this case[17, 36, 96]. It is also possible that the rates of hydrocarbon oxidation are dependent on Ag particle size with higher rates being associated with larger particles. Hence, Ag/Al_2O_3 (II) was expected to yield a better activity for C_3H_6 oxidation.

It was observed from Figure 31b and Table 13 that large amounts of O_2 were desorbed from amorphous Al_2O_3 (III) at 688 °C compared to Al_2O_3 (I and II). Additionally, a significant O_2 desorption (194 μmol/g) occurred over Ag/Al_2O_3 (III) along with CO_2 and H_2O desorption at T>600 °C. Interestingly, Ag/Al_2O_3 (IV) having a γ-phase Al_2O_3 structure, obtained by means of calcining the amorphous Ag/Al_2O_3 (III) at 800 °C, shows a comparable O_2 desorption (9.4 μmol/g) with those for Ag/Al_2O_3 (I and II) (Figure 31b). Therefore, this abundant O_2 desorption in O_2-TPD occurring over Ag/Al_2O_3 (III) was related to the depletion of lattice oxygen during the phase transformation of amorphous hydroxy alumina to gamma type transition alumina. In addition, metallic Ag appearing in Ag/Al_2O_3 (IV and V) was confirmed by XRD (not shown). These data indicate that the structure and nature of the silver and amorphous support phases of Ag/Al_2O_3 (III) undergo drastic changes upon high temperature calcination (≥800 °C).

10.2.4. Activity Test in Excess O_2

The temperature dependence of NO conversion to N_2 in a feed flow of 1000 ppm NO + 3000 ppm C_3H_6 + 10% O_2 over Al_2O_3 (I) and Ag/Al_2O_3 (I-V) is depicted in Figure 32a. For alumina alone, N_2 yield was marginal until T>400 °C and then increased rapidly at increasing temperature reaching a maximum of 76% at 500 °C, eventually, declining to 23% at 600 °C. With the addition of silver to alumina, the best performance for NO selective reduction to N_2 was obtained over Ag/Al_2O_3 (I) giving a N_2 yield of 74% at a temperature as low as 300 °C and achieving values of more than 90% over a wide temperature range with a plateau from 350 to 550 °C. A moderate N_2 yield over Ag/Al_2O_3 (II) was observed starting at 200 °C and passing through the maximum of 84% at 450 °C, before gradually declining at relatively higher temperature. A plot of N_2 yield similar to that for pure Al_2O_3 (I) was found for Ag/Al_2O_3 (III) displaying significant SCR activity solely at T>400 °C and showing a maximum N_2 yield of 60% at 500 °C. In the case of Ag/Al_2O_3 (IV) synthesized by calcining Ag/Al_2O_3 (III) at 800 °C, a good N_2 yield at low temperature (300-400 °C) and its progressive diminution at raising temperature (400-600 °C) were observed. The worst N_2 yield was obtained over Ag/Al_2O_3 (V) with values around 20% in the range of 300-450 °C and level off at >600 °C.

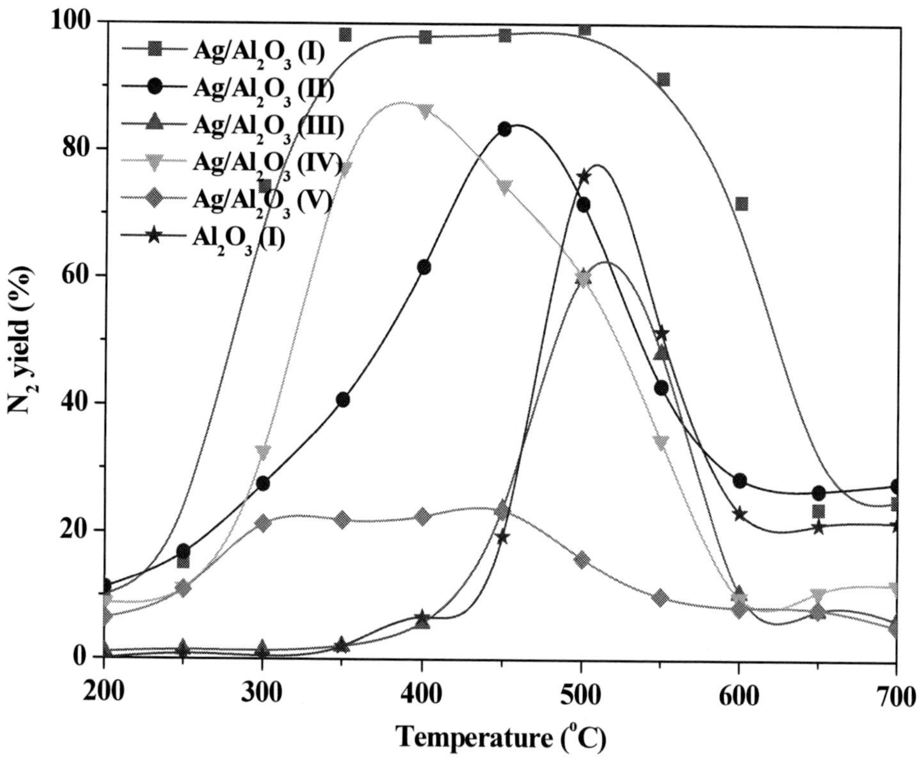

Figure 32a. N_2 yield in C_3H_6-SCR of NO over alumina-based samples [110]. Conditions: GHSV=30,000 h^{-1}, 1000 ppm NO, 3000 ppm C_3H_6, 10% O_2.

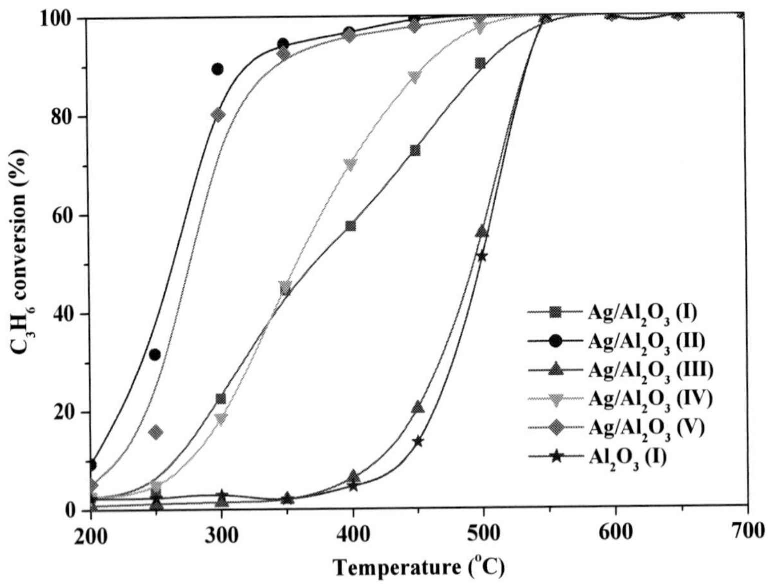

Figure 32b. C_3H_6 conversion in C_3H_6-SCR of NO over alumina-based samples [110]. Conditions: GHSV=30,000 h^{-1}, 1000 ppm NO, 3000 ppm C_3H_6, 10% O_2.

The temperature dependence of C_3H_6 conversion on temperature during NO-SCR is shown in Figure 32b. Relatively lower C_3H_6 conversions were achieved over Al_2O_3 (I) and Ag/Al_2O_3 (III) with a light-off at ca. 400 °C. These conversions keep going on with temperature and up to almost 100% at 600 °C. On the contrary, the best conversion of C_3H_6 occurred over Ag/Al_2O_3 (II) showing a 90% C_3H_6 conversion even at the relatively lower temperature of 300 °C followed by a complete conversion at temperature higher than 450 °C. A C_3H_6 conversion comparable to that for Ag/Al_2O_3 (II) was also obtained in the case of Ag/Al_2O_3 (V). In addition, the moderate C_3H_6 conversion among the tested six samples was found over Ag/Al_2O_3 (IV) and Ag/Al_2O_3 (I) with values of 70% and 57% at 400 °C, respectively.

Attempts to correlate the catalytic performance of Ag-promoted alumina in NO-SCR reaction with the local structure of Ag cluster and alumina support have also been made by several authors[97, 112]. Jen ascribed the larger deNO$_x$ activity for 2% Ag/Al_2O_3 to the support having a greater fraction of pore volumes in the range 25-100 Å or a narrower pore-size distribution[113]. It is known that the state of Ag is strongly depending on its loading[112, 114]. Although all detected silver species are thought to be active in the SCR of NO_x, large silver oxide particles tend to give a lower SCR specific activity simultaneously with a higher oxidation of the hydrocarbon reductants[97, 115].

The ratio of N_2 yield to C_3H_6 conversion ($Y(N_2)/X(C_3H_6)$) determined from data collected during the C_3H_6-SCR of NO as a function of temperature over the above discussed alumina-based catalysts is shown in Figure 33. This ratio decreases gradually due to a decline in N_2 yield together with an enhancement in C_3H_6 conversion at increasing temperature. In the low-temperature region of 300-500 °C Ag/Al_2O_3 (I and IV) showed a higher rate for NO selective reduction than propene oxidation giving $Y(N_2)/X(C_3H_6) > 1$, the opposite occurred for Ag/Al_2O_3 (II and V) showing $Y(N_2)/X(C_3H_6) < 1$ due to a largely non-selective combustion. The ionic Ag bound with the alumina forming an Ag-O-Al structure is believed to be vital for NO_x reduction since the SCR reaction rate is proportional to the number of Ag-O-Al sites[112]. Both Ag/Al_2O_3 (I) and Ag/Al_2O_3 (IV) are prepared by an impregnation method but using Al sources containing -OH groups (such as AlOOH and $Al(OH)_3$), which also facilitates the formation of an Ag-O-Al local structure. Their good selectivity towards the formation of N_2 can thus correspond to the high density of Ag-O-Al species. On the contrary, the excellent oxidative abilities of Ag/Al_2O_3 (II) and Ag/Al_2O_3 (V) related to the formation of large Ag_mO (m>2) crystallites are responsible for their high C_3H_6 conversion. The Ag-O-Al species are also believed to exist in Ag/Al_2O_3 (III) sample but they may be immersed in the alumina substrate thus not showing their SCR activities. The low surface area and amorphous structure of Ag/Al_2O_3 (III) limit its activity for both NO reduction and C_3H_6 oxidation resulting in a $Y(N_2)/X(C_3H_6)$ around 1.

A γ-Al_2O_3 structure was obtained over Ag/Al_2O_3 (IV) after calcination at 800 °C of amorphous Ag/Al_2O_3 (III) leading to a clear enhancement in both N_2 yield and C_3H_6 conversion. This suggests that the high surface area γ-Al_2O_3 acting as a support is necessary for lean NO reduction. Ag/Al_2O_3 (V) was obtained by calcining Ag/Al_2O_3 (III) at 950 °C. It is surprising that N_2 yield reached over Ag/Al_2O_3 (V) was much lower while its C_3H_6 conversion was fully developed compared to Ag/Al_2O_3 (IV). This result may be explained by the formation of large-size Ag clusters during calcination at 950 °C as verified by XRD (not shown), yielding more readily reducible entities with a higher activity for nonselective propene oxidation.

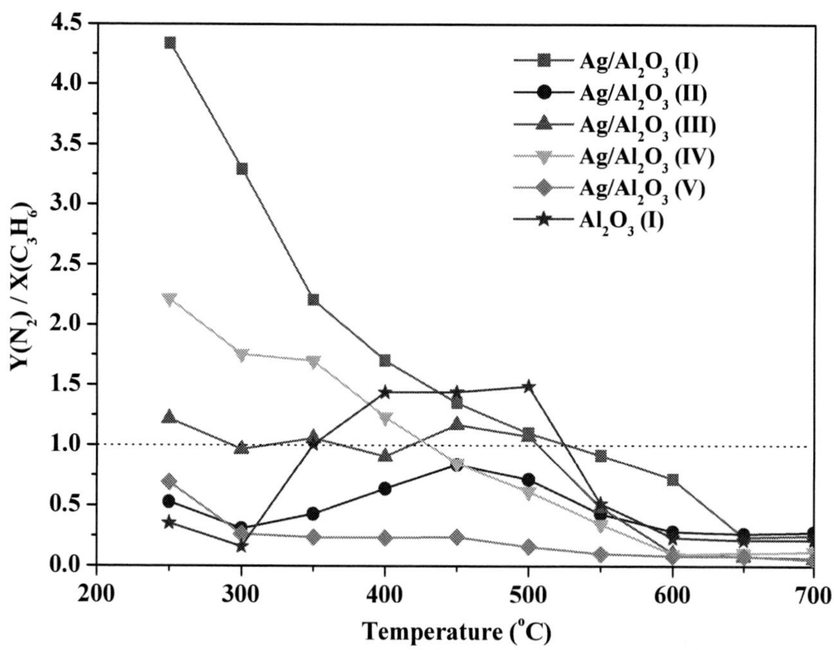

Figure 33. Ratio of N_2 yield to C_3H_6 conversion in C_3H_6-SCR of NO over alumina-based samples [110].Conditions: GHSV=30,000 h^{-1}, 1000 ppm NO, 3000 ppm C_3H_6, 10% O_2.

10.3. Cu-MCM-41

Ever since Iwamoto et al.[116] reported that transition-metal ion exchanged zeolites, in particular Cu-ZSM-5, showed high HC-SCR activity, the copper-containing zeolite catalysts have been comprehensively investigated[114-116]. Copper exchanged ZSM-5, mordenite and ferrite were found active, whereas zeolites X and Y were shown to be poor supports[116, 117]. Cu-zeolites have received much attention as a possible means of controlling the emissions from vehicles, the engines of which operate in lean-burn conditions (e.g. diesel and lean-burn engine)[118, 119]. The catalytic reduction process is however inhibited by the presence of steam and by the low thermal stability of the catalyst[120, 121]. Progress remains therefore to be made for the development of active and selective catalysts over the desired range of temperature. It has been found that copper-exchanged Al-MCM-41 has activity similar to Cu-ZSM-5, but contrary to Cu-ZSM-5, the presence of 10% water induces only a minor inhibition[122].

The present contribution reports a study of NO-SCR with propylene under lean-burn conditions over Cu-exchanged Al-MCM-41 catalysts and of the nature of the copper species present in these catalysts. The Al-MCM-41 samples were synthesized according to the procedure given in the literature[123]. The copper-containing catalysts were prepared by ion-exchange of Al-MCM-41 with an aqueous solution of copper (II) acetate.

Figure 34 shows the NO and propylene conversions as functions of temperature in the case of Cu-Al-MCM-41-10-61 (Si/Al = 10; Cu^{2+} exchange 61%). The maximum conversions observed for the formation of N_2 and of NO_2 are around 370 and 450 °C, respectively. The latter product appears only at temperatures higher than 370 °C. Propylene is essentially oxidized to carbon dioxide and water.

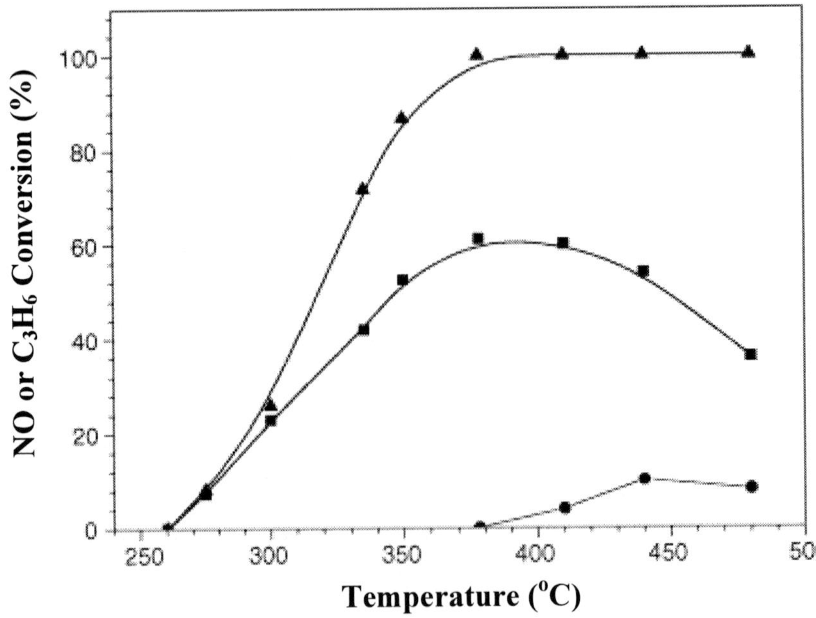

Figure 34. Effect of temperature on NO conversion to N_2 (■), NO conversion to NO_2 (●) and propylene conversion (▲) over Cu-Al-MCM-41-10-61 catalyst (reaction conditions: GHSV = 100,000 h^{-1}, 0.1 vol.% NO, 0.1 vol.% C_3H_6, 2% O_2) [124].

It is found that the complete conversion of C_3H_6 is obtained at the temperature corresponding to the maximum for N_2 formation and to the start of NO_2 formation. This suggests that the reduction of NO could involve a chemisorbed NO_2 as a very reactive intermediate.

Since water was shown to be a poison for Cu-ZSM-5, it is especially important to examine the effect of water vapor on the activity of Cu-Al-MCM-41 in NO-SCR. A full study is currently undertaken in our laboratory. Figure 35 shows the dynamic effect of introducing 10% H_2O in the feed, on catalytic activity at 400 °C. The decrease in NO conversion was sensible but still not very high and by sequential suppression of the water in the gas feed the effect was found to be reversible. The low and reversible deactivation by water vapor during NO-SCR is also significant for potential commercial application of these new catalysts. The surface chemistry of the support, its hydrophilicity, will require investigations before any definitive conclusion can be made.

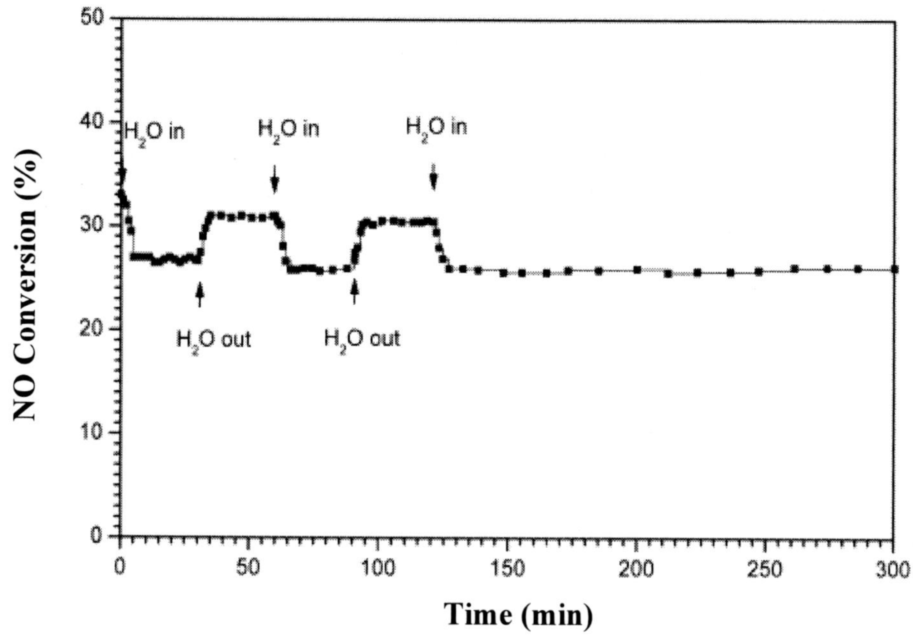

Figure 35. Effect of water vapor on NO conversion over Cu-Al-MCM-41 at 400 °C [124].

Table 14. Characteristics of NO and oxygen adsorbed over Cu-Al-MCM-41-10-61 [124]

Feed gas	NO-TPD					NO + O_2-TPD		
	Temperature of desorption peak (°C)		NO desorbed[a] (µmol/g)		$A_L/(A_L + A_H)$[b]	Temperature of desorption peak (°C)	NO_2 desorbed[a] (µmol/g)	NO_2/NO_x
	First	Second	First	Second				
NO	149	422	34.8	1.64	0.95	86	8.2	0.19
NO + O_2	135	411	38.8	4.1	0.9	80	44.4	0.53

[a] Based on the NO desorption peak area at low temperature.[b] A_L: peak area at low temperature; A_H: peak area at high temperature.

TPD of Cu-Al-MCM-41 (after NO adsorption under 0.8% NO in He) was conducted (Table 14). NO and NO_2 are the species detected coming off the surface as the temperature of the catalyst is increased. Two features were observed in the NO desorption profile: a principal peak at 149 °C and a second NO desorption feature at higher temperature (440 °C). This indicates that there are at least two types of NO adsorption sites available. The presence of two types of adsorbed NO species over Cu catalysts has been reported earlier in the literature[45]. These have been proposed to be the desorption of NO from Cu^{2+} ions and nitrate (NO_3^-), nitrite (NO_2^-) or NO_2^+ adsorbed species, respectively. Assuming the sensitivity factors of the peaks at low and high temperature are equivalent, the areas can be used to estimate the normalized desorption of NO. As listed in Table 14, the amount of NO desorbed at low temperature is close to the total amount of desorbed NO. This feature indicates that copper is mainly as isolated Cu^{2+} in the catalyst. During NO desorption, a small amount of NO_2 (8.2 μmol/g) desorbed at 80 °C.

TPD experiments were also performed after adsorption from a mixture of 0.8% NO and 1.6% O_2 in helium over the catalyst. From Table 14, it is clear that the amount of desorbed NO increases and the temperature decreases with the addition of oxygen into the feed gas. Additionally, a different trend appeared in the NO_2 desorption curves. In an oxygen-containing atmosphere, the desorbed amount of NO_2 increased to 44.4 μmol/g. This suggests that the presence of oxygen enhances the NO adsorption and the oxidation state of the adsorbed species.

The hydrogen TPR profiles of Cu-Al-MCM-41 catalysts before and after reaction are illustrated in Figure 36. The fresh catalyst shows two reduction peaks at about 190 and 543 °C, which reflects the two-step reduction process of isolated Cu^{2+} species [124, 125]. The low temperature peak is assigned to the reduction of isolated Cu^{2+} to Cu^+ ions. The other peak at a higher temperature is attributed to the reduction of Cu^+ ion to Cu^0, as was done for Cu-ZSM-5[124].

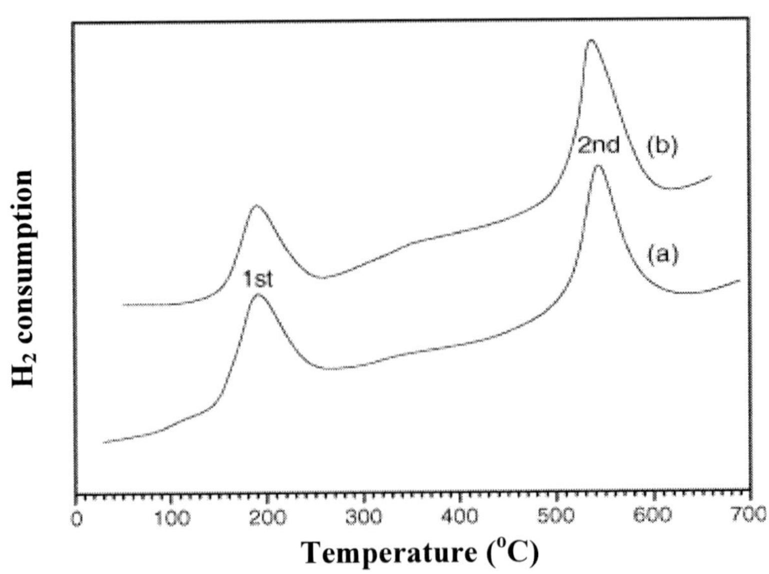

Figure 36. H_2-TPR profiles of Cu-Al-MCM-41-10-61 catalysts before (a) and after (b) reaction [124].

Since the CuO aggregates would be reduced to Cu^0 by H_2 in one step at about 350 °C[124], the peak at 348 °C in TPR profile of the catalyst after reaction indicates the presence of traces of CuO. It is suggested that copper in the catalysts is mainly in the form of isolated copper ions. This is in accordance with the results by Delahay et al.[126]. Comparing the H_2/Cu ratios of the two reduction peaks, we observe that in the fresh catalyst Cu-Al-MCM-41, copper is mainly present as Cu^{2+} ion, since the areas of the two peaks are very close from each other. After reaction, a decrease in the H_2/Cu ratio is observed for the first peak. By comparison, the H_2/Cu ratio for the second peak increases. This indicates that some Cu^{2+} is reduced to Cu^+ during the reaction.

Figure 37 shows the difference IR spectra of Cu-Al-MCM-41 after exposure to the reaction gases mixtures at 100 °C after 1 h under flow conditions, and subsequent flushing with He for 1 h. Spectrum (a) in Figure 37 concerns the adsorption of NO in the presence of O_2. Several peaks are observed in the range between 1400 and 2000 cm^{-1}, with pronounced peaks at 1625 and 1455 cm^{-1}. The band at 1625 cm^{-1} was assigned to the asymmetric stretch of NO_2 adsorbed on Cu^{2+} [127, 128], and the other strong band 1455 cm^{-1} may be caused by the vibration of the Cu^{2+}-ON=O species[129]. ESR and TPD experiments[129] also indicated the existence of the last species on Cu-ZSM-5 catalysts. The peak at 1905 cm^{-1} can be attributed to NO linearly adsorbed on Cu^{2+} [64, 127, 129], while the band at 1820 cm^{-1} is associated to NO adsorbed on Cu^+ [64, 127]. The results indicate that a significant amount of oxidized NO species is formed in the presence of a NO + O_2 gas mixture.

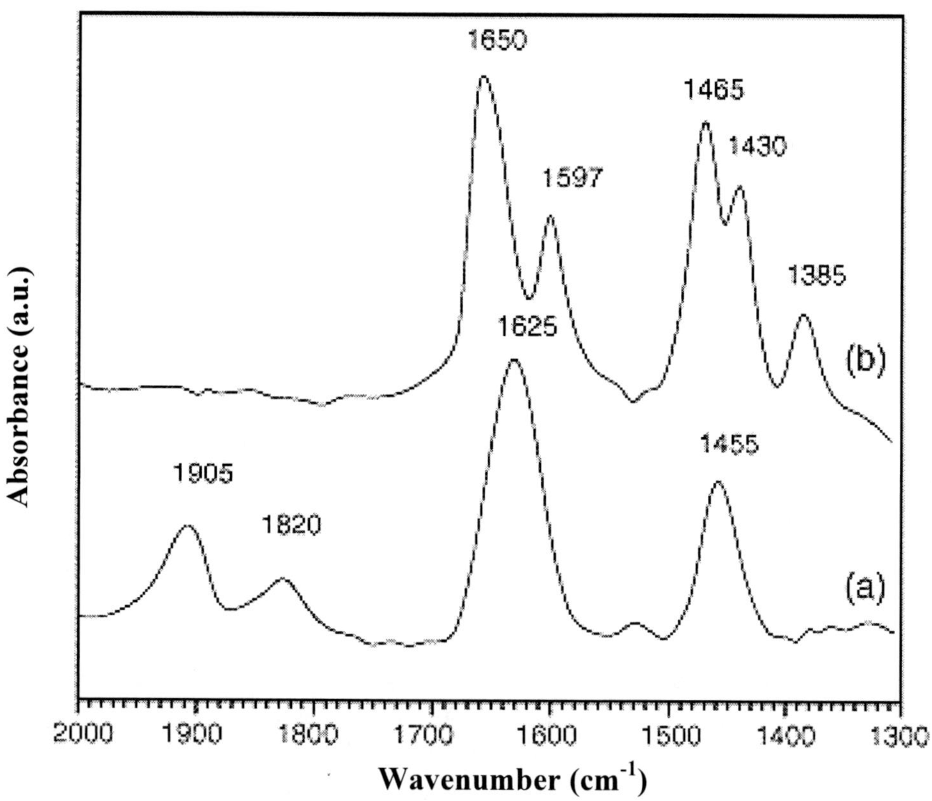

Figure 37. IR spectra of NO + O_2 (a) and C_3H_6 + O_2 (b) adsorbed on Cu-Al-MCM-41-10-61 [124].

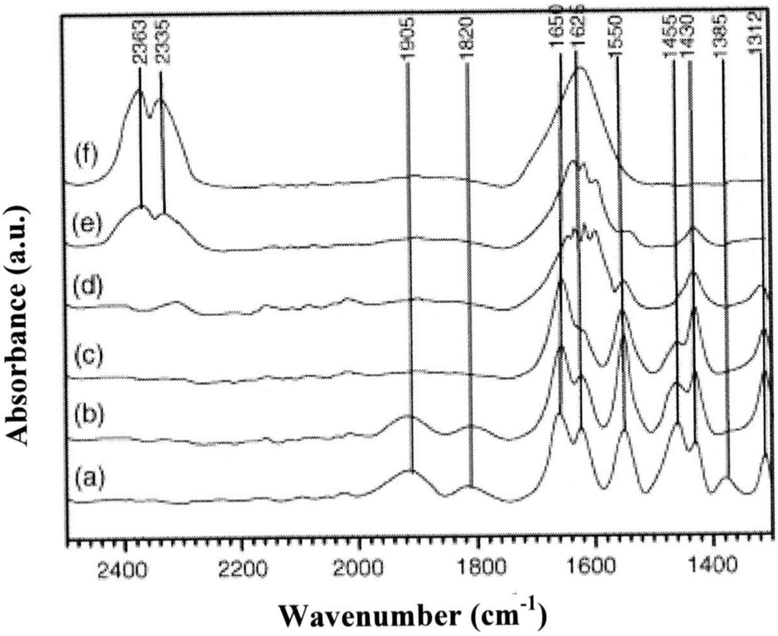

Figure 38. In situ IR spectra of Cu-Al-MCM-41-10-61 exposed to C_3H_6-SCR reaction gases at 100 °C (a), 150 °C (b), 200 °C (c), 250 °C (d), 300 °C (e) and 350 °C (f) [124].

After 1 h of $C_3H_6 + O_2$/He exposure, adsorbed species with characteristic IR bands in the range between 1700 and 1350 cm^{-1} are observed (Figure 37 (b)). By referring to the literature[130-132], the absorption bands can be assigned as follows: the band at 1650 cm^{-1} is resulting from acrolein species; the bands at 1465 and 1385 cm^{-1} are due to π-allyl complex (π-C_3H_5) and allylic species; and the bands at 1597 and 1430 cm^{-1} are caused by carboxylate adspecies. The above observations revealed that in the presence of O_2 the adsorption of C_3H_6 on Cu-Al-MCM-41 resulted not only in different adspecies but also in the formation of oxygen-containing intermediates.

Figure 38 (a)-(f) shows the changes in the IR spectra during exposure of Cu-Al-MCM-41 to a gas mixture containing 0.8% NO, 0.8% C_3H_6 and 16% O_2 in He between 100 and 350 °C. At the lowest temperature, intense bands from surface nitrogen dioxide, nitrites, carboxylate, acrolein and allylic species between 2000 and 1400 cm^{-1} are observed. Besides these, some new surface species, indicated by bands at 1550 and 1312 cm^{-1} are formed. According to literature, these bands can be assigned to the vibration of organic nitro species (R-NO_2)[133-135]. The intensity of organic nitro species increases, reaches a maximum at 150 °C and then declines with further temperature increase. Meanwhile, the intensities of acrolein and carboxylate bands reach their maximum at a temperature of 200 °C. Increasing temperature leads to a decrease in concentration of adsorbed species.

At 250 °C, a split band around 1610 cm^{-1} appears, showing the formation of -HCOOH adspecies. As temperature exceeds 300 °C, new bands ascribed to adsorbed CO_2 occur at 2363 and 2335 cm^{-1}. Only the bands of CO_2 and H_2O caused by the decomposition of HCOOH adspecies[136] remain as the temperature is raised to 350 °C.

As indicated by TPD of NO (Table 14) and FT-IR (Figure 37) at room temperature, NO is adsorbed on both Cu^{2+} and Cu^{+} and the corresponding adsorbed species are both partially oxidized to adsorbed NO_2 (lines at 1625 and 1455 cm^{-1} in Figure 37). It seems that in this process Cu^+ may be oxidized:

$$Cu^+ + NO_2 \longrightarrow Cu^{2+}\text{-}ON{=}O \qquad (27)$$

or

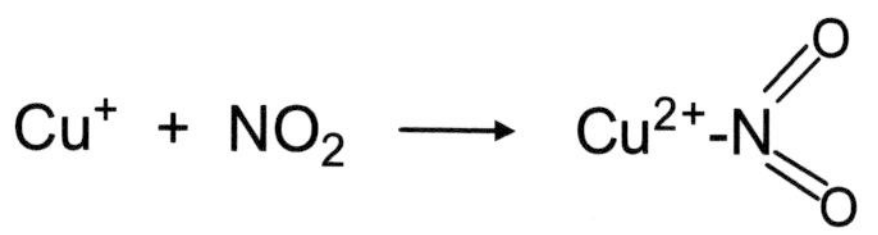

(28)

Propylene is also adsorbed as an allylic species (Figure 38 (a), lines at 1465 and 1385 cm^{-1}) followed by successive oxidation to acrolein, carboxylate species (lines at 1650, 1597 and 1430 cm^{-1}) and eventually $CO_2 + H_2O$.

$$CH_3\text{-}CH{=}CH_2 \longrightarrow CH_2{=}CH{=}CH_2\,(ads) + H_2\,(ads) \qquad (29)$$

$$CH_3{=}CH{=}CH_2 \xrightarrow{O_2} CH_2{=}CH\text{-}CHO \xrightarrow{O_2} RCOO^- \xrightarrow{O_2} CO_2 + H_2O \qquad (30)$$

where R=H-, CH_3-, C_2H_3-, or C_2H_5-.

Similar successive oxidations may be performed with adsorbed NO_2 as the oxidizing agent:

$$CH_2{=}CH{=}CH_2 \xrightarrow{NO_{2\,ads}} CH_2{=}CH{=}CHO \xrightarrow{NO_{2\,ads}} R\text{-}NO_2 \xrightarrow{NO_{2\,ads}\text{ or }O_2} N_2 + CO_2 + H_2O \qquad (31)$$

and the intermediate organic nitro species may also be formed from the carboxylates:

$$RCOO^- \xrightarrow{NO_{2\,ads}} R\text{-}NO_2 \xrightarrow{NO_{2\,ads}\text{ or }O_2} N_2 + CO_2 + H_2O \qquad (32)$$

Reaction (31) suggests acrolein as a key intermediate in SCR of NO by propylene. The formation of nitro species by this reaction was already discussed in the literature and evidenced by a significant reduction in the surface concentration of propylene adspecies in the presence of a $NO_2 + O_2$ mixture[133, 134]. Note that the role of organic nitro species as active intermediate in the SCR of NO over Cu-ZSM-5 was already discussed by Hayes et al.[135]. In addition, in TPR experiments, we observe Cu^+ forming predominantly on the surface of Cu-Al-MCM-41 after exposure to C_3H_6, and a redox of Cu^{2+} and Cu^+ during the reaction. It is therefore possible to postulate that the divalent copper ion is reduced to monovalent in the conditions of the reaction between adsorbed propylene and NO_2 species.

In Figure 38, the band intensity of the adsorbed nitro species reduces gradually at increasing temperature above 200 °C. The phenomenon may be caused by the decomposition of the species promoted by oxygen (reactions (31) and (32)). Li and Armor[137] investigated the selective decomposition of nitromethane over Cu-ZSM-5 and H-ZSM-5 and found that the presence of oxygen can significantly enhance the decomposition. Simultaneously, O_2 plays an essential role in promoting the oxidation of Cu^+ to Cu^{2+}, thus completing the catalytic cycle of the active center. In addition, the presence of oxygen promotes the HC-SCR by oxidizing NO to NO_2.

11. CONCLUSIONS

In this chapter, our continued work on the application of perovskites for the purification of exhaust gases including NO, C_3H_6 and CO of motor vehicles was investigated other recent catalysts for SCR of NO_x were also discussed including different mixed oxides, alumina supported silver and copper exchanged mesostructured alumino-silicate MCM-41. The following conclusions can be made based on the experimental results.

- High-energy-grinding is an ideal method to synthesize perovskite-type mixed-oxides having extraordinarily high BET surface area (even on the order of 100 m^2/g when grinding additives are used) and nanoscale grain size (below 20 nm). Perovskites generated by this novel reactive grinding exhibit more grain boundary and excellent catalytic performance for various redox reactions. Moreover, this preparation method is suitable to a large scale continuous production. Furthermore, the preparation of perovskite by reactive grinding is environmental friendly avoiding the production of waste liquids like the other methods (e.g. citric method) involved. Hence, high-energy-grinding is a promising technology for perovskite synthesis.
- The redox properties of series of perovskites were studies by means of O_2-TPD, H_2-TPR, XPS, and isotopic exchange experiments. Various oxygen species as well as their role in the corresponding reactions were discussed. α-O_2 was desorbed usually at temperature below 700°C and can be further classified as α_1-O_2 and α_2-O_2. α_1-O_2 was described as molecular oxygen adsorbed upon anion vacancies of perovskite as O_2^-. α_2-O_2 is possibly generated via a futher dissociation of molecular oxygen into atomic oxgen (O^-). The desorption of β-O_2 is due to the liberation of lattice oxygen from mixed-oxides, which occurs only at high temperature (>700 °C). Overstoichiometric oxygen may also come from the lattice, however, associated with a cation deficiency. The existence of overstoichiometric oxygen commonly appears in the case of lanthanum manganites, while lanthanum cobaltites or divalent-ion substituted perovskite always present large anion deficiency displaying as a condense α-O_2 desorption. Although both α-O_2 and overstoichiometric oxygen can make a contribution for redox process, α-O_2 seems to be easier used. In addition, the grain boundary oxygen, which is formed due to the oxygen diffusion and adsorption taking place around the interface between individual grains, was verified to play an important role on hydrocarbon oxidation. Reactive grinding can facilitate the generation of more grain boundary oxygen.
- Propene and carbon monoxide are two kinds of exhaust gases for gasoline engine, which can be utilized as reducing agents for NO reduction. A mechanism was proposed for NO reduction by propene which involves organo nitrogen compounds

generated from the interaction between nitrate species and adsorbed C_3H_6. Subsequently, an isocyanate intermediate formed from organo nitrogen intermediates can react with NO and/or O_2 to get the desired products. In NO + CO reaction, the chemisorbed NO (nitrosyl species) was dissociated over perovskites with the formation of both N_2 and N_2O as well as an oxidized surface. This oxidized surface can be subsequently reduced by CO with a regeneration of anion vacancies for the continuous reaction at low temperature. N_2O was further reduced by CO and this transformation becomes pronounced as temperature increases. The higher efficiency of CO (compared to propene) in reducing the perovskite surface gives the explanation for the difference between these two systems. Cu substation in B site can obviously improve the catalytic performance of NO reduction by C_3H_6 or CO. Besides the essential nature of copper ions in the transformation of nitrogen oxides, Cu substitution promoted the generation of anion vacancies, which can facilitate the adsorption of reactants and the formation of some crucial intermediates for deNO$_x$ process. The outstanding ability of $LaFe_{0.97}Pd_{0.03}O_3$ for the low-temperature NO removal was ascribed to the excellent redox properties of palladium. In short, perovskites with B-site substitution by Pd, Cu are suitable material for simultaneous purification of NO, C_3H_6, and CO at an O_2 content close to stoichiometric.

- Facing up with the global energy crisis, the present gasoline engine will tend to introduce more oxygen to achieve a more complete combustion of fuel, resulting into a "lean-burn" condition. This requires the autocatalysts can endure higher O_2 concentration during the NO reduction. Unfortunately, a higher gas phase O_2 concentration reduces the N_2 yield over the highly oxidative perovskites because of a lack of reducing agent due to unfavorable combustion of reductants. This problem was solved by adding small amount of perovskite into alumina, yielding satisfactory efficiency of NO reduction at higher O_2 contents as well as a good transformation of HCs and CO.
- A moderate deactivation upon addition of 10% H_2O in the feed in NO reduction by propene over these perovskites was found, which is well recovered upon removal of H_2O from the feed. This steam deactivation is believed to involve a competitive adsorption between steam and the reactants (NO, C_3H_6 and O_2). More oxygen vacancies of perovskites are beneficial for a better resistance to water vapor. Similar reversible H_2O poisoning was also observed in the case of Cu-Al-MCM-41. Two regimes depending on the SO_2 poisoning extents were announced during both NO reduction and methane oxidation over perovskites. The reversible deactivation occurring at the first regime was mainly ascribed to the competitive adsorption between SO_2 and reactant molecules. The severe poisoning at the second stage was caused by partial destruction of perovskite structure accompanying with the generation of new phase of $La_2(SO_4)_3$ and simple B-site oxides when sufficient amounts of sulfate were formed in the bulk.
- Diesel engine with good fuel economy is popularly used in the daily life. It just emits minor amounts of CO and HCs due to complete combustion at excess oxygen (approximately 8-10%). However, NO$_x$ abatement under such oxidizing atmosphere is a great challenge because it requires using the limited reductant to selectively react with NO rather than O_2. Success was achieved by using alumina supported silver. It was proven that highly isolated Ag cations, strongly bonded to alumina, are the

active sites for NO-SCR process, whereas metallic silver clusters dominate in the high-silver content alumina-based catalysts. The hydroxyl groups of AlOOH precursors can facilitate the anchoring of Ag^+ ions on the alumina support through an exchange process in a colloid solution, finally, achieving a Ag-O-Al local structure. The high density of Ag-O-Al species in this Ag/Al_2O_3 catalyst is deemed to be crucial for NO selective reduction into N_2, showing values above 90% over a wide temperature range with a plateau from 350 to 550 °C.

- In summary, perovskites are desired materials for the reduction of NO from gasoline engine at stoichiometric oxygen or under "rich-burn" condition. The lean reduction of NO from gasoline engine at O_2 contents of 4-6% can be performed over $ZnGa_2O_4$ spinel, Cu-Al-MCM-41, as well as a perovskite-alumina mixture. Ag/Al_2O_3 is still the best choice for the catalytic removal of NO from diesel engine at excess oxygen (8-10%). It is interesting that the structure of Cu-Al-MCM-41 was maintained in the presence of steam, while the dealumination will happen for Cu-ZSM-5 as reported in the literature. This special properties imply mesoporous materials such as MCM-41, SBA-15, SBA-16 are possible not only to act as a support for high dispersion of ions with excellent ability for NO transformation (such as Cu, Pd, and Rh) due to its high specific surface area but also to have an ideal resistance to water vapor during practical application.

References

[1] Raveau, B. *J. Res. Natl. Bur. Stand.*, 1957, Volume 58, 75.

[2] Goldschmidt, V. M. *Skr. Nor. Videnk.-Akad., Kl. 1: Mat.-Naturvidensk. Kl.*, 1926, Volume 8, 56.

[3] Megaw, H. D. Crystal structure of double oxides of the perovskite type. *The Proc. Phys. Soc.*, 1946, Voulme 58, 133-152.

[4] Voorhoeve, RJH; Remeika, JP; Johnson, DWJr. Rare-earth manganites: catalysts with low ammonia yield in the reduction of nitrogen oxides. *Science*, 1973, Volume 180, 62-64.

[5] Voorhoeve, RJH; Johnson, DWJr; Remeika, JP; Gallagher, PK. Perovskite oxides: materials science in catalysis. *Science*, 1997, Voulme 195, 827-833.

[6] Kirchnerova J; Klvana D. Preparation and characterization of high surface perovskite electrocatalysts. *Int. J. Hydrogen Energy.*, 1994, Volume 19, Issue 6, 501-506.

[7] Cui X. and Liu Y. New methods to prepare ultrafine particles of some perovskite-type oxides, *Chem. Eng. J.*, 2000, Volume 78, Issues 2-3, 205-209.

[8] Giannakas, AE; Vaimakis, TC; Ladavos, AK; Trikalitis, PN; Pomonis, PJ. Variation of surface properties and textural features of spinel $ZnAl_2O_4$ and perovskite $LaMnO_3$ nanoparticles prepared via CTAB-butanol-octane-nitrate salt microemulsions in the reverse and bicontinuous states. *Journal of Colloid and Interface Science*, 2003, Volume 259, Issue 2, 244-253.

[9] Li, X; Zhang, HB; Liu, XX; Li., SJ; Zhao, MY. XPS study on O(1s) and Fe(2p) for nanocrystalline composite oxide $LaFeO_3$ with the perovskite structure. *Mater. Chem. Phys.,* 1994, Volume 38, Issue 4, 355-362.

[10] Leanza, R; Rossetti, I; Fabbrini, L; Oliva, C; Forni, L. Perovskite catalysts for the catalytic flameless combustion of methane: Preparation by flame-hydrolysis and characterisation by TPD–TPR-MS and EPR. *Appl. Catal., B: Environmental*, 2000, Volume 28, Issue 1, 55-64.

[11] Johnson, DWJr; Gallaguer, PK; Schrey, F; Rhodes, WW. Preparation of high surface area substituted $LaMnO_3$ catalysts, *Am. Cer. Soc. Bull.*, Annual meeting of the American Ceramic Society; Washington, DC, USA, 1976, Volume 55, Issue 5, 520-523.

[12] Tascón, JMD; Mendioroz, S; Tejuca, LG. Preparation, characterization and catalytic properties of $LaMeO_3$. *Z. Phys. Chem. Neue Folge*, 1981, Volume 124, 109-115.

[13] Wang, J; Yasuda, H; Inumaru, K; Misono M. Catalytic decomposition of dinitrogen oxide over perovskite-related mixed oxides, *Bull. Chem. Soc. Japan*, 1995, Volume 68, Issue 4, 1226-1231.

[14] Skoglundh, M; Löwendahl, L; Jansson, K; Dahl, L; Nygren, M. Characterization and catalytic properties of perovskites with nominal compositions $La_{1-x}Sr_xAl_{1-2y}Cu_yRu_yO_3$. *Appl. Catal., B: Environmental*, 1994, Volume 3, Issue 4, 259-274.

[15] Öcal, M; Oukaci, R; Marcelin, G. Steady state isotopic transient kinetic analysis (SSITKA) investigation of NO reduction with CO over perovskite catalysts. *Ind.Eng. Chem. Res.*, 1994, Volume 33, 2930-2934.

[16] Kaliaguine, S., and van Neste, A., Process for synthesizing perovskites using high energy milling. US patent, 6,017,504 (2000).

[17] Kaliaguine, S., van Neste, A., Szabo, V., Gallot, J.E., Bassir, M., Muzychuk, R., Perovskite-type oxides synthesized by reactive grinding: Part I. Preparation and characterization. *Appl. Catal., A: General*, 2001, Volume 209, Issues 1-2, 345-358.

[18] Royer, S., van Neste, A., Davidson, R., McIntyre, S., Kaliaguine, S., Methane oxidation over nanocrystalline $LaCo_{1-x}Fe_xO_3$: Resistance to SO_2 poisoning. *Ind. and Eng. Chem. Res.*, 2004, Volume 43, 5670-5680.

[19] Zhang, RD; Villanueva, A; Alamdari, H; Kaliaguine, S. Catalytic reduction of NO by propene over $LaCo_{1-x}Cu_xO_3$ perovskites synthesized by reactive grinding, *Appl. Catal., B: Environmental*, 2006, Volume 64, Issues 3-4, 220-233.

[20] Zhang, RD; Villanueva, A; Alamdari, H; Kaliaguine, S. SCR of NO by propene over nanoscale $LaMn_{1-x}Cu_xO_3$ perovskites, *Appl. Catal., A: General*, 2006, Volume 307, Issue 1, 85-97.

[21] Zhang, RD; Villanueva, A; Alamdari, H; Kaliaguine, S. Cu- and Pd-substituted nanoscale Fe-based perovskites for selective catalytic reduction of NO by propene. *J. Catal.*, 2006, Volume 237, Issue 2, 368-380.

[22] Seiyama, T., Yamazoe, N., Eguchi, K., Characterization and activity of some mixed metal oxide catalysts. *Ind. Eng. Chem., Product Research and Development*, 1985, Volume 24, 19-27.

[23] Lisi, L; Bagnasco, G; Ciambelli, P; Rossi, Sde; Porta, P; Russo, G; Turco, M. Perovskite-type oxides II. Redox properties of $LaMn_{1-x}Cu_xO_3$ and $LaCo_{1-x}Cu_xO_3$ and methane catalytic combustion. *J. Sol. State Chem.*, 1999, Volume 146, Issue 1, 176-183.

[24] Echchahed, B; Kaliaguine, S; Alamdari, H. Well dispersed Co^0 by reduction of $LaCoO_3$ perovskite, *International Journal of Chemical Reactor Engineering*, 2006, Volume 4, A 29.

[25] Fierro, J.L.G., Pena, M.A., Tejuca, L.G., An XPS and reduction study of $PrCoO_3$, *J. Mat. Sci.*, 1988, Volume 23, 1018-1023.
[26] Simonot, L., Garrin, F., Maire, G., A comparative study of $LaCoO_3$, Co_3O_4 and $LaCoO_3$-Co_3O_4 1. Preparation, characterization and catalytic properties for the oxidation of CO, *Appl. Catal., B: Environmental*, 1997, Volume 11, 167-179.
[27] Bedel, L. Roger, A.C., Estournes, C., Kiennemann, A., Co^0 from partial reduction of La(Co,Fe)O_3 perovskites for Fischer Tropsch synthesis, *Catal. Today*, 2003, Volume 85, 207-218.
[28] Royer, S., Alamdari, H., Duprez, D., Kaliaguine, S., Oxygen storage capacity of $La_{1-x}A_xBO_3$ perovskites (with A=Sr, Ce; B=Co, Mn) – Relation to the catalytic activity in CH_4 oxidation reaction, *Appl. Catal., B: Environmental*, 2005, Volume 58, 273-288.
[29] Baiker, A; Marti, PE; Keusch, P; Fritsch, E; Reller, A. Influence of the A-site cation in $ACoO_3$ (A = La, Pr, Nd, and Gd) perovskite-type oxides on catalytic activity for methane combustion. *J. Catal.,* 1994, Volume 146, Issue 1, 268-276.
[30] Wan, Y; Ma, JX; Wang, Z; Zhou, W; Kaliaguine, S. Selective catalytic reduction of NO over Cu-Al-MCM-41. *J. Catal.,* 2004, Volume 227, Issue 1, 242-252.
[31] Varma, S; Wani, BN; Gupta, NM. Redox behavior and catalytic activity of La-Fe-V-O mixed oxides. *Appl. Catal., A: General*, 2003, Volume 241, Issues 1-2, 341-348.
[32] Ciambelli, P; Cimino, S; Lisi, L; Faticanti, M; Minelli, G; Pettiti, I; Porta, P. La, Ca and Fe oxide perovskites: preparation, characterization and catalytic properties for methane combustion. *Appl. Catal., B: Environmental*, 2001, Volume 33, Issue 3, 193-203.
[33] Porta, P; Cimino, S; Rossi, Sde; Faticanti, M; Minelli, G; Pettiti, I. $AFeO_3$ (A=La, Nd, Sm) and $LaFe_{1-x}Mg_xO_3$ perovskites: structural and redox properties. *Mater. Chem. Phys.,* 2001, Volume 71, Issue 2, 165-173.
[34] Shen, WJ; Okumura, M; Matsumura, Y; Haruta, M. The influence of the support on the activity and selectivity of Pd in CO hydrogenation. *Appl. Catal., A: General*, 2001, Volume 213, Issue 2, 225-232.
[35] Tejuca, LG; Fierro, JLG; Tascón, JMD. Structure and reactivity of perovskite-type oxides. *Adv. Catal.*, 1989, Volume 36, 237-328.
[36] Fierro, J. L. G and Tejuca, L. G. Non-stoichiometric surface behaviour of $LaMO_3$ oxides as evidenced by XPS. *Appl. Surf. Sci.*, 1987, Volume 27, Issue 4, 453-457.
[37] Royer, S; Duprez, D; Kaliaguine, S. Role of bulk and grain boundary oxygen mobility in the catalytic oxidation activity of $LaCo_{1-x}Fe_xO_3$, *J. Catal.,* 2005, Volume 234, Issue 2, 364-375.
[38] Van Roosmalen JAM; Cordfunke, EHP; Helmholdt, RB; Zandbergen, HW. The defect chemistry of $LaMnO_{3\pm\delta}$: 2. Structural aspects of $LaMnO_{3+\delta}$. *J. Sol. State Chem.,* 1994, Volume 110, Issue 2, 100-105.
[39] Iwamoto, M; Yahiro, H; Tanda, K; Mizuno, N; Mine, Y; Kagawa, S. Removal of nitrogen monoxide through a novel catalytic process. 1. Decomposition on excessively copper-ion-exchanged ZSM-5 zeolites. *J., Phys. Chem.,* 1991, Volume 95, 3727-3730.
[40] Isupova, LA; Budneva, AA; Paukshtis, EA; Sadykov, VA. Nature of the perovskites surface centers as studied by the Infrared spectroscopy of adsorbed NO test molecule. *J. Mol. Catal., A: Chem.,* 2000, Volume 158, Issue 1, 275-280.
[41] Valyon, J. and Hall, WK. Studies of the surface species formed from nitric oxide on copper zeolites. *J. Phys. Chem.,* 1993, Volume 97, 1204-1212.

[42] Tabata, T; Ohtsuka, H; Kokitsu, M; Okada O; IR investigation of selective reduction of NO by ethene on Cu-ZSM-5, *Bull. Chem. Soc. Japan*, 1995, Volume 68, Issue 7, 1905-1914.

[43] Tanaka, T; Okuhara, T; Misono, M. Intermediacy of organic nitro and nitrite surface species in selective reduction of nitrogen monoxide by propene in the presence of excess oxygen over silica-supported platinum. *Appl. Catal., B: Environmental*, 1994, Volume 4, Issue 1, L1-L9.

[44] Burch, R. and Scire, S. Selective catalytic reduction of nitric oxide with ethane and methane on some metal exchanged ZSM-5 zeolites. *Appl. Catal., B: Environmental*, 1994, Volume 3, Issue 4, 295-318.

[45] Coq, B; Tachon, D; Figuéras, F; Mabilon, G; Prigent, M. Selective catalytic reduction of nitrogen monoxide by decane on copper-exchanged mordenites. *Appl. Catal., B: Environmental*, 1995, Volume 6, Issue 3, 271-289.

[46] Centi, G. and Perathoner, S. Nature of active species in copper-based catalysts and their chemistry of transformation of nitrogen oxides. *Appl. Catal., A: General*, 1995, Volume 132, Issue 2, 179-259.

[47] Otsuka, K; Takahashi, R; Yamanaka, I. Oxygenates from light alkanes catalyzed by NO_x in the gas phase. *J Catal.*, 1999, Volume 185, Issue 1, 182-191.

[48] Cowan, AD; Cant, NW; Haynes, BS; Nelson, PF. The catalytic chemistry of nitromethane over Co-ZSM-5 and other catalysts in connection with the methane-NO_x SCR reaction. *J. Catal.*, 1998, Volume 176, Issue 2, 329-343.

[49] Smits, R. H. H. and Iwasawa, Y. Reaction mechanisms for the reduction of nitric oxide by hydrocarbons on Cu-ZSM-5 and related catalysts. *Appl. Catal., B: Environmental*, 1995, Volume 6, Issue 3, L201-L207.

[50] Haneda, M; Kintaichi, Y; Inaba, M; Hamada, H. Infrared study of catalytic reduction of nitrogen monoxide by propene over Ag/TiO_2-ZrO_2, *Catal. Today*, 1998, Volume 42, 127-135.

[51] Yokoyama, C. and Misono, M. Catalytic reduction of nitrogen oxides by propene in the presence of oxygen over cerium ion-exchanged zeolites: II. Mechanistic study of roles of oxygen and doped metals. *J. Catal.*, 1994, Volume 150, 9-17.

[52] Haj, KO; Ziyade, S; Ziyad, M; Garin, F. DeNO$_x$ reaction studies: Reactivity of carbonyl or nitro-compounds compared to C_3H_6: influence of adsorbed species in N_2 and N_2O formation. *Appl. Catal., B: Environmental*, 2002, Volume 37, Issue 1, 49-62.

[53] Blower, C. J., Smith, T. D. The gas-phase decomposition of nitromethane over metal ion-exchanged sodium Y zeolite and sodium X zeolite. *Zeolites*, 1993, Volume 13, Issue 5, 394-398.

[54] Witzel, F; Sill, GA; Hall, WK. The selective reduction of NO and combustion of paraffins over MFI zeolites. *Stud. Surf. Sci. Catal.*, 1994, Volume 84, 1531-1536.

[55] Parravano, G. Ferroelectric transitions and heterogenous Catalysis, *J. Chem. Phys.*, 1952, Volume 20, 342-343.

[56] Libby, W. F. Promising catalyst for auto exhaust. *Science*, 1971, Volume 171, 499-500.

[57] Giannakas, AE; Ladavos, AK; Pomonis, PJ. DeNO$_x$ reaction studies: Reactivity of carbonyl or nitro-compounds compared to C_3H_6: influence of adsorbed species in N_2 and N_2O formation. *Appl. Catal., B: Environmental*, 2004, Volume 49, Issue 3, 147-158.

[58] Leontiou, AA; Ladavos, AK; Vaimakis, TC; Pomonis, PJ. Catalytic NO reduction with CO on $La_{1-x}Sr_x(Fe^{3+}/Fe^{4+})O_{3\pm\delta}$ perovskite-type mixed oxides (x = 0.00, 0.15, 0.30, 0.40, 0.60, 0.70, 0.80, and 0.90). *Appl. Catal., A: General*, 2003, Volume 241, Issues 1-2, 133-141.

[59] Belessi, VC; Costa, CN; Bakas, TV; Anastasiadou, T; Pomonis, PJ; Efstathiou, AM. Catalytic behavior of La-Sr-Ce-Fe-O mixed oxidic/perovskitic systems for the NO+CO and $NO+CH_4+O_2$ (lean-NO_x) reactions. *Catal. Today*, 2000, Volume 59, Issues 3-4, 347-363.

[60] Peter, SD; Garbowski, E; Perrichon, V; Primet, M. NO reduction by CO over aluminate-supported perovskites. *Catal. Lett.*, 2000, Volume 70, 27-33.

[61] Zhang, RD; Villanueva, A; Alamdari, H; Kaliaguine, S. Reduction of NO by CO over nanoscale $LaCo_{1-x}Cu_xO_3$ perovskites, *J. Mol. Catal., A: Chemical*, 2006, Volume 258, 22-34.

[62] Zhang, RD; Alamdari, H; Kaliaguine, S. Fe-based perovskites substituted by copper and palladium for No + CO reaction, *J. Catal.*, 2006, Volume 242, 241-253.

[63] Shin, S; Arakawa, H; Hatakeyama, Y; Ogawa, K; Shimomura, K. Absorption of NO in the lattice of an oxygen-deficient perovskite $SrFeO_{3-x}$ and the infrared spectroscopic study of the system NO-$SrFeO_{3-x}$. *Mater. Res. Bull.*, 1979, Volume 14, Issue 5, 633-639.

[64] Matyshak, V. A. and Krylov, O. V. In situ IR spectroscopy of intermediates in heterogeneous oxidative catalysis. *Catal. Today*, 1995, Volume 25, Issue 1, 1-87.

[65] Tascón, JMD; Tejuca, LG; Rochester, CH. Surface interactions of NO and CO with $LaMO_3$ oxides. *J. Catal.*, 1985, Volume 95, Issue 2, 558-566.

[66] Viswanathan, B. CO oxidation and NO reduction on perovskite oxides. *Catal. Rev.*, 1992, Volume 34, 337-354.

[67] Chien, MW; Pearson, IM; Nobe, K. Reduction and absorption kinetics of nitric oxide on colbalt perovskite catalysts. *Ind. Eng. Chem., Prod. Res. Dev.*, 1975, Volume 14, 131-134.

[68] Forni, L; Oliva, C; Barzetti, T; Selli, E; Ezerets, AM; Vishniakov, AV. FT-IR and EPR spectroscopic analysis of $La_{1-x}Ce_xCoO_3$ perovskite-like catalysts for NO reduction by CO. *Appl. Catal., B: Environmental*, 1997, Volume 13, Issue 1, 35-43.

[69] Cooper, W. J. and Hollway, L. A. Recipes for cleaner air, *Nature*, 1996, Volume 384, Issue 6607, 313-315.

[70] Taylor, H. and O'Leary, R. A study of uranium oxide based catalysts for the oxidative destruction of short chain alkanes, *Appl. Catal., B: Environmental*, 2000, Volume 25, Issues 2-3, 137-149.

[71] Heck, R. M. and Farrauto, R. J. The automobile catalyst Its present and future family tree, *CATTECH*, 1997, Volume 1, Issue 2, 117-124.

[72] Hodjati, S; Vaezzadeh, K; Petit, C; Pitchon, V; Kiennemann, A; Absorption/desorption of $NO_{(x)}$ process on perovskites: Performances to remove $NO_{(x)}$ from a lean exhaust gas, *Appl. Catal., B: Environmental*, 2000, Volume 26, Issue 1, 5-16.

[73] Paulis, M; Gandía, LM; Gil, A; Sambeth, J; Odriozola, JA; Montes, M. Influence of the surface adsorption–desorption processes on the ignition curves of volatile organic compounds (VOCs) complete oxidation over supported catalysts, *Appl. Catal., B: Environmental*, 2000, Volume 26, Issue 1, 37-46.

[74] Voorhoeve, RJH; Remeika, JP; Freeland, PE; Mathias, BT. Rare-earth oxides of manganese and cobalt rival platinum for the treatment of carbon monoxide in auto exhaust. *Science*, 1972, Volume 177, 353-354.

[75] Szabo, V; Bassir, M; Gallot, JE; van Neste, A; Kaliaguine, S. Perovskite-type oxides synthesized by reactive grinding Part III. Kinetics of n-hexane oxidation over $LaCo_{1-x}Fe_xO_3$. *Appl. catal., B: Environmental*, 2003, Volume 42, Issue 3, 265-277.

[76] Szabo, V; Bassir, van Neste, A; Kaliaguine, S. Perovskite-type oxides synthesized by reactive grinding Part II. Catalytic properties of $LaCo_{1-x}Fe_xO_3$ in VOC oxidation. *Appl. Catal. B: Environmental*, 2002, Volume 37, 175-180.

[77] Royer, S; Berube, F; Kaliaguine, S. Effect of the synthesis conditions on the redox and catalytic properties in oxidation reactions of $LaCo_{1-x}Fe_xO_3$, *Appl. Catal. A: General*, 2005, Volume 282, 273-284.

[78] Iwamoto, M. Heterogeneous catalysis for removal of NO in excess oxygen. Progress in 1994. *Catal. Today,* 1996, Volume 29, Isuues 1-4, Second Japan-EC joint workshop on the frontiers of catalytic science and technology for energy, environment and risk prevention, pages 29-35.

[79] Grinsted, RA; Jen, HW; Montreuil, CN; Rokosz, MJ; Shelef, M. The relation between deactivation of Cu-ZSM-5 in the selective reduction of NO and dealumination of the zeolite. *Zeolites*, 1993, Volume 13, Issue 8, 602-606.

[80] Forni, L. and Rossetti, I. Catalytic combustion of hydrocarbons over perovskites. *Appl. Catal., B: Environmental*, 2002, Volume 38, Issue 1, 29-37.

[81] Zhang, RD; Alamdari, H; Kaliaguine, S. Water vapor sensitivity of nanosized La(Co, Mn, $Fe)_{1-x}(Cu, Pd)_xO_3$ perovskites during NO reduction by C_3H_6 in the presence of oxygen. *Appl. Catal. B: Environmental*, 2007, Volume 72, 331-341.

[82] Crespin, M. and Hall, W. K. The surface chemistry of some perovskite oxides. *J. Catal.,* 1981, Volume 69, Issue 2, 359-370.

[83] Takita, Y; Tashiro, T; Saito, Y; Hori, F. The effects of water coadsorption on the adsorption of oxygen over metal oxides: I. Temperature-programmed desorption study of Co_3O_4. *J. Catal.,* 1986, Volume 97, Issue 1, 25-35.

[84] Peri, J. B. and Hannan, R. B. Surface hydroxyl groups on γ-alumina. *J. Phys. Chem.*, 1960, Volume 64, 1526-1530.

[85] Guilhaume, N; Peter, SD; Primet, M. Palladium-substituted lanthanum cuprates: application to automotive exhaust purification. *Appl. Catal., B: Environmental*, 1996, Volume 10, Issue 4, 325-344.

[86] Zhang, RD; Alamdari, H; Kaliaguine, S. SO_2 poisoning of $LaFe_{0.8}Cu_{0.2}O_3$ perovskite prepared by reactive grinding during NO reduction by C_3H_6, *Appl. Catal. A: General*, 2008, Volume 340, 140-151.

[87] Alifanti, M; Auer, R; Kirchnerova, J; Thyrion, F; Grange, P; Delmon, B. Activity in methane combustion and sensitivity to sulfur poisoning of $La_{1-x}Ce_xMn_{1-y}Co_yO_3$ perovskite oxides. *Appl. Catal., B: Environmental*, 2003, Volume 41, Issues 1-2, 71-81.

[88] Waqif, M; Bazin, P; Saur, O; Lavalley, JC; Blanchard, G; Touret, O. Study of ceria sulfation. *Appl. Catal., B: Environmental*, 1997, Volume 11, Issue 2, 193-205.

[89] Hodjati, S; Petit, C; Pitchon, V; Kiennemann, A. Absorption/desorption of NO_x process on perovskites: impact of SO_2 on the storage capacity of $BaSnO_3$ and strategy to develop thioresistance. *Appl. Catal., B: Environmental*, 2001, Volume 30, Issues 3-4, 247-257.

[90] Choisnet, J; Abadzhieva, N; Stefanov, P; Klissurski, D; Bassat, JM; Rives, V; Minchev, L. X-ray photoelectron spectroscopy, temperature-programmed desorption and temperature-programmed reduction study of $LaNiO_3$ and $La_2NiO_{4+\delta}$ catalysts for methanol oxidation. *J. Chem. Soc., Faraday Transactions*, 1994, Volume 90, 1987-1991.

[91] Brion, D. Photoelectron spectroscopic study of the surface degradation of pyrite (FeS_2), chalcopyrite ($CuFeS_2$), sphalerite (ZnS), and galena (PbS) in air and water, *Appl. Surf. Sci.,* 1980, Volume 5, 133-52.

[92] Briggs, D.; Seah, M. P. Practical Surface Analysis; Wiley: New York, 1993.

[93] Pischinger, S. The future of vehicle propulsion-combustion engines and alternatives. *Top. Catal.* 2004, Volume 30/31, 5-16.

[94] Chin, YH; Alvarez, WE; Resasco, DE. Comparison between methane and propylene as reducing agents in the SCR of NO over Pd supported on tungstated zirconia. *Catal. Today*, 2000, Volume 62, Issue 4, 291-302.

[95] Miyadera, T. and Yoshida, K. Alumina-supported catalysts for the selective reduction of nitric oxide by propene. *Chem. Lett.,* 1993, Volume 22, 1483-1486.

[96] Meunier, FC; Zuzaniuk, V; Breen, JP; Olsson, M; Ross, JRH. Mechanistic differences in the selective reduction of NO by propene over cobalt- and silver-promoted alumina catalysts: kinetic and in situ DRIFTS study. *Catal. Today* 2000, Volume 59, Issues 3-4, 287-304.

[97] Bethke, K. A. and Kung, H. H. Supported Ag catalysts for the lean reduction of NO with C_3H_6. *J. Catal.,* 1997, Volume 172, Issue 1, 93-102.

[98] Wichterlova, B. Structural analysis of potential active sites in metallo-zeolites for selective catalytic reduction of NO_x. An attempt for the structure versus activity relationship. *Top. Catal.*, 2004, Volume 28, 131-140.

[99] Seiyama, T. Total oxidation of hydrocarbons on perovskite oxides. *Catal. Rev.-Sci. and Eng.*, 1992, Volume 34, 281-300.

[100] Meunier, FC; Ukropec, R; Stapleton, C; Ross, JRH. Effect of the silver loading and some other experimental parameters on the selective reduction of NO with C_3H_6 over Al_2O_3 and ZrO_2-based catalysts. *Appl. Catal., B: Environmental*, 2001, Volume 30, Issues 1-2, 163-172.

[101] Zhang, RD; Alamdari, H; Bassir, M; Kaliaguine, S. Optimization of mixed Ag catalysts for catalytic conversions of NO and C_3H_6, *International Journal of Chemical Reactor Engineering*, 2007, Volume 5, A 70.

[102] Armor, J. N. NO_χ/hydrocarbon reactions over gallium loaded zeolites: A review, *Catal. Today*, 1996, Volume 31, Issues 3-4, 191-198.

[103] Shimizu, K; Satsuma, A; Hattori, T. Selective catalytic reduction of NO by hydrocarbons on Ga_2O_3/Al_2O_3 catalysts, *Appl. Catal., B: Environmental*, 1998, Volume 16, Issue 4, 319-326.

[104] Shimizu, K; Takamatsu, M; Nishi, K; Yoshida, H; Satsuma, A; Tanaka, T; Yoshida, S; Hattori, T. Alumina-supported gallium oxide catalysts for NO selective reduction: Influence of the local structure of surface gallium oxide species on the catalytic activity, *J. Phys. Chem., B*, 1999, Volume 103, Issue 9, 1542-1549.

[105] Haneda, M; Kintaichi, Y; Hamada, H. Enhanced activity of metal oxide-doped Ga_2O_3-Al_2O_3 for NO reduction by propene, *Catal. Today*, 1999, Volume 54, Issue 4, 391-400.

[106] Takahashi, M; Nakatani, T; Iwamoto, S; Watanabe, T; Inoue, M. Effect of the composition of spinel-type Ga_2O_3-Al_2O_3-ZnO catalysts on activity for the CH_4-SCR of NO and optimization of catalyst composition, *Ind. Eng. Chem. Res.*, 2006, Volume 45, Issue 10, 3678-3683.

[107] Zahir, MH; Katayama, S; Awano, M. Synthesis and de-NO_x properties of ZnO-Ga_2O_3-Al_2O_3 spinel, *Mater. Chem. Phys.*, 2004, Volume 86, Issue 1, 99-104.

[108] Zhang, RD; Alamdari, H; Kaliaguine, S. Crystal structure, redox properties and catalytic performance of Ga-based mixed oxides for NO reduction by C_3H_6, *Catalysis Communications*, 2008, Volume 9, 111-116.

[109] Patcas, F; Buciuman, FC; Zsako, J. Oxygen non-stoichiometry and reducibility of B-site substituted lanthanum manganites, *Thermochimica Acta*, 2000, Volume 360, Issue 1, 71-76.

[110] Zhang, RD and Kaliaguine, S. Lean reduction of NO by C_3H_6 over Ag/alumina derived from Al_2O_3, AlOOH and $Al(OH)_3$, Appl. Catal. B: Environ. 2008, Volume 78, 275-287.

[111] Ren, LP; Dai, WL; Yang, XL; Gao, Y; Xie, ZK; Fan, KN. Transformation of various oxygen species on the surface of electrolytic silver characterized by in situ raman spectroscopy, *Chinese J. Catal.*, 2006, Volume 27, Issue 2, 115-118.

[112] She, X. and Flytzani-Stephanopoulos, M. Mechanistic differences in the selective reduction of NO by propene over cobalt- and silver-promoted alumina catalysts: kinetic and in situ DRIFTS study, *Catal. Today*, 2000,Volume 59, Issues 3-4, 287-304.

[113] Jen, H.-W. Study of nitric oxide reduction over silver/alumina catalysts under lean conditions: Effects of reaction conditions and support, *Catal. Today*, 1998, Volume 42, Issues 1-2, 37-44.

[114] Iglesias-Juez, A; Hungría, AB; Martínez-Arias, A; Fuerte, A; Fernández-García, M; Anderson, JA; Conesa, JC; Soria, J. Nature and catalytic role of active silver species in the lean NO_x reduction with C_3H_6 in the presence of water, *J. Catal.*, 2003, Volume 217, Issue 2, 310-323.

[115] Furusawa, T; Seshan, K; Lercher, JA; Lefferts, L; Aika, K. Selective reduction of NO to N_2 in the presence of oxygen over supported silver catalysts, *Appl. Catal., B: Environmental*, 2002, Volume 37, Issue 3, 205-216.

[116] Iwamoto, M. Proceedings of the Meeting on Catalytic Technology for Removal of Nitrogen Monoxides, Tokyo. 1990, p. 17.

[117] Iwamoto, M. and Mizuno, N. NO_x emission control in oxygen-rich exhaust through selective catalytic reduction by hydrocarbon, Proceedings - IMechE: Part D, *J. Automobile Eng.*, 1993, Volume 207, Issue D1, 23-33.

[118] Long, R. Q. and Yang, R. T. Selective Catalytic reduction of nitric oxide with ethylene on copper ion-exchanged Al-MCM-41 catalyst, *Ind. Eng. Chem. Res.*, 1999, Volume 38, Issue 3, 873-878.

[119] Trong, OD; Desplantier-Giscard, D; Danumah, C; Kaliaguine, S. Perspectives in catalytic applications of mesostructured materials, *Appl. Catal., A: General*, 2003, Volume 253, Issue 2, 545-602.

[120] Kharas, K. C. C. Performance, selectivity, and mechanism in Cu-ZSM-5 lean-burn catalysts, *Appl. Catal., B: Environmental*, Volume 2, Issues 2-3, 1993, 207-224.

[121] Ciambelli, P; Corbo, P; Migliardini, F. Potentialities and limitations of lean de-NO_x catalysts in reducing automotive exhaust emissions, *Catal. Today*, 2000, Volume 59, Issues 3-4, 279-286

[122] Wan, Y; Wang, Z; Ma, JX; Zhou, W; Zhang YQ; Wu MZ, Effect of water vapor on selective catalytic reduction of NO by propylene over Cu-Al-MCM-41, *Chinese J. Catal.*, 2002, Volume 23, Issue 3, 257-261.

[123] Wan, Y; Wang, Z; Jin, XM; Ma, JX; Zhou, W. Synthesis of MCM-41 with high content of framework aluminum using mixed templates, *Micropor. Mesopor. Mater.*, 2004, Volume 76, Issues 1-3, 35-40.

[124] Wan, Y; Ma, JX; Wang, Z; Zhou, W; Kaliaguine, S. On the mechanism of selective catalytic reduction of NO by propylene over Cu-Al-MCM-41, *Appl. Catal. B: Environmental*, 2005, Volume 59, Issues 3-4, 235-242.

[125] Bulanek, R; Wichterlova, B; Sobalik, Z; Tichy, J. Reducibility and oxidation activity of Cu ions in zeolites Effect of Cu ion coordination and zeolite framework composition, *Appl. Catal., B: Environmental*, 2001, Volume 31, Issue 1, 13-25.

[126] Delahay, G; Coq, B; Broussous, L. Selective catalytic reduction of nitrogen monoxide by decane on copper-exchanged beta zeolites, *Appl. Catal., B: Environmental*, 1997, Volume 12, Issue 1, 49-59.

[127] Zhang, B; Yamaguchi, T; Kawakami, H; Suzuki, T. Selective reduction of nitric oxide over platinum catalysts in the presence of sulfur dioxide and excess oxygen, *Appl. Catal., B: Environmental*, 1992, Volume 1, Issue 3, L15-L20.

[128] Wichterlova, B; Dedecek, J; Sobalik, Z. Cu coordination in high silica zeolites. Effect of the framework Al local siting, *Stud. Surf. Sci. Catal.*, 1995, Volume 94, 641-648, Catalysis by Microporous Materials, Proceedings of ZEOCAT "95.

[129] Wichterlova, B; Sobalik, Z; Vondrova, A. Differences in the structure of copper active sites for decomposition and selective reduction of nitric oxide with hydrocarbons and ammonia, Catal. Today, 1996, Volume 29, Issues 1-4, 149-153, Second Japan-EC Joint Workshop on the Frontiers of Catalytic Science and Technology for Energy, Environment and Risk Prevention

[130] Inui, T; Zeolites as the key matrix for superior deNO$_x$ catalysts, Studies in Surface Science and Catalysis, 1995, Volume 97, Zeolites: A refined tool for designing catalytic sites, Proceedings of the International Zeolite Symposium, Pages 277-285.

[131] Dedecek, J; Sobalik, Z; Tvaruzkova, Z; Kaucky, D; Wichterlova, B. *J. Phys. Chem.*, 1994, Volume 99, 1632.

[132] Torre-Abreu, C; Ribeiro, MF; Henriques, C; Delahay, G. Characterisation of CuMFI catalysts by temperature programmed desorption of NO and temperature programmed reduction. Effect of the zeolite Si/Al ratio and copper loading, *Appl. Catal., B: Environmental*, 1997, Volume 12, Issues 2-3, 249-262.

[133] Raj, A; Le, THN; Kaliaguine, S; Auroux, A. Involvement of nitrate species in the SCR of NO by NH_3 at ambient conditions over TS-1 catalysts, *Appl. Catal., B: Environmental*, 1998, Volume 15, Issues 3-4, 259-267.

[134] Petunchi, JO; Gustave, S; Hall, WK. Studies of the selective reduction of nitric oxide by hydrocarbons, *Appl. Catal., B: Environmental*, 1993, Volume 2, Issue 4, 303-321.

[135] Hayes, NW; Joyner RW; Shpiro, ES. Infrared spectroscopy studies of the mechanism of the selective reduction of NO$_x$ over Cu-ZSM-5 catalysts, *Appl. Catal., B: Environmental*, 1996, Volume 8, Issue 3, 343-363.

[136] Li, C; Domen, K; Maruya K; Onishi, T. Spectroscopic identification of adsorbed species derived from adsorption and decomposition of formic acid, methanol, and formaldehyde on cerium oxide, *J. Catal.*, 1990, Volume 125, Issue 2, 445-455.

[137] Li, Y. and Armor, J. N. Temperature-programmed desorption of nitric oxide over Cu-ZSM-5, *Appl. Catal.*, 1991, Volume 76, Issue 2, L1-L8.

In: Heterogeneous Catalysis Research Progress
Editor: Mathias B. Gunther, pp. 81-110
ISBN 978-1-60456-979-7

Chapter 2

NEW SYNTHETIC METHODS TO EFFICIENT NANOPOROUS AND NANOSTRUCTURED CATALYSTS: SELF-ASSEMBLY AND CO-ASSEMBLY OF MULTIPLE SITE-ISOLATED CATALYTIC SITES ON MESOPOROUS MATERIALS

Tewodros Asefa*, Abhishek Anan, Cole Duncan and Youwei Xie
Department of Chemistry, 111 College Place, Department of Chemistry, Syracuse University, Syracuse, NY 13244, USA

ABSTRACT

The introduction of organic groups into nanoporous materials results in functionalized nanostructured heterogeneous catalysts. By various self-assembly routes including a new solvent-assisted grafting and co-assembly of multiple types of organic groups, we and others have recently demonstrated the synthesis of a broad range of nanoporous catalysts containing high surface areas, site-isolated catalytic sites, nanometer pores and efficient and selective catalytic properties for various types of reactions. The grafting of site-isolated organic groups with a polar solvent such as ethanol or isopropanol is among the newest synthetic strategies that allow the synthesis of novel heterogeneous catalysts with enhanced catalytic efficiency. While the distributions and site-isolation of the catalytic sites in these materials can be controlled simply by changing the solvents used for grafting the catalytic functional groups, the degree of site-isolation of the catalytic sites can be probed by a simple new colorimetric method involving measuring of the d-d electronic spectra of metal complexes with the catalytic sites. By introducing appropriate secondary functional groups into the nanoporous catalyst structure, selective catalysts for specific reactants can also be produced. Furthermore, by a judicious choice of the active catalytic sites, selectivity for a specific product by these catalysts has been demonstrated recently.

* Rm 1-014, CST Bldg., Department of Chemistry, 111 College Place, Department of Chemistry, Syracuse University, Syracuse, NY 13244, USA

1. Introduction

Catalysis in biological systems by various enzymes, perhaps the first catalytic processes ever, before any knowledge whatsoever of catalytic process was available to mankind has been playing a crucial role in evolution of various organisms. The indispensable role of catalysts in various biological and chemical processes has been reiterated and emphasized time and again. Numerous catalytic processes simultaneously taking place in various life forms are the basis for their sustained development. For example, without carbonic anhydrase the conversion of carbon dioxide to bicarbonate anion in our body would have been impossible. Nucleoside monophosphate (NMP) kinase catalyzed dephosphorylation of adenosine triphosphate (ATP) in animals is absolutely critical to meet energy requirements.

While Berzelius coined the term catalysis in 1836, first well-documented recognition accorded to the importance of catalysts in chemical systems is very recent and attributed to the pioneering work of Wilhelm Ostwald for which he was awarded the Nobel Prize in chemistry for the year 1909. Enormous literature has since accumulated on the role of catalysts in biological and chemical systems. While on one hand, thousands of enzymes present in various life forms have been characterized and their mechanism of action were reported, on the other hand, a staggering number of chemicals are used presently in the catalytic production of many fine chemicals in industries. Both homogeneous as well heterogeneous catalysts are used extensively in many of these various industrial catalytic processes. Interest in heterogeneous catalytic systems is primarily due to the advantage of recovery and recyclability offered by heterogeneous catalysts. Hence a lion's share of the current research in catalysis is taken up in development of ever more efficient and economically profitable heterogeneous catalysts. Conventional homogeneous catalytic systems, although efficient, offer little or no freedom with variations in catalytic sites and thus have a limited scope of application. Further, their tedious and extravagant recovery processes increase the process cost. Recent advances in hybrid heterogeneous catalysts allow modification of a single solid support with multiple functional groups capable of acting as catalytic sites. Active involvement of the solid support itself in some cases accelerates or decelerates the catalytic process as well as affects the outcome (reaction products) of reactions. Thus the chemical nature of solid support provides another tool to judiciously tailor the catalytic processes. Easy recovery often with little or no loss in activity over multiple catalytic cycles makes them desirable for catalysis applications both in terms of efficiency and processing cost.

Hybrid heterogeneous catalysts based on porous supports such as aerogels, zeolites, metal-organic framework (MOF) materials, silica and metal-oxide based mesoporous materials and polymers have attracted enormous attention owing to the ease of modification of solid support with suitable organic functionality and other desirable properties attributed to the high surface area and tunable porosity of these materials. Mesoporous silica materials by virtue of easily modifiable extremely high surface area and tunable pores are among the most studied porous supports for heterogeneous catalytic systems in recent years. The present chapter focuses primarily on recent developments in synthesis, organic modification with catalytic groups, characterization of heterogeneous mesoporous catalysts and their application to catalysis of different organic reaction.

1.1. Mesoporous Silica

As defined by the IUPAC Compendium of Technology for mesopore in catalysis, mesopores are pores of intermediate size, i.e. those with widths in the range of 2 nm and 50 nm diameter. Thus inorganic silica framework materials reported by Kresge et al, [1] for the first time classify as mesoporous silica, MCM-41. Other groups have also reported synthesis of variations of mesoporous silica materials such as MCM-41 [2], MCM-48 [3], SBA-15 [4], FSM-16 [5, 6], FDU [7], etc. principally based on differences in the template and synthetic conditions used for the synthesis of mesophase. For example, while MCM-41 and SBA-15 both use tetraethoxysilane as the silicate source, cetyltrimethylammonium bromide is used as the structure directing reagent in the former and poly(ethyleneoxide)-block-(polypropyleneoxide)-block-poly(ethyleneoxide), a triblock copolymer is used in later. Also FSM-16 uses polysilicate Kanemite as the silicate source compared to tetralkoxysilane in MCM-41 and SBA-15. Differences in synthetic strategies apart, these materials are characterized by very large surface area (typically > 500 – 1,000 m^2/g), relatively narrow distribution of pore widths, high pore volumes, and high thermal stability. These properties make them promising candidates for applications in catalysis or as catalytic supports [8]. Apart from high catalytic efficiency, the high surface area and large pore volume of mesopores allows for functionalization with multiple functional groups without severely restricting the diffusion of the reactants into the pores as in zeolites. Furthermore, relatively narrow distribution of pore widths makes them ideal candidates to achieve selective access of reactants to catalytic sites.

1.1.1. Synthesis of Mesoporous Silica

All major synthetic routes to mesoporous silica involve use of a template to achieve mesoporosity. The templates that are most commonly used are surfactants or block co-polymers. Below we describe the synthetic strategies for mesoporous silica MCM-41[1] and SBA-15[4], which are representative of this class of materials.

MCM-41

As originally described by Kresge et al.[1,2] the synthesis of MCM-41 involves the formation of a liquid crystalline phase with a cationic surfactant hexadecyltrimethylammonium bromide (Figure 1). The hydrolysis and condensation of tetraethylorthosilicate (or a sol-gel process) on these liquid crystalline templates leads to a hexagonal framework structure of networked silicates with surface area between 500 to 1000 m^2/g and pore widths ranging from 2 to 4 nm depending on the synthetic conditions. MCM-41 belongs to the *p6mm* space group with d-spacings ranging between 4 to 5 nm.

The mechanism for the formation of hexagonal array of cylindrical micelles and condensation of tetraethylorthosilicate with regards to the order in which they take place has been a matter of extensive research [9, 10]. While liquid crystal template mechanism proposes the formation of silicate anion mediated liquid crystal phase followed by condensation and growth of the silicate framework [2], co-operative templating mechanism precludes the formation of a pre-organized array of surfactants, followed by nucleation and growth of the silicate framework as the possible mechanism. Instead the latter proposes the formation of highly ordered mesophase [9], a result of cooperative organization of inorganic precursor and surfactants based on their charge, coordination and steric requirements. The nucleation, growth and transition between different phases are attributed to interaction of surfactant and inorganic precursor at the interface. The model successfully explains transition of mesophases between hexagonal, lamellar and cubic with subtle changes in reaction parameters such as inorganic-surfactant ratio, temperature, pH and ionic strengths. Significant deviations from these mechanisms can be expected based on changes in the synthetic conditions and precursors.

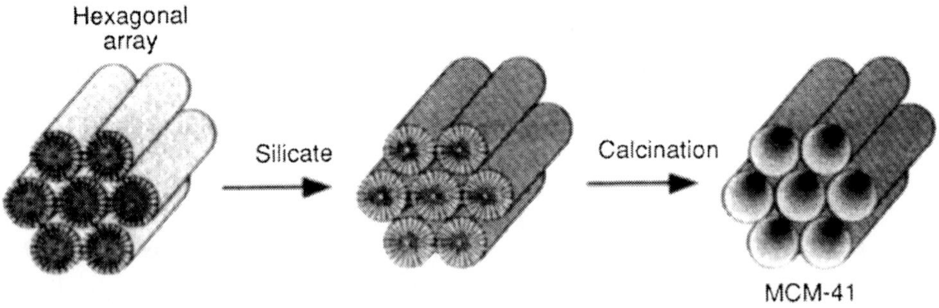

Figure 1. Scheme for the liquid crystalline templating mechanism proposed by Kresge et al1 for synthesis of mesoporous silica MCM-41. Formation of a hexagonal array of cylindrical micelles possibly mediated by silicate anions followed by condensation of the silicate anions from the silicate source (tetraethylorthosilicate) leads to templated framework structure. Calcination or extraction of the template produces hexagonally ordered mesoporous silica.

1.2. Mesoporous SBA-15

While mesoporous silica MCM-41 has pores in the range of 2 to 4 nm, it can be extended up to 10 nm using swelling agents such as 1,3,5-trimethylbenzene [11]. However, such swelling usually can also lead to less ordered materials. An interesting report by Zhao et al[4] described the synthesis of highly ordered mesoporous silica SBA-15 with pores in the range of 5 to 30 nm and wall thickness of 3.1 to 6.4 nm, templated by triblock copolymer polyethylene glycol. SBA-15 belongs to the space group p6mm with unusually high inter-lattice d-spacing approaching 10 nm. In a typical synthesis, an aqueous solution of triblock copolymer poly(ethylene oxide)-*block*-poly(propylene oxide)-*block*-poly(ethylene oxide) (PEO-PPO-PEO) is mixed with inorganic silicate precursor, teraalkoxyorthosilicate, under acidic conditions to form SBA-15 mesostructured silica. Pore size and pore wall thickness is tuned by varying the solution temperature between 35 and 80 °C. By virtue of its thick wall thickness, SBA-15 material has robust structure and high hydrothermal stability. The extraction of the template is done by a variety of methods, solvent wash and calcination being the most widely practiced. While extraction with mixture of diethyl ether, acetone and ethanol removes most of the co-polymer, calcination are more effective at removing the templates albeit with loss of much of the surface silanols crucial for post-modification of the mesopores. Microwave digestion technique recently described by Tian B.Z. et al. [12] also lead to removal of most of the template without significant loss in surface silanols.

2. ORGANIC FUNCTIONALIZED MESOPOROUS MATERIALS AND SITE-ISOLATION OF FUNCTIONAL GROUPS

The functionalization of mesoporous silica materials with organic groups is an important avenue to introduce functional groups especially catalytic sites for catalysis. Co-condensation, post-modification and assembly of various organosilanes are the most commonly used methods to incorporate organic functionality in the inorganic silicate mesoporous materials. Briefly, co-condensation involves the addition of organosilane precursors with the desired organic functionality into the micellar solution of surfactant (and tetralkoxysilane). The as-synthesized material thus obtained is extracted to yield organic functionalized mesoporous silica [13, 14]. Post-modification often referred to as grafting involves stirring the organic precursor with previously synthesized mesoporous silica in an appropriate solvent. The proposed mechanism involves the nucleophilic attack of surface silanols or silanolate on the silicon center of the alkoxysilane in a bimolecular fashion to anchor organic functional groups on the mesoporous silica surface [15]. The newest method involves the self-assembly of bridging organosilanes or silosesquioxanes with various types of surfactants and it produce organic functionalized mesoporous materials called periodic mesoporous organosilica (PMO) material that contain bridge-bonded organic groups on the walls of mesoporous materials [16, 17, 18].

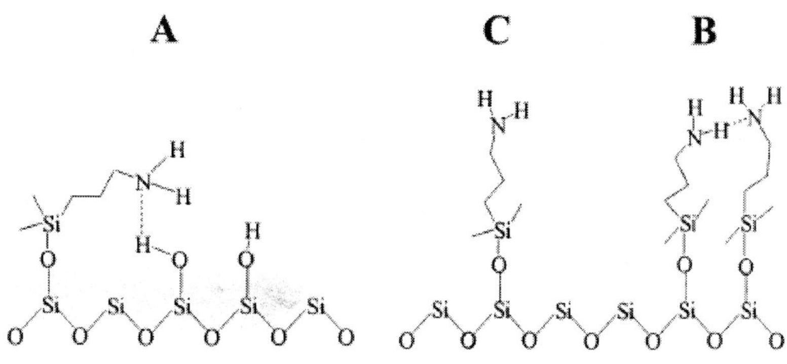

Figure 2. Scheme illustrating (A) amine-silanol and (B) amine-amine interactions and (C) single site or site-isolated 3-aminopropyl groups grafted onto mesoporous silica [20].

The various methods used to incorporate organic functional groups into mesoporous silica differ in terms of the loading of functional groups, their spatial distribution within the mesopores and the interaction of the organic functional groups with the silica framework. It is counter-intuitive that minimization of interaction of the organic functional groups among themselves and with the surface silanols would increase the potency of the organic functionality towards catalysis. While enormous research has been directed towards development of new mesoporous materials with a broad spectrum of organic functional groups, relatively little effort has been put into the understanding of interactions among the different components of the inorganic-organic hybrid mesoporous silica. However, research undertaken by Katz A. [19]'Jones C., [20] Walcarius A. [21] and more recently our group [22] has provided new methodologies to obtain organically functionalized mesoporous silica with "site-isolated" functional groups. By strictest definition, site-isolated functional groups in mesoporous silica would mean functional groups placed a distance apart so as to minimize interaction of the organic functional groups among themselves [20] (Figure 2). However, the tendency of the acidic surface silanols to act as general acid catalysts [23], which may or may not assist the catalytic process also requires synthetic methodologies aimed at achieving site-isolated functional groups with tunable interaction with surface silanols (Figure 2).

Site isolation of catalytic groups has been obtained by a variety of techniques including a molecular imprinting approach, by using bulky spacer groups, and more recently, by solvent assisted site-isolated grafting (SASIG) [24]. Although these three methods achieve site-isolation of functional groups, they differ significantly in terms of loading of the functional groups obtained. Most of these studies were conducted on obtaining site-isolation of 3-aminopropyl groups supposedly due to the fact that primary amine functionalized mesoporous materials are highly efficient heterogeneous catalysts in a variety of carbon-carbon bond forming reactions such as nitroaldol condensation [25], knoevenagel condensation [26], aldol condensation [27], and Michael addition [28]. Each of the aforementioned techniques to obtain site-isolation of aminopropyl groups on silica based heterogeneous catalysts is discussed below in brief with particular emphasis on methodology developed for obtaining single site 3-aminopropyl functional groups on silica support and experimental probes for site-isolation.

2.1. Molecular Imprinting

Molecular imprinting is a powerful technique to control the spatial orientation and relative distribution of functional groups in solid-state materials [29]. A pre-designed template with desired functional groups and monomer molecules with propensity to cross-link are polymerized to form an amorphous networked matrix. Removal of the template leads to a cavity with spatially controlled functional groups covalently bound to the matrix. Since the cavity thus generated is complementary to the template or imprint used, it is possible to exercise selectivity in terms of access to the cavity or the active site. Katz, A. and Davis, M. E [29].reported molecular imprinting approach to amorphous silica with one to three spatially controlled 3-aminpropyl groups (Figure 3). An imprint with one to three 3-aminopropyltriethoxysilane groups covalently attached to a benzene ring through carbamate linkage was hydrolyzed with tetraethoxysilane under acidic conditions to form amorphous silica gel. Removal of the template resulted in heterogeneous catalysts with amine functional groups covalently bound to the silica support. While amine-amine interaction was reduced by isolation of functional groups through the use of imprint, amine-silanol interaction was diminished by capping of surface silanols with 1,1,3,3-hexamethyldisilazane (HMDS). The imprinted amorphous silica was shown to be an efficient and selective base catalyst for the Knoevenagel condensation of malononitrile and isopthalaldehyde (74 turnovers per amine sites per hour). Spatial distribution of amines within the imprint sites was probed with two different experiments. In the first experiment, the imprinted silica was reacted with azelaoyl chloride to form amide bonds. While imprinted silica containing one amine per imprint formed single amide bonds with azelaoyl chloride, those with two amines per imprint formed two amide bonds as demonstrated by FT-IR spectroscopy [29]. Formation of single or double

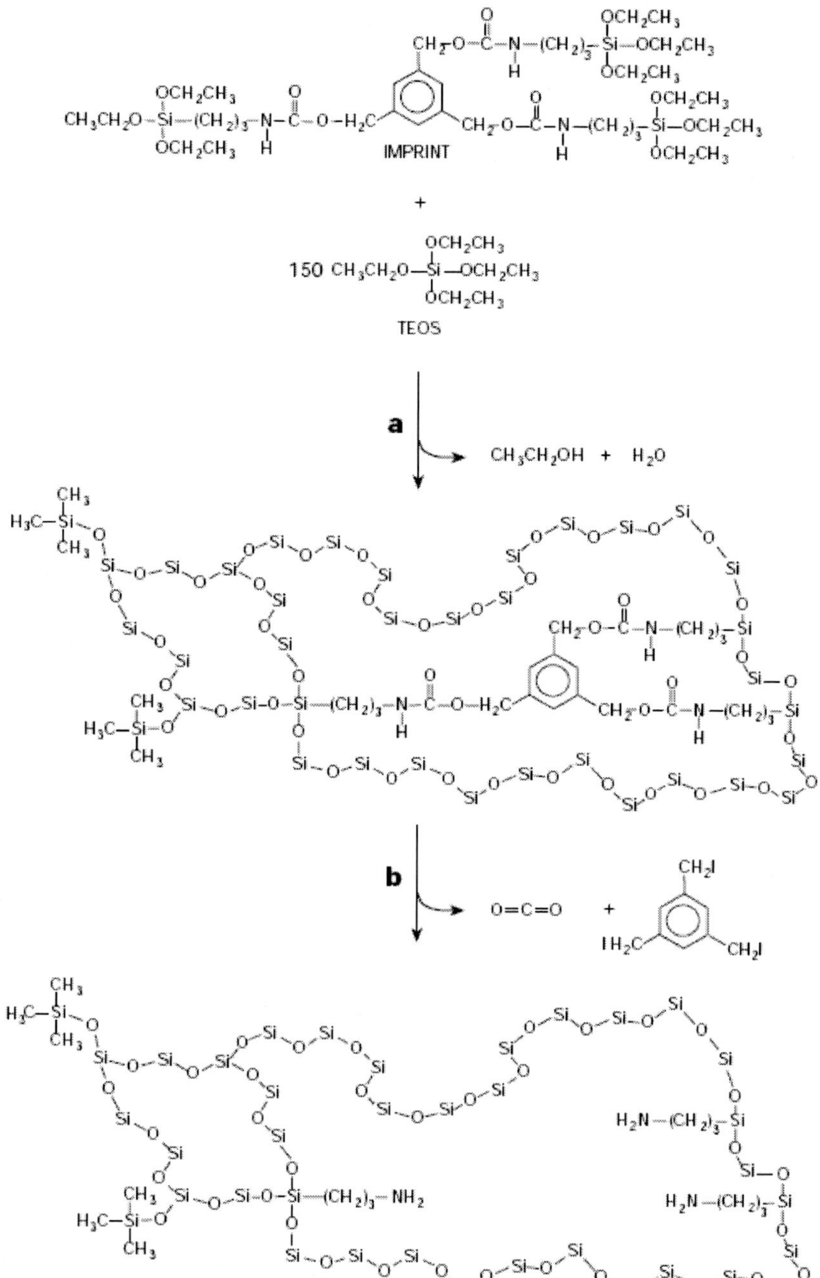

Reproduced with permission from Katz A., Davis M.E., *Nature*, Vol. 403, 286-289.

Figure 3. Synthetic scheme to produce silica containing site-isolated aminopropyl groups through molecular imprinting approach [29].

amide bond is an indicator of the relative spatial separation of the amine groups. Site isolation was also demonstrated by fluorescence emission spectra of materials obtained by reaction of one and three aminopropyl groups per imprint containing silica with pyrenebutyric acid [29]. The method is based on difference in the fluorescence emission spectra of the probe molecule pyrenebutyric acid when present as a monomer or excimer. While an emission maximum around 400nm is observed for the monomer, the excimer shows a maximum around 470 nm (Figure 4). While materials containing one aminopropyl group per imprint showed principally monomer emission spectra, excimer emission spectra was observed for imprinted silica with three aminopropyl groups per imprint thus confirming single site nature of the amine in the former and clustering of the amine sites in the later. Although site isolated catalytic sites were successfully obtained by molecular imprinting approach, this was accompanied with a low aminopropyl loading of 0.07 to 0.23 mmol/g of the imprinted material as demonstrated by benzoic acid adsorption experiments [29].

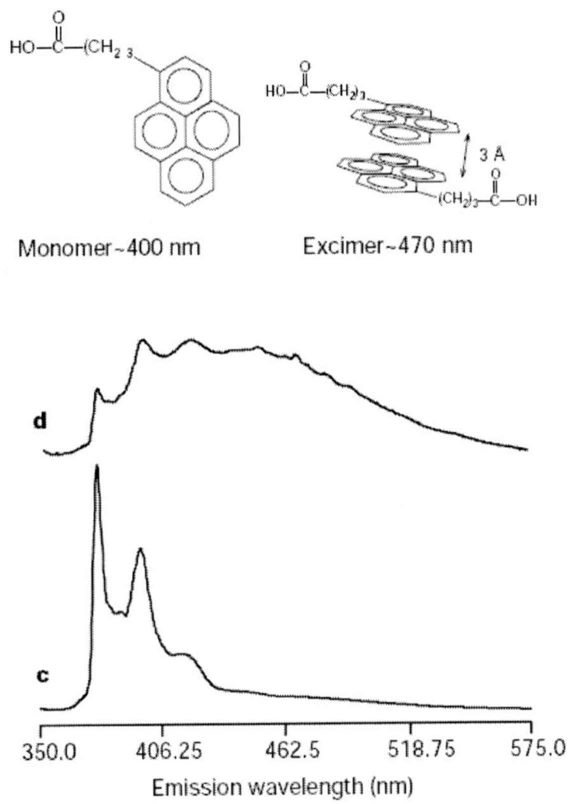

Reproduced with permission from Katz A., Davis M.E., *Nature*, Vol. 403, 286-289.

Figure 4. Fluorescence emission spectra obtain for aminopropyl-pyrenebutyric acid salt functionalized on imprinted silica. The excimer emission at ~ 470 nm serves as probe to investigate clustering of amines [29].

Furthermore, the molecular imprinting methodology is lengthy and specific to a few functional groups and extension to other functional groups would require pre-designing new imprint molecules and multi-step synthetic procedures. Thus other more versatile and less tedious techniques that could manipulate the loading and site-isolation of a variety of functional groups simultaneously are desirable.

2.3. Spacer-Mediated Site-Isolation

A more commonly used technique to obtain site-isolation of functional groups on silica is the use of bulky spacers. Typically, organosilane with the desired functional group is covalently attached to an easily removable spacer. Post-modification of the silica surface with the spacer attached organosilane and subsequent removal of the spacer results in isolated functional groups. Varying the steric bulk of spacer molecule can control the degree of isolation of the functional groups. McKittrick, M. W. et al. [20] reported single site 3-aminopropyl functional groups on mesoporous silica SBA-15 (Figure 5). The 3-aminopropyl groups protected with bulky trityl groups were functionalized on mesoporous silica SBA-15. The steric bulk of trityl group served to functionalize 3-aminopropyl groups at a minimum distance. Hydrolysis and removal of trityl group resulted in mesoporous silica with single site aminopropyl groups. The relative spatial distribution of the amines was probed by reaction with acid chloride to form amide bonds. Reaction of functionalized silica with diacid chloride, isopthaloyl chloride led to the formation of only one amide bond confirming single site or site-isolated nature of immobilized functional groups (Figure 6).

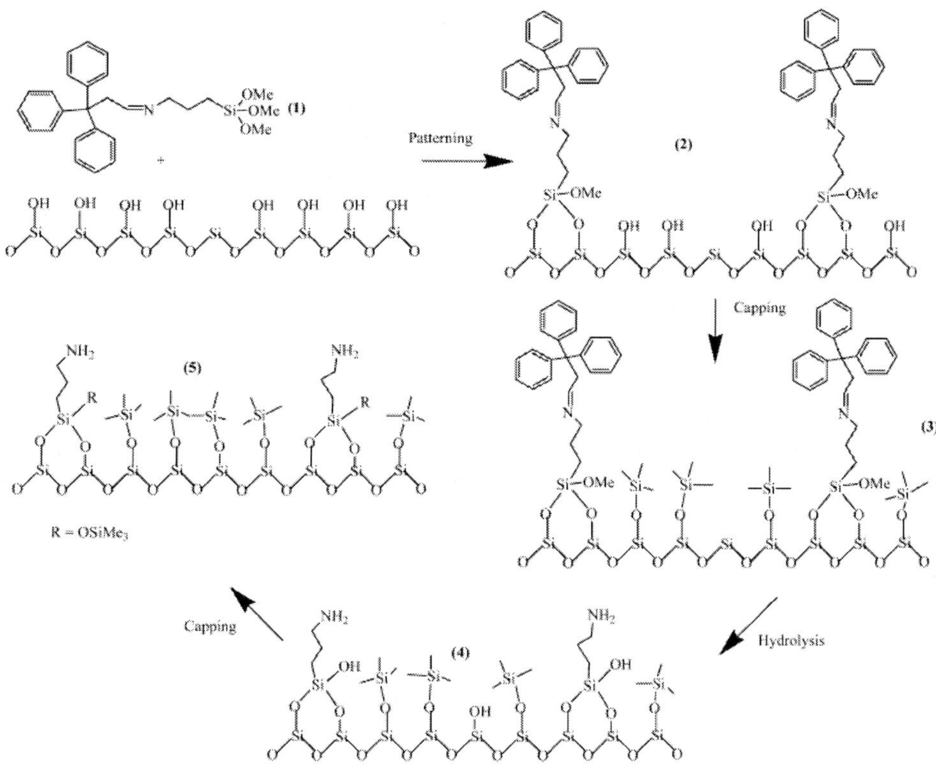

Reproduced with permission from Michael W. McKittrick and Christopher W. Jones. *Chemistry of Materials*, 2003, 15, 1132-1139.

Figure 5. Synthetic route to site-isolated 3-aminopropyl groups functionalized on mesoporous silica through use of 3-aminopropyltrimethoxysilane with bulky trityl group protected amine that serves as spacer [20].

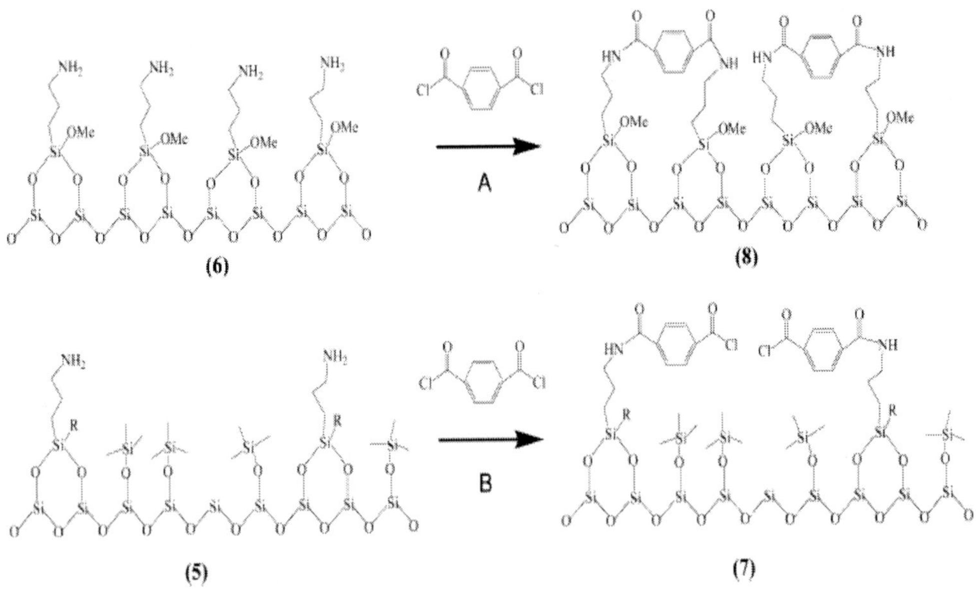

Reproduced with permission from Michael W. McKittrick and Christopher W. Jones. *Chemistry of Materials*, 2003, 15, 1132-1139.

Figure 6. Formation of a single or double amide bond by reaction of site-isolated or otherwise aminopropyl functionalized mesoporous silica with isopthaloyl chloride [20].

The use of bulky spacer groups to obtain site-isolated functional groups on solid supports is a powerful approach to tailor the relative separation between amine groups by simply varying the steric bulk of the spacer. Furthermore, higher loading of functional groups can be obtained by the use of spacer molecules of relatively smaller steric bulk. While a loading of 0.345 mmol/g of aminopropyl was reported by McKittrick, et al. [20] with trityl spacer, another report by Hicks J.C., et al. [30] obtained a loading of 0.99 mmol/g of aminopropyl with the use of less bulky benzyl spacer. While it would be expected that a loading as high as 0.99 mmol/g would lead to significant decrease in relative separation of aminopropyl groups, steady state fluorescence emission studies of pyrenecarboxylic acids conducted on functionalized SBA-15 showed emission spectrum largely corresponding to monomers [30]. Control experiments conducted on SBA-15 with slightly higher loading of aminopropyl groups (1.22 mmol/g), however, showed excimer fluorescence emission indicating clustering of the amine sites. Thus a higher loading of aminopropyl groups was achieved without compromising on the single site nature of the catalytic sites. While this approach is useful in obtaining high loading of aminopropyl groups accompanied by site-isolation, the methodology requires lengthy, usually two to three, synthetic steps until the material could be finally put to use in catalysis and also the probes used to investigate site isolation are labor-intensive. Therefore, a simple methodology to tailor the loading and relative distribution of multiple functional groups is required.

2.4. Solvent Assisted Site-Isolated Grafting (SASIG)

In a very detailed and comprehensive study, Hicks, J. C., et al. [31] presented different synthetic techniques to obtain single site aminopropyl groups and used pyrenecarboxylic acid or pyrenebutyric acid fluorescence emission spectra as a probe to investigate site-isolation. Apart from use of bulky spacer molecules such as trityl and benzyl in obtaining site-isolation, it was demonstrated that grafting of 3-aminpropyltrimethoxysilane done by traditional approach of stirring silica with organosilane in toluene would also result in site isolated aminopropyl group provided the organosilane is used in small concentration. It was proposed that the traditional approach of grafting in toluene leads to clustering of aminopropyl groups in solution (Figure 7) and hence they are grafted in "bunches". Using a dilute solution of organosilane reduces clustering and site-isolation is achieved.

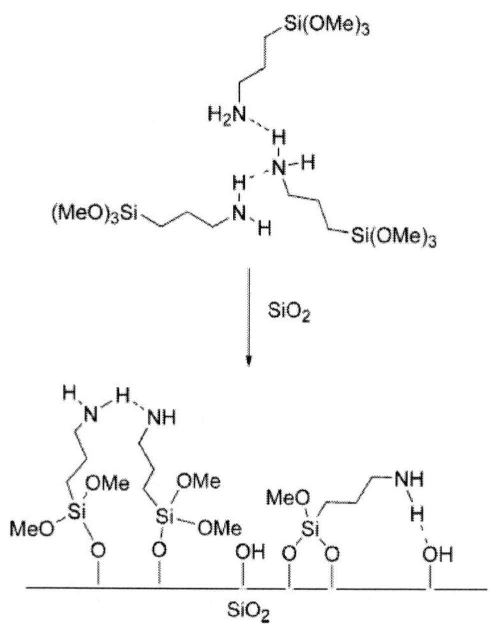

Figure 7. Clustering of amine groups through H-bonding of amino groups of 3-aminopropyltrimethoxysilane in toluene [31].

Based on the hypothesis that diminishing clustering of organosilanes would lead to site-isolated grafting Sharma K.K., et al. [22] recently reported site-isolated grafting of the organosilanes in polar protic solvents such as ethanol and isopropanol. It was proposed that use of polar protic solvents prevents H-bonding amongst amino groups of the organosilane. A detailed study of the effect of dielectric constant and polarity of the solvents on the loading and site-isolation of organosilanes with polar as well as non-polar functional groups was presented.

Table 1. Grafting of 3-aminopropyl groups on mesoporous silica obtained in polar-protic, polar-aprotic and non-polar solvents, surface area and catalytic efficiency for Nitroaldol condensation of 4-hydroxybenzaldehyde and nitromethane [22].

Solvent	Dielectric Constant, *e*	wt % N, Elemental Analysis	%T[a]	wt% (100-600°C)	Surface Area (m^2/g)	Catalytic Efficiency (% yield in 12 min)[b]
			Polar-Protic Solvents			
Isopropanol	18	2.92	11.7	12.0	902 (±7)	92
Ethanol	24	1.71	7.2	12.5	905 (±3)	89
Methonal	33	1.39	6.8	8.0	864 (±3)	72
Formamide	84	2.90[c]	na	na	659 (±4)	5
			Nonpolar and Dipoloar-Aprotic Solvents			
Toluene	2.4	3.82	23.4	17.8	259 (±5)	42
Chloroform	4.8	3.24	22.0	15.2	628 (±7)	99
Ethyl Acetate	6.0	3.55	24.0	18.5	153 (±4)	28
THF	7.5	3.75	17.1	15.4	293 (±7)	36
Acetone	21	3.38	16.6	13.5	617 (±7)	94
DMF	38	3.82[c]	20.4	17.5	501 (±11)	58
Acetonitrile	38	2.52	18.6	14.0	618 (±6)	83
DMSO	47	3.47	17.8	13.8	620 (±11)	99

Remarkably high loading (~ 1-2 mmol/g) of functional groups in polar solvents and (~ 3-4 mmol/g) of functional groups in non-polar solvents was obtained. Aminopropyl loadings comparable to the traditional methods of grafting were obtained accompanied by site-isolation. The method is simple and provides excellent control over the loading of functional groups without compromising on single site nature of the catalytic group. Table 1 presents the loading of aminopropyl groups obtained in a variety of different solvents. Further, a new and simple method to elucidate site-isolation is demonstrated. The technique is based on the colorimetric detection of different copper-amine complexes formed for site-isolated or clustered amines (Figure 8). While multiple amines in an aggregated grafting could bind to copper ions, only one amine could bind to copper in single site grafting. UV-vis reflectance measurements on these samples provided an indication of the relative distribution of amines within the mesoporous channels.

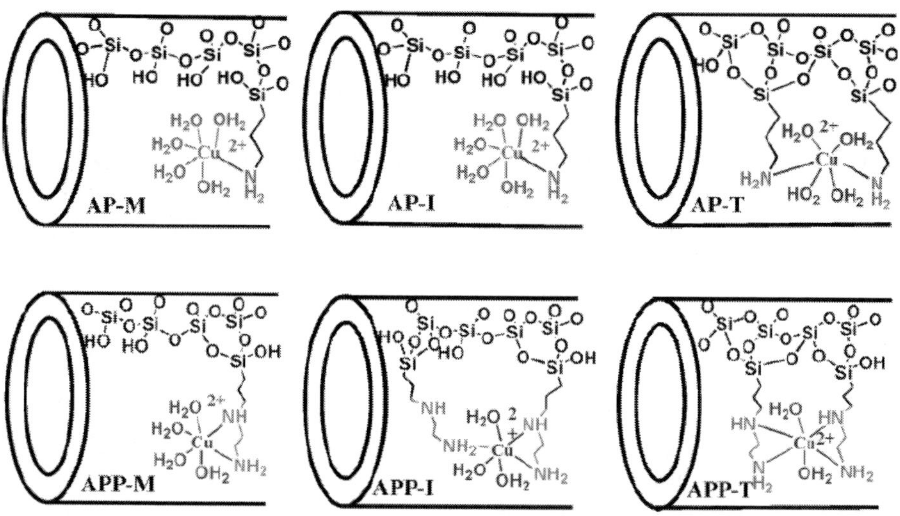

Figure 8. Colorimetric determination of site isolation in 3-aminopropyl functionalized mesoporous silica by complexation to copper ions [22].

3. Nanoparticles Supported in Mesoporous Materials as Heterogeneous Catalysts

3.1. Introduction

Nanoparticles, which often show enhanced catalytic abilities [32, 33] unusual optical properties [34], and novel quantum size effects [35], have been widely used in fields such as catalysis [36, 37], sensing [38], optoelectronics [39], and microelectronics [40]. Nanoparticle catalysis is industrially and experimentally important because a large variety of C-C coupling [41] and alcohol oxidation [32] can be effectively catalyzed by nanoparticles. In this part, we will present a brief review on recent advances in supported nanoparticle heterogeneous catalysts on various mesoporous materials. Heterogeneous nanoparticle catalysts have several advantages such as ease of separation and recycling, less extent of metal leaching and enhanced catalytic performance, etc. Especially with the tightening of federal legislation regulating chemical disposal, catalyst chemist are obliged to develop more catalytically active and less environmentally demanding catalysts, thus heterogeneous nanoparticle catalysts supported mesoporous materials have become a good choice and a lot of promising results have already been obtained. However, we regret that we can not acknowledge each outstanding work and contribution because of the consideration of the length of article. Instead, focus will be put on recent development in methods for preparing composites of nanoparticles and supporting mesoporous materials and catalytic performance of these materials.

3.2. Enhanced Catalytic Activity of Nanoparticles Supported on Mesoporous Materials

Up to now, nanoparticles that have been successfully incorporated into mesoporous materials span a large area, such as metallic Au [42], Pt [43], Pd [44], Ni [45], Co [46], metal oxide TiO_2 [47], ZrO_2 [48], WO_3 [49], Mn_3O_4 [50], compound $MnFe_2O_4$ [51], $NiFe_2O_4$ [51], $NaAlH_4$ [52], or even complex structures such as core-shell Pt-C [53] and bimetallic nanoparticles [54]. The mesoporous materials that are used for the supporting materials include mesoporous carbon [54] and other mesoporous materials based on oxides such as SiO_2 [55, 56, 48], TiO_2 [57, 58], MgO [59], ZnO [60], CeO_2 [59, 60], ZrO_2 [60], $A12O_3$ [60], etc. The reason that all possible combinations of different nanoparticles and various mesoporous materials are being applied to catalysis is that not only different mesoporous materials have different enhancement effects on the catalytic properties of nanoparticles, but also nanoparticles have some influence on the catalytic performance of mesoporous materials. For example, Denkwitz and colleagues[58] found that Au nanoparticles with a size of ~3 nm supported on mesoporous TiO_2 exhibited similar or higher catalytic ability toward low temperature CO oxidation compared to standard Au/TiO_2 catalyst, while their tendency towards deactivation was much lowered. On the other hand, Li and coworkers developed a new method for synthesizing mesoporous Au/TiO_2 nanocomposites with the Au nanoparticles sitting on the walls of mesoprous TiO_2 (See Figure 9) [57]. They found that the photocatalytic acivity was significantly improved by choosing the right loading of Au; catalysts with medium loading of Au were found to be the most efficient (See Figure 10). It was believed that the improved photocatalytic activity was a combination of enhanced light absorption and improved quantum efficiency [57]. Besides the mutual enhancement effects of the nanoparticles and the supporting mesoporous materials, there are several other avenues that could be adopted to further improve the catalytic performance of these heterogeneous systems. For example, Wang, et al. found that with the addition of alumina as a promoter, the catalytic activity of sulfated zirconia supported on mesoporous molecular sieves MCM-41 toward butane isomerization was dramatically improved, and the promotion effects of alumina on this material is not monotonic, for high activity the level of alumina loading should be optimized [48].

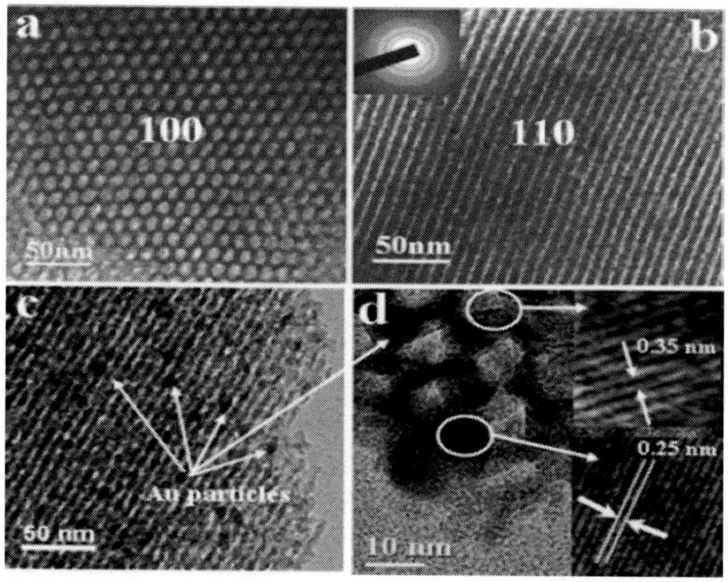

Reprint from Reference 56.

Figure 9. Representative TEM images of 0.5 mol % TiO2 along (a) [100] and (b) [110] Plane, (c) 1.0 mol % TiO2 along [110] plane and (d) high-resolution image of 2.0 mol % TiO2 [57].

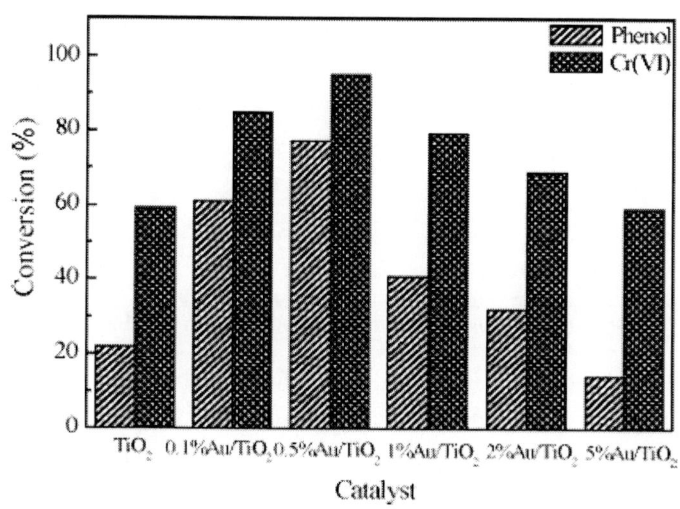

Reprint from Ref. 56 with Permission.

Figure 10. Photocatalytic activity of Au/titania nanocomposites containing 0 to 5 % Au in phenol-oxidation and chromium-reduction reactions [57].

3.3. Methods toward Synthesizing Nanoparticle/Mesoporous Materials Composites

The traditional method for generating nanoparticles in mesoporous materials includes a wet impregnation process, followed by treatment with heat, reducing reagents or oxidation. This method could be applied to a lot of metals [43, 61] or metal oxides [50, 59]. Other methods such as vapor grafting [62], ultra-sound assisted sono-chemical process [42], and one-pot co-assembly method [63], etc. have also been reported. Besides these conventional methods, other avenues for generating these materials were tried for specific metals. For example, Li and co-workers developed a novel and simple in-situ reduction route for the synthesis of ultra-thin metal nanocoating in the channels of mesoporous silica materials. Mesoporous material was first grafted with Si-H groups, which was then used as a reducing species to reduce Pd^{2+} to form a layer of ultra-thin layer of Pd nanoparticle catalytic sites in the mesoporous channel of SBA-15 (Figure 11) [64].

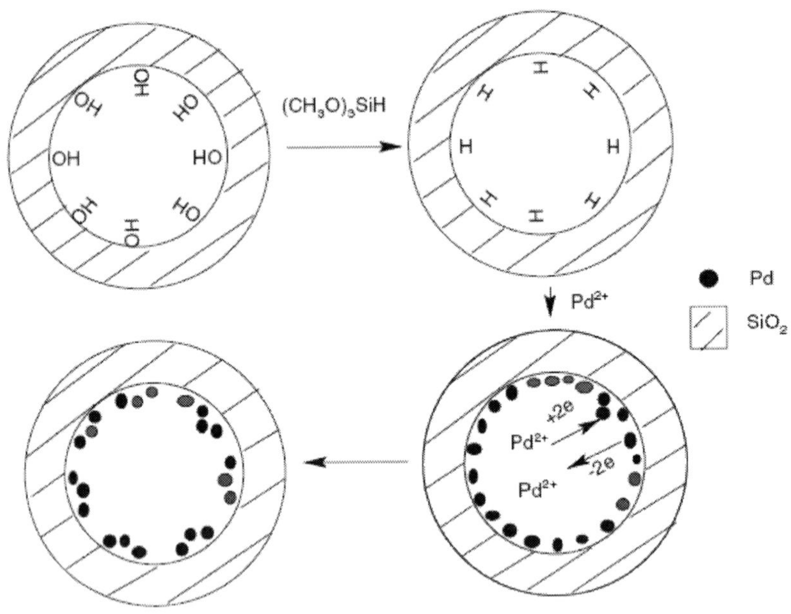

Reprint from Ref. 64 with permission.

Figure 11. Formation mechanism of an ultra-thin metal nanocoating in the channel of SBA-15 [64].

More advanced catalytic application could be achieved if we can tailor the materials in the way that we want, such as the selective deposition of nanoparticles in a specific region of the mesoporous materials. Yuranov et al. achieved selective synthesis of Pd nanoparticles in the complementary micropores of SBA-15 by modifying the post heat treatment after the wet impregnation [61]; SBA15 was treated in two ways: (a) oxidative treatment which involves direct calcination after impregnation of metal ions and (b) reductive-oxidative treatment which involves immobilization of the precursor and subsequent reduction with hydrogen before the material is calcined. The materials obtained from both methods (a) and (b) were then reduced in hydrogen. The former method resulted in Pd nanoparticles sitting in the mesopores of SBA-15 while the later one produced Pd nanoparticles in the complementary micropores (See Figure 12) [61]. It was proposed that this is because palladium atoms can be incorporated into the silica support lattice resulting in palladium silicide (Pd_nSi) under reductive conditions and due to the strong interaction of Pd with SiO_2, pd nanoparticles are stabilized within SBA-15 micropores [61].

Reprint from Ref. 57 with permission.

Figure 12. TEM images of (a) 1.85% Pd/SBA-15 (oxidation) and (b) 1.85% Pd/SBA-15 (reduction-oxidation) [61].

Another method for generating nanoparticles selectively in micropores or mesopores developed by Yang and co-workers [65] reported treatment of as-synthesized SBA-15 with 45 wt.% H_2SO_4 to selectively remove the poly(ethylene oxide) or PEO block of block copolymer, Pluronic123 followed by functionalization of the mesopores with trimethylsilyl groups. PEO block from the micropores was subsequently removed and functionalized with trivinylsilyl groups further complexed to Pd compound. After reduction of the complex, Pd nanoparticles were deposited in the micropores (Figure 13). The catalytic activity of this catalyst was tested for C-C coupling reactions with poorer catalytic performance compared to normal Pd/SBA15 because of the less accessibility of the Pd nanoparticles, which were buried in micropores [65]. However, these new materials might provide great insights for more advanced catalytic application in more complicated systems. All in all, this field of catalysis is still of great challenges and opportunities.

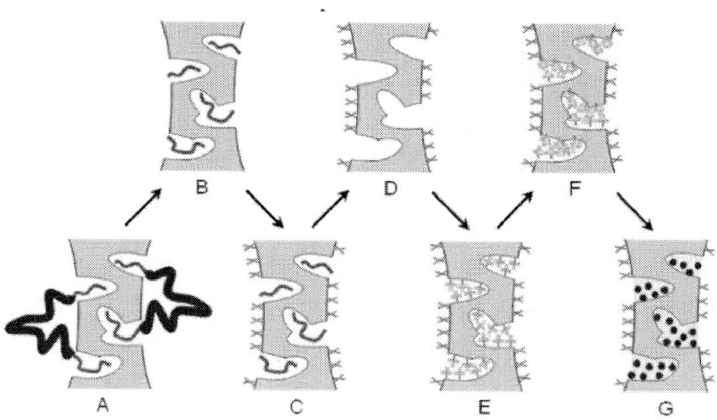

Reprint from Ref. 62 with permission.

Figure 13. Selective Surface Functionalization and Selective deposition of Pd Nanoparticles in the micropores of SBA-15.

4. Mesoporous Materials Supported Organometallic Catalysts

4.1. Introduction

The use of mesoporous materials as scaffolding for organometallic catalysts has become an increasingly popular, and diverse field of investigation. Organometallic catalysis has shown utility in many various synthetic applications, but industrial use is generally hampered by cost, product separation, recyclability, and other large scale concerns. Mesoporous materials, as has been discussed, have many potential advantages compared homogeneous catalysis such as facile separation, recyclability, and stability in a wide range of solvents and pH. In addition, single site or site isolated catalytic sites can be generated through one of the many strategies discussed before. Independent catalytic sites can diminish undesirable inter-site interactions and affects the catalytic efficiency of the material. Also, the site isolated catalytic sites in these highly ordered materials allow a better understanding of the intrinsic and mechanistic properties of the catalyst. Studies performed by Jones et al. have even shown the concave natures of the interior of mesoporous materials are capable of enantioselective enhancement [66]. Due to the overwhelming amount of literature available pertaining to homo/heterogeneous organometallic catalysis, this section can not possibly cover all of the systems realized to date an their properties and can be found elsewhere [67, 68, 69]. This section merely seeks to illuminate common motifs toward the incorporation of organometallic catalysts in mesoporous materials, and their catalytic success compared to homogeneous systems in various syntheses.

4.2. Mesoporous Silica Supported Organometallic Catalysts by Co-Condensation

The synthesis of organic functionalized mesoporous materials and periodic mesoporous organosilicas has been a very promising synthetic scheme utilized in the production of heterogeneous organometallic catalysts. Co-condensation has been shown to produce a more uniform surface chemistry [70], in comparison to grafting, ion-exchange, impregnation, which has been shown to have non-uniform substrate loading [71]. Other synthetic strategies have been cited to suffer from limited diffusion, pore blockage, lower catalytic activity/selectivity [72, 73, 74]. For example, Ru-complex **1** containing mesoporous materials (Figure 14) were synthesized via co-condensation of 2-(diphenylphosphino)ethyltriethoxysilane (DPPES) and subsequent reaction with $RuCl_2(PPh_3)_3$ and $RhCl(PPh_3)_3$ respectively. **1** was found to catalyze the isomerization of homoallylic alcohols in water comparable to homogeneous catalysis with only a minor reduction in catalytic efficiency after multiple runs (Table 2) [75].

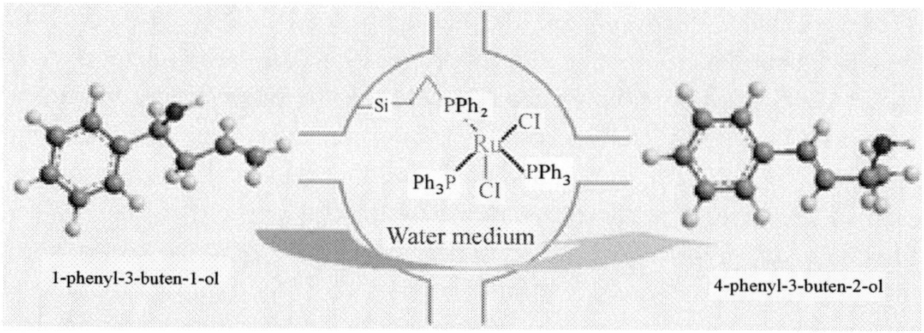

Reproduced with permission from Ref. 73.

Figure 14. Isomerization of 1-phenyl-3-buten-1-ol to 4-phenyl-3-buten-2-ol using 1 immobilized in cage-like FDU-12 [75].

Table 2. Catalytic results for water medium isomerization of 1-phenyl-3-buten-1-ol to 4-phenyl-3-buten-2-ol with homogeneous $RuCl_2(PPh_3)_3$ and immobilized Ru–PPh$_2$-FDU-12 heterogeneous catalyst [75]

Catalyst	Recycling times	Conversion (%)	Selectivity (%)	Yield (%)
$RuCl_2(PPh_3)_3$	1	79	95	75
Ru–PPh$_2$-FDU-12	1	71	96	68
Ru–PPh$_2$-FDU-12	2	71	96	68
Ru–PPh$_2$-FDU-12	3	69	95	66
Ru–PPh$_2$-FDU-12	4	65	95	62
Ru–PPh$_2$-FDU-12	5	60	94	57

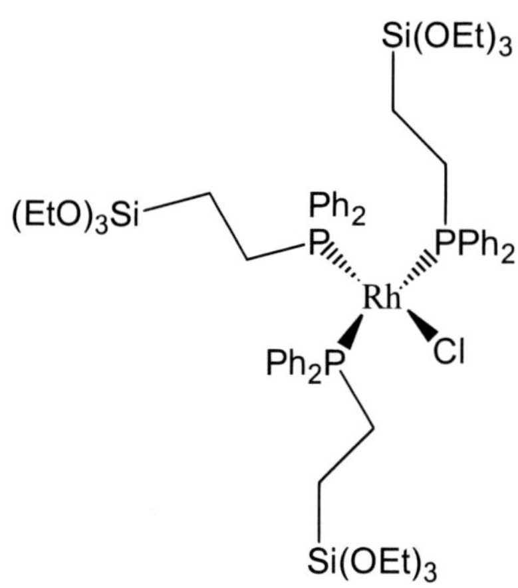

2

Figure 15. Rh-containing organosilane precursor 2 incorporated into mesoporous silica through post-modification and one-pot synthesis [76].

While Ru-dppes incorporated mesoporous silica was used in isomerization reactions [75], hydrogenation of various olefins was accomplished by the use of Rh-dppes complex containing organosilanes functionalized into mesoporous silica **2** (Figure 15) with catalytic efficiency and chemoselectivity for the hydrogenation of carbon-carbon single bonds comparable to corresponding homogeneous catalyst [76]. Interestingly, contrary to usual large differences observed in catalytic efficiency for metal-complex containing mesoporous materials prepared by co-condensation and post-modification, catalysts prepared by both co-condensation of precursor **2** with tetraethoxysilane in one pot and those obtained through post-modification of pre-synthesized mesoporous silica with the precursor by wet-impregnation showed similar catalytic efficiency in terms of the turnover frequency [TOF (h^{-1})]. From these results it can be conjectured that catalytic sites in these two materials should be intrinsically similar and equally accessible.

Li, et al. reported ethyl-bridged PMOs with Pd(II) complexed to 3-aminopropyl-ltrimethoxysilane grafted onto the mesoporous walls to be an efficient catalysts for Barbier reaction of benzaldehyde and allyl bromide (Figure 16) [74]. Use of water as the reaction medium combined with the presence of ethyl moiety in the framework (which increased hydrophobicity of the pores) enhanced diffusion of the organic substrates. As can be seen in Table 3 the PMO material showed superior catalytic efficiency compared to grafted SBA-15 and MCM-41 materials with values comparable to homogeneous trials.

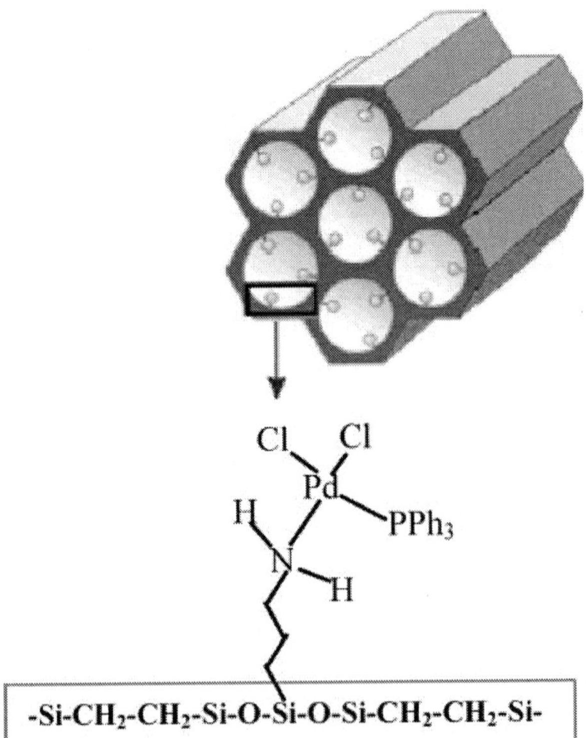

Reproduced with permission from Li, et al.

Figure 16. Pictorial representation of Pd(II) metal ions complexed to the 3-aminopropyl groups grafted onto mesoporous silica wall [74].

Table 3. Catalysis results for the Barbier reaction of benzaldehyde and allyl bromide in presence of homogeneous palladium catalyst (entry 1) and those prepared by complexing palladium to amine functionalized mesoporous silica (entry 2-4) [74]

Entry	Catalyst	Conversion (%)	Selectivity (%)	Yield (%)
1	$PdCl_2(PPh_3)_2$	96	94	90
2	Pd(II)-PMO(Et)-4	88	94	83
3	Pd(II)-SBA-15	76	94	71
4	Pd(II)-MCM-41	74	92	68

4.3. Mesoporous Silica Supported Organometallic Catalysts by Grafting

As discussed in previous sections, the addition/modification of the interior surface functionality (or exterior, for that matter) is intimately dependent upon pore dimensions, hydrophobicity/hydrophilicity, and most importantly, surface silanol concentration. Surface silanols, which are slightly acidic in nature [77], serve as sites for the addition of organic/organometallic moieties. Some have shown these groups to be catalytically active or serve commutatively with catalytically active materials incorporated into the mesoporous materials, a so-called bi-functional catalyst [24, 23]. The methods of grafting siloxy, halosilane, and other related materials have been discussed previously and will not be reviewed here. This section will focus on the methodologies successfully undertaken in the chemisorption, either ionic or covalent in nature, of organometallic functionalities toward heterogeneous catalysis.

The immobilization of organometallic species may be accomplished by two different modes. The previously mentioned acidic silanol groups have been shown useful in ion-exchange reactions with catalysts containing negatively charged weak coordination ligands (i.e. halogens, borates). This methodology, also known as direct grafting or impregnation, has been applied to systems such as MCM-41, MCM-48, and SBA-15 [78], but has proven to be most efficient in aluminum substituted mesoporous material due increased siloxane nucleophilicity upon deprotonation [79]. Kühn et al. have utilized aluminated mesoporous material extensively in the immobilization of $[M(NCCH_3)_6]X_2$ complexes (M= Mn^{II} or Cu^{II}), as seen in Figure 17, for use in polymerization of isobutene [80] and the aziridination of olefins [81].

The Kühn group has also reported extensively on the direct grafting of $Cp^*Mo(CO)_3Cl$ (**3**) and $[(-)\text{-menthylCp}]Mo(CO)_3Cl$ (**4**) complexes in aluminated mesoporous materials for use in olefin epoxidation reactions (Figure 18) [79, 82, 83].

The immobilization of samarium(II) complexes analogous to Kagan's reagent, $SmI_2(THF)_2$, in MCM-41 have shown to catalyze the formation of polymethyl methacrylate (PMMA) [84]. Catalytic efficiency, however, could not be determined due to pore blockage and reduced diffusion from PMMA sorption onto the mesoporous material as the reaction progressed. Crosman, A., et al. [85] have also taken advantage of ionic interaction in the direct grafting of chiral rhodium diphosphine complexes (**5**) in aluminated mesoporous material (Figure 19 and 20).

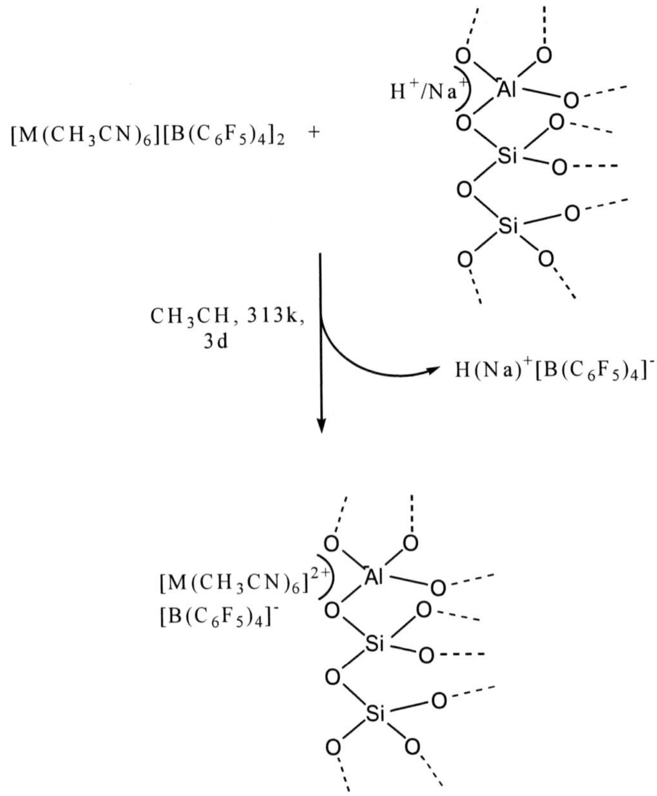

Figure 17. Immobilization of [M(NCCH3)6]X2 complexes (M= MnII or CuII) in aluminated mesoporous materials.

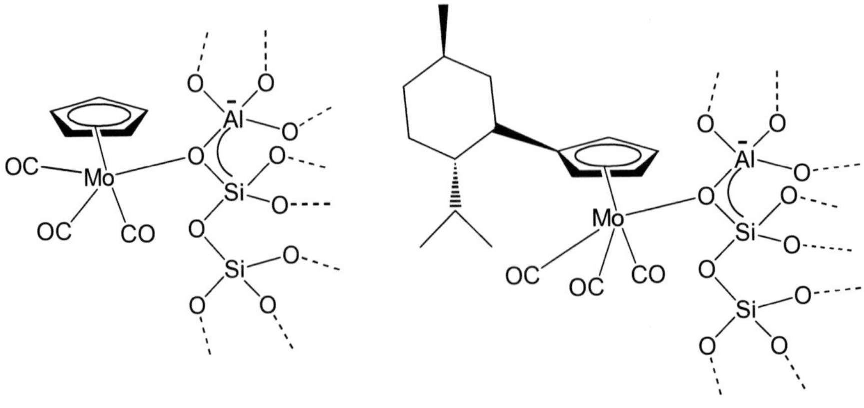

3 4

Figure 18. The synthesis of heterogeneous catalyst by grafting of Cp*Mo(CO)3Cl (3) and [(-)-menthylCp]Mo(CO)3Cl (4) complexes in aluminated mesoporous materials for olefin epoxidation reactions.

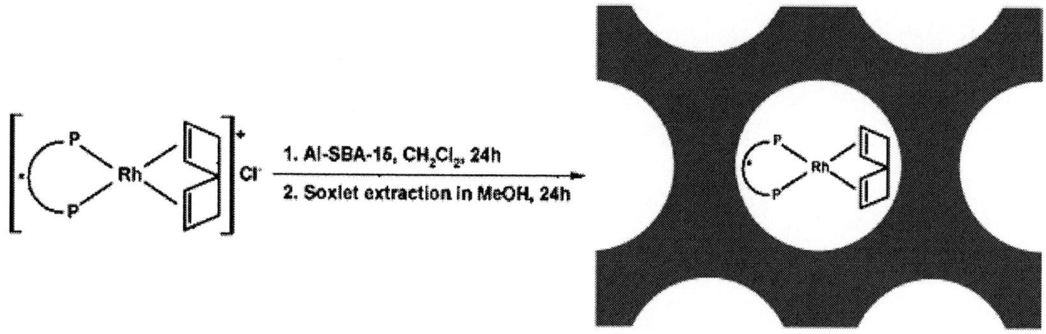

Reproduced with permission from reference 85.

Figure 19. Immobilization of chiral rhodium diphosphine catalysts (5) to Al-SBA-15/Al-MCM-41 [85].

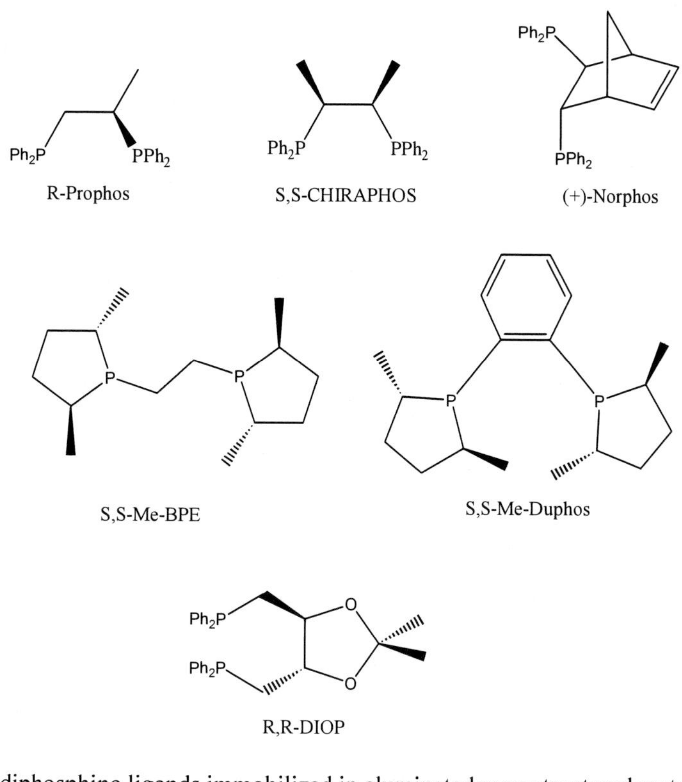

Figure 20. Chiral diphosphine ligands immobilized in aluminated mesostructured materials with rhodium to form catalytic centers [85, 86, 87].

These heterogeneous catalysts have shown excellent reactivity in the enantioselective hydrogenation of prochiral olefins with enantiomeric excess (ee's) $\geq$ 92% and regioselectivities ($\geq$99%) with quantitative conversion.

The second method of catalyst incorporation in mesoporous materials involves, generally, multi-step syntheses where organic moieties (i.e. amines, phosphines, thiols) are immobilized followed by post modification toward the final product. Alternatively, the catalyst may be synthesized as the corresponding alkoxysilane complex followed by immobilization into the mesostructured materials. For instance, Kühn and co-workers demonstrated covalent modification of mesoporous materials with molybdenum complexes **6-9** to catalyze olefin epoxidation (Figure 21) [88, 89, 90].

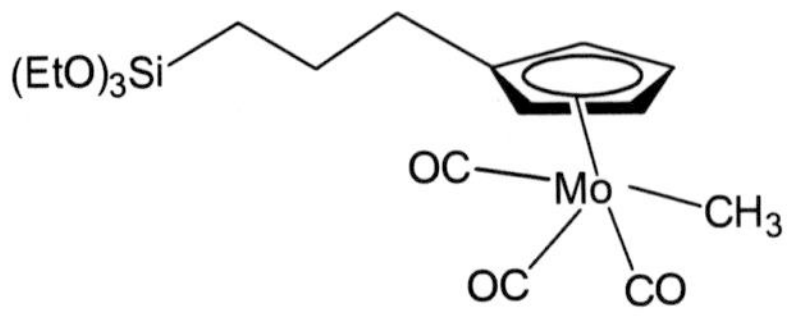

6

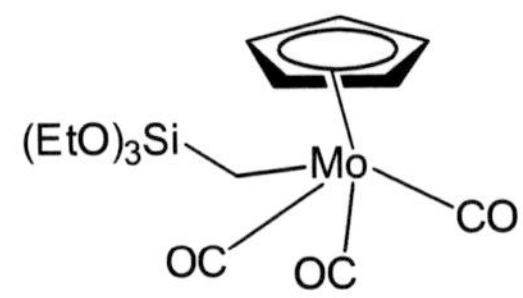

7

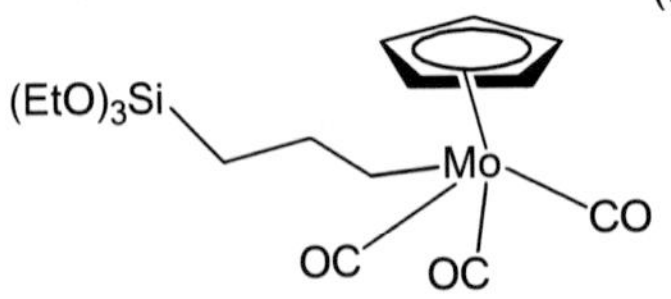

8

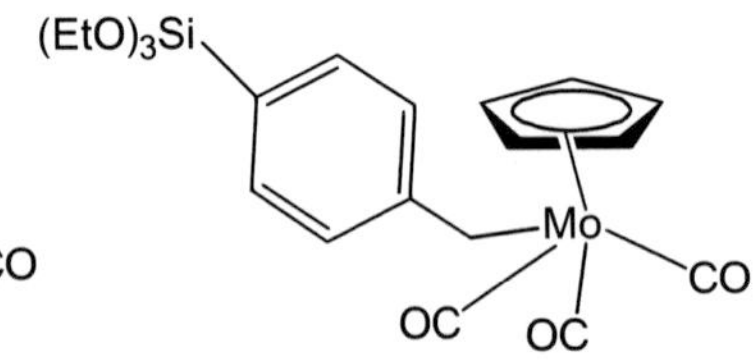

9

Figure 21. Examples of or Mo-containing organosilanes synthesized by modification of Kühn's molybdenum complexes [88, 89, 90].

These immobilized moieties were found to react somewhat slower than corresponding homogeneous catalytic trials, as seen in Figure 22 [88], but still exhibited good yields, selectivities, and TOF's (h^{-1}) in excess of 10000 with little loss in efficiency after multiple trials.

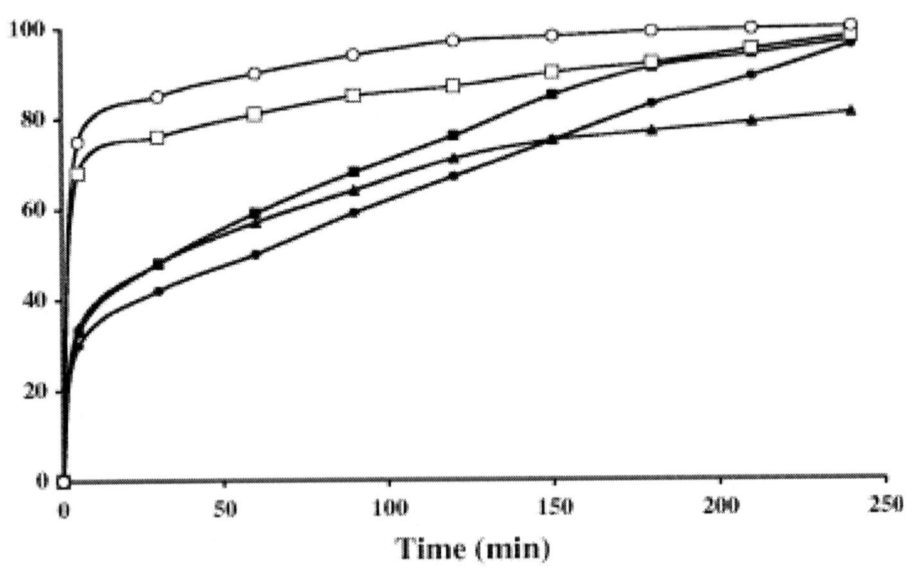

Reproduced with Permission from reference Ref. 89.

Figure 22. Time-dependent yield of cyclooctene epoxide in the presence of 1 mol% 6(triangles), 7 (squares), 8 (circles), CpMo(CO)3Me (open squares), and CpMo(CO)3Cl (open circles) at 55°C [89].

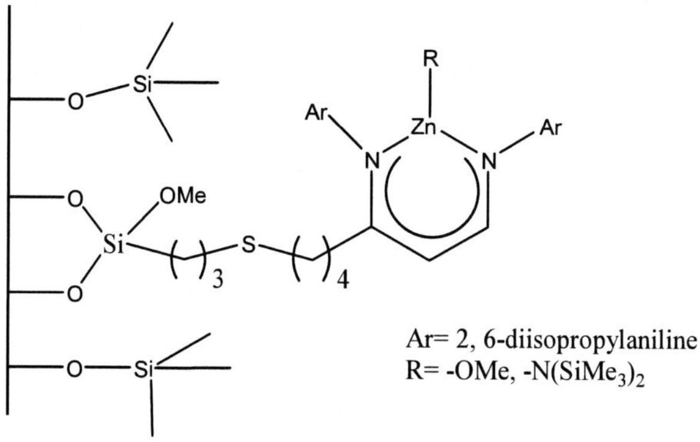

10

Figure 23. Mesoporous silica supported Zn-complex catalysts [91].

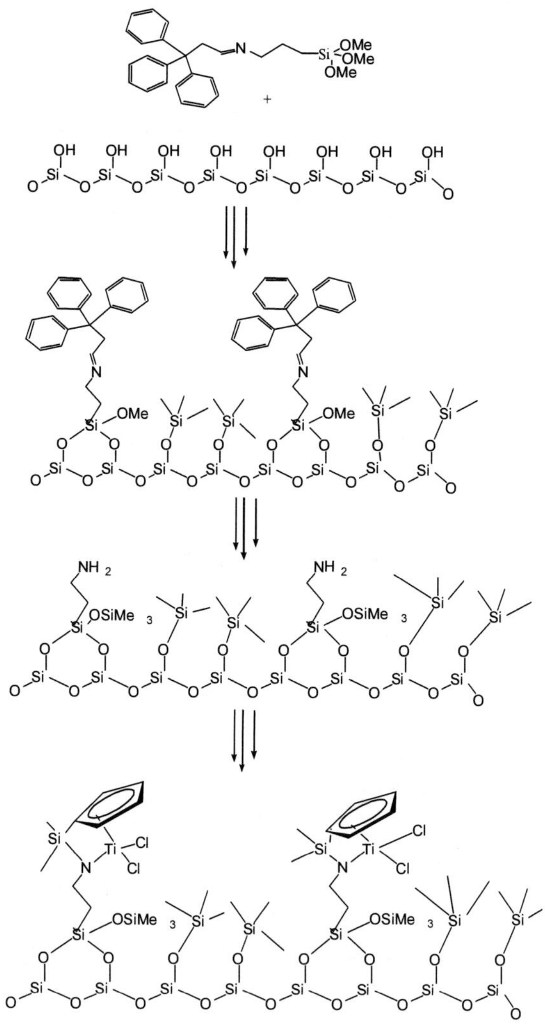

Reproduced with permission from Ref. 93.

Figure 24. Synthetic scheme to obtain site-isolated Ti-ethylene polymerization catalytic sites [93].

Besides olefin epoxidation, CO_2/epoxide copolymerization reactions were catalyzed with Zn-complex catalyst **10** (Figure 23) [91] as well as a silsesquinoxane-based Zn-complex [92] synthesized via multi-step *in situ* synthetic strategies.

Previously mentioned spacer mediated technique to obtain single site catalytic groups for enhanced catalytic efficiency was put to use by McKittrick M.W. et al. for the successful synthesis of a variety of site-isolated group 4 metal-olefin polymerization catalysts (Figure 24) [93]. This methodology was developed as an alternative to co-condensation and performed much more favorably than catalysts formed from grafting APTS and subsequent synthesis.

5. Conclusion and Future Outlooks

The design and assembly of nanostructured materials for efficient heterogeneous catalysts of various reactions is being explored by many researchers currently. In recent years, the introduction of organic groups into nanoporous materials has been widely used as versatile way to result in efficient functionalized and multifunctionalized nanostructured catalysts. By various self-assembly routes including a new solvent-assisted grafting and co-assembly of multiple types of organic groups, we and others have also recently demonstrated the synthesis of a broad range of nanoporous catalysts containing high surface areas, site-isolated catalytic sites, nanometer pores and efficient and selective catalytic properties for various types of reactions. The grafting of site-isolated organic groups with a polar solvent such as ethanol or isopropanol is among the newest synthetic strategies that allow the synthesis of novel catalysts with enhanced catalytic efficiency. By introducing appropriate secondary functional groups into the nanoporous catalyst structure, selective catalysts for specific reactants can also be produced. Furthermore, by a judicious choice of the active catalytic sites, selectivity for a specific product by these catalysts can be accomplished.

Acknowledgments

We gratefully acknowledge the financial assistance by the US National Science Foundation (NSF), CAREER Grant No. CHE-064534 for our work in nanostructured catalysis.

References

[1] Kresge, C. T.; Leonowicz, M. E.; Roth, W. J.; Vartuli, J. C.; and Beck, J. S. *Nature,* 1992, *359*, 710-712.

[2] Beck, J. S.; Vartuli, J. C.; Roth, W. J.; Leonowicz, M. E.; Kresge, C. T.; Schmitt, K. D.; Chu, C. T. W.; Olson, D. H.; Sheppard, E. W.; McCullen, S. B.; Higgins, J. B.; and Schlenker, J. L. *Journal of the American Chemical Society,* 1992, *114*, 10834-10843.

[3] Vartuli, J. C.; Schmitt, K. D.; Kresge, C. T.; Roth, W. J.; Leonowicz, M. E.; McCullen, S. B.; Hellring, S. D.; Beck, J. S.; Schlenker, J. L.; Olson, D. H.; and Sheppard, E. W. *Chemistry of Materials,* 1994, *6*, 2317-2326.

[4] Zhao, D. Y.; Feng, J. L.; Huo, Q. S.; Melosh, N.; Fredrickson, G. H.; Chmelka, B. F.; and Stucky, G. D. *Science,* 1998, *279*, 548-552.

[5] Inagaki, S.; Fukushima, Y. and Kuroda, K. *Journal of the Chemical Society-Chemical Communications,* 1993, 680-682.

[6] Inagaki, S.; Fukushima, Y. and Kuroda, K. In *Zeolites and Related Microporous Materials: State of the Art, 1994* 1994; Vol. 84, p 125-132.

[7] Yu, C. Z.; Yu, Y. H.; Miao, L.; and Zhao, D. Y. *Microporous and Mesoporous Materials,* 2001, *44*, 65-72.

[8] Wight, A. P. and Davis, M. E. *Chemical Reviews,* 2002, *102*, 3589-3613.

[9] Firouzi, A.; Kumar, D.; Bull, L. M.; Besier, T.; Sieger, P.; Huo, Q.; Walker, S. A.; Zasadzinski, J. A.; Glinka, C.; Nicol, J.; Margolese, D.; Stucky, G. D.; and Chmelka, B. F. *Science, (Washington D C)* 1995, *267*, 1138-1143.

[10] Nakamura, T.; Mizutani, M.; Nozaki, H.; Suzuki, N.; and Yano, K. *Journal of Physical Chemistry C,* 2007, *111*, 1093-1100.

[11] Bambrough, C. M.; Slade, R. C. T. and Williams, R. T. *Journal of Materials Chemistry,* 1998, *8*, 569-571.

[12] Tian, B. Z.; Liu, X. Y.; Yu, C. Z.; Gao, F.; Luo, Q.; Xie, S. H.; Tu, B.; and Zhao, D. Y. *Chemical Communications,* 2002, 1186-1187.

[13] Lim, M. H. and Blanford, C. F.; Stein, A. *Chemistry of Materials,* 1998, *10*, 467.

[14] Moller, K. and Bein, T. *Chemistry of Materials,* 1998, *10*, 2950-2963.

[15] Corriu, R. J. P.; Guerin, C.; Henner, B. J. L.; and Wang, Q. *Organometallics,* 1991, *10*, 3200-3205.

[16] Asefa, T.; MacLachan, M. J.; Coombs, N.; and Ozin, G. A. *Nature,* 1999, *402*, 867-871.

[17] Melde, B. J.; Holland, B. T.; Blanford, C. F.; and Stein, A. *Chemistry of Materials,* 1999, *11*, 3302-3308.

[18] Inagaki, S.; Guan, S.; Fukushima, Y.; Ohsuna, T.; and Terasaki, O. *Journal of the American Chemical Society,* 1999, *121*, 9611-9614.

[19] Bass, J. D. and Katz, A. *Chemistry of Materials,* 2006, *18*, 1611-1620.

[20] McKittrick, M. W. and Jones, C. W. *Chemistry of Materials,* 2003, *15*, 1132-1139.

[21] Walcarius, A. and Delacote, C. *Chemistry of Materials,* 2003, *15*, 4181-4192.

[22] Sharma, K. K.; Anan, A.; Buckley, R. P.; Ouellette, W.; and Asefa, T. *Journal of the American Chemical Society,* 2008, *130*, 218-228.

[23] Bass, J. D.; Solovyov, A.; Pascall, A. J.; and Katz, A. *Journal of the American Chemical Society,* 2006, *128*, 3737-3747.

[24] Sharma, K. K. and Asefa, T. *Angewandte Chemie-International Edition,* 2007, *46*, 2879-2882.

[25] Suzuki, T. M.; Nakamura, T.; Fukumoto, K.; Yamamoto, M.; Akimoto, Y.; Akimoto, Y.; and Yano, K. *Journal of Molecular Catalysis a-Chemical,* 2008, *280*, 224-232.

[26] Angeletti, E.; Canepa, C.; Martinetti, G.; and Venturello, P. *Tetrahedron Letters,* 1988, *29*, 2261-2264.

[27] Shimizu, K.; Hayashi, E.; Inokuchi, T.; Kodama, T.; Hagiwara, H.; and Kitayama, Y. *Tetrahedron Letters,* 2002, *43*, 9073-9075.

[28] Ballini, R. and Bosica, G. *Journal of Organic Chemistry,* 1997, *62*, 425-427.

[29] Katz, A. and Davis, M. E. *Nature,* 2000, *403*, 286-289.

[30] Hicks, J. C. and Jones, C. W. *Langmuir,* 2006, *22*, 2676-2681.

[31] Hicks, J. C.; Dabestani, R.; Buchanan, A. C.; and Jones, C. W. *Chemistry of Materials,* 2006, *18*, 5022-5032.

[32] Tsunoyama, H.; Sakurai, H.; Negishi, Y.; and Tsukuda, T. *Journal of the American Chemical Society,* 2005, *127*, 9374-9375.

[33] Tian, N.; Zhou, Z. Y.; Sun, S. G.; Ding, Y.; and Wang, Z. L. *Science,* 2007, *316*, 732-735.

[34] Murphy, C. J. and Jana, N. R. *Advanced Materials,* 2002, *14*, 80-82.

[35] Pethkar, S.; Aslam, M.; Mulla, I. S.; Ganeshan, P.; and Vijayamohanan, K. *Journal of Materials Chemistry,* 2001, *11*, 1710-1714.

[36] Aiken, J. D. and Finke, R. G. *Journal of Molecular Catalysis a-Chemical,* 1999, *145*, 1-44.

[37] Narayanan, R. and El-Sayed, M. A. *Journal of Physical Chemistry B,* 2005, *109*, 12663-12676.

[38] Luo, X. L.; Morrin, A.; Killard, A. J.; and Smyth, M. R. *Electroanalysis,* 2006, *18*, 319-326.

[39] Trindade, T.; O'Brien, P.; and Pickett, N. L. *Chemistry of Materials,* 2001, *13*, 3843-3858.

[40] Schmid, G.; Baumle, M.; Geerkens, M.; Helm, I.; Osemann, C.; and Sawitowski, T. *Chemical Society Reviews,* 1999, *28*, 179-185.

[41] Kogan, V.; Aizenshtat, Z.; Popovitz-Biro, R.; and Neumann, R. *Organic Letters,* 2002, *4*, 3529-3532.

[42] Perkas, N.; Zhong, Z.; Grinblat, J.; and Gedanken, A. *Catalysis Letters,* 2008, *120*, 19-24.

[43] Li, X. H.; Shen, Y. L.; Xing, R.; Liu, Y. M.; Wu, H. H.; He, M. Y.; and Wu, P. *Catalysis Letters,* 2008, *122*, 325-329.

[44] Behrens, S. and Spittel, G. *Dalton Transactions,* 2005, 868-873.

[45] Sietsma, J. R. A.; Meeldijk, J. D.; Versluijs-Helder, M.; Broersma, A.; van Dillen, A. J.; de Jongh, P. E. and de Jong, K. P. *Chemistry of Materials,* 2008, *20*, 2921-2931.

[46] Lira, E.; Lopez, C. M.; Oropeza, F.; Bartolini, M.; Alvarez, J.; Goldwasser, M.; Linares, F. L.; Lamonier, J. F.; and Zurita, M. J. P. *Journal of Molecular Catalysis a-Chemical,* 2008, *281*, 146-153.

[47] Luan, Z. H.; Maes, E. M.; van der Heide, P. A. W.; Zhao, D. Y.; Czernuszewicz, R. S.; and Kevan, L. *Chemistry of Materials,* 1999, *11*, 3680-3686.

[48] Wang, J. H. and Mou, C. Y. *Microporous and Mesoporous Materials,* 2008, *110*, 260-270.

[49] Yang, X. L.; Gao, R. H.; Dai, W. L.; and Fan, K. N. *Journal of Physical Chemistry C,* 2008, *112*, 3819-3826.

[50] Han, Y. F.; Chen, F.; Ramesh, K.; Zhong, Z.; Widjaja, E.; and Chen, L. *Applied Catalysis B-Environmental,* 2007, *76*, 227-234.

[51] Du, Y. C.; Liu, S.; Ji, Y. Y.; Zhang, Y. L.; Xiao, N.; and Xiao, F. S. *Journal of Magnetism and Magnetic Materials,* 2008, *320*, 1932-1936.

[52] Zheng, S.; Fang, F.; Zhou, G.; Chen, G.; Ouyang, L.; Zhu, M.; and Sun, D. *Chemistry of Materials,* 2008, ASAP.

[53] Wen, Z. H.; Liu, J. and Li, J. H. *Advanced Materials,* 2008, *20*, 743-+.

[54] Liu, S. H.; Yu, W. Y.; Chen, C. H.; Lo, A. Y.; Hwang, B. J.; Chien, S. H.; and Liu, S. B. *Chemistry of Materials,* 2008, *20*, 1622-1628.
[55] Han, P.; Wang, X. M.; Qiu, X. P.; Ji, X. L.; and Gao, L. X. *Journal of Molecular Catalysis a-Chemical,* 2007, *272*, 136-141.
[56] Prasetyanto, E. A.; Sujandi; Lee, S. C.; and Park, S.E. *Bulletin of Korean Chemical Society,* 2007, *28*, 2359-2362.
[57] Li, H. X.; Bian, Z. F.; Zhu, J.; Huo, Y. N.; Li, H.; and Lu, Y. F. *Journal of the American Chemical Society,* 2007, *129*, 4538-+.
[58] Denkwitz, Y.; Geserick, J.; Hormann, U.; Plzak, V.; Kaiser, U.; Husing, N.; and Behm, R. J. *Catalysis Letters,* 2007, *119*, 199-208.
[59] Tsoncheva, T.; Roggenbuck, J.; Tiemann, M.; Ivanova, L.; Paneva, D.; Mitov, I.; and Minchev, C. *Microporous and Mesoporous Materials,* 2008, *110*, 339-346.
[60] Yuan, Z. Y.; Idakiev, V.; Vantomme, A.; Tabakova, T.; Ren, T. Z.; and Su, B. L. *Catalysis Today,* 2008, *131*, 203-210.
[61] Yuranov, I.; Kiwi-Minsker, L.; Buffat, P.; and Renken, A. *Chemistry of Materials,* 2004, *16*, 760-761.
[62] Mehnert, C. P.; Weaver, D. W. and Ying, J. Y. *Journal of the American chemical Society,* 1998, *120*, 12289-12296.
[63] Chen, A.; Zhang, W.; Li, X.; Tan, D.; Han, X.; and Bao, X. *Catalysis Letters,* 2007, 119-164.
[64] Li, L.; Shi, J. L.; Zhang, L. X.; and Xiong, L. M.; Y, J. N. *Advanced Materials,* 2004, *16*, 1079-1082.
[65] Yang, C.M.; Lin, H.A.; Zibrowius, B.; Spliethoff, B.; Schuth, F.; Liou, S.C.; Chu, M.W.; and Chen, C.H. *Chemistry of Materials,* 2007, *19*, 3205-3211.
[66] Jones, M. D.; Raja, R.; Thomas, J. M.; Johnson, B. F. G.; Lewis, D. W.; Rouzaud, J.; and Harris, K. D. M. *Angewandte Chemie-International Edition,* 2003, *42*, 4326-4331.
[67] Crabtree, R. H. *The organometallic chemistry of transition metals*; (4 Ed.); John Wiley & Sons Inc.: New York, 2005.
[68] Thomas, J. M. and Raja, R. *Journal of Organometallic Chemistry,* 2004, *689*, 4110-4124.
[69] Song, C. E.; Kim, D. H. and Choi, D. S. *European Journal of Inorganic Chemistry,* 2006, 2927-2935.
[70] (70) Huh, S.; Wiench, J. W.; Yoo, J. C.; Pruski, M.; and Lin, V. S. Y. *Chemistry of Materials,* 2003, *15*, 4247-4256.
[71] Yoshitake, H.; Aoki, Y. and Hemmi, S. *Microporous and Mesoporous Materials,* 2006, *93*, 294-303.
[72] Li, H. X.; Zhang, F.; Wan, Y.; and Lu, Y. F. *Journal of Physical Chemistry B,* 2006, *110*, 22942-22946.
[73] Lin, V. S. Y.; Radu, D. R.; Han, M. K.; Deng, W. H.; Kuroki, S.; Shanks, B. H.; and Pruski, M. *Journal of the American Chemical Society,* 2002, *124*, 9040-9041.
[74] Li, H. X.; Xiong, M. W.; Zhang, F.; Huang, J. L.; and Chai, W. *Journal of Physical Chemistry C,* 2008, *112*, 6366-6371.
[75] Li, H. X.; Zhang, F.; Yin, H.; Wan, Y.; and Lu, Y. F. *Green Chemistry,* 2007, *9*, 500-505.
[76] Dufaud, V.; Beauchesne, F. and Bonneviot, L. *Angewandte Chemie-International Edition,* 2005, *44*, 3475-3477.

[77] Bascom, W. D. and Timmons, R. B. *Journal of physical chemistry,* 1972, *76*, 3192-3200.

[78] Jain, K. R. and Kuhn, F. E. *Dalton Transactions,* 2008, 2221-2227.

[79] Freund, C.; Abrantes, M. and Kuhn, F. E. *Journal of Organometallic Chemistry,* 2006, *691*, 3718-3729.

[80] Sakthivel, A.; Hijazi, A. K.; Yeong, H. Y.; Kohler, K.; Nuyken, O.; and Kuhn, F. E. *Journal of Materials Chemistry,* 2005, *15*, 4441-4445.

[81] Sakthivel, A.; Hijazi, A. K.; Hanzlik, M.; Chiang, A. S. T.; and Kuhn, F. E. *Applied Catalysis a-General,* 2005, *294*, 161-167.

[82] Sakthivel, A.; Zhao, J.; Hanzlik, M.; and Kuhn, F. E. *Dalton Transactions,* 2004, 3338-3341.

[83] Abrantes, M.; Sakthivel, A.; Romao, C. C.; and Kuhn, F. E. *Journal of Organometallic Chemistry,* 2006, *691*, 3137-3145.

[84] Anwander, R.; Nagl, I.; Zapilko, C.; and Widenmeyer, M. *Tetrahedron,* 2003, *59*, 10567-10574.

[85] Crosman, A. and Hoelderich, W. E. *Journal of Catalysis,* 2005, *232*, 43-50.

[86] Wagner, H. H.; Hausmann, H. and Holderich, W. F. *Journal of Catalysis,* 2001, *203*, 150-156.

[87] Crosman, A. and Hoelderich, W. F. *Catalysis Today,* 2007, *121*, 130-139.

[88] Sakthivel, A.; Zhao, J.; Hanzlik, M.; Chiang, A. S. T.; Herrmann, W. A.; * and Kühn, F. E. *Advanced synthesis and catalysis,* 2005, *347*, 473-483.

[89] Zhao, J.; Sakthivel, A.; Santos, A. M.; and Kuhn, F. E. *Inorganica Chimica Acta,* 2005, *358*, 4201-4207.

[90] Sakthivel, A. and Zhao, J.; Raudaschl-Sieber, G.; Kuhn, F. E. *Journal of Organometallic Chemistry,* 2005, *690*, 5105-5112.

[91] Yu, K. Q. and Jones, C. W. *Organometallics,* 2003, *22*, 2571-2580.

[92] Duchateau, R.; van Meerendonkt, W. J.; Huijser, S.; Staal, B. B. P.; van Schilt, M. A.; Gerritsen, G.; Meetsma, A.; Koning, C. E.; Kemmere, M. F.; and Keurentjes, J. T. F. *Organometallics,* 2007, *26*, 4204-4211.

[93] McKittrick, M. W. and Jones, C. W. *Journal of the American Chemical Society,* 2004, *126*, 3052-3053.

In: Heterogeneous Catalysis Research Progress
Editor: Mathias B. Gunther, pp. 111-143
ISBN 978-1-60456-979-7

Chapter 3

COBALT(III)-SUPPORTED CHEMICALLY MODIFIED MESOPOROUS SILICAS AS HETEROGENEOUS OXIDATION CATALYSTS

Birinchi K. Das*, Rajesh Chakrabarty and Purabi Sarmah
Department of Chemistry, Gauhati University, Guwahati 781 014, India

ABSTRACT

Mesoporous silica containing -$(CH_2)_2CN$ groups on the surface has been prepared from tetraethylorthosilicate (TEOS) and (2-cyanoethyl)triethoxysilane (CTES) in aqueous ethanol in the presence of n-dodecylamine as the templating agent. Unlike in the cases of zeolitic materials or other porous variants of silica such as MCM-41, no high temperature calcinations are required for removing the templating molecules because it is possible to obtain the functionalized silica material of high porosity by either washing at an elevated temperature or soxhlet extraction. The cyanopropyl groups on the chemically modified silica (CMS) are converted to -$(CH_2)_2CO_2H$ groups via acid hydrolysis. The resultant materials, CMS-$(CH_2)_2CN$ and CMS-$(CH_2)_2CO_2H$, are found to be highly porous displaying very high BET surface areas of 1400 and 1071 m^2/g respectively. The tetranuclear cobalt(III) oxo clusters of the type $Co_4(\mu_3\text{-}O)_4(\mu\text{-}O_2CR)_4L_4$ with R = CH_3 or Ph and L = pyridine or a substituted pyridine have been chemically immobilized on CMS-$(CH_2)_2CO_2H$. N_2 adsorption data on CMS-$(CH_2)_2CO_2H$ and the supported reagent obtained after immobilization of the cubane-like clusters of cobalt(III) indicate retention of high surface area and the porous nature of the organosilica support. Most pores in the supported reagent are found to have diameters in the 3-6 nm range. More importantly, both the silica support and also the immobilized cobalt(III) complexes are hydrolytically stable. The supported materials, which have been characterized by various methods including spectroscopy and electron microscopy, are suitable as heterogeneous oxidation catalysts for substrates such as alkylaromatic hydrocarbons, α-pinene and alcohols using air, oxygen and TBHP as oxidants. The high reaction yields, improved selectivities, and possibilities for reusing the catalysts suggest that the cobalt(III) based catalysts developed by us may be seen as a promising class of supported reagents.

* Corresponding author, E-mail: das_bk@rediffmail.com

1. INTRODUCTION

Porous materials are important as heterogeneous catalysts. Mesopores in catalytic and other materials are normally considered to have diameters in the 2–50 nm range. Various forms of silica display mesopores and thus these catalysts and catalyst support materials continue to be the subject of intense activity in the materials community [1]. The discovery of M41S class of mesoporous silicas, of which MCM-41 is a member, by Beck and co-workers [2] in 1992, and subsequent preparation of a wide variety of mesoporous silicas including hexagonal mesoporous silica (HMS), which has much less internal order, has proven to be an important development in the area of catalysis involving supported catalysts [3,4]. Since these forms of silica are highly porous and also because their surface modification is rather facile, they may be conveniently exploited as supports for transition metal based catalysts to prepare useful heterogeneous catalysts. In the present context, heterogeneous catalysts may be viewed as supported reagents, in which the activity of one chemical entity is amplified (or moderated) through dispersal on another non-interacting material. Supported reagents are in use since the 19th century; nonetheless, it is also an important area of current research directed toward organic transformations [1]. Through our interests in synthesizing novel transition metal complexes for catalytic oxidation, we have recently made use of a few supported reagents in which cobalt(III)-oxo clusters have been immobilized on an HMS-type silica having carboxyl functionalities as surface modifiers [5]. For want of a more descriptive term, we call these materials 'chemically modified silica' (CMS) which we are not inclined to prefix with the term *hexagonal* because of its symmetry connotations.

In this chapter we shall focus on the synthesis and adsorption characteristics of a CMS, prepared by a co-condensation or sol-gel route following the S^oI^o (S^o, a neutral amine; I^o, a neutral inorganic precursor) pathway [3,6]. Immobilization of some cobalt(III) oxo clusters on CMS support, characterization of the resultant supported materials, and the use of these Co(III)-CMS materials in catalytic oxidation under environmentally friendly conditions are also described. Related results available in the published literature are also included at appropriate places with a view to broadening the scope of our discussion.

2. HYBRID INORGANIC-ORGANIC MESOPOROUS SILICATES

Hybrid inorganic-organic mesoporous silicates with uniform channel structures constitute an important group of catalytic materials. These materials are produced by attaching organic functionalities on the surface of mesoporous silica. Attachment of suitable functional groups may enable anchoring of active homogenous catalysts, including transition metal complexes, on solid supports. This indeed has proven to be an ideal tool in achieving better and more eco-friendly catalysts with added advantages of ease of separation, low toxicity as well as simple work-up and recycle procedures. Among strategies employed for immobilization of metal complexes ion exchange, physisorption, covalent binding between metal and ligands, entrapment of complexes in porous matrices, and incorporation of metals into lattices are the most important [7]. As will be seen later, we have made use of one of the three primary methods of heterogenization of metal complexes on solid supports [8], namely the direct reaction of the complex catalyst or catalyst precursor with the solid support. Owing to the inert nature of silica surface, mesoporous materials have suffered from lack of active sites, which are necessary for practical applications, especially in catalysis. However, availability active groups in the form of immobilized metal complexes can buoy up the prospects considerably [9-11].

The modification of silicas and related materials by attachment of organic functionalities onto their surfaces is an important area of research in heterogeneous catalysis directed toward achieving green chemistry goals. Some of the major aims of 'Green Chemistry' are: (i) to increase process selectivity, (ii) to maximize the utilization of starting materials (aiming for 100% atom efficiency), (iii) to replace stoichiometric reagents with catalysts and (iv) to facilitate easy separation of the final mixture including the efficient recovery of the catalyst [12]. The use of efficient solid catalysts may thus be highly relevant in trying to achieve one or more of these goals [13]. In view of the greater stability of inorganic solids, catalysts based on high surface area inorganic support materials, e.g., MCM-41, HMS, SBA-15, etc. should have greater thermal stability and primarily due to this reason these catalytic reagents have attracted a lot of interest as solid catalysts and reagents in liquid phase organic reactions calling for activation support by way of the use of elevated reaction temperature and/or catalyst reusability [14]. It is important to be able to achieve high conversion and selectivity of organic transformations via liquid phase reactions under atmospheric conditions in view of green chemistry benefits accruable through the use of such processes [15].

Surface functionalization of mesoporous silica has been widely practiced in recent years for generating active sites necessary for heterogeneous catalytic activity. Numerous strategies have been described for immobilizing inorganic and/or organic active species on or within the silica matrices [16]. Functionalization of mesoporous silica surface via tethering of an active group generally can be done in two different ways: through the so-called "post-synthesis" method, which involves grafting an organotrialkoxysilane onto the pore surfaces after mesoporous material synthesis, and through the "direct synthesis" method, in which the functional groups are introduced during the synthesis of the mesoporous material. The pore functionalization is achieved through the co-condensation reaction occurring between the organotrialkoxysilane and the silica source in the presence of a structure-directing agent [17]. The direct synthesis method is considered more predictable and hence valuable, because it can avoid several shortcomings of the post-grafting method, such as reduction in pore size, pore blocking at the aperture, and difficulties in controlling the loadings as well as the distribution of active sites [18]. The resulting hybrid inorganic-organic mesoporous silicates may further be functionalized through the chemical immobilization of metal-organic or organometallic compounds via ligation of organic groups present on the surface as mentioned above and to be discussed in detail below.

3. Silica-Supported Cobalt(III) Catalysts

Cobalt salts and their complexes have been widely used as homogeneous oxidation catalysts in organic syntheses as well as in the chemical industry [19]. Various complexes of cobalt are useful as catalysts for oxidative organic transformations [20-25]. In view of the ability of cobalt to cycle between the commonly encountered II and III oxidation states, complexes of cobalt find application as catalysts in the oxidation of a variety of substrates

[19]. However, because of the kinetic lability of cobalt(II), heterogenized catalysts based on Co^{2+} are susceptible to metal leaching during liquid phase reactions and thus repeated use of such catalysts is not practical from a chemical point of view. In order to avoid this problem, use of catalysts based on cobalt(III), which is substitutionally inert, may be expected to show more attractive catalytic properties for the same reactions. As expected, substitutionally inert cobalt(III) complexes have been shown to be catalytically very active, and hence attractive, for alkylaromatic oxidation [26]. Also, as we shall see later, a series of tetrameric cobalt(III) complexes capable of cycling oxidation states between III and IV has also been found to be effective as catalysts for the oxidation of alkyl aromatics, alcohols and alkenes [5,15,27].

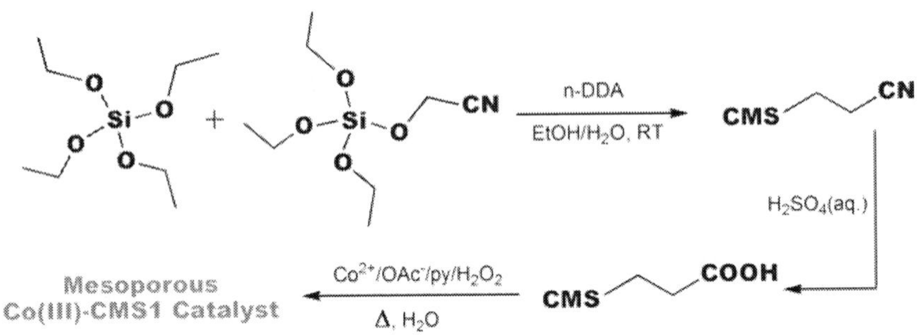

Figure 1. Preparation of Co(III)-CMS by the sol-gel route.

An olive green catalyst containing an as yet unknown Co(III) complex, immobilized on a chemically modified silica (CMS), was proven to be an active catalyst for the selective oxidation of alkylaromatics using air as the source of oxygen under solvent free conditions [26]. The CMS support having carboxyl groups as the surface functionality was prepared by the method originally developed by Macquarrie [28]. This novel cobalt(III)-supported reagent was prepared (Figure 1) by adding pyridine slowly to a preformed mixture of Co(II) salt and sodium acetate along with the CMS particles dispersed uniformly in water. Complexation started only after the addition of pyridine as indicated by the purple-coloured CMS particles. A transformation in the oxidation state of the metal complex from Co(II) to Co(III) was achieved by treating the supported Co(II) complex with H_2O_2 . This olive green supported reagent, called Co(III)-CMS1 in this chapter, was found to retain a fairly high surface area (528 m^2/g) even after the loading of the metal complex. The supported cobalt(III) reagent was tested as a catalyst in the aerobic oxidation of neat ethylbenzene, chlorotoluene and toluene as representative alkylbenzenes and the results are summarized in Table 1.

The greatest advantage of the above catalytic system was the elimination of the unwelcome induction period which otherwise occurs commonly for cobalt(II)-based oxidation catalysts. This has been possible because through direct use of a cobalt(III)-based heterogeneous system it has been possible to eliminate the time required to change Co(II) to Co(III). The isolated yield of acetophenone was found to be 70% at 94% selectivity during aerobic oxidation of ethylbenzene under atmospheric liquid phase conditions.

Table 1. Catalytic oxidations of alkylaromatics using Co(III)-CMS1 in neat substrate at atmospheric pressure for 22 h [26]

Substrate	*T*/°C	Product (isolated yield, %)
Ethylbenzene	130	Acetophenone (70)
4-Chlorotoluene	130	4-Chlorotoluic acid (25)
Toluene	100	Benzoic acid (6)

Sujandi *et al.* [29] in 2006 reported a Co(III) system immobilized on SBA-15 in the form of a Co(III)-cyclam complex, which was covalently tethered onto the SBA-15 surface via the propyl group (Figure 2). Cyclam was chosen because it is a well-known tetraazamacrocycle that can form and stabil˙ e various high-valent transition metal complexes [30]. Thus, the use of cyclam as the macrocyclic ligand to chemically bind Co(III) apparently provides a stable environment from which metal leaching becomes less likely. This catalyst was evaluated for the oxidation of cycloolefins, such as cyclopentene, cyclohexene, and cycloheptene. It an affective catalyst in cycloolefin oxidation with hydrogen peroxide as the terminal oxidant to give allylic hydroperoxides as major products at selectivities generally exceeding 50% after 12 h of reaction. Other products forming in the reactions included the corresponding epoxides, allylic alcohols, and enones. It was not possible to reuse the catalyst for a second. It was surmised that ligand decomposition, which in turn would cause catalyst degradation, was primarily responsible for this catalyst deactivation.

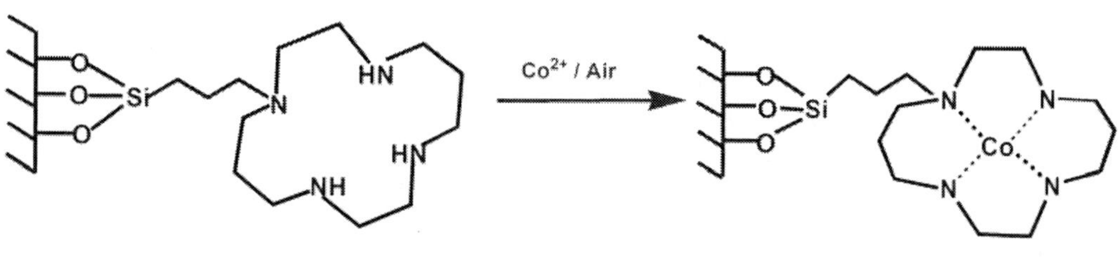

Figure 2. Immobilization of Co(III) as its cyclam complex [29].

In contrast to the above heterogeneous catalyst based on a Co(III)cyclam complex, in the catalyst reported earlier by Das & Clark [26], pyridine which was used as an ancillary ligand to stabilize the cobalt(III) complex in the immobilized state was found to escape from reaction mixtures heated between 110-130°C. However, no metal leaching was observed. Catalysis reuse was possible for this catalyst after catalyst regeneration achieved by addition of pyridine to the substrate-used catalyst reaction systems.

It was also reported by Pruβ *et al.* [31] that *in situ* formed cobalt(III) complexes of pyridine-4-ylmethyl-propyl-amine (PYPA) on preformed organomodified HMS are active as catalysts in the aerobic oxidation of styrene and also 1-decene (Figure 3). Incorporation of PYPA may be achieved by following several routes viz. sol-gel synthesis, post modification of sol-gel AMP-HMS, and grafting. The authors proposed that all materials are able to act as hosts for cobalt-acetato complexes to varying extents and also that metal-ligand interactions occur mainly via the aromatic N atoms present in the system. They also studied immobilization of cobalt complexes derived from various metal salts on PYPA-organomodified HMS. Use of peroxides (TBHP or H_2O_2) as oxidants in the preparation of the catalysts leads to different complexes including cobalt-peroxo or an oxygen bridged dinuclear species on the surface.

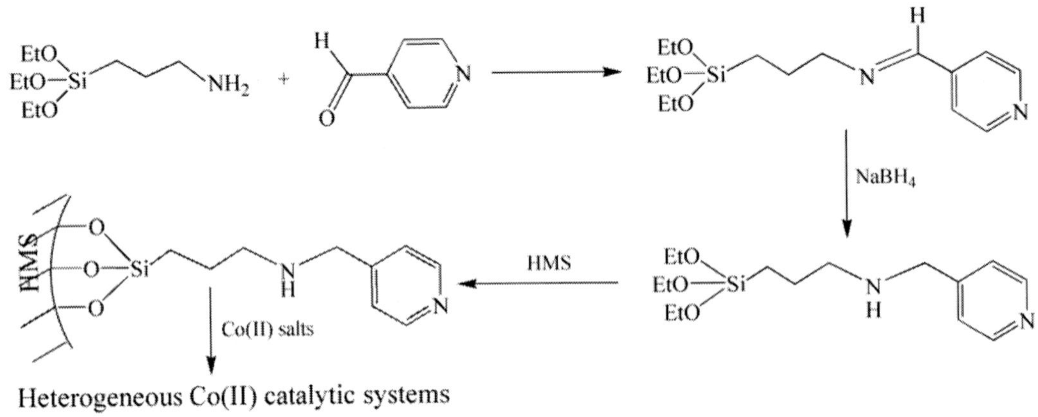

Figure 3. Preparative scheme of the catalyst reported by Pruβ *et al.* [31].

Burri et al. [32] reported yet another Co(III) catalyst heterogenized on mesoporous SBA-15 and used it in the liquid phase oxidation of *p*-xylene under both solvent as well as bromide ion free environment. Organomodified SBA-15 was synthesized by sol-gel technique using 4-triethoxysilylbutyronitrile as the surface modifier by following the method of Stucky *et al.* [33] The solid product containing –CN groups (SBA-15-CN) was converted to SBA-15-COOH on which a cobalt(III) complex containing pyridine and acetate groups as ligands was immobilized by following the method of Das et al. [26]. The resulting Co(III)-SBA-15 material was examined as a catalyst for the oxidation of *p*-xylene under solvent free conditions with molecular oxygen at moderate pressure. This catalyst exhibited comparable activity with the existing industrial homogeneous catalyst ($CoBr_2/Mn(OAc)_2$) at 130 °C and 2 atm pressure in terms of *p*-toluic and terephthalic acid yield. The Co(III)-SBA-15 catalyst however, showed the formation of a substantial amount of terephthalaldehyde, production of which is low with $CoBr_2/Mn(OAc)_2$, under the above conditions. The performance of the Co(III)-SBA-15 catalyst is however inferior compared to that of the Co-Mn catalyst at 190 °C and 20 atm pressure.

4. Preparation of Cobalt(III)-Based Catalysts

The variety of finely divided silica known as hexagonal mesoporous silica (HMS) [6,34] is generally prepared by the copolymerization of a silica precursor such as tetraethylorthosilicate (TEOS), in presence of a suitable template. This acid or base catalyzed reaction follows the sol-gel route and it has been found [28] that during silica preparation by this technique, it is also possible to add an organotriethoxysilane [$RSi(OEt)_3$] that would stamp the silica surface with R, where R could be a group like $-CH_2CH_2CN$, $-CH_2CH_2NH_2$, $-CH_2CH_2Cl$ etc. Such chemical modification allows precise control over the surface properties and pore sizes for specific applications (including use for noncatalytic ones such as chromatographic separation [1] while at the same time stabilizing the structurally disordered chemically modified silica (CMS) materials against hydrolysis. One of the prerequisites for the incorporation of the organic tethering groups into the synthesis gel is that the template should be removable without damaging the incorporated groups during the removal process. Unlike ionic templates which are used in the synthesis MCM-41 or its analogues, which requires a very high calcination temperature for their complete removal, neutral templates can easily be removed by simply washing with boiling ethanol or by solvent extraction and thus suitable for one-step process.

Of the various methods utilized for surface modification we have adopted the sol-gel technique where co-polymerization of tetraethoxysilane (TEOS) and (2-cyanoethyl)triethoxysilane using n-dodecylamine as the neutral templating agent to give an inorganic-organic hybrid network. Subsequent hydrolysis of the –CN group provides the –CO_2H functionalized supported material.

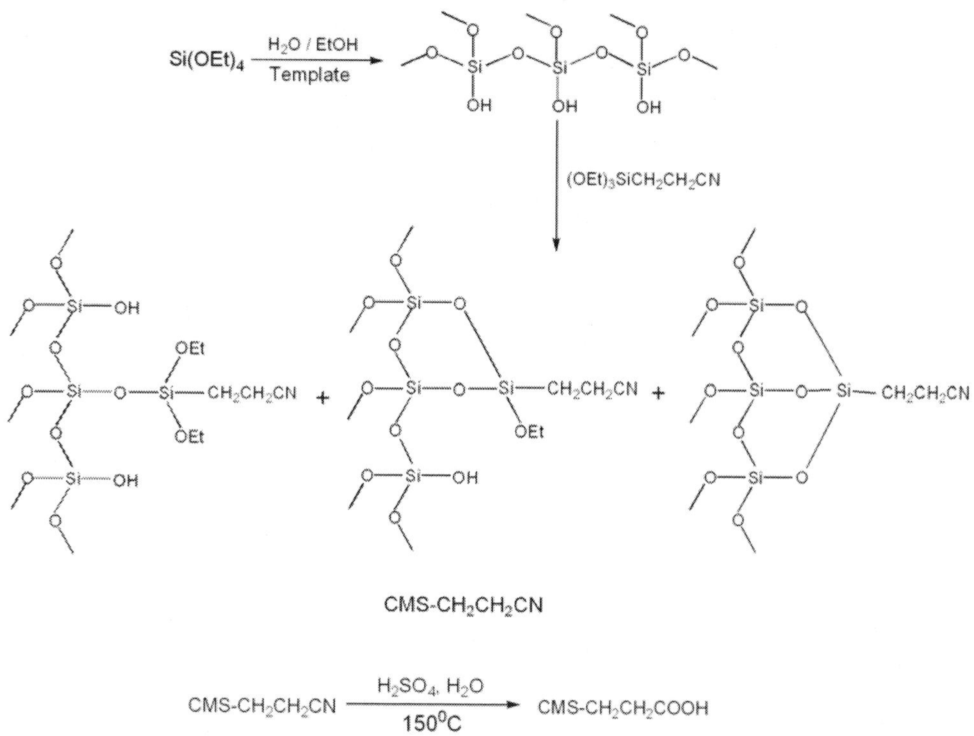

Figure 4. Preparation of CMS-$(CH_2)_2COOH$ by the method of Macquarrie et al. [28].

4.1. Preparation of $CMS\text{-}CH_2CH_2CO_2H$

The carboxylate functionalized CMS support has been synthesized by the method originally developed by Macquarrie *et al.* [28] The CMS-$(CH_2)_2CN$ is formed by employing the sol-gel technique (Figure 4) in aqueous ethanol in the presence of n-dodecylamine (DCA) as the templating agent via co-polymerization reaction involving tetraethoxysilane (TEOS) and (2-cyanoethyl)triethoxysilane (CTES). The templating agent may be removed either by washing the material with hot ethanol or by soxhlet extraction. After removal of the template, the resulting white solid is treated with hot aqueous H_2SO_4 so as to hydrolyze the CN group to CO_2H.

The free-flowing white powders of the organosilicas, CMS-$(CH_2)_2CN$ and CMS-$(CH_2)_2COOH$, are finely divided but readily filterable materials, hence the washed products may be isolated in near-quantitative yield and they are dried subsequently under vacuum at >100 °C. The infrared spectra recorded in the mid-IR region for CMS-$(CH_2)_2CN$ and CMS-$(CH_2)_2COOH$ show bands for the organic functional groups. Elemental analyses give values corresponding to their expected C-H-N content. More importantly, the observed C:N ratios preclude the presence of any significant amount of n-dodecylamine which may be expected to remain as an impurity. The powder X-ray diffraction pattern of CMS-$(CH_2)_2COOH$ has been recorded (Figure 5). The observed pattern does not show well resolved diffraction peaks. The broad peaks seen for this CMS material suggest the lack of long-range order in the structure. Hybrid inorganic-organic materials including the CMS substances under discussion in this chapter are generally thought to be amorphous [35]. In view of the disordered nature of the carboxyl-functionalized silica material prepared by us, we have considered it pertinent to call it CMS instead of HMS in spite of the S^oI^o route followed by us in preparing the material. Elsewhere, these materials have been christened micelle templated silicas (MTS) which are known to display pore sizes up to 10 nm [1].

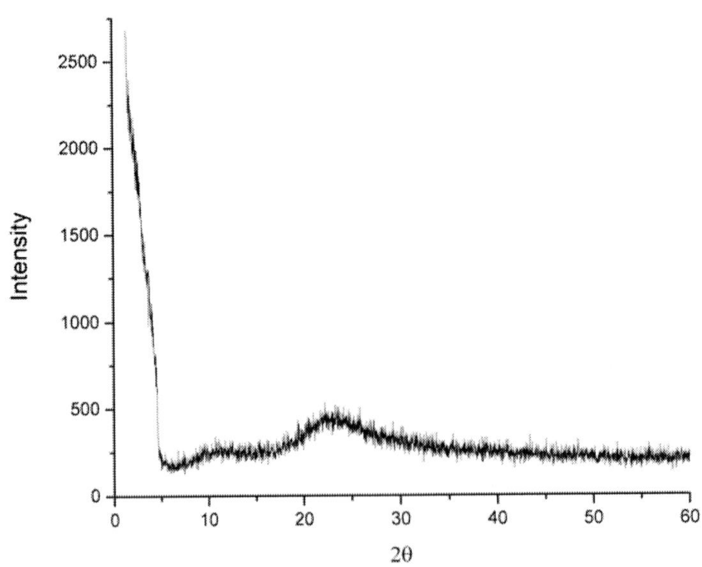

Figure 5. XRD pattern of CMS-$(CH_2)_2COOH$ prepared by the DCA-templated sol-gel technique.

4.2. Synthesis and Characterization of Oxo-Cobalt(III) Carboxylate Clusters

The synthetic strategy for the preparation of tetranuclear Co(III)-oxo complexes of the type $Co_4(\mu_3\text{-}O)_4(\mu\text{-}O_2CR)_4L_4$, with R = CH_3, C_6H_5, or a 4-substituted aryl group and L = pyridine or a 4-substituted pyridine, involves the reaction of $Co(NO_3)_2{\cdot}6H_2O$ with two equivalents of a sodium salt of carboxylic acid, NaO_2CR and one equivalent of the chosen N-donor ligand – pyridine (py), 4-cyanopyridine (4-CNpy), 4-methylpyridine (4-Mepy) or 4-ethylpyridine (4-Etpy) – followed by H_2O_2-oxidation of Co^{2+} at an elevated temperature in commercial grade methanol. Oxidation of the pink solution forming at the beginning of the reaction with excess H_2O_2 leads to an olive green solution from which dark olive green Co(III) complexes of the above type may be isolated. Figure 6 summarizes the synthetic route followed by us to obtain the cobalt(III)-oxo cluster in good to very good yield. Most of the complexes may be isolated in pure form as $Co_4(\mu_3\text{-}O)_4(\mu\text{-}O_2CR)_4L_4$ by following a general procedure involving extraction of the species into CH_2Cl_2 followed by precipitation of the complexes via addition of petroleum ether into the extract.

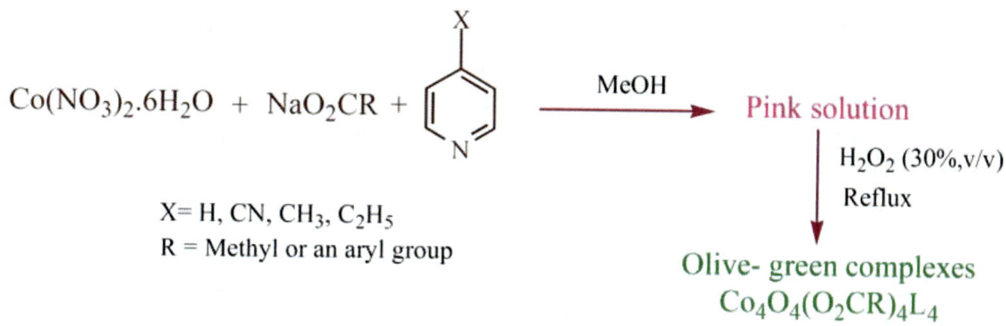

Figure 6. Preparation cubane clusters of Co(III).

The identity and structure of the olive green complexes have been investigated by various spectroscopic and analytical techniques as well as single crystal X-ray diffraction. Since these tetranuclear metal complexes are diamagnetic (octahedral complex having $^1A_{1g}$ ground state for the d^6-Co^{3+}), ^1H NMR spectroscopy provides a useful tool for characterizing the neutral compounds which are soluble to varying degrees in various polar as well as apolar organic media. A few complexes including $Co_4(\mu_3\text{-}O)_4(\mu\text{-}O_2CMe)_4(py)_4$, $Co_4(\mu_3\text{-}O)_4(\mu\text{-}O_2CPh)_4(py)_4$ and $Co_4(\mu_3\text{-}O)_4(\mu\text{-}O_2CMe)_4(4\text{-Mepy})_4$ have been characterized X-ray crystallography. All of them contain cocrystallizing solvent molecules in the crystal lattice and that is why crystals of the complexes are seldom stable in absence of the mother liquor. The molecular structure of complexes of the type $Co_4(\mu_3\text{-}O)_4(\mu\text{-}O_2CR)_4L_4$, which has been established on the basis of structures determined by the single crystal method [20] is shown in Figure 7. By drawing analogy between cubane (C_8H_8) and the cube-like $[Co_4(\mu_3\text{-}O)_4]^{4+}$ core in our complexes, we shall have occasions to call $Co_4O_4(O_2CR)_4L_4$ cubane complexes or even cubane clusters because of the clustering of four Co^{3+} ions by four O^{2-} ions. Presence of the triply bridging oxide ions in these complexes is evidenced by the appearance of a moderate intensity band at ~635 cm^{-1}. In fact the four-band pattern observed at ~760, ~695, ~635 and ~580 cm^{-1} for the cubane complexes is a characteristic feature of the IR spectra of $Co_4O_4(O_2CR)_4L_4$.

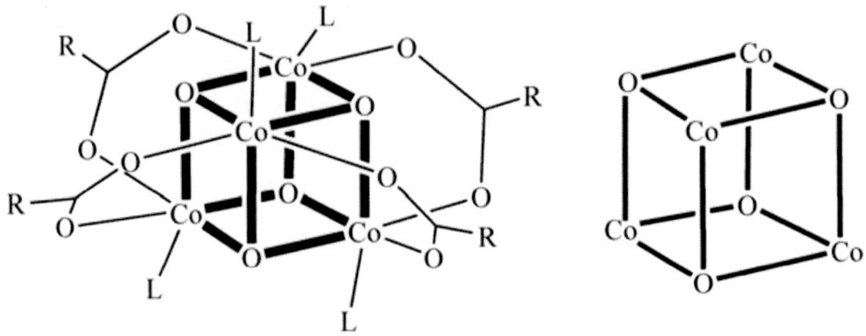

cubane cluster *cubane core*

Figure 7. Structure of cubane clusters of Co(III).

The electrically neutral complexes are generally stable over indefinite periods of time in air. The olive green complexes (those having O_2N-4-$C_6H_4CO_2^-$ as ligands are in general brownish) dissolve in CH_2Cl_2 and hydrocarbon solvents to produce solutions of the same colour. Of particular note is the solubility of $Co_4(\mu_3\text{-}O)_4(\mu\text{-}O_2CMe)_4(py)_4$, which in a rather curious manner dissolves in such solvents of widely differing dielectric constants as water, DMF, DMSO, acetonitrile, ethanol, methanol acetone, diethyl ether and dichloromethane. The cubane clusters in which acetate ligands are present good solubility in water is observed.

The ^{1}H NMR spectra of these complexes are characterized by sharp signals due their diamagnetic nature. The cobalt(III) ions with a low-spin d^6 electron configuration are expected to be diamagnetic, particularly in the presence of pyridine as a ligand [36]. ^{1}H and ^{13}C NMR spectra of $Co_4(\mu_3\text{-}O)_4(\mu\text{-}O_2CMe)_4(py)_4$ are shown in Figure 8 and Figure 9 respectively. In the ^{1}H NMR spectrum recorded in D_2O, all the characteristic resonances due to the coordinated pyridine ligand and $CH_3CO_2^-$ moiety appear at expected positions. A sharp singlet at 2.04 ppm indicates the presence of the acetate ligand, while protons of the pyridine ligand resonate at 8.16, 7.68 and 7.16 ppm in 2:1:2 integration ratio. The simplicity of the spectrum clearly indicates the presence of only one type of environment for pyridine and acetate ligands and is thus consistent with virtual T_d symmetry of the molecule in solution. This indicates the retention of the structure on dissolution. In ^{13}C NMR also peaks due to the carbon atoms of the complex appear at expected positions. It is also possible to obtain expected NMR spectra for the other complexes in the series. The NMR spectra thus provide evidence for the bulk purity of the product as well as the substitutional inertness of the tetrameric Co(III) complex in water.

UV-visible spectra of the tetranuclear cobalt(III)-oxo clusters have been studied in the solution phase. The spectral data for the complex $Co_4(\mu_3\text{-}O)_4(\mu\text{-}O_2CMe)_4(py)_4$ obtained using several common solvents are presented in Table 2. The electronic spectra for this compound in CH_2Cl_2 and H_2O at three different concentrations are shown in Figure 10. As can be seen from the data, the spectra are slightly solvent- as well as concentration-dependent. The spectral bands found in MeOH occur at 617 nm ($\varepsilon = 360\ M^{-1}cm^{-1}$), 330 nm ($\varepsilon = 5900\ M^{-1}cm^{-1}$) and 246 nm ($\varepsilon = 21{,}900\ M^{-1}cm^{-1}$). The lowest energy band at 617 nm can be assigned to the *d-d* transition involving either $^1A_1 \rightarrow {}^1T_1$ or $^1A_1 \rightarrow {}^1T_2$ for the approximately octahedral low-spin d^6-Co(III) centers. As judged from the observed intensities, the other two bands are attributable to ligand to metal charge transfer transitions (LMCT). While the first LMCT band at 330 nm is likely to be due to a transition involving the μ_3-O-Co(III) moiety present in the complex, the latter occurring at 246 nm most probably involves a transition between a molecular orbital from the carboxylato ligands and a vacant cobalt(III) e_g* orbital. The above spectral results are typical of complexes of the type $Co_4(\mu_3\text{-}O)_4(\mu\text{-}O_2CR)_4L_4$ which have been studied in some detail by us. As will be seen later in this chapter, the characteristic spectral signature of the tetranuclear species is quite useful in specifying the nature of the complexes present in the immobilized state over mesoporous silica.

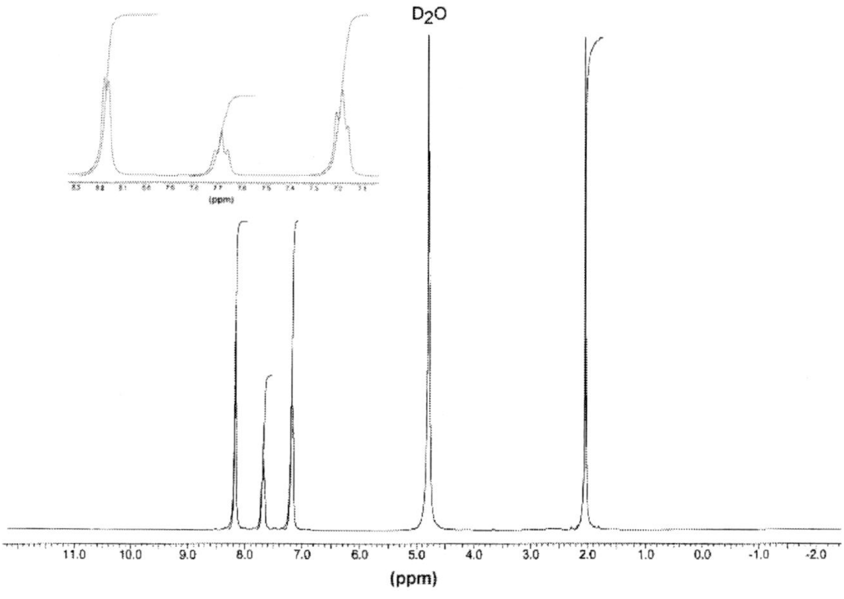

Figure 8. ^{1}H NMR spectrum of $Co_4(\mu_3\text{-}O)_4(\mu\text{-}O_2CCH_3)_4(py)_4$ in D_2O [27c].

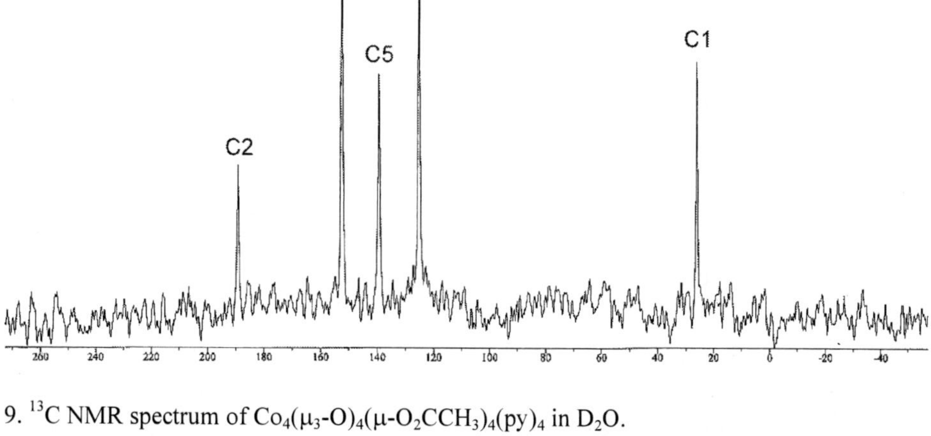

Figure 9. ^{13}C NMR spectrum of $Co_4(\mu_3\text{-}O)_4(\mu\text{-}O_2CCH_3)_4(py)_4$ in D_2O.

Table 2. Electronic spectral data on $Co_4(\mu_3\text{-}O)_4(\mu\text{-}O_2CMe)_4(py)_4$ in various solvents

Solvent	$\square_{max}$, nm (ε, $M^{-1}cm^{-1}$)		
	$^1A_1 \rightarrow {}^1T_1$	LMCT ($\square_3$-oxo-bridge)	LMCT
MeOH	617 (360)	330 (5,900)	246 (21,900)
CH_2Cl_2	638 (sh)	362 (7,428)	227 (26,537)
MeCN	650 (330)	355 (7,149)	–
H_2O	618 (388)	354 (4,742)	257 (42,879)

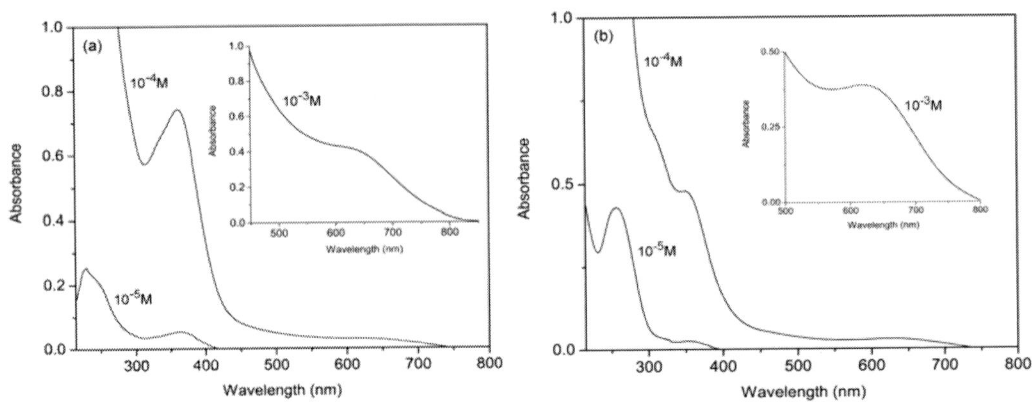

Figure 10. UV-visible spectra of $Co_4(\mu_3\text{-}O)_4(\mu\text{-}O_2CMe)_4(py)_4$ in (a) CH_2Cl_2 and (b) H_2O obtained through measurements on 10^{-3}, 10^{-4} and 10^{-5} M solutions [20].

As noted above, the spectral records depend to some extent on the solvent and concentration used. For example, the above complex shows bands in the visible region in CH_2Cl_2 and MeCN at 638 and 650 nm respectively indicating a significant shift in band maxima. Again, the 362 nm (ε = 7,428 $M^{-1}cm^{-1}$) band in CH_2Cl_2 becomes blue-shifted to 355 nm (ε = 7,149 $M^{-1}cm^{-1}$) in MeCN. The characteristic visible peak due to the d-d transition remains similar in shape, although the band position undergoes considerable shift from the values found in CH_2Cl_2 and MeCN. On the other hand, the visible band at 618 nm (ε = 388 $M^{-1}cm^{-1}$) observed in H_2O occurs almost at the same energy with the band observed in MeOH (617 nm), although the lower energy LMCT bands in these two solvents appear at widely separated wavelengths (354 nm for H_2O and 330 nm for MeOH). From these results, it may be said that the $[Co_4O_4]^{4+}$ core of the complex undergoes deformation to varying degrees in different solvents. At the same time, it is not possible to rule out the possibility of H_2O and MeOH entering into hydrogen bonding interactions with the oxo ligands so as to change the shape of the Co_4O_4 cube further. This could perhaps explain why the d-d band positions in these two solvents appear at nearly the same energies. The highest energy LMCT transition is not observed in MeCN for as yet unknown reasons.

The corresponding absorptions for other complexes also appear at similar positions. For example, in the spectrum of $Co_4(\mu_3\text{-}O)_4(\mu\text{-}O_2CPh)_4(py)_4$ in CH_2Cl_2 also three bands are observed at 634 nm (ε = 490 $M^{-1}cm^{-1}$), 361 nm (ε = 10,700 $M^{-1}cm^{-1}$) and 236 nm (ε = 80,000 $M^{-1}cm^{-1}$).

The redox behaviour of the cubane clusters of cobalt(III) has been studied by cyclic voltammetry. Figure 11 shows the CV recorded for a 10^{-3} M solution of complex $Co_4(\mu_3\text{-}O)_4(\mu\text{-}O_2CMe)_4(py)_4$ in MeCN containing 0.2 M TBAP as the supporting electrolyte (scan rate = 10 mVs^{-1}) and ferrocene as the internal standard using platinum electrode. The electrochemical studies reveal a reversible oxidation at 0.73 V for this particular complex. Other complexes also show similar $E_{½}$ values, some of which are listed in Table 3. However, change of solvent from MeCN to CH_2Cl_2 makes the electrochemical process quasireversible. It is imperative that the tetrameric $[Co_4(\mu_3\text{-}O)_4]^{4+}$ core undergoes a one-electron oxidation to $[Co_4(\mu_3\text{-}O)_4]^{5+}$ to form a mixed valence tetranuclear complex that contains three Co^{III} and one Co^{IV} ions, as had been observed for a related complex. Such a favourable electrochemical redox potential for the cubane complexes makes it an interesting candidate for use as a catalyst for the oxidation of organic substrates.

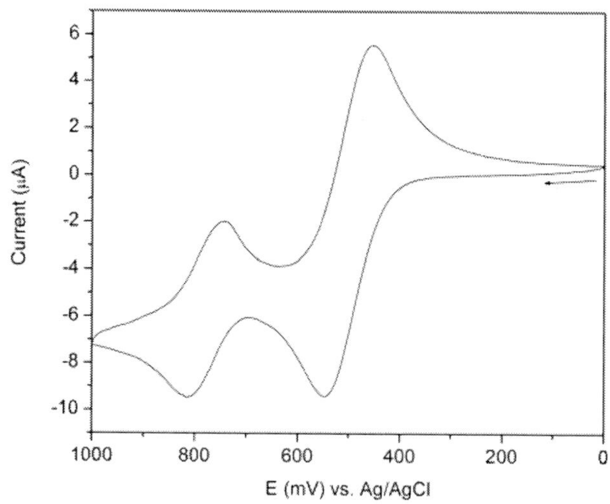

Figure 11. Cyclic voltammogram at 20 mVs^{-1} on Pt electrode (Ag/AgCl reference electrode) of $Co_4(\mu_3\text{-}O)_4(\mu\text{-}O_2CMe)_4(py)_4$ in MeCN containing 0.2M TBAP and ferrocene as an internal standard. $E_{½}$ (Co^{3+}/Co^{4+}) = 0.73 V, ΔE_p = 66 mV[20].

Table 3. CV^a and DPV^b data for a few $Co_4(\mu_3\text{-}O)_4(\mu\text{-}O_2CR)_4(L)_4$ complexes in CH_2Cl_2-0.1M TBAP

Complex	CV data		DPV data (V)
	$E_{½}$ (V)	E_p (mV)	
$Co_4(_3\text{-}O)_4(\text{-}O_2CMe)_4(py)_4$	0.75	170	0.77
$Co_4(_3\text{-}O)_4(\text{-}O_2CMe)_4(4\text{-}Mepy)_4$	0.71	138	0.73
$Co_4(_3\text{-}O)_4(\text{-}O_2CMe)_4(4\text{-}Etpy)_4$	0.72	140	0.74
$Co_4(_3\text{-}O)_4(\text{-}O_2CMe)_4(4\text{-}CNpy)_4$	1.0	128	1.02

[a]scan rate, 20 mVs^{-1}. [b]scan rate, 5 mVs^{-1}. Working electrode, Pt; reference electrode, Ag/AgCl.

The stoichiometry of electron transfer determined from the peak current measurements using ferrocene in equimolar concentration as an internal standard shows that the ratio of observed peak currents for ferrocene and $Co_4(\mu_3\text{-}O)_4(\mu\text{-}O_2CMe)_4(py)_4$ is ~4 (Figure 11). This suggests that only one Co^{III} centre per '$(Co^{III})_4O_4$' cluster undergoes oxidation. Thus the electron transfer process involves a one electron oxidation of the cubane core from $[(Co^{III})_4(\mu_3\text{-}O)_4]^{4+}$ to $[(Co^{III})_3Co^{IV}(\mu_3\text{-}O)_4]^{5+}$ (Figure 12). A literature report [37] on a structurally related complex $[Co_4(\mu_3\text{-}O)_4(\mu\text{-}O_2CMe)_2(bpy)_2]^{2+}$ in MeCN revealed a similar oxidation process where the cubane core $[Co_4(\mu_3\text{-}O)_4]^{4+}$ containing $4Co^{III}$ was oxidized to a $3Co^{III}$, Co^{IV} form.

$$[Co_4^{III}O_4(\mu_3\text{-}O)_4(\mu\text{-}O_2CMe)_4(L)_4] \underset{+e}{\overset{-e}{\rightleftharpoons}} [Co_3^{III}Co^{IV}O_4(\mu_3\text{-}O)_4(\mu\text{-}O_2CMe)_4(L)_4]^+$$

Figure 12. The redox couple involved in the oxidation of cubane complexes.

The electrochemical results described above indicate that unlike in the cases of other cobalt-catalyzed oxidation processes where the Co^{2+}/Co^{3+} redox couple is invariably involved [19b,38], in the present case where cubane clusters of the general formula $Co_4(\mu_3\text{-}O)_4(\mu\text{-}O_2CR)_4(L)_4$ are to be employed as catalysts for the air/O_2 or TBHP oxidation of alkylaromatics, alcohols, etc., we have a catalytic system wherein the oxidation states of cobalt cycle between +3 and +4. The kinetic inertness of Co(III) coupled with the inadequately explored reactivity of Co(IV) thus make the catalysts based on 'Co_4O_4' cubanes quite interesting [36]. We shall now discuss the resulting materials prepared by supporting the cubane-like cobalt(III)-oxo clusters discussed above in this section by following the chemical route in which the carboxylate anion derived from CMS-$CH_2CH_2CO_2H$ binds the in situ or preformed cobalt(III)-oxo tetramers at elevated temperatures.

4.3. Immobilization of Co(III) Species on CMS

Cobalt(II) and chromium(III) complexes were earlier immobilized on porous silica support with a view to preparing supported reagents meant for use as heterogeneous catalysts for oxidation of organic substrates [39]. It has been mentioned above that Co(II) catalyzes the side-chain oxidation of aromatic hydrocarbon compounds. However, such catalytic systems are reported to be affected by induction periods necessitated by the Co(II)→Co(III) oxidation involved in aerobic oxidation processes. Since the catalytic cycle clearly includes Co^{3+}, it was found instructive to try to prepare heterogeneous silica-based catalysts incorporating the Co(III) complexes [26] with a view to preparing heterogeneous oxidation catalysts. For this purpose, the organofunctional group $-CH_2CH_2CO_2H$ present on CMS surface may act as a ligand for Co^{3+} ions which are commonly known to form coordination complexes with the carboxylate anions. To incorporate the cobalt(III) species discussed in the previous section on to the CMS material, we have adopted two different preparative routes.

4.3.1. The In Situ Complex Formation Method

In the in situ complex formation method, the desired complex of Co(III) ions has been allowed to form in presence of CMS-$CH_2CH_2CO_2H$. For this purpose $Co(OAc)_2\cdot 4H_2O$ and 4-cyanopyridine are stirred in water having suspended particles of CMS-$CH_2CH_2CO_2H$. Complexation is indicated by the change in the colour of the solid particles to purple. The resultant mixture is then treated with H_2O_2 after raising the temperature to *ca.* 100 °C. The pink-purple solid particles convert to green particles immediately, but stirring under reflux is continued for 4h to ensure complete oxidation of Co(II). The olive-green residue obtained after filtration is washed thoroughly with water and methanol. The preparative method followed here is analogous with that described in Figure 1. In the present instance $Co(OAc)_2{\cdot}4H_2O$ has been used in place of a mixture of $Co(NO_3)_2{\cdot}6H_2O$ and $Na(O_2CCH_3).3H_2O$.

Apart from Co(III)-CMS1 which was reported earlier [26], two other supported reagents have been prepared by this technique. In the first case, 1 mmol of $Co(OAc)_2{\cdot}4H_2O$ was used for 1 g of the CMS-$CH_2CH_2CO_2H$ support while keeping the Co:4-CNpy ratio at 1.0. In the second instance 2 mmol of $Co(OAc)_2{\cdot}4H_2O$ has been used for 1 g of the support, Co:4-CNpy ratio has been kept at 1.0 again. In both cases 30% H_2O_2 has been used in excess (molar ratio of Co^{2+} and $H_2O_2 \geq 1{:}10$). The dried catalysts obtained as above are called Co(III)-CMS2 and Co(III)-CMS3 respectively for our purposes in this chapter (Figure 13).

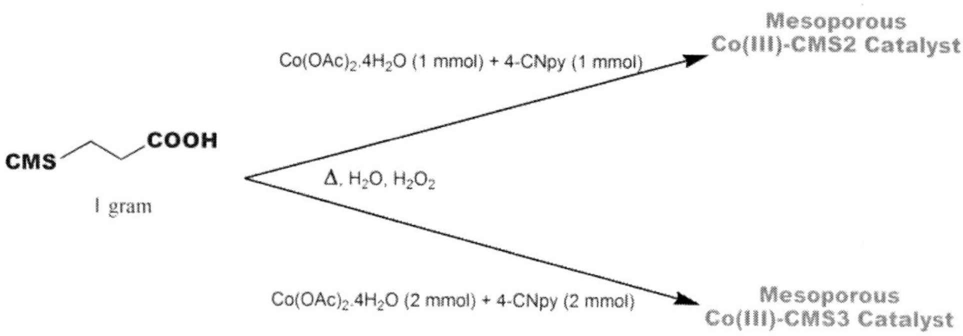

Figure 13. In-situ preparation of Co(III)-CMS1 and Co(III)-CMS2.

4.3.2. The Ligand Exchange Method

In this method of immobilization of cobalt(III) complexes on CMS-$CH_2CH_2CO_2H$ is achieved by carboxylate exchange reaction following reported procedure in the literature [40]. For the purpose of preparing cobalt(III)-based supported reagents from CMS-$CH_2CH_2CO_2H$ we have made use of the series of complexes having the general formula $Co_4(\mu_3\text{-}O)_4(\mu\text{-}O_2CR)_4(L)_4$ which have been described in the previous section. A suspension of CMS-$(CH_2)_2CO_2H$ in a solution of $Co_4(\mu_3\text{-}O)_4(\mu\text{-}O_2CR)_4L_4$ with R = CH_3 or Ph and L = pyridine or 4-cyanopyridine (4-CNpy) is stirred under reflux for 12h. The olive green powders so obtained have been named as given below:

Co(III)-CMS4: $Co_4(\mu_3\text{-}O)_4(\mu\text{-}O_2CMe)_4(py)_4$ supported on CMS-$CH_2CH_2CO_2H$
Co(III)-CMS5: $Co_4(\mu_3\text{-}O)_4(\mu\text{-}O_2CMe)_4(4\text{-}CNpy)_4$ supported on CMS-$CH_2CH_2CO_2H$
Co(III)-CMS6: $Co_4(\mu_3\text{-}O)_4(\mu\text{-}O_2CPh)_4(4\text{-}CNpy)_4$ supported on CMS-$CH_2CH_2CO_2H$.

While the reagents Co(III)-CMS4 and Co(III)-CMS5 were prepared by using water as the solvent, Co(III)-CMS6 was prepared in MeCN (Figure 14). The change in solvent was necessitated by the insolubility of $Co_4(\mu_3\text{-}O)_4(\mu\text{-}O_2CPh)_4(4\text{-}CNpy)_4$ in water. Prior to drying the catalysts for 12h at 105 °C, the first two catalysts were washed with water, methanol and acetone; on the other hand the last catalyst was washed with water, acetone and acetonitrile.

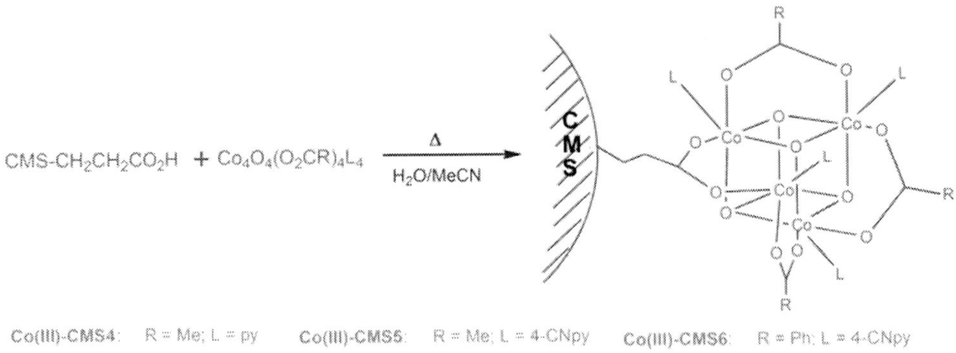

Figure 14. Immobilization cubane complexes by ligand exchange method on CMS.

Although, we are not certain regarding the true nature of the supported complex species, using spectroscopic results to be discussed below, we propose the following scheme to show the immobilization of the above complexes on CMS-$CH_2CH_2CO_2H$. It is believed that the formation of the olive green catalysts has been possible due to the exchange of one of the carboxylato ligands in the cubane clusters with the carboxyl functional group present on the surface of CMS-$CH_2CH_2CO_2H$. The powder X-ray diffraction pattern (Figure 15) recorded for Co(III)-CMS4 indicates that no crystalline material originating in the complex incorporated into the support. Close similarity of the powder diffraction trace with the corresponding pattern for CMS-$CH_2CH_2CO_2H$ suggests the retention of the disordered nature of the support.

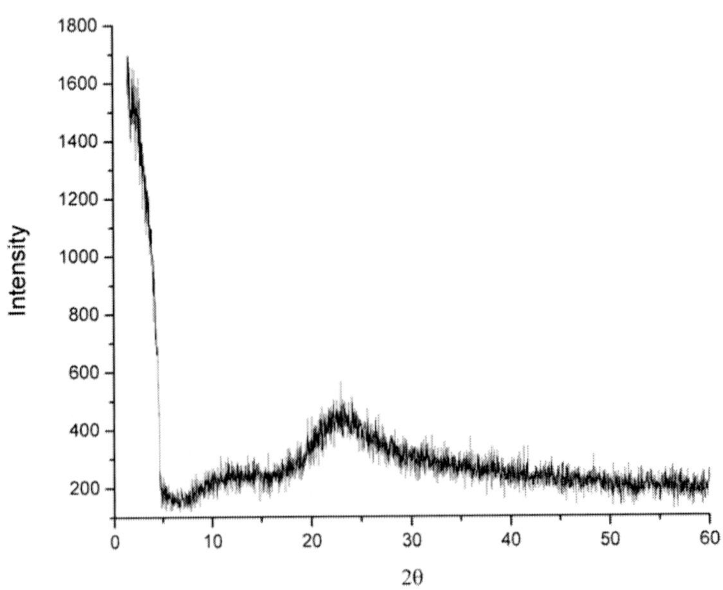

Figure 15. XRD pattern of Co(III)-CMS4.

5. Characterization of Catalysts

5.1. N_2 Adsorption Studies

Although no direct correlation between the surface area or the porosity and the catalytic activity of the supported material can be made, N_2 adsorption experiments provide valuable information about the structure of the support materials. Isotherms of the solid material CMS-$CH_2CH_2CO_2H$ and the supported material obtained after immobilization of cubane complexes are of similar shape indicating that the porous structure has been retained after the immobilization. The isotherms for CMS-$CH_2CH_2CO_2H$ (and also CMS-CH_2CH_2CN, from which the former was obtained) and the supported reagents prepared by us are of Type IV according to the scheme of classification recommended by IUPAC [41]. These isotherms are generally amenable to BET analysis. The characteristic hysteresis loop for Type IV can be observed in all the isotherms. The adsorption isotherms of CMS-$CH_2CH_2CO_2H$, and Co(III)-CMS4-6 are presented in Figure 16 and Figure 17. Table 4 lists the BET surface areas and pore volumes of the support materials as well as the heterogeneous catalysts obtained via immobilization of cobalt(III) complexes. The cobalt loadings measured by atomic absorption spectroscopy are also presented in the extreme right-hand column of the table.

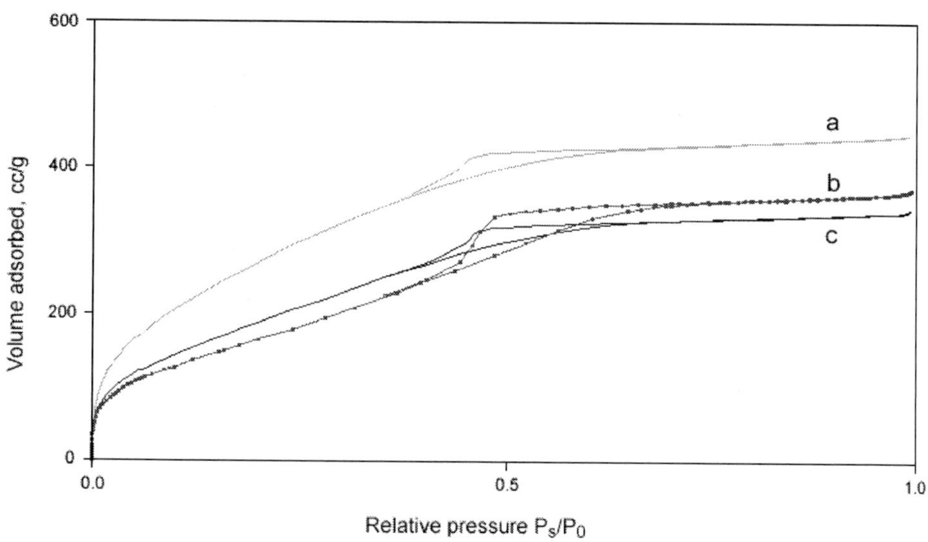

Figure 16. N_2 adsorption isotherms for (a) CMS-CH_2CH_2COOH and (b) Co(III)-CMS5 and (c) Co(III)-CMS6.

The BET surface area of *ca.* 1400 m^2/g for CMS- CH_2CH_2CN decreases to 1071 m^2/g for CMS- $CH_2CH_2CO_2H$. The corresponding pore volumes are 0.75 and 0.69 cm^3/g, respectively at P_s/P_o = 0.9814 for adsorption. While only 60% of the pores in CMS-CH_2CH_2CN have diameters under 6 nm, 84.4% pores in CMS- $CH_2CH_2CO_2H$ have diameters below 6 nm. The pore size distribution is quite narrow with majority of the pores having diameters between 3.48 and 3.82 nm. Clearly, upon conversion of the –CN group to the –COOH there is a drop in both surface area and average pore size. In CMS-$CH_2CH_2CO_2H$, about 74% of desorption BJH pore volume is accounted for by pores having diameters in the range of 3.2–3.8 nm.

Pores of such diameter are likely to allow passage of bulky metal complexes such as $Co_4O_4(O_2CPh)_4(4\text{-CNpy})_4$ that is estimated to have the largest molecular dimension of *ca.* 15 Å (= 1.5 nm). Thus, cubane-like cluster complexes may be expected to bind on the inner walls of the pores of CMS-$CH_2CH_2CO_2H$. The BET surface areas and pore volumes for the supported materials decrease considerably following immobilization of the tetrameric metal complexes.

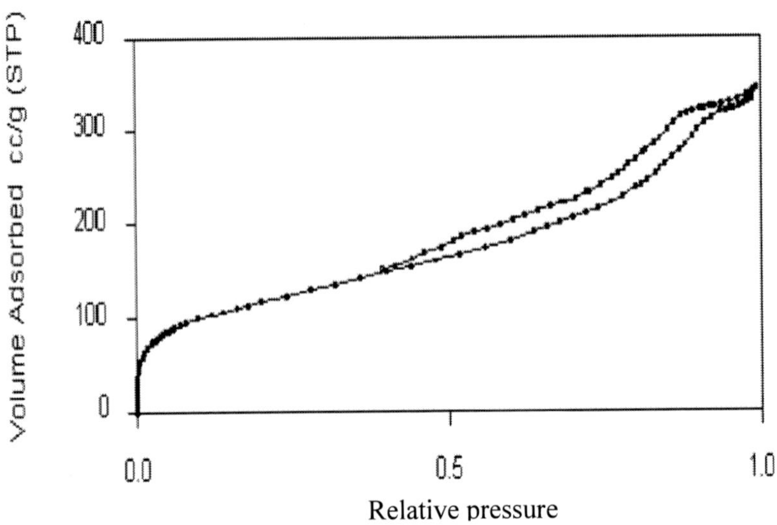

Figure 17. N_2 adsorption isotherm for Co(III)-CMS4 [27c].

Table 4. Porosimetry and cobalt-loading data of CMS-CH_2CH_2CN, CMS-$CH_2CH_2CO_2H$ and the supported catalysts

	BET surface area (m^2/g)	Total pore volume (ml/g)	Cobalt loading (mmol/g)
CMS-CH_2CH_2CN	1400	0.75	-----
CMS-$CH_2CH_2CO_2H$	1071	0.69	-----
Co(III)-CMS1	528	0.31	0.68
Co(III)-CMS2	403	0.21	0.83
Co(III)-CMS3	622	0.60	1.39
Co(III)-CMS4	396	0.48	0.87
Co(III)-CMS5	620	0.57	0.99
Co(III)-CMS6	714	0.53	1.08

From the data presented in Table 4 it may be concluded that the porous nature of the chemically modified silica remains more or less the same after immobilization of the cobalt(III) complexes. In addition, there is a decrease in the surface area of the support following the incorporation of the metal complexes. The AAS data on Co(III)-CMS2 and Co(III)-CMS3 appear to suggest that the extent of cobalt loading is dependent upon the initial amount of cobalt used. The cobalt loadings obtained for the catalysts prepared by H_2O_2 oxidation of CMS suspensions in 1 and 2 mmol cobalt(II) solutions (in presence of 1 and 2 mmol of 4-CNpy) respectively are 0.83 and 1.39 mmol/g. Such high loading of cobalt is not achievable from solutions of cobalt(II) salts. It is thus likely that higher amount of loading achieved upon addition of H_2O_2 can only be caused by the changes in the composition of the complex. A possible explanation that may be given by stating that the formation of multi-nuclear, oxo-bridged complexes, which are likely to remain anchored to the silica surface by interaction with one or more $-CO_2H$ groups present on the silica surface, has facilitated the high loadings of cobalt found the supported reagents prepared by us. This assumption is also corroborated by the results of the DRIFT experiments on the materials (*vide infra*).

5.2. Vibrational Spectroscopy

Diffuse reflectance infrared spectroscopy (DRIFTS) provides useful information about the degree of incorporation and nature of the immobilized complex in supported reagents. We have studied the supports as well as the supported reagents containing the immobilized complexes by this form of infrared spectroscopy where samples in the solid state are examined to determine the nature species under examination. We discuss here the infrared spectral evidence of two supported materials – Co(III)-CMS3, prepared by the in-situ method and Co(III)-CMS4 which has been prepared by the ligand substitution route.

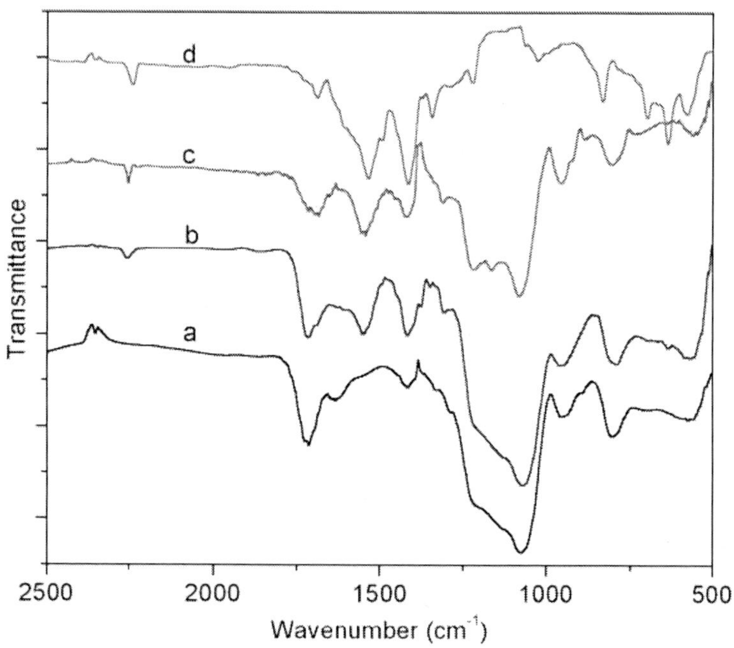

Figure 18. DRIFT spectra of (a) HMS-$CH_2CH_2CO_2H$; (b) Co(III)-CMS3; (c) Co(III)-CMS5; (d) $[Co_4(\mu_3\text{-}O)_4(\mu\text{-}O_2CCH_3)_4(4\text{-CNpy})_4]$ [15].

Figure 18 shows the DRIFTS spectra of the support material (i.e., CMS-$CH_2CH_2CO_2H$) and the immobilized catalyst Co(III)-CMS3 along with the spectrum of the tetrameric cobalt(III) complex $[Co_4(\mu_3\text{-}O)_4(\mu\text{-}O_2CCH_3)_4(4\text{-CNpy})_4]$, and Co(III)-CMS5, i.e., the supported reagent prepared by immobilizing this complex on CMS-$CH_2CH_2CO_2H$ via ligand exchange. Appearance of a band for the nitrile group (2255 cm^{-1}) and of two strong bands at 1534 and 1415 cm^{-1} corresponding to the ν_{asym} and ν_{sym} vibrations of bridging -COO^- groups respectively is suggestive of the presence of a cobalt(III) complex closely related to [$Co_4(\mu_3$-$O)_4(\mu$-$O_2CCH_3)_4$(4-CNpy)$_4$] in Co(III)-CMS3. The striking similarity of the spectra of the Co(III)-CMS3 and Co(III)-CMS5 (Figure 18c) leaves no doubt that in Co(III)-CMS3 also we indeed have anchored a tetrameric cobalt(III) complex having a $[Co_4(\mu\text{-}O)_4]^{4+}$ core as in the case of the latter. Immobilization of an analogous oligonuclear oxo-bridged complex was indicated earlier [26] for the cobalt(III) based heterogeneous catalyst, Co(III)-CMS1.

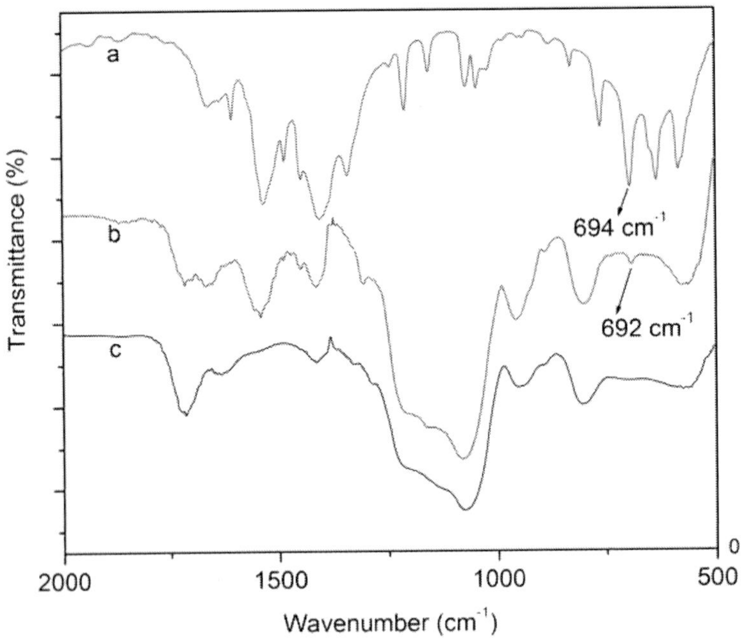

Figure 19. Infrared spectra of (a) $Co_4(\mu_3$-$O)_4(\mu$-$O_2CMe)_4(py)_4$; (b) Co(III)-CMS4; (c) CMS-$CH_2CH_2CO_2H$ [27c].

In order to see how vibrational spectra may be useful in characterizing the heterogeneous systems prepared by immobilizing the cubane complexes on CMS-$CH_2CH_2CO_2H$ by the ligand exchange method we present some more results in Figure 19, in which the infrared spectra of $Co_4(\mu_3$-$O)_4(\mu$-$O_2CMe)_4(py)_4$, CMS-$(CH_2)_2CO_2H$ and the catalyst, Co(III)-CMS4 in the 500-2000 cm^{-1} range are shown. In the KBr phase IR spectrum of the pure complex, it is possible to identify bands due to the bridging acetate ions as well as for the pyridine ligands. The carboxyl stretching vibrations ν_{asym}(COO) and ν_{sym}(COO) for the acetate groups appear as strong bands at 1538 cm^{-1} and 1410 cm^{-1} respectively. The separation of the asymmetric and symmetric stretching vibrations (ν_{asym}-ν_{sym} = $\Delta\nu$ = 128 cm^{-1}) of the carboxyl group is consistent with this ligand bonding in a *syn-syn* bridging mode. The breathing vibration for the $(\mu_3$-$O)Co_3$ groups present in [$Co_4(\mu_3$-$O)_4(\mu$-$O_2CCH_3)_4(py)_4$] shows up as a band at 694 cm^{-1}. A moderately strong band occurring at ~690 cm^{-1} has been found to be characteristic of cubane-like cobalt(III) complexes of general formula [$Co_4(\mu_3$-$O)_4(\mu$-$O_2CR)_4(L)_4$] [5,20]. In fact an infrared absorption at such an energy value was also noted earlier [42] by other authors while reporting catalytic properties of trinuclear complexes of cobalt(III) which also contain the (μ_3-O)Co_3 core that has a flat geometry as compared to the pyramidal one observed for the cubane clusters. Appearance of this vibration as a weak band at 692 cm^{-1} in the DRIFT spectrum of the supported reagent under study suggests the existence of a tetrameric cobalt(III) complex closely related to the cubane complex $Co_4(\mu_3\text{-}O)_4(\mu\text{-}O_2CMe)_4(py)_4$ in Co(III)-CMS4 as well. The ν_{asym} and ν_{sym} vibrations of bridging carboxyl groups appear at 1540 and 1417 cm^{-1} respectively in the spectrum of Co(III)-CMS4. The conclusion to be drawn from the infrared spectral evidence is that we have successfully anchored the cubane-like metal complex by replacing one of its μ-acetato ligands with the carboxylate anions derived from CMS-$CH_2CH_2CO_2H$ as illustrated in Figure 19.

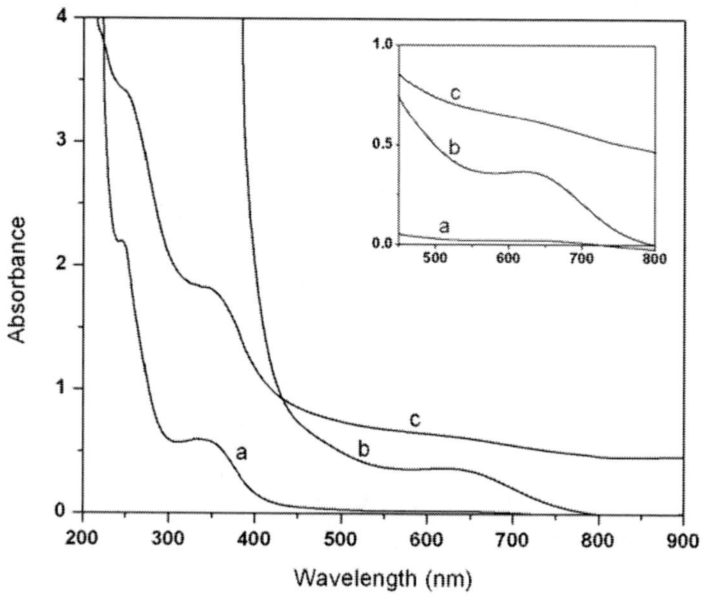

Figure 20. UV-visible spectra of complex $Co_4(\mu_3\text{-}O)_4(\mu\text{-}O_2CMe)_4(py)_4$ as (a) 10^{-4} M and (b) 10^{-3}M solutions in MeOH and (c) Co(III)-CMS4 in nujol mull [27c].

5.3. Electronic Spectroscopy

The electronic spectra of the cubane-like clusters have been discussed earlier in this chapter. The spectra are characterized by two LMCT bands in the UV region and a *d-d* band of moderate intensity around 620 nm. Figure 20 shows the solution-phase UV-visible spectra recorded for $Co_4(\mu_3\text{-}O)_4(\mu\text{-}O_2CMe)_4(py)_4$ at two different concentrations as well as the nujol mull spectrum of Co(III)-CMS4. As can be seen in the figure, all three spectral bands of the cobalt complex are observed in the spectrum recorded for the supported catalyst as well. The LMCT transitions for Co(III)-CMS4 are observed at 251 and 354 nm while the band due to the *d-d* transition appears at 620 nm. These positions and their relative intensities are comparable with the bands for $Co_4(\mu_3\text{-}O)_4(\mu\text{-}O_2CMe)_4(py)_4$, which occur at 257, 354 and 618 nm respectively in aqueous medium. This close spectral similarity thus clear suggests that the nature of the complex present in Co(III)-CMS4 is nearly the same as the tetrameric complex

$Co_4(\mu_3\text{-}O)_4(\mu\text{-}O_2CMe)_4(py)_4$. In the present as well as the other cases where supported reagents have been obtained by immobilizing cubane clusters on CMS-$CH_2CH_2CO_2H$ the colour of the finely ground complexes as well as the supported reagents are identical. This fact strongly suggests that the nature of the cubane complexes remains intact after immobilization. However, more studies involving UV-visible spectra recorded in the diffuse reflectance mode may be useful in elucidating the identity of complexes present in the immobilized state.

5.4. Scanning Electron Microscopy

SEM pictures recorded for Co(III)-CMS3 (Figure 21) show that the average particle diameters are ~1 μm or smaller, and that the particles are spherical in shape.

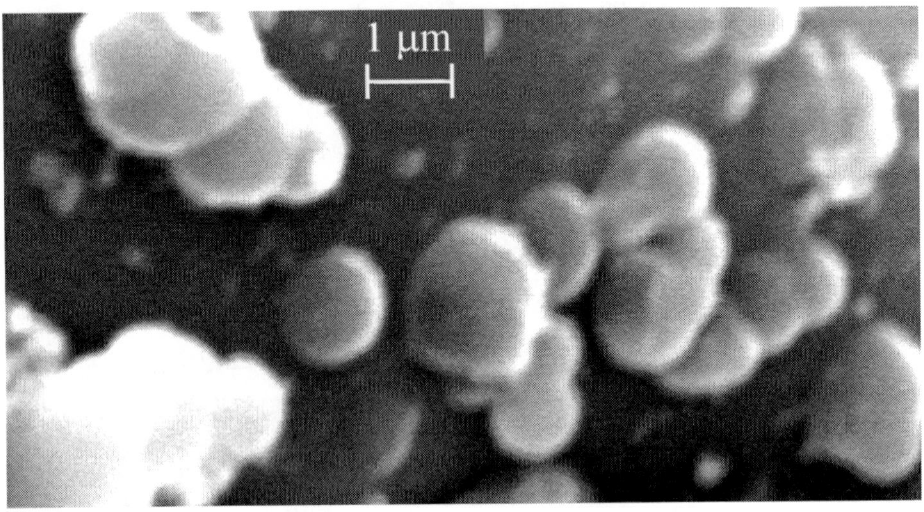

Figure 21. Scanning electron micrograph of Co(III)-CMS3 particles [15].

The SEM pictures recorded for Co(III)-CMS4 (Figure 22) also show that the particles are once again spherical in shape. It may also be said that the distribution of particle size is narrow. . EDAX measurements of the catalyst gives a Co:Si ratio of ~1:8 which is consistent with the cobalt loading determined by AAS. This study also shows that the cobalt complex is uniformly dispersed in the support. The uniform distribution of the cobalt(III) cluster complex in the chemically modified silica matrix is al clearly indicated by the TEM image recorded for Co(III)-CMS4 (Figure 23). As noted earlier while discussing the N_2 adsorption behaviour of the supported heterogeneous reagents prepared by immobilization of tetranuclear cobalt(III) complexes on CMS-$CH_2CH_2CO_2H$, the immobilized complexes are likely to be tethered to the inside walls of the disordered channels present in the functionalized silica. Figure 23 corroborates this expectation.

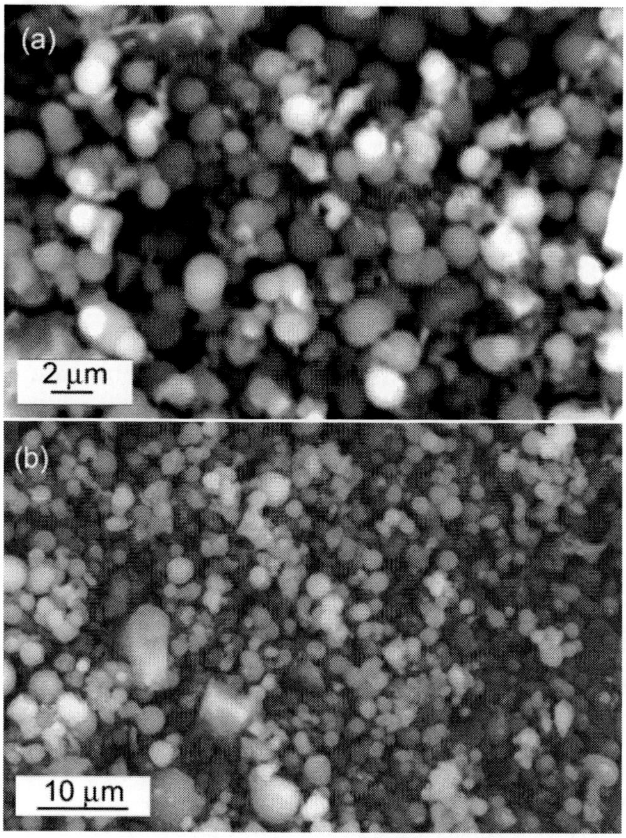

Figure 22. Scanning electron micrographs of Co(III)-CMS4 at two different magnifications [27c].

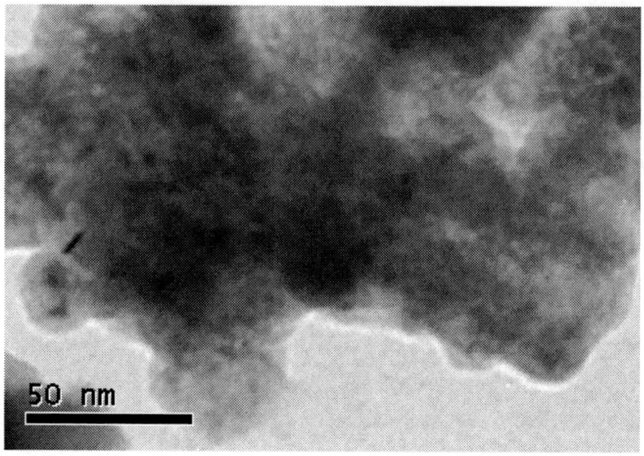

Figure 23. Transmission electron micrograph of Co(III)-CMS4.

6. Catalytic Use of Co(III)-CMS Materials

It has been found that the cubane clusters, $Co_4(\mu_3\text{-}O)_4(\mu\text{-}O_2CR)_4(L)_4$ themselves are capable of catalyzing the aerobic oxidation of α-pinene [27a], and ethylbenzene [20]. Rather remarkably, the tetrameric Co(III)-oxo cubane complex *viz.* $Co_4(\mu_3\text{-}O)_4(\mu\text{-}O_2CCH_3)_4(4\text{-}CNpy)_4$ was proven to be a very good catalyst for the oxidation of *p*-xylene [27b] under mild pressure with dioxygen in aqueous media. The main oxidation product in the catalyzed reaction is *p*-toluic acid, with terephthalic acid forming in traces. The reaction rate and product yield (maximum *p*-toluic acid yield ~36% at 20 bar oxygen pressure at 130 °C) are linearly dependent upon the applied pressure and reaction temperature. In this backdrop it is only to be expected that the heterogeneous CMS materials which have been prepared by anchoring the cubane clusters will also display catalytic activity for the above reactions. The catalysts prepared by anchoring Co(III) cubane clusters on CMS-$CH_2CH_2CO_2H$ both by in-situ complex formation as well as ligand exchange methods are indeed found to act as efficient catalysts for the oxidation of the above organic substrates. In addition, notable TBHP-oxidation activity for benzylic and other alcohols, catalyzed by Co(III)-CMS4, has been an important finding for these heterogenized materials [27c]. In the present section we shall discuss our results on the use of Co(III)-CMS materials as heterogeneous catalysts in the oxidation of a few organic substrates under environmentally friendly conditions.

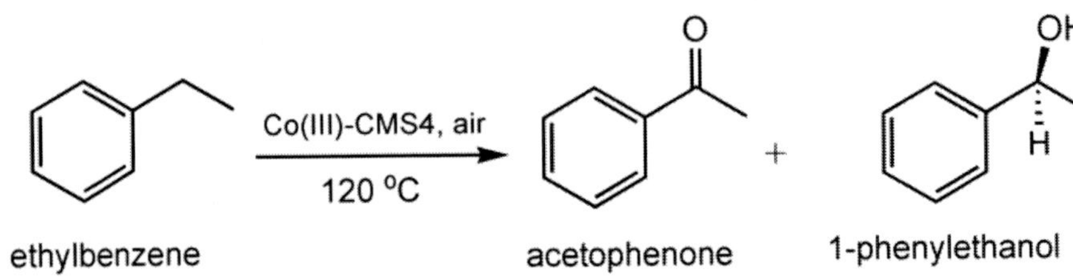

Figure 24. Aerobic oxidation of ethylbenzene catalyzed by Co(III)-CMS4.

6.1. Oxidation of Ethylbenzene

Catalytic oxidation of ethylbenzene by using air as the oxidant and Co(III)-CMS1 as the catalyst has been mentioned earlier in this chapter. Although the nature of the metal complex involved remained unknown at the time, it is now reasonably justified to state in view of our recent results that a cubane complex of cobalt(III) could be involved in the catalyzed process. This belief is confirmed by our studies using Co(III)-CMS4 as the catalyst for the aerobic oxidation of ethylbenzene under liquid phase conditions that do not call for the use of any solvents.

The Co(III) immobilized catalyst (Co(III)-CMS4) prepared by ligand exchange involving CMS-$CH_2CH_2CO_2H$ and the preformed complex $Co_4(\mu_3\text{-}O)_4(\mu\text{-}O_2CCH_3)_4(py)_4$ has been used for the aerobic oxidation of ethylbenzene to produce acetophenone and 1-phenylethanol as the products (Figure 24). The experimental findings suggest that in terms of activity as well as selectivity the present catalyst prepared by conscious immobilization of the cubane cluster $Co_4(\mu_3\text{-}O)_4(\mu\text{-}O_2CCH_3)_4(py)_4$ is comparable to Co(III)-CMS1 which was previously reported to covert 76% ethylbenzene at 94% selectivity. The present reaction using Co(III)-CMS4 has been carried out at 120 °C instead of the higher 130 °C which was used earlier. Perhaps due to this difference, although the observed selectivity for the formation of acetophenone is superlative at 99%, the ethylbenzene conversion is lower at 68.5%. The only other product that is found to form is 1-phenylethanol for which selectivity stands at 1%.

Formation of benzoic acid has not been observed. These results clearly suggest that in the aerobic oxidation of neat ethylbenzene the cubane complexes cobalt(III) are involved as the catalytic species.

Selectivities greater than 90% for acetophenone observed with immobilized catalysts can be considered significantly high for a reaction involving free radicals. In contrast to our results, Mal *et al.* [43] obtained only 40-50% selectivity for acetophenone using Ti-, V- and Sn-silicalites, while obtaining other aromatic ring hydroxylated products *ortho-* and *para-*hydroxy ethylbenzene in the oxidation of ethylbenzene. In addition, and rather remarkably, in our reaction we observed no induction period prior to the onset of oxidation. Absence of induction period is presumably due to the presence of cobalt in higher oxidation state because the initial time loss in this reaction is believed to occur due the required Co(II) to Co(III) oxidation, which is kinetically hindered. A recent study [44] showed that induction period indeed disappears when Co^{3+} is electrogenerated in the medium during or before the autoxidation of alkylaromatics in the liquid phase. Furthermore, in the present instance the reactions have been carried out in absence of any promoter and also under solvent-free conditions. This makes our reaction attractive from the environmental point of view as well. The plots of GC results are shown in Figure 25.

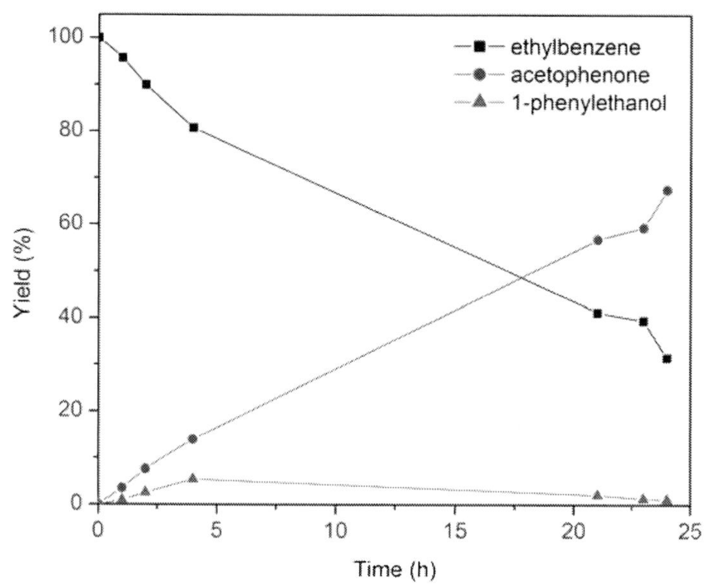

Figure 25. Oxidation of ethylbenzene with air under atmospheric pressure using Co(III)-CMS4.

6.2. Oxidation of α-Pinene

The second reaction that we wish to describe is that of the aerobic oxidation of α-pinene, which is a bicyclic monoterpene with an oxidizable double bond. The use of such a renewable feedstock in the production of chemicals is considered as the first step towards greening the life cycle of chemical products. In this context, oxidation α-pinene, which occurs widely in the plant kingdom, may be viewed as an important reaction because its oxidation products find use as the starting materials for fragrance, flavour and therapeutic agents including taxol. Oxidation of α-pinene proceeds through two competing pathways: epoxidation and allylic oxidation. While the former involves an electrophilic attack on the double bond, the latter results from allylic H-abstraction. Due to the radical nature of the reaction pathway, autoxidation in such processes is generally difficult to control and thus the possibility to achieve greater conversions while maintaining high selectivity is a challenging proposition.

In search for efficient and greener processes over the past few years various heterogeneous catalysts such as titanium incorporated mesoporous molecular sieves [45,46], Schiff-base complexes supported on zeolite [47] and Zn(II)-Al(III) layered double hydroxide (LDH) [48], oxomolybdenum(VI) complexes supported on MCM-41 and MCM-48 [49], polyoxometallate supported materials [50], Co and Mn-AlPO's [51] etc. have been developed and studied for the catalytic epoxidation of α-pinene. Many of these processes suffer from drawbacks and limited applicability due to the poor conversion and also the selectivities. Sacrificial aldehydes are often used as an oxygen acceptor in such processes to achieve reasonable yield and selectivities; but, this procedure leads to an increase in the E-factors and decrease in the atom economy [51].

The Co(III) supported material, Co(III)-CMS3 has been tested as a catalyst in the aerobic oxidation of α-pinene. It has been found to show very high activity and selctivity in the epoxidation of α-pinene to α-pinene oxide. Results are summarized in Table 5 and Figure 26 shows a representative kinetic plot. In a typical reaction, 100 mg (0.09 mol%) of the catalyst was added to the reaction mixture consisting of α-pinene (3.97 mL, 25 mmol) and 40 mL of 1,4-dioxane with air flowing at the rate of 15 mL/min through the mixure. As observed earlier [52], absence of solvent leads to rearranged and hence selectivity is lowered. It has been observed that an increase in the reaction temperature accelerates aerobic oxidation of α-pinene. Similar acceleration in air oxidation of α-pinene is in conformity with literature reports [53,54]. At 100°C an excellent conversion of 81.5% (Entry 3, Table 5) has been observed. The TOF (h^{-1}) values calculated range from *ca.* 9 to 37. Rearrangement of the products is also found to occur at higher temperatures. The observed selectivity (61%) for α-pinene oxide at 100°C (at 81.5% conversion) is rather significant because the yield for the epoxidized product is also reasonably high. The increased selectivity for α-pinene oxide at higher temperatures is likely to be an outcome of the suppression of the allylic oxidation route of α-pinene oxidation.

In contrast to our results, previous reports have shown that selectivities >50% for α-pinene oxide are difficult to achieve [55]. More recently Co/SiO_2 composite was shown [56] to catalyze α-pinene oxidation with molecular oxygen giving both epoxidation and allylic oxidation products with substrate conversion of around 40% with selectivities for the former and the latter being 24-30% and 44-52% respectively. A higher selectivity of 84% for α-pinene epoxide could be observed by Thomas *et al.* [51] at 56% conversion in a study on the aerobic oxidation of α-pinene using a microporous Co(III) containing aluminophosphate (CoAlPO-36) catalyst. This reaction however required the use of a sacrificial aldehyde co-reactant and 30 bar air pressure. On the other hand, formation of only trace amounts of α-pinene oxide was observed in α-pinene oxidation by TBHP and H_2O_2 using Cu, Mn and Pd based catalysts [57]. Finally, it may be mentioned here that the observed activity and selectivity shown by Co(III)-CMS3 are comparable to those observed for the homogeneous reaction catalysed by the complex $[Co_4O_4(O_2CC_6H_5)_4(4\text{-CNpy})_4]$ [27a]. More importantly, the α-pinene conversion as well as epoxide selectivity observed by us using the Co(III) based homogeneous and heterogeneous catalysts stand among the highest reported for the reaction. Moreover, our results are significant from a green chemistry point of view because no sacrificial aldehyde, other additives or applied pressure have been used by us in the aerobic oxidation of α-pinene. Furthermore, in addition to being a catalyst that doesn't show cobalt leaching during and at the end of the reaction, our studies show that it may be a reusable catalyst.

Table 5. Catalytic aerobic oxidation of α-pinene by Co(III)-CMS3 under atmospheric pressure [a]

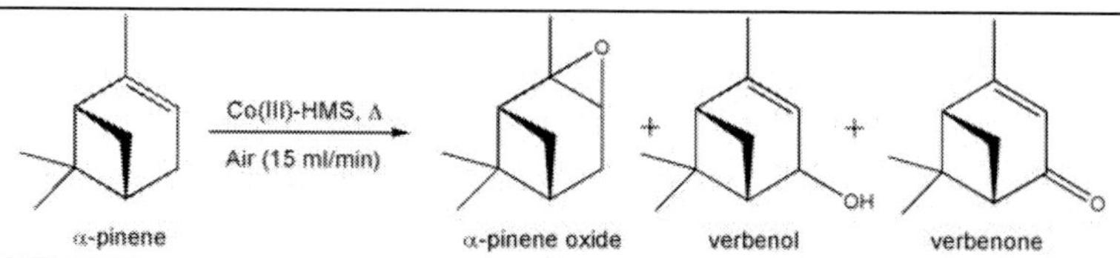

Entry	Catalyst (mol%)	Temp. (°C)	Conv. (%)	Product[b] yield (selectivity), %		
				-Pinene oxide	Verbenol	Verbenone
1	0.09	60	6.6	3.5 (53)	1.1 (17)	2.0 (30)
2	0.09	80	34.2	11.6 (34)	5.4 (16)	7.5 (22)
3	0.09	100	81.5	50.1 (61)	2.9 (4)	16.9 (21)
4	0.005	100	66.0	40.1 (61)	5.8 (9)	13.4 (20)
5	0.02	100	73.0	41.1 (56)	8.0 (11)	9.5 (13)

[a] Reaction conditions: α-pinene: 3.97 mL (25 mmol); 1,4-dioxane: 40 mL; oxidant: air (15 mL/min); reaction time = 24h. Conversion and product composition were determined by GC and identified by GC-MS,[b] Minor amounts of rearranged products of α-pinene oxide also formed.

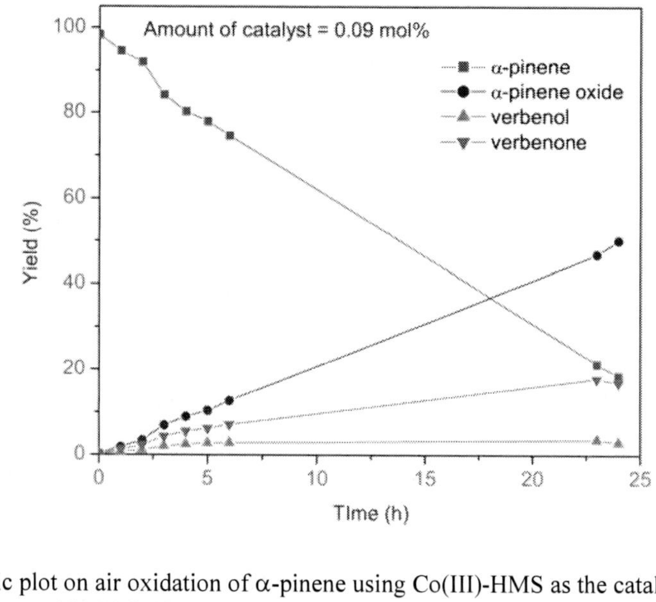

Figure 26. A kinetic plot on air oxidation of α-pinene using Co(III)-HMS as the catalyst under atmospheric pressure at 100 °C [15].

The catalysts Co-CMS5 and Co-CMS6 have also been tested for the air oxidation of α-pinene. In both the cases air oxidation commences quickly with the formation of α-pinene oxide, verbenol and verbenone as the main products along with other rearranged products as observed in the reactions described above. A conversion of 50.6% and 48.0% are observed for Co-CMS5 and Co-CMS6 respectively after 24h (Table 6).

Table 6. Results of air oxidation of α-pinene catalyzed by Co-CMS5 and Co-CMS6[a]

	Conv. (%)	TON[b]	Product yield (selectivity) (%)			
			-Pinene oxide	Verbenol	Verbenone	others[c]
Co-CMS5	50.67	1023	12.67 (37)	9.01 (26)	13.71 (37)	15.73
Co-CMS6	48.24	893	12.04 (36)	7.8 (24)	13.35 (40)	15.05

[a]**Conditions:** α-pinene: 15.9 mL (100 mmol); reaction temperature: 60°C; oxidant: air (15 mL/min); reaction time = 24h. Conversion and product composition were determined by GC; [b] Turnover number (TON), mole of α-pinene converted per mole of cobalt. [c]Rearranged products of α-pinene oxide were the main products.

Although the mechanistic pathway of the reaction is not clear at this stage, the reaction probably proceeds via peroxo-intermediates. Chavez *et al.* [58] have shown that the formation of a transient Co(III)-alkylperoxo complex, which acts as a key intermediate in autoxidation of hydrocarbons, is crucial for the sustained production and decomposition of the alkyl hydroperoxides. Cobalt(III)-alkylperoxo species then undergo homolytic cleavage of either the cobalt-oxygen or the oxygen-oxygen bond, generating either alkoxy or peroxoalkyl radical, which then would react with the substrate.

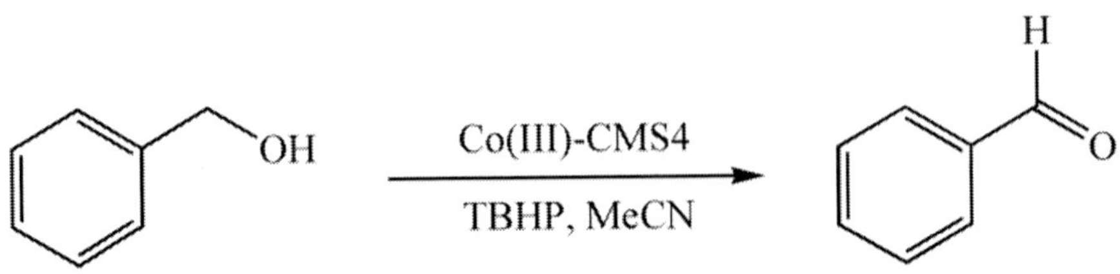

Figure 27. TBHP oxidation of benzyl alcohol catalyzed by Co(III)-CMS4.

6.3. Oxidation of Alcohols

Oxidation of alcohols to carbonyl compounds is an important reaction. Stoichiometric oxidants such as chromates, permanganates and MO_4 (M = Ru, Os) are the commonly used reagents [19a,59,60]. However, they are going out of favour increasingly because they create 'heavy metal wastes'. In view of this, development of environmentally friendly heterogeneous catalysts for alcohol oxidation is very important. In the use of catalytic amounts of transition metal salts or complexes as homogeneous catalysts for the oxidation of alcohols [61-64], separation of the catalyst from the reaction mixture and its subsequent recovery in active form is cumbersome. Heterogeneous catalysts for this kind of reaction are therefore necessary [65]. Clearly, encapsulation and/or immobilization of known homogeneous catalysts on the surface and in the cavities of porous supports appears to address this requirement [66,67]. Such an approach facilitates the separation of catalysts from products and as a result, catalyst reuse may also become possible. It is also to be noted that in many cases, the heterogenization of catalytically active transition metal complexes is known to yield catalysts with improved activity and selectivity [68-70]. Various cobalt(II) compounds have been used as homogeneous catalysts in the oxidation of alcohols to its corresponding carbonyl compounds [71]. We have indicated earlier in this chapter and that the use of cobalt(III) complexes in these reactions may be advantageous due to their substitutional inertness. In this context it may be said that cobalt(III) based heterogeneous catalysts are by and large unexplored for alcohol oxidation.

Table 7. Oxidation of various alcohols by TBHP using Co-CMS4 as the catalyst [27c] [a]

Entry	Substrate	Product	Time (h)	Yield[b]
1	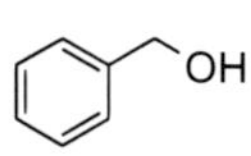	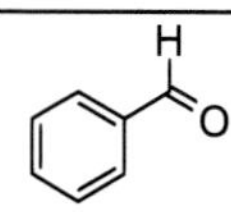	4.5	93
2	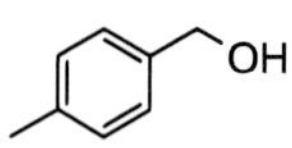	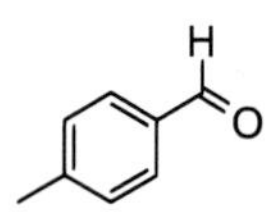	1	69
3	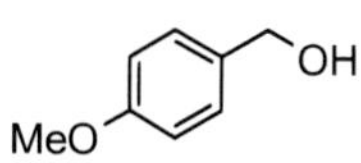	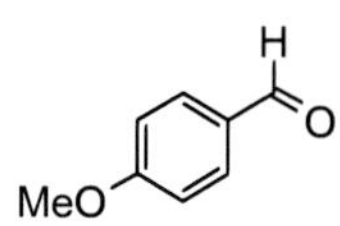	4.5	89
4	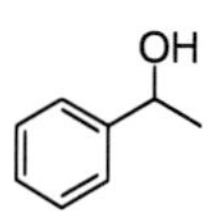	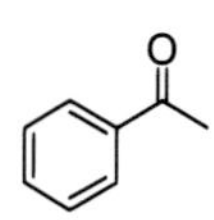	1	78
5	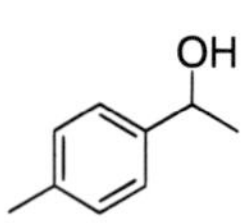	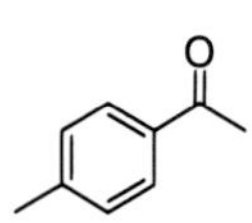	2	86
6	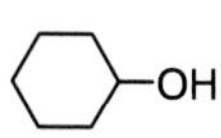	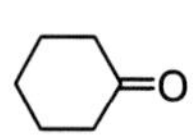	5	64
7	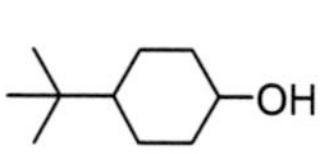	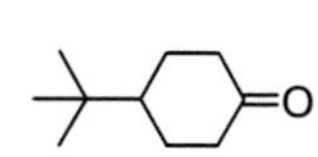	3	73
8	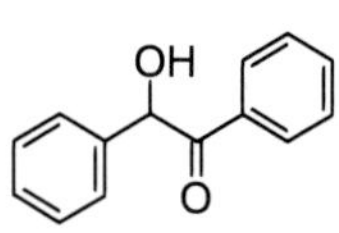	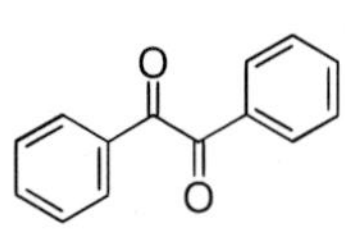	12	89

[a] Reaction condition: substrate, 10 mmol; amount of catalyst, 50 mg; acetonitrile, 5 mL; reaction temperature, 82 °C. [b] Isolated yields after chromatographic purification.

We have investigated [27c] the use of Co(III)-CMS4 as a heterogeneous catalyst for the oxidation alcohols using *tert*-butyl hydroperoxide (TBHP). To begin with, a systematic study has been carried out for catalytic evaluation of Co(III)-CMS4 for the oxidation of benzyl alcohol (Figure 27). Optimization of the reaction conditions has been done by varying the amounts of the catalyst and TBHP. In a typical reaction, benzyl alcohol (10 mmol), catalyst (50 mg), and 80% TBHP (4 mL) are heated to reflux in 5 mL acetonitrile. The progress of the reaction has been monitored by TLC. After chromatographic separation, the isolated yield of benzaldehde is found to be 93%. However, the reaction is slow when carried out at room temperature with only 37% benzaldehyde obtained after 24h. The essential role played by the catalyst is evident from the extremely low (2%) yield of benzaldehyde obtained in a blank reaction carried out in absence of the catalyst.

The heterogenized catalyst Co(III)-CMS4 shows enhanced catalytic activity compared to the homogeneous catalytic efficiency of $Co_4(\mu_3\text{-}O)_4(\mu\text{-}O_2CMe)_4(py)_4$ in the dissolved state. The amount of complex (10 mg) used in the homogeneous reaction was approximately the same as the amount of the cobalt complex estimated to be present in 50 mg of Co(III)-CMS4. Benzaldehyde was the only product observed in both the cases. Studies on catalyst reuse show that there is a slight decrease in product yield in the second run. However, no detectable leaching of cobalt was observed in the first as well as the second run of the reaction. Furthermore, the amount of cobalt estimated for the catalyst obtained after first use remained unchanged. This appears to indicate that during the course of the reaction the nature of the catalytically active complex also has undergone some change. Clearly, more studies on catalyst regeneration and recycling will be beneficial.

Our results on the catalytic conversion of other alcohols are given in Table 7. The data indicate that Co(III)-CMS4 acts as an efficient catalyst for the oxidation of other benzylic as well as secondary alcohols to the corresponding carbonyl compounds. In the process these results compare very well with other cobalt-based catalysts [71].

CONCLUSION

Preparation and properties of a disordered mesoporous silica having $-CH_2CH_2COOH$ as surface functional groups have been described. A series of chemically robust complexes of general formula $Co_4O_4(O_2CR)_4L_4$ has been synthesized by following a general procedure and protocols for immobilization of these complexes on the surface of the above chemically modified silica (CMS) have been developed. The resulting supported reagents, the so-called Co(III)-CMS materials, have been examined as catalysts for the oxidation ethylbenzene, α-pinene and a few alcohols in the liquid phase under atmospheric conditions. It is an important finding since the Co(III)-CMS catalysts are effective as heterogeneous catalysts in the oxidation of three different types of organic substrates, viz. alkylaromatics, alkenes and alcohols. Performance of the catalysts used and also the perceived catalyst stability suggest that one or more of these heterogeneous catalysts may prove practically important. The prospect of utilizing Co(IV) in the catalytic cycle involving these catalysts is quite novel in itself. Furthermore, possibility of using mesoporous but disordered CMS materials displaying high surface areas for heterogeneous catalysis is believed to be a viable one.

REFERENCES

[1] Clark, J. H.; Macquarrie, D. J. and Tavener, S. J. *Dalton Trans*. 2006, 4297-4309.

[2] (a) Kresge, C. T.; Leonowicz, M. E.; Roth, W. J.; Vartuli, J. C.; and Beck, J. S. *Nature* 1992, *359*, 710-712. (b) Beck, J. S.; Vartuli, J. C.; Roth, W. J.; Leonowicz, M. E.; Kresge, C. T.; Schmitt, K.D.; Chu, C. T.W.; Olson, D.H.; Sheppard, E.W.; McCullen, S.B.; Higgins, J. B.; and Schlenker, J. L. *J. Am. Chem. Soc.* 1992, *114*, 10834-10843.

[3] Tanev, P. T. and Pinnavaia, T. J. *Science* 1995, *267*, 865-867.

[4] Macquarrie, D. J. *Green. Chem.* 1999, *1*, 195-198.

[5] Chakrabarty, R. *Ph. D. Thesis* Gauhati University, 2006.

[6] Zhang, W.; Fröba, M.; Wang, J.; Tanev, P. T.; Wong, J.; and Pinnavaia, T. J. *J. Am. Chem. Soc.* 1996, *118*, 9164-9171.

[7] De Vos, D. E.; Sels, B. F. and Jacobs., P. A. *Adv. Catal.* 2001, *46*, 1-87.

[8] Quignard, F. and Choplin, A. In *Comprehensive Coordination Chemistry*; J. A. McCleverty, T. J. Meyer (Eds.); Pergamon Press: Oxford, 2003; Vol. 9, pp. 445-470.

[9] Corma, A. *Chem. Rev.* 1997, *97*, 2373-2419.

[10] Ying, J. Y.; Mehnert, C. P. and Wong, M. S. *Angew. Chem. Int. Ed.* 1999, *38*, 56-77.

[11] Stein, A.; Melde, B. J. and Schroden, R. C. *Adv. Mater.* 2000, *12*, 1403-1419.

[12] Clark, J. H. (Ed.) *Chemistry of Waste Minimization*, Chapman and Hall, London, 1995.

[13] Clark, J. H. and Macquarrie, D. J. *Chem. Soc. Rev.* 1996, 303-310.

[14] Clark, J. H. and Macquarrie, D. J. *Chem. Commun.* 1998, 853-860.

[15] Chakrabarty, R.; Das, B. K. and Clark, J. H. *Green Chem.* 2007, *9*, 845-848.

[16] Arends, I. W. C. E. and Sheldon, R. A. *Appl. Catal.* 2001, *A212*, 175-188.

[17] Corriu, R. J. P.; Mehdi, A. and Reye, C. *J. Mater. Chem.* 2005, *15*, 4285-4294.

[18] Chong, A. S. M. and Zhoa, X. S. *J. Phys. Chem.* 2003, *B107*, 12650-12657.

[19] (a) Sheldon, R. A. and Kochi, J. A. *Metal Catalysed Oxidations of Organic Compounds*; Academic Press: New York, 1981. (b) Suresh, A. K.; Sharma, M. M. and Sridhar, T. *Ind. Eng. Chem. Res.* 2000, *39*, 3958-3997. (c) Łukasiewicz, M. Ciunik, Z.; Mazurek, J.; Sobczak, J.; Staroń, A.; Wołowiec, S.; and Ziółkowski, J. *Eur. J. Inorg. Chem.* 2001, 1575-1579. (d) Saussine, L.; Brazi, E.; Robine, A.; Mimoun, H.; Fischer, J.; and Weiss, R. *J. Am. Chem. Soc.* 1985, *107*, 3534-3540.

[20] Chakrabarty, R.; Bora, S. J. and Das, B. K. *Inorg. Chem.* 2007. *46*, 9450-9462.

[21] Chavez, F. A. and Mascharak, P. K. *Acc. Chem. Res.*, 2000, *33*, 539-545.

[22] Chavez, F. A.; Briones, J. A.; Olmstead, M. M.; and Mascharak, P. K. *Inorg. Chem.* 1999, *38*, 1603-1608.

[23] Sharma, V. B.; Jain, S. L. and Sain, B. *J. Mol. Catal. A: Chem.* 2004, *212*, 55-59.

[24] Das, S. and Punniyamurthy, T. *Tetrahedron Lett.* 2003, *44*, 6033-6035.

[25] Qi, J.-Y.; Ma, H.-X.; Li, X.-J.; Zhou, Z.-Y.; Choi, M. C. K.; Chan, A. S. C.; and Yang, O.Y. *Chem. Commun.* 2003, 1294-1295.

[26] Das, B. K. and Clark, J. H. *Chem. Commun.* 2000, 605-606.

[27] (a) Chakrabarty, R. and Das, B. K. *J. Mol. Catal. A: Chemical* 2004, *223*, 39-44. (b) Chakrabarty, R.; Kalita, D.; and Das, B. K. *Polyhedron* 2007, *26*, 1239-1244.(c) Sarmah, P.; Chakrabarty, R.; Phukan, P.; and Das, B. K. *J. Mol. Catal. A: Chem.* 2007, *268*, 36-44.

[28] Macquarrie, D. J. *Chem. Commun.* 1996, 1961-1962.

[29] Sujandi, S.C. H.; Han, D.S.; Jin, M.J.; and Park, S.E. *J. Catal.* 2006, *243*, 410-419.

[30] Lindoy, L. F. *The Chemistry of Macrocyclic Ligand Complexes*, Cambridge Univ. Press, Cambridge, 1989.

[31] (a) Pruβ, T.; Macquarrie, D. J. and Clark, J. H. *Appl. Catal. A: General* 2004, *276*, 29-37. (b) Pruβ, T.; Macquarrie, D. J. and Clark, J. H. *J. Mol. Catal. A: Chemical* 2004, *211*, 209-217.

[32] Burri, D. R.; Jun, K.W.; Kim, Y.H.; Park, S.E.; and Yoo, J. S. *Chem. Lett.* 2002, 212-213.

[33] Zhao, D.; Feng, J.; Huo, Q.; Melosh, N.; Fredrickson, G. H.; Chmelka, B. F.; and Stucky, G. D. *Science* 1998, *279*, 548-552.

[34] Macquarrie, D. J.; Jackson, D. B.; Mdoe, J. E. G.; and Clark, J. H. *New J. Chem.* 1999, *23*, 539-544.

[35] Corriu, R. J. P. *Eur. J. Inorg. Chem.* 2001, 1109-1122.

[36] Cotton, F. A.; Wilkinson, G.; Murillo, C. A.; and Bochmann, M. *Advanced Inorganic Chemistry*; John Wiley & Sons: New York, 6th Ed., 1999.

[37] Dimitrou, K.; Brown, A. D.; Concolino, T. E.; and Rheingold, A. L.; Christou, G. *Chem. Commun.* 2001, 1284-1285.

[38] (a) Partenheimer, W. *Catal. Today* 1995, 69-158. (b) Raghavendrachar, P. and Ramachandran, S. *Ind. Eng. Chem. Res.* 1992, *23*, 453-462.

[39] Price, P. M.; Clark, J. H. and Macquarrie, D. J. *J. Chem. Soc. Dalton Trans.* 2000, 101-110.

[40] (a) Das, B. K. and Chakravarty, A. R. *Polyhedron* 1988, *7*, 685-687. (b) Das, B. K. and Chakravarty, A. R. *Inorg. Chem.* 1991, *30*, 4978-4986.

[41] Sing, K. S. W.; Everett, D. H.; Haul, R. A. W.; Moscou, L.; Pierotti, R. A.; Rouquérol, J.; and Siemieniewska, T. *Pure Appl. Chem.* 1985, *57*, 603-619.

[42] Chavan, S. A.; Srinivas, D. and Ratnasamy, P. *J. Catal.* 2001, *204*, 409-419.

[43] Mal, N. K. and Ramaswamy, A. V. *App. Catal. A: General* 1996, *143*, 75-85.

[44] Bejan, D.; Lozar, J.; Falgayrac, G.; and Savall, A. *Catal. Today* 1999, *48*, 363-369.

[45] Kapoor, M. P.; Bhaumik, A.; Inagaki, S.; Kuraoka, K.; and Yazawa, T. *J. Mater. Chem.* 2002, *12*, 3078-3083.

[46] (a) Chiker, F.; Nogier, J. P.; Launay, F.; and Bonardet, J. L. *Appl. Catal. A: General* 2003, *243*, 309-321. (b) Chiker, F.; Launay, F.; Nogier, J. P.; and Bonardet, J. L. *Environ. Chem. Lett.* 2003, *1*, 117-120.

[47] Joseph, T.; Sawant, D. P.; Gopinath, C. S.; and Halligudi, S. B. *J. Mol. Catal. A: Chemical* 2002, *184*, 289-299.

[48] Bhattacharjee, S.; Dines, T. J. and Anderson, J. A. *J. Catal.* 2004, *225*, 398-407.

[49] Nunes, C. D.; Pillinger, M.; Valente, A. A.; Rocha, J.; Lopes, A. D.; and Gonçalves, I. S. *Eur. J. Inorg. Chem.* 2003, 3870-3877.

[50] Sakamoto, T. and Pac, C. *Tetrahedron Lett.* 2000, *41*, 10009-10012.

[51] Raja, R.; Sankar, G. and Thomas, J. M. *Chem. Commun.* 1999, 829-830.

[52] Rothenberg, G.; Yatziv, Y. and Sasson, Y. *Tetrahedron* 1998, *54*, 593-598.

[53] Lajunen, M. K.; Myllykoski, M. and Asikkala, J. *J. Mol. Catal. A: Chemical* 2003, *198*, 223-229.

[54] Lajunen, M. K.; Manula, T. and Koskinen, A. M. P. *Tetrahedron Lett.* 1994, *55*, 4461-4464.

[55] Lajunen, M. K. *J. Mol. Catal. A: Chemical* 2001, *169*, 33-40.

[56] Guo., C.C.; Yang, W.J. and Mao, Y.L. *J. Mol. Catal. A: Chemical* 2005, *226*, 279-284.

[57] Allal, B. A.; Firdoussi, L. E.; Allaoud, S.; Karim, A.; Castanet, Y.; and Mortreux, A. *J. Mol. Catal. A: Chemical*, 2003, *200*, 177-184.

[58] Chavez, F. A. and Mascharak, P. K. *Acc. Chem. Res.* 2000, *33*, 539-545.

[59] Larock, R. C. *Comprehensive Organic Transformations*, Wiley-VCH, New York, 1999, pp. 1646.

[60] March, J. *Advanced Organic Chemistry*, John Wiley, New York, 1992, pp. 1167.

[61] Müller, P. and Godoy, J. *Tetrahedron Lett.* 1981, *22*, 2361-264.

[62] Kanemoto, S.; Oshima, K.; Matsubara, S.; Takai, K.; and Nozaki, H. *Tetrahedron Lett.* 1983, *24*, 2185-2188.

[63] Kaneda, K.; Kawanishi, Y. and Teranishi, S. *Chem. Lett.* 1984, 1481-1482.

[64] Zondervan, C.; Hage, R. and Feringa, B. L. *Chem. Commun.* 1997, 419-420.

[65] Zhan, B-Z. and Thompson, A. *Tetrahedron Lett.* 2004, *60*, 2917-2935 and references therein.

[66] Karimi, B.; Abedi, S.; Clark, J.H.; and Budarin, V. *Angew. Chem. Int. Ed.* 2006, *45*, 4776-4779.

[67] Rafelt, J.S. and Clark, J.H. *Catal. Today* 2000, *57*, 33-44.

[68] Kumar, K. R.; Choudary, B. M.; Jamil, Z.; and Thyagarajan, G. *J. Chem. Soc. Chem Commun.* 1986, 130-131.

[69] Barloy, L.; Battioni, P. and Mansuy, D. *J. Chem. Soc. Chem Commun.* 1990, 1365-1366.

[70] Balkas, K. J.; Gabrielor, A. G.; Bell, S. L.; Bedioui, F.; Roue, L.; and Devyank, J. *Inorg. Chem.* 1994, *33*, 67-72.

[71] Punniyamurthy, T.; Velusamy, S. and Iqbal, J. *Chem. Rev.* 2005, *105*, 2329-2363.

In: Heterogeneous Catalysis Research Progress
Editor: Mathias B. Gunther, pp. 145-191

ISBN 978-1-60456-979-7

Chapter 4

SELECTIVE ALKYLATION OF AROMATIC SUBSTRATES ON SPINELS

Munusamy Vijayaraj, Thomas Mathew and Chinnakonda S. Gopinath[1]
Catalysis Division, National Chemical Laboratory, Dr. Homi Bhabha Road, Pune 411 008, India

ABSTRACT

The purpose of the present chapter is to review the status of spinels for selective alkylation of aromatic compounds with functional groups such as –OH, -NH_2. Alkylation of aromatic compounds is an important industrial reaction due to the usage of alkylated products in a wide variety of applications. Traditional Friedel-Crafts alkylation suffers from many disadvantages, and solid acid catalysts are preferred for several known advantages. Most of the known catalysts systems, such as zeolites, MgO, various metal oxides, lead to non-selective alkylation or involves high operating temperature (≥400^{0}C). Among the metal oxide catalysts used for aromatic alkylation, spinel (AB_2O_4) compounds have been effective and emerging as one of the potential systems for selective alkylation of aromatic compounds. Main advantages associated with spinels are the selective alkylation of aromatic compounds with good to very good yield and stable activity, with molecules such as phenol, aniline, pyridine in fixed-bed reactors around 300^{0}C. Methanol was employed as the methylating agent in most of the cases, and higher alcohols for higher alkylation. Few detailed characterization studies reported in the literature reveals the heterogeneous distribution of metal ions on the surface, and hence the associated synergism might be responsible for high catalytic activity. In situ infrared spectroscopic studies reveal the reaction mechanism aspects of selective alkylation of phenol and aniline. Medium range acid-base property of spinels is another important factor that makes them indispensable for different reactions with different cations. Indeed inverse spinel systems $B(AB)O_4$ are the most interesting due to the versatile nature of cation distribution in tetrahedral (T_d) and octahedral (O_h) sites, which contributes to the structural integrity and enhanced activity apart from the ability to tune cation site distribution with different metal ions. Although there are many positive factors with

---

[1] Author for correspondence; Fax: 0091-20-2590 2633; Email: cs.gopinath@ncl.res.in; http://www .ncl.res.in/csgopinath.

spinels, such as selective alkylation, more needs to be explored to exploit the spinel system for commercial applications.

1. INTRODUCTION

Historically alkylation by Friedel-Crafts catalysis is of great significance, since a wide variety of fine and bulk chemical products are obtained employing this route [1]. Alkylations, acylations, benzoylations and sulfonylations are types of this name reaction, respectively leading to useful products like alkyl aromatics, ketones, alcohols and sulfones. Petrochemical industry is also a beneficiary of this class of reaction as number of alkyl hydrocarbons are synthesized by alkylation reactions. Acids (HF, H_2SO_4) and anhydrous metal halides, such as, $AlCl_3$, BF_3 are the conventionally practiced catalyst systems that are very reactive. Since they are effective the reaction can be carried out at low temperature in homogeneous liquid phase conditions. Many reactions are carried out in batch processes using anhydrous $AlCl_3$ as a soluble catalyst, which is inexpensive and very reactive. However, there are numerous disadvantages associated with these catalysts; very often such catalysts are required in stoichiometric quantities, and difficult to handle as they get readily hydrolyzed; corrosive and toxic nature of these catalysts lead to potential environment hazards and associated operational problems, including downstream processes such as difficulty in separation, recovery and reutilization that result in higher capital costs. Thus attention is devoted to the development of environmentally benign catalysts for the last several decades towards the production of important chemical intermediates and value added products [2].

Utilization of solid acids as catalysts in place of traditional Friedel-Crafts catalysts and mineral acids has been developed [3]. Although, these solid catalysts show low to comparable activity as that of Friedel-Crafts catalysts, possess more advantages such as alleviating product separation, non-corrosiveness, reduces handling problems, and recycling ability. The acid-base and/or redox properties of these solid catalysts renders prospects to study these reactions by varying their catalytic characteristics. Numerous reports on the production of alkyl aromatics are available in the literature and some of them are in the patents domain. Several varieties of catalysts used for the alkylation reactions of aromatic compounds have been examined over mixed (metal) oxides, supported oxides, zeolites and mesoporous materials were discussed in the recent reviews [3-6]. Alkylating agents such as C_1-C_4 alcohols, various olefins, dimethyl carbonate etc., are widely employed for alkylation. The reaction is sensitive to acid-base property of the catalysts, reaction parameters such as temperature, mole ratio of reactants and type of alkylating agent. In general, reactions are carried out at vapor phase conditions in the conventional fixed bed catalytic reactors under atmospheric pressure. Alumina, silica, transition metal oxides, magnesium oxide, zeolites and silica-alumina in their various modified forms have been evaluated for alkylation reactions [7-12]. Nonetheless, alkylation reactions of aromatic compounds using spinels as catalysts have been found quite attractive, and recent research work from our group as well as number of other groups are also precedent for the use of ferrospinels for alkylation of aromatic substrates, like phenol, aniline, and heterocyclic compounds [13, 14]. Synergistic interaction of metal ions in the spinel, tunable acid-base character, and structural and electronic integrity of the spinel system under reaction conditions are some of the salient features of the spinels. Specifically, it is not a single metal ion, but a combination of metal ions, and hence the heterogeneity of the surface, decides the course of the reaction and the stability of the spinel catalysts. Since the reaction is mainly effected on the surface for any catalysts, a critical evaluation of the surface structure is also inevitable. This is quite important to identify the positive (and negative) aspects of active (and not-so-active) catalyst, and hence in this chapter special attention is given to include the surface properties in detail through phenol and aniline as model candidates on ferrospinel surfaces.

2. SPINELS

2.1. Structural Aspects of Spinels

The term 'spinel' has been originated from the mineral $MgAl_2O_4$ called spinel, which has a well defined crystal structure. General formula for any simple binary spinel may be written as AB_2O_4. Ternary spinel system can be represented as $(A'A'')B_2O_4$ or $A(B'B'')O_4$. In the above formula A and B are the divalent tetrahedral (T_d) and trivalent octahedral (O_h) cations, respectively. If A type cations occupy T_d and B type cations occupy O_h sites then it is a normal spinel. If half of the B cations occupy T_d sites and A cations along with other half of B occupy O_h sites then it is called a fully inverse spinel. There are other types of spinels called random or intermediate spinels, where cations distribution will be in between the normal spinel and fully inverse spinel arrangement. These intermediate spinels are normally labeled in terms of the percent inverse character that they exhibit. Table 1 illustrates some examples of normal, inverse and intermediate spinels. Spinels having iron as B site ions are known as ferrospinels or ferrites, similarly Cr and Al containing spinels are chromites and aluminates. It is the difference in bond lengths of two types of cations with the anions is an important factor that decides the electronic character of spinels, as explained in the following.

Table 1. Examples of some normal and inverse spinels

Type	Structure	Examples
Normal	$(A^{2+})[B_2^{3+}]O_4$	$ZnFe_2O_4$, $ZnCrFeO_4$, $ZnCr_2O_4$, $MgCr_2O_4$
Inverse	$(B^{3+})[A^{2+}B^{3+}]O_4$	$MgFe_2O_4$, $NiFe_2O_4$, $CoFe_2O_4$
Random	$(A_{1-x}^{2+}B_y^{3+})[A_x^{2+}B^{3+}_{2-y}]O_4$	$MgCrFeO_4$

The unit cell of an ideal spinel contains 8 formula units represented as $8[AB_2O_4]$, which forms a face centered cubic *(fcc)* arrangement with 32 O^{2-} ions, while metal cations occupy interstitial sites. Close packing arrangement of 32 O^{2-} ions form 64 T_d and 32 O_h interstices, out of which 8 T_d and 16 O_h sites are occupied by the cations [15]. The unit cell of an ideal spinel structure is shown in Figure 1. From the figure, it is evident that there are two types of cubic building units inside a big fcc O-ion lattice, filling all 8 octants. The positions of the ions in the spinel lattice are not perfectly regular and some distortion does occur. The T_d sites are often too small for the metal ions so that the oxygen ions move slightly to accommodate them. The oxygen ions connected with the O_h sites move in such a way as to shrink the size of the O_h cell by the same amount as T_d site expands.

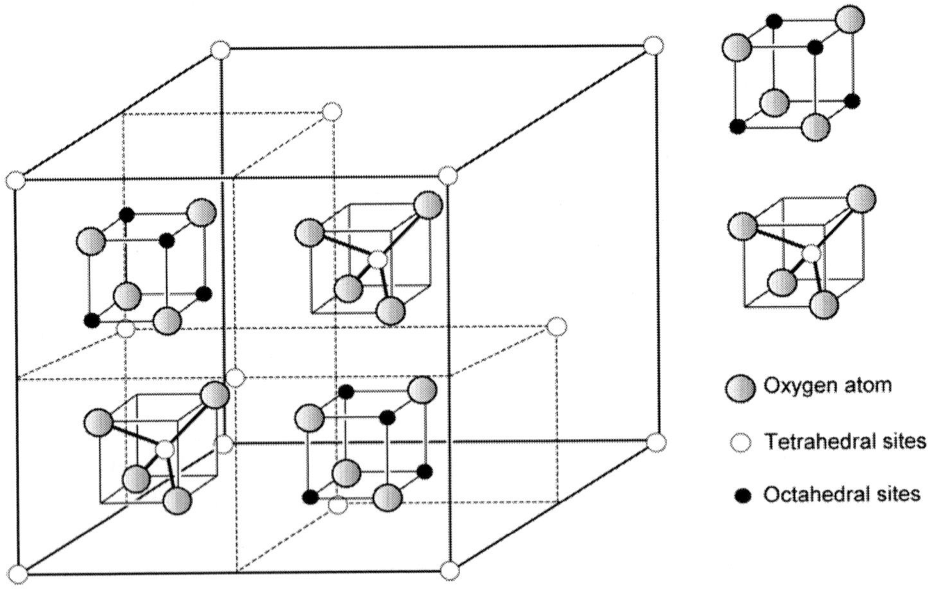

Figure 1. The unit cell of an ideal spinel structure.

Thus the position of the metal ions are fixed by the symmetry of the structure but the anion positions are variable and are specified by the parameter ‘u’, called the oxygen parameter, which is the distance between the oxygen ion and the face of the cube edge along the cube diagonal of the spinel sub cell. In the ideal spinel the parameter ‘u’ has a value in the neighborhood of 0.375. However, in actual spinel lattices this ideal pattern is slightly deformed, usually, corresponding to $u > 0.375$ [15,16], in such a way that the oxygen tetrahedron in the A-cubes is somewhat expanded and the oxygen octahedron in the B-cubes is slightly contracted. Accordingly, in that case the octahedrons formed by the 6 oxide ions directly surrounding the positive ions in these B-cubes deviate somewhat from regular octahedrons. Each anion (O^{2-} ion) in the spinel structure is surrounded by one A and three B cations. If ‘a’ denotes the cell edge of the spinel unit cell, the AX distance is a(u-1/4) and BX distance is a(5/8-u) for small deviations from the ideal lattice. Where AX is the distance between anion (O^{2-}) and A cation and BX is the distance between B cation and anion. The angles AXB and BXB are about 125^0 and 90^0, respectively.

$$AX = a(u-1/4)$$
$$BX = a(5/8-u)$$

For an ideal spinel, u is 0.375 and hence, AX = 0.125a, and BX = 0.25a. This difference in the bond length of two types of cations with the anions is an important factor in determining their electrical and catalytic properties [15].

2.2. Catalytic Activity of Spinels

As well documented in the literature, spinels are basically known for its magnetic properties. Generally, high temperature preparation methods (ceramic methods) have been employed to prepare these materials. Ceramic method produces materials with high purity, however with very low surface area. For the last two decades, a wide variety of synthesis procedures, which include co-precipitation, hydrothermal synthesis, sol-gel process, has been introduced for preparing spinels with high surface area. The high surface area spinel materials provide enough surface space for organic molecule adsorption-desorption process, which provoked many researchers to utilize these materials for different organic transformation reactions such as oxidation, dehydrogenation, acylation, alkylation etc.

The ample diversity of properties that the spinel compounds exhibit is derived from the fact that the possibility of synthesis of multi-component spinels by partial substitution of cations in position A and B giving rise to compounds of formula $(A_xA'_{1-x})(B_yB'_{2-y})O_4$. This accounts for the variety of reactions in which they have been used as catalyst. Moreover, partial substitution of A and B ions giving rise to complex oxides is possible, while maintaining the spinel structure.

Spinels show interesting catalytic properties, in which the properties are controlled by the nature of ions, their charge and site distribution between T_d and O_h sites. Bautiles and Barboax [17] investigated the surface composition of normal spinel oxides using differential neutron diffraction (DND) techniques. This method suggested the surface of normal spinel consists of mixture of (110) and (111) planes. However, in the case of distorted spinels like γ-alumina, 80% of the exposed crystallites have the (110) facets. They observed that such planes contain only octahedral cation with oxygen anion. Probing the surface layer of the materials has become more accurate after introducing low energy ion scattering (LEIS) technique in surface science. The most characteristic feature of the LEIS is that the peaks in the spectra results exclusively from the scattering by top-most atomic layer. Jacobs et al. [18] applied this technique to evaluate spinel surface composition. Their work revealed that O_h sites are exposed almost exclusively at the surface of the spinel oxide and the catalytic activity of many such systems is mainly due to O_h cations. Several other workers also believed that T_d site ions are either inactive or contributing only a little to the catalytic properties [19]. Another possible reason for the lower activity of T_d cations comes from the fact that, due to lower coordination number, the metal oxygen bonds will be stronger and hence such cations are not easily accessible to the reactants. Nonetheless, under the catalytic reaction conditions what are the facets are exposed on the surface, and how does the surface composition changes were not studied in detail. There are some indications from our earlier work [13,14] suggesting that both A and B cations might have equal significance towards the final catalytic activity.

Catalytic activities of different spinel compounds have been well established for various reactions. They are reported to be thermally stable and maintain increasingly sustained activity for a variety of industrially important reactions such as decompositions of N_2O and H_2O_2 [20-23], hydrodesulfurization of petroleum crude [24], oxidation of CO, H_2, methane and methanol [25-29], oxidative dehydrogenation reaction [30-37], hydrogenation of organic compounds [38,39], treatment of automobile exhaust gases [40,41], and dehydrogenation of ethyl benzene and alcohols etc [42-46]. In addition to the above-mentioned reactions, a variety of organic transformation reactions are carried out over spinel oxides. For example, Roesky etal [47] reported an improved synthesis method for indenes (from indanones) and styrenes (from acetophenones) by use of a ZnO/Al_2O_3 spinel catalyst. Cobalt modified $MgAl_2O_4$ used for the production of ethene and other industrial gases by diesel fuel pyrolysis [48]. Spinel catalysts are also employed for reducing NO by using NH_3 and propane as reductants [49,50]. Another synthetically important reaction is the synthesis of methyl formate. Sato et al [51] reported the vapour phase dehydrocoupling reaction of methanol to methyl formate using $CuAl_2O_4$ catalyst. Reactions such as hydrogenation of maleic anhydride to γ-butyrolactone and tetrahydrofuran [52], oxidation of benzoic acid to phenol [53] and styrene to benzaldehyde [54,55], reduction of fatty esters into higher alcohols [56], methanation of CO and CO_2 [57-59], synthesis of higher alcohols [60] etc were successfully applied over various types of spinels. Other miscellaneous and environmental catalytic studies using spinels were also found in the literature [61-65] Among the spinel compounds ferrospinels have been used as effective catalysts because of the ease with which iron can exchange its oxidation state between 2 and 3. Another important feature attributed with these materials, from the commercial viewpoint, is that spinel structure provides high structural stability so that these materials can withstand reducing conditions to a reasonable extent. Even if reduction of Fe^{3+} to Fe^{2+} occurs, spinel structure remains unaltered and upon reoxidation Fe returns to the original oxidation state [40].

Spinels used in the catalytic applications are generally synthesized by low temperature co-precipitation method [66] or sol-gel route [67] are sufficient to overcome the drawbacks such as low surface area (<5 m^2/g), and inhomogeniety at atomic level, which are generally associated with high temperature solid-state preparation. Uniform contrast observed throughout the surfaces of ferrospinels, prepared by co-precipitation (Figure 2) suggesting the distribution of metal-ions in a uniform manner, without any preferential segregation. However, particle size varied between 20 nm to few micrometers with a relatively high surface area between 30-50 m^2/g. Nevertheless, in general, surface morphology, do not seem to have a significant role towards the alkylation reactions reported [13,14].

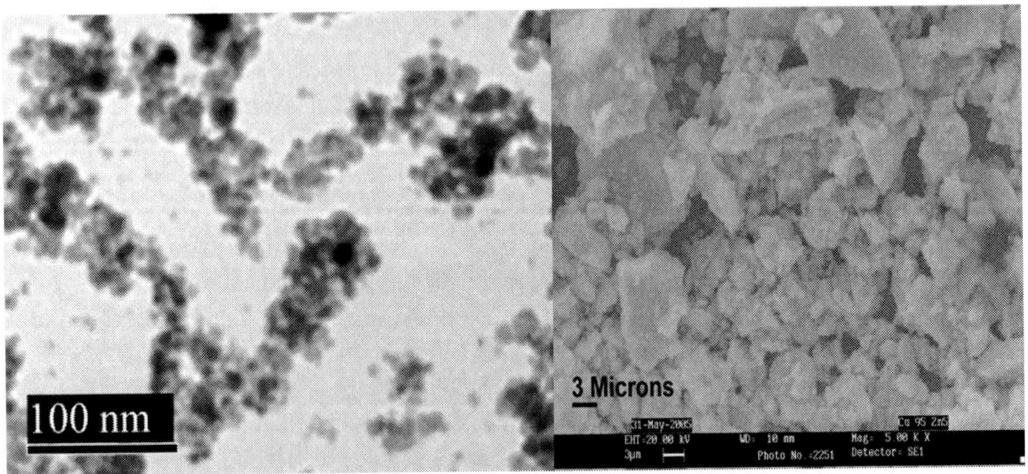

(a) (b)

Figure 2. Representative TEM and SEM images obtained from (a) CoFe2O4 and (b) Cu0.95Zn0.05Fe 2O4, respectively.

3. Alkylation of Aromatic Substrates

The alkylation of aromatic compounds is widely used in the large-scale synthesis of fine chemicals and valuable synthetic intermediates [68]. The reaction is essentially replacing a hydrogen atom of an aromatic compound by an alkyl group derived from an alkylating agent. The substitution for hydrogen by an alkyl group can be on the aromatic ring (C-alkylation), and/or on the side chain of aromatic compounds, or on the heteroatom (N-, O- alkylation) present in the aromatic compounds like anilines, phenols, etc. Nevertheless, various experimental factors decide the center for substitution of hydrogen by an alkyl group. Thus the nature of the alkylated product formed depends on the catalysts employed, and the reaction conditions applied. However, the former influences the products formation to a larger extent than the latter. In general, acidic catalyst favors aromatic ring alkylation, whereas basic catalyst favors side chain or heteroatom alkylation [69]. An overwhelming majority of alkylation reactions are acid catalyzed. The base-catalyzed alkylation of aromatic compounds has received much less attention than that catalyzed by acids.

3.1. Phenol Alkylation

Alkyl phenols are industrially important [68,70], either directly or as chemical intermediates and the important and representative applications of some of the alkyl phenols are given in Table 2. The stringent specifications and the demand of these chemicals necessitate the development of catalytic systems and processes towards highly selective production of various alkyl phenols.

Table 2. Industrial applications of various alkyl phenols

Alkyl phenol	Applications
o-Cresol	In the production of novolak resins, 4,6-dinitro – 2 – methyl phenol (a herbicide) and antioxidants. o-Cresol is a component of crysilic acids, which are used as solvents in a number of coating applications.
2,6-xylenol	Poly(phenylene oxide) resin, epoxy resin.
Anisole	Additives in gasoline to boost octane, used for the production of dyes, agricultural chemicals and antioxidants.
2-Ethyl phenol	Starting material for photochemicals. Intermediate for the synthesis of benzofuran.
2-Isopropyl phenol	An efficient antiskinning aid in air-drying coatings.
2-tert-butyl phenol	Starting material for the synthesis of antioxidants and agrochemicals.
2,4-di-tert-butyl phenol	In the manufacture of its triphosphite that is employed as a co-stabilizer for poly (vinyl chloride), and of its benzotriazole derivatives that are used as UV absorbers in polyolefins.
2,6-di-tert-butyl phenol	An indispensable building block in the synthesis of higher molecular mass antioxidants and light-protection resins for plastics, especially polyolefins.

3.1a Phenol Methylation

Phenol methylation to 2,6-xylenol has been widely studied for the past few decades owing to the room for improvisation from the viewpoint of product selectivity. Generally during phenol methylation to 2,6-xylenol, occurs via sequential methylation of phenol to *o*-cresol to 2,6-xylenol, various reaction parameters mediate the selectivity between the two. For instance, when the reactants stoichiometry of methanol to phenol molar ratio > 2, and significant residence time of o-cresol may favor 2,6-xylenol selectivity. However, excess methanol is often used, since some amount of methanol tend to undergo oxidation into various reformate products [71] under vapor phase condition. Similarly, reaction temperature, catalyst acid-base property, and space velocity of the reactant are the parameters that govern the selectivity to 2,6-xylenol.

Sreekumar et al. [72] reported the methylation of phenol with methanol to o-cresol and 2,6-xylenol over Co and Ni containing ternary ferrospinel system ($Ni_{1-x}Co_xFe_2O_4$). The total *ortho* selectivity (*o*-cresol+2,6-xylenol) was 94%, and only traces of anisole, and hardly other xylenol or cresol isomers were detected regardless of the catalyst composition at a reaction temperature 350°C. Phenol conversion and individual selectivities for *o*-cresol and 2,6-xylenol depend strongly on the catalyst composition rather than any other parameter. Up on increasing Co-content 2,6-xylenol selectivity increases with concomitant decrease in the *o*-cresol selectivity. It was demonstrated that substitution of Co^{2+} for Ni^{2+} ions increases the acidity of the catalyst system that brings out high phenol conversion and 2,6-xylenol selectivity; relatively stronger acid sites were required for the sequential methylation of *o*-cresol to yield 2,6-xylenol. A plausible reaction mechanism proposed [13, 73] is shown in a simple manner in Figure 3. Since phenol is highly polar, even weak basic sites can easily abstracts a H^+ from the phenolic OH group. The phenolate anion is resonance stabilized, and a stronger acid site is required to interact with the same. Thus, it is the catalyst acidity, rather than the basicity, that controls the catalytic activity.

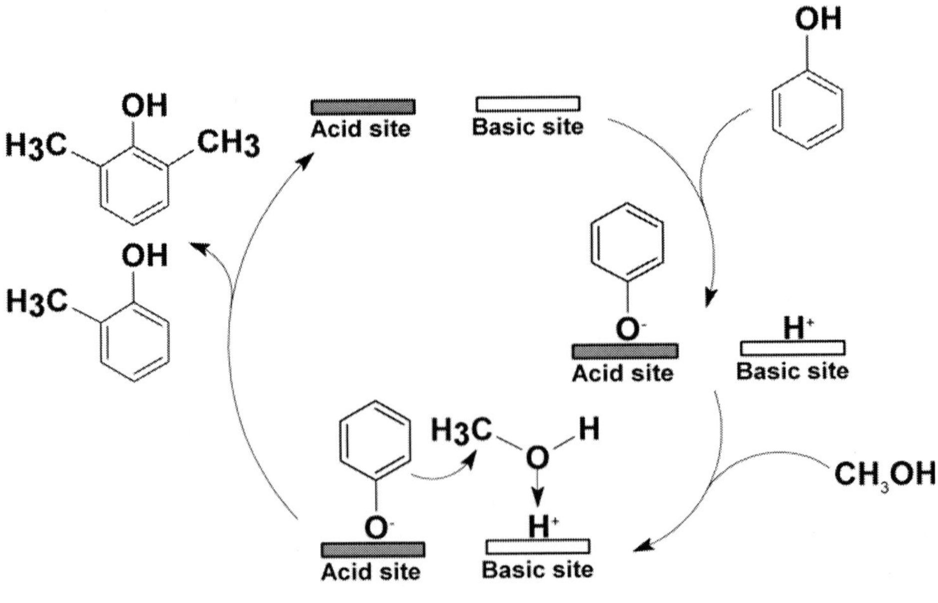

Figure 3. Reaction mechanism of ortho methylation of phenol using methanol.

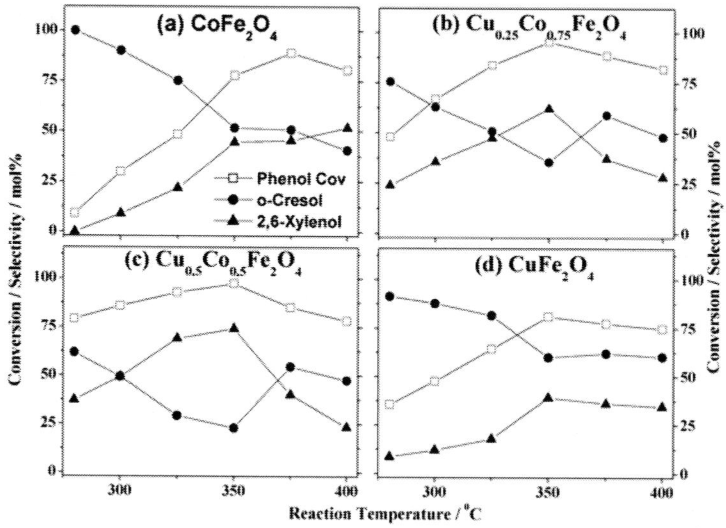

Figure 4. Phenol conversion, and o-cresol and 2,6-xylenol selectivity dependence on catalyst composition and reaction temperature with a feed composition of MeOH;PhOH = 5:1, at time on stream = 3h. Note that 2,6-xylenol selectivity increases with phenol conversion, and at the expense of o-cresol with all catalyst compositions, indicating the first order conversion dependence and the sequential methylation, respectively.

Mathew et al. [74] demonstrated that phenol methylation to 2,6-xylenol over Cu-Co-ferrospinel ($Cu_{1-x}Co_xFe_2O_4$) is sequential and the analysis of the kinetic data indicates that under most reaction conditions the rate of the overall process might be limited by the first methylation step to *o*-cresol. As shown in Figure 4, phenol conversion over all of the $Cu_{1-x}Co_xFe_2O_4$ catalyst compositions increases with temperature up to 350°C. The first order conversion dependence of 2,6-xylenol selectivity is evident from the linear relation between phenol conversion and 2,6-xylenol selectivity. 2,6-xylenol selectivity increases from the end compositions ($CuFe_2O_4$ and $CoFe_2O_4$) to a maximum at intermediate composition, x = 0.5 ($Cu_{0.5}Co_{0.5}Fe_2O_4$), in contrast to the *o*-cresol selectivity that decreases. This concurrent increase in 2,6-xlyenol selectivity at the expense of *o*-cresol shows the mode of phenol methylation is undoubtedly sequential. It is also inferred from the absence of other cresol and xylenol isomers that the conversion of *o*-cresol to 2,6-xylenol is relatively a fast reaction. The better catalytic results observed with x = 0.5 is attributed to an optimum distribution of Cu-species with heteroatom neighbors (Co and Fe) on the catalyst surface under reaction conditions. Further, the importance of 1:1 ratio of Cu:Co combination in $Cu_{0.5}Co_{0.5}Fe_2O_4$ is evident from the high catalytic activity observed at 280^{0}C, and comparable to that of the end compositions at T ≥350^{0}C.

Cu-Mn mixed-oxide binary spinel catalysts ($Cu_xMn_{3-x}O_4$, where x = 0, 0.25, 0.5, 0.75 and 1) prepared through co-precipitation method exhibit phenol methylation activity under vapor phase conditions [75]. All of the catalysts, irrespective of the compositions, produced only C-methylated phenols. However, a total ortho selectivity of 100% with 2,6-xylenol selectivity of 74% was observed over x = 0.25 compositions at 400°C. This composition was found to be relatively stable under reaction conditions compared with the other compositions studied. The catalysts with high copper content suffered severe reduction under methylation conditions whereas, catalysts with low copper content had a hausmannite phase (Mn_3O_4) that sustained with minimal reduction. Cu^{2+} reduces to Cu-metal from Cu-rich compositions under reaction conditions indicate the less effectiveness of the system, mainly due to the disintegration of $Cu_xMn_{3-x}O_4$ to Cu^0 and Mn_3O_4. This is further supported by XPS and X-ray diffraction studies [75]. It was also described that the strong basicity and weak acidity associated with these catalysts were responsible for the selective formation of 2,6-xylenol.

Very recently [76], the phenol methylation mechanism was compared between basic catalysts (Mg-Fe oxides) and Bronsted acid catalyst (H-Mordenite). The catalyst having spinel phase ($MgFe_2O_4$) before and after reaction is more active than individual oxides MgO and Fe_2O_3 which possess comparable but low activity. The main product was *o*-cresol (<350°C), however, its selectivity decreased in favor of 2,6-xylenol when the reaction temperature was increased. The key reaction step was identified as the dehydrogenation of methanol to formaldehyde, and generation of the active methoxy species for the methylation of phenol to *o*-cresol and 2,6-xylenol. The generation of active methoxy species through formaldehyde formation is attributed as Cannizaro-type reaction with intermolecular disproportionation of formaldehyde. Briefly, a nucleophilic attack of the O^{2-} surface species by the carbonyl (with development of the formate), is accompanied by a hydride transfer to a second molecule of adsorbed formaldehyde, which is converted to an active methoxy species. In contrast, Bronsted acid type catalyst activation of methanol led to the development of electrophilic methyl species and to the formation of anisole and cresols. This is in contrast to the mechanism shown in Figure 3, in which the Lewis acid sites play a significant role. Indeed the above differences in reaction mechanism is due to the deployment of acidic or basic catalysts. It is interesting to note that product selectivity varied with respect to catalysts preparation and source of metal ion in the catalyst.

Grabowska et al [77] accounted phenol methylation over zinc aluminate ($ZnAl_2O_4$) prepared through hydrothermal conditions from zinc acetate and aluminum isopropoxide (catalyst A) and basic aluminium nitrate (catalyst B) as metal ions source. Both the catalysts were showing phenol methylation activity, however, they show different product selectivity trend depending on the aluminium source used for the catalyst preparation. Very high ortho selectivity was obtained with catalyst A, whereas catalyst B was active in O- as well as C-alkylation. The reason is noteworthy that catalyst A contains pure $ZnAl_2O_4$, and catalyst B contains γ-Al_2O_3 as undesired component besides $ZnAl_2O_4$, and the reason for O-alkylation to anisole.

3.1b. Efficient Methylation Through Synergism

It is evident from the literature results that multi-component metal oxide systems show better activity and selectivity towards phenol methylation than individual metal oxides [72-82]. The mixture of metal oxides brings out a combined effect, in other words a synergistic behavior, enhancing the catalytic activity. In general, such multi-component systems require relatively low temperature for better catalytic performance than the individual components. For instance, mixed oxides show catalytic activity at the temperature range between 300 and 400°C, and they are more active and selective for 2,6-xylenol production. Thus the synergetic or combined effect of metal-ion components has considerable influence on catalytic performance. Due to synergism, the combined effect of metallic ions shows more homogeneity in their property towards a particular reaction and hence high selectivity of a specific product. This is possible only when the valence band (VB) orbitals of the metallic ions of mixed oxide system have comparable energy. If VB orbitals energy are widely separated, significant overlapping between the metal-ion VB orbitals is energetically unfavorable and hence reduces the catalytic effectiveness. For example, binary oxide systems MO-Fe_2O_3 (M = Mg, Ca, Ba, Co, Ni, Zn and Cu) were found to be active and selective for 2,6-xylenol production at 280-350°C, whereas the individual oxides components require more than 400°C for achieving good conversion and high 2,6-xylenol selectivity [7].

It was clearly demonstrated how the phenol methylation activity was strongly influenced by the metal orbitals overlap [74, 75, 80-82]. XPS and XAES investigation on fresh $Cu_{1-x}Co_xFe_2O_4$ (where x = 0, 0.25, 0.5, 0.75 and 1) indicated the existence of Cu^{2+}, Fe^{3+}, Co^{2+} and Co^{3+} ions. However, the spent catalysts after phenol methylation exhibit composition dependent oxidation states. Cu $2p_{3/2}$ XPS results from spent catalysts are interesting compared to fresh Cu-containing catalysts, which displays Cu^{2+} at 934.2±0.2 eV and a satellite at 942 eV (Figure 5). Spent $CuFe_2O_4$ shows a peak at 932.6 eV with a full-width at half maximum (FWHM) of 1.4 eV and without any satellite. Nonetheless, for x ≥0.25, there is a large broadening observed with a satellite peak at high binding energy. Deconvolution reveals the contribution of different species. For simplicity, it is shown only for $Cu_{0.5}Co_{0.5}Fe_2O_4$ in Figure 5. Important points to be noted are the following: Cu-species are reduced during methylation reaction and the extent of copper reduction decreases with increasing Co-content (inset to Figure 5), and the binding energy of Cu^{2+} decreases by 0.4 eV for x = 0.5. The above points clearly demonstrate that there is a strong influence from Co on Cu and its reducibility. The inset to figure 5 shows the reducibility of Cu and it is calculated as the amount of reduced Cu to total Cu-content. Reducibility of Cu decreases with increasing Co-content. High polarizability induced by the

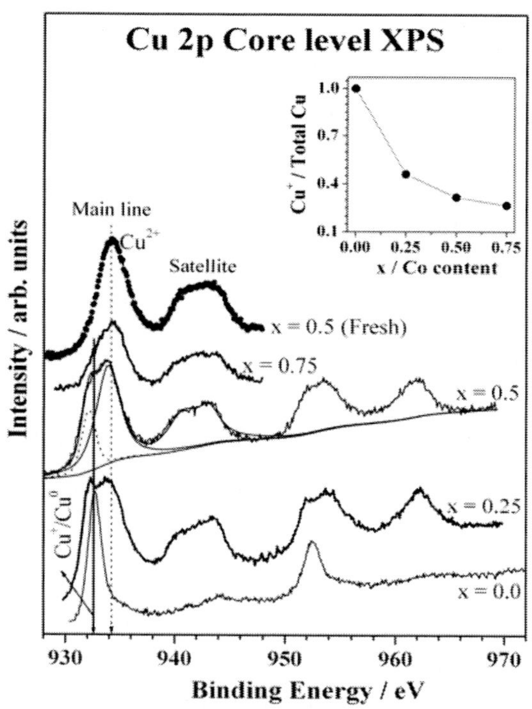

Figure 5. Cu 2p core level XPS measured on spent Cu1-xCoxFe2O4 catalysts. Cu 2p3/2 core level spectrum obtained from x = 0.5 calcined catalyst is shown for comparison.

Inset shows the extent of surface Cu-reduction from the ratio of Cu^{+}/total Cu. Note a decrease in copper reducibility with increasing Co-content.

formation of Co-containing Cu-spinel would change the nature of the Cu-O bond [83]. A complete reduction of Cu^{2+} to Cu and Co^{3+} to Co^{2+} was witnessed at the end compositions x = 0 and 1, respectively. At intermediate compositions, the reducibility of Cu^{2+} decreases with increasing x along with a partial reduction of Fe^{3+} to Fe^{2+} and a decrease in the Fe 3d binding energy was observed.

Valence band shows (Figure 6) a change from marginal overlap between 3d bands of Cu and Co on fresh catalysts to a large overlap on spent catalysts. The changes observed in the VB region on spent catalysts at intermediate compositions offer a picture on the overlap of 3d bands of all metal ions. Partial reduction of Cu^{2+} (to Cu^{+}) and Fe^{3+} (to Fe^{2+}) brings in the energy overlap between 3d levels of Co, Cu and Fe, highlights the Cu-Co synergism especially in the intermediate composition (x = 0.5) and that was attributed to very high stability and activity associated with x = 0.5 system towards phenol methylation. Further, Co 3d level is bridging the energy gap between 4s and 3d bands of Cu and helps to integrate the material electronically.

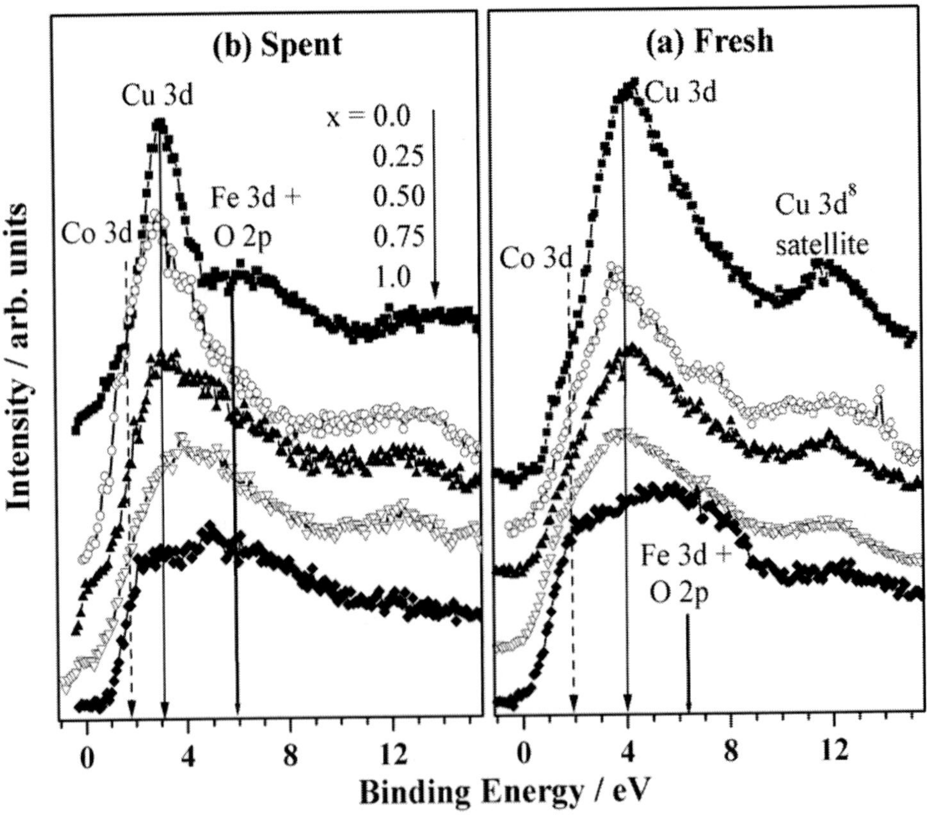

Figure 6. Valence band XPS of fresh and spent catalysts. A significant decrease in the energy gap between the 3d levels of Cu and Co and a decrease in the overall bandwidth are observed on spent catalysts. Fe 3d level decreases on spent catalysts at $0.25 \leq x \geq 0.75$. Reprinted from Journal of Catalysis, 210, Mathew T., et al., 2002, 405-417 with permission from Elsevier.

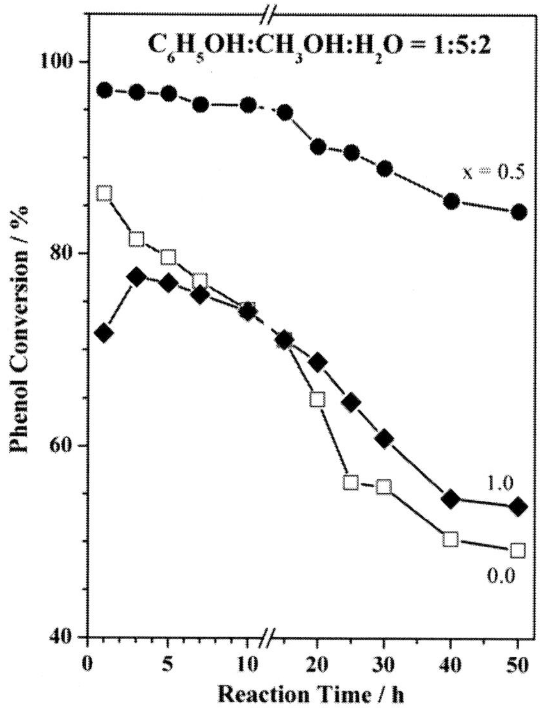

Figure 7. Effect of synergism on phenol conversion over three selected compositions of Cu1-xCoxFe2O4, x = 0.0, 0.50 and 1.0. Phenol methylation was carried out with phenol:methanol:water composition of 1:5:2 at 3500C and at WHSV of 0.869 h-1. 2,6-xylenol selectivity was maintained at 75 mol% throughout the 50 h period.

The importance of Cu-Co synergism and its influence over the phenol methylation activity is shown in Figure7. PhOH:MeOH:H_2O ratio of 1:5:2 was used in this study. Generally, a prolonged stability is observed with considerably low deactivation compared to the feed without H_2O (see, Figure 4) [74]. Further, 2,6-xylenol selectivity was maintained at 75 mol% throughout the 50 h period. It is evident that x = 0.5 system is more stable and active than other two catalyst compositions (x = 0 and 1) that deactivates fast with time on stream. The prolonged activity and the stability observed with x = 0.5 system is related to the combined role of all the three metals towards phenol methylation. Advantage of water in the feed is that the steam would suppress the coke formation and maintains catalytic activity for longer reaction time. Further, phenol methylation carried out in hydrogen atmosphere [13,74] showed fast deactivation of the catalyst with low PhOH conversion demonstrate the large reduction of catalyst suppresses the methylation, and hints the metallic species are not the methylation centers.

The activity trend obtained with $Cu_{1-x}Co_xFe_2O_4$ catalysts is well supported by the surface metal ions composition determined from XPS analysis. Figure 8 displays the Cu/(Co+Fe) (Co/Fe for x = 1) ratio calculated from XPS results in the left panel and phenol conversion with products selectivity for all catalyst compositions in the right panel. This exercise is mainly to understand the distribution of metal ions and their heterogeneity on the surface, as it directly influences catalytic activity. On fresh catalysts, relative Cu-content decreases linearly with decreasing Cu-content and it is in good correlation with bulk Cu-content measured by x-ray fluorescence. A high Cu/Fe ratio is found on spent catalysts at $0.25 \leq x \leq 0.75$. It is to be emphasized here that the predominant Fe-enriched surface at x = 0 changes to a Cu+Co-enriched surface at x = 0.25 and 0.5 on spent catalysts.

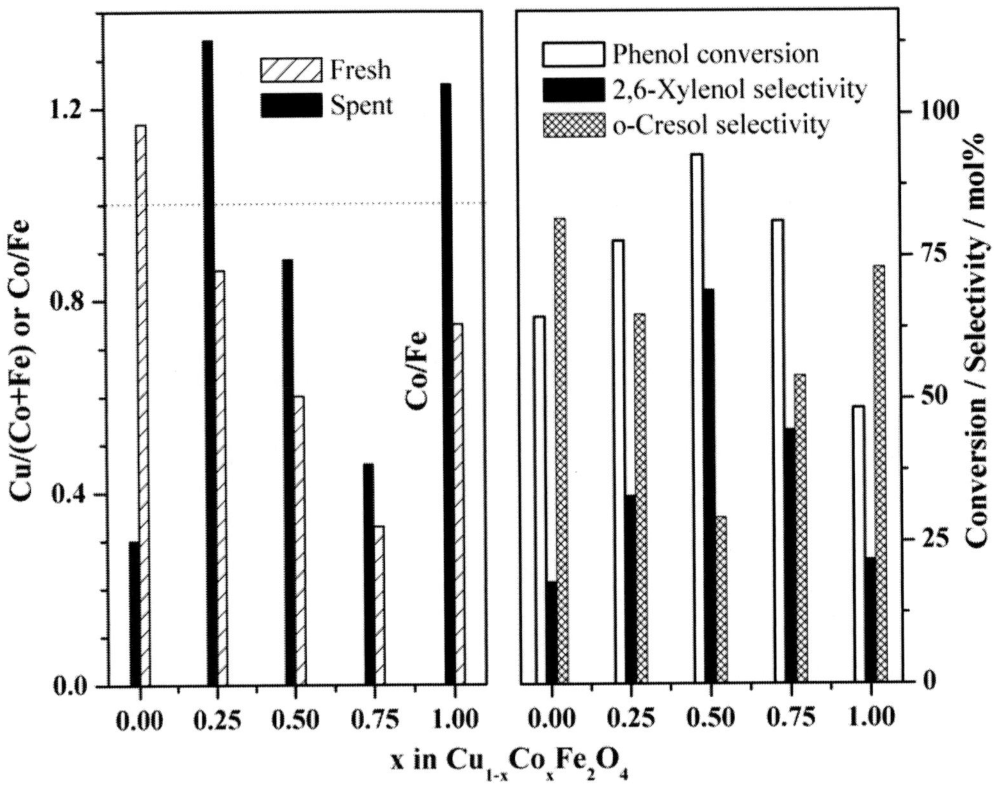

Figure 8. Comparison of phenol conversion and ortho products selectivity performance of Cu1-xCoxFe2O4 catalysts at 3250C, TOS = 3h (right panel) and atomic ratio of Cu/(Co+Fe) (or Co/Fe) on Cu-containing (Cu-less) catalysts (left panel). Note the large production of desired 2,6-xylenol and Cu/(Co+Fe) = 0.9 at x = 0.5 composition on spent catalyst. Reprinted from Journal of Catalysis, 210, Mathew T., et al., 2002, 405-417 with permission from Elsevier.

Cu/(Co+Fe) ratio decreases linearly with increasing Co-content on fresh catalysts; however there is no such trend observed on spent catalysts. However, on spent catalyst the ratio Cu/(Co+Fe) is close to one for x = 0.5 that shows maximum phenol conversion and 2,6-xylenol selectivity. This correlation hints that no single ion, but a combination of metal ions, specifically the heterogeneity of the surface, decides the course of the reaction towards high 2,6-xylenol selectivity. Further, the above observation hints that it might be the pair of heteroatoms, say Cu-Co or Cu-Fe, that is preferred for two different reactants to be adsorbed on nearby sites for the desired high selectivity. Large production of 2,6-xylenol and a ratio of 0.9 for Cu/(Co+Fe) at x = 0.5 requires the optimum distribution of active metal ions to have a highly heterogeneous surface. When the surface is dominated by a single metal ion (Fe, Cu or Co on x = 0.0, 0.25 or 1, respectively), desired reactivity is not achieved. Above drastic change in surface composition on spent catalyst is due to contribution from significant extent of side reactions, like methanol reforming on x = 0.0, which leads to the reduction of Cu^{2+} to Cu-metal and hence a loss in activity at TOS > 10 h. However, a high catalytic activity maintained on x = 0.5 even at TOS = 50 h clearly demonstrates the heterogeneity of the surface.

3.1c Acid-Base Properties of $Cu_{1-x}co_xfe_2o_4$

A systematic characterization of surface acid-base property of $Cu_{1-x}Co_xFe_2O_4$ was carried out through FTIR spectroscopy of adsorbed probe molecules, such as pyridine and CO_2 [13,78]. For oxide like ferrospinel, the acid-base properties can be very decisive in determining their catalytic activities. Temperature dependent IR studies of pyridine adsorbed spinels, and on deliberately added metal oxide to the spinel phase exemplifies the contribution of metal ions and its coordination state towards Lewis acidity. Predominant IR bands, due to pyridine coordination to an electron deficient metal atom, at ν_{8a} (1606 cm^{-1}) and ν_{19b} (1446 cm^{-1}) confirm the major contribution of acidity is due to Lewis acid sites on all $Cu_{1-x}Co_xFe_2O_4$ compositions. Relative acidity calculated from the above two bands increases linearly with increasing Co-content. Intensity of various bands, due to CO_2 adsorption on $Cu_{1-x}Co_xFe_2O_4$, observed between 1700 cm^{-1} and 1260 cm^{-1} (symmetric OCO stretching of bidentate and unidentate carbonates at 1430-1260 cm^{-1}, asymmetric OCO stretching of unidentate carbonate and symmetric OCO stretching of bicarbonate at 1580-1435 cm^{-1} and asymmetric OCO stretching of bicarbonate and asymmetric OCO stretching of bidentate carbonates at 1680-1600 cm^{-1}) generally decreases with increasing x, indicates a decrease in basicity from x = 0 to 1.

It was found that the acidity (basicity) of the $Cu_{1-x}Co_xFe_2O_4$ spinel system increases (decreases) from x = 0 to 1. A correlation between acidity and catalytic performance reveals that an intermediate acidic character enhances the phenol methylation activity. Figure 9 displays a comparison between the catalytic performance in terms of 2,6-xylenol yield obtained at a reaction temperature 325^{0}C with a phenol:methanol = 1:5, and relative acidity and basicity obtained from pyridine and CO_2 adsorption studies. In this reaction, methylation is sequential and 2,6-xylenol is produced at the expense of o-cresol and for simplicity o-cresol detail is not given in Figure 9. A steep increase in relative acidity with Co concentration clearly supports an increase in acidity both in terms of strength and number. It is clear from Figure 9 that neither highly acidic nor basic catalyst at x = 1.0 and 0.0, respectively, is active towards phenol conversion to produce 2,6-xylenol. However $0.25 \geq x \leq 0.75$ show better catalytic performance and x = 0.5 is superior and demonstrates an intermediate acidity character. This clearly indicates that an optimum acid-base property favors the overall reaction. It is also interesting to note that usually Friedel-Crafts alkylation reactions require highly acidic catalysts, which is not entirely true in the present case.

3.1d Phenol Methylation Mechanism on $Cu_{1-x}co_xfe_2o_4$

Kotanigawa and co-workers [7-9] studied the adsorption behavior of phenol and methanol on a ZnO–Fe_2O_3 system and highlighted the importance of acid–base sites for the selective *ortho* methylation. However, detailed studies of the adsorption behavior of possible products and reactants of phenol methylation on catalytic systems and their interaction among them are not available widely.

The mechanism of phenol methylation on Cul-xCoxFe2O4 and the participation of Lewis acid-base pair site in the methylation step were recently demonstrated in detail [79].

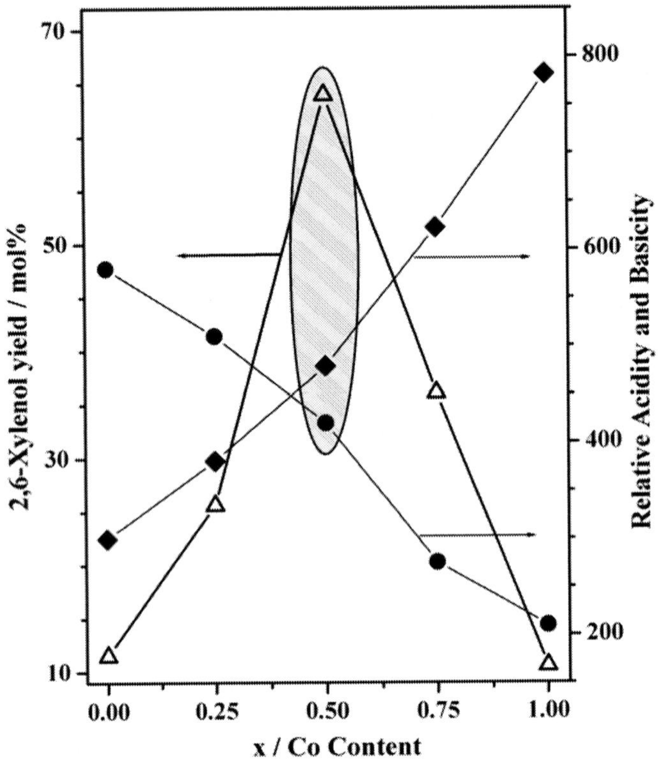

Figure 9. A comparison between the catalytic performance through 2,6-xylenol yield, and relative acidity (solid diamonds) obtained from the area of ν8a and ν19b bands from pyridine adsorbed on catalyst surface at 1000C and the relative basicity (solid circles) obtained from the area of the bands between 1700 and 1250 cm-1 from CO2 adsorbed on Cu1-xCoxFe2O4 catalysts at 250C.

The interaction of adsorbed reactants (phenol and methanol adsorbed separately and co-adsorbed) and possible reaction products of phenol methylation with the Cu1-xCoxFe2O4 system has been studied at temperatures between 100oC and 350oC and probed by in situ FTIR spectroscopy. The spectra of adsorbed methanol, phenol and methylated products on catalyst surface, at 100oC, did not possess much changes compared to the spectra of pure components that indicated the molecular adsorption of species on catalyst surface. The remarkable changes in the spectra occur, above 100°C due to the chemisorption of substrates, were observed and correlated with the observed reaction trend.

The spectra obtained from the chemisorption of methanol onto catalyst above 100°C indicated the progressive oxidation of methoxy species to formate via dioxymethylene/HCHO and finally to CO, CO_2, and H_2. Phenol adsorbed on the surface Lewis acid-base pair site and dissociated to phenolate anion and proton. The formation of phenolate anion and proton were discerned from the strong intense C-O stretching vibration and the disappearence of phenolic O-H stretching vibration, respectively. Importantly, there were series of definite low intensity bands between 2050 and 1780 cm^{-1} that were identified as the out-of-plane aromatic C-H bending vibrations [79, 84-85]. These bending vibrations are possible only if the phenyl ring of phenol is perpendicular to the catalyst surface.

Figure 10 shows the temperature-dependent FTIR spectra of $Cu_{1-x}Co_xFe_2O_4$ catalysts adsorbed with 1:5 mole ratio of phenol:methanol at 100^0C and subsequently heated up to

350^0C. Results obtained on $Cu_{0.5}Co_{0.5}Fe_2O_4$ adsorbed at 300^0C with phenol and different products are shown in Figure 10d for comparison as well as to demonstrate the ortho-methylated products formation around 300^0C. Very interesting features were observed due to the above reaction mixture undergoes chemical changes with temperature and catalyst composition too. The main points worth highlighting from the co-adsorption of PhOH and MeOH are as follows. (1) First, phenol features dominate all the catalyst surfaces despite a low phenol content in the reaction mixture. (2) Low-intensity ν_{C-O} bands of methoxy moieties are observed around 1067 cm^{-1} at 100^0C, and its intensity increases with increasing Co-content on the catalyst. At 200^0C the methoxy moiety disappears on $x<1$, however a weak ν_{C-O} still appears on $x=1$. (3) No bands due to MeOH oxidation were observed at any temperature on any catalyst composition, demonstrating that the MeOH oxidation is prohibited in the presence of phenol. (4) ν_{C-H} stretching vibrations of phenol disappear with increasing temperature on $x<1$; however, the intensity of the above band at any temperature increases with Co content. Further $x=1$ does not show considerable desorption of phenol up to 350^0C. (5) A sharp ν_{C-O} band appearing at 1250 cm^{-1} at 100^0C on all catalyst compositions broadens at high temperatures on $x<1$ and a shoulder, characteristic of ortho-methylated phenols, is visible around 1210 cm^{-1}; note the large valley between bands at 1250 and 1166 cm^{-1} on $x = 1$ and a distinct shoulder at 1280 cm^{-1} at 200^0C. (6) The C–O–H bending band at 1166 cm^{-1} disappears below 300^0C on $x<1$; however, it strongly persists on $x = 1$ at 350^0C; $O–H_{str}$ bands also persists at 350^0C on $x = 1$ and disappear gradually with increasing temperature for $x < 1$. (7) The aromatic ring vibration of phenol at 1585 cm^{-1} is shifted to 1593 cm^{-1} and exhibits a shoulder at 1570 cm^{-1}. The narrow band at 1481 cm^{-1} observed on phenol adsorption is shifted to 1487 cm^{-1} and overlapped with an additional band at 1472 cm^{-1}. The above four bands are characteristic bands of *o*-cresol and 2,6-xylenol. (8) Out-of-plane C–H (overtone) vibrations between 2050 and 1700 cm^{-1} are seen nicely for all catalyst compositions; however, it disappears for $x = 0$ and 0.5 at >200 and >300^0C, respectively; It remains observed on $x = 1$ at 350^0C. (9) A good correspondence between o-cresol and/or 2,6-xylenol, and reaction mixture on x = 0.5 at 300^0C suggesting the products formation between 200-300^0C. All the above points hint that the co-adsorption of methanol and phenol leads to a good interaction among them above 100^0C on $x<1$ and there are indications toward methylated product formation at 200^0C. Contrarily, $CoFe_2O_4$ indicates a low profile interaction among the reactants, even at 300^0C, and in combination with the above points indicates that the phenol dissociation and methyl cation availability could be the limiting factors to the overall methylation reaction; however, the *ortho* methylation step is somewhat fast and occurs between 200^0C and 300^0C on Cu-containing compositions, suggesting the importance of Cu^{2+} in the spinels.

On co-adsorbing phenol and methanol, the protonation of methanol occurs on the active acid sites as the labile protons released from the phenol reacted with methanol. Thus protonated methanol became electrophilic methyl species, which undergo electrophilic substitution. The ortho position of phenol, which is close to the catalyst surface, has eventually become the substitution reaction center to form the ortho methylated products (Figure 3). This mechanism was also supported by the competitive adsorption of reactants with acidity probe pyridine [79]. A sequential adsorption of phenol and pyridine has shown the formation of phenolate anion and pyridinium ion that indicated the protonation of pyridine.

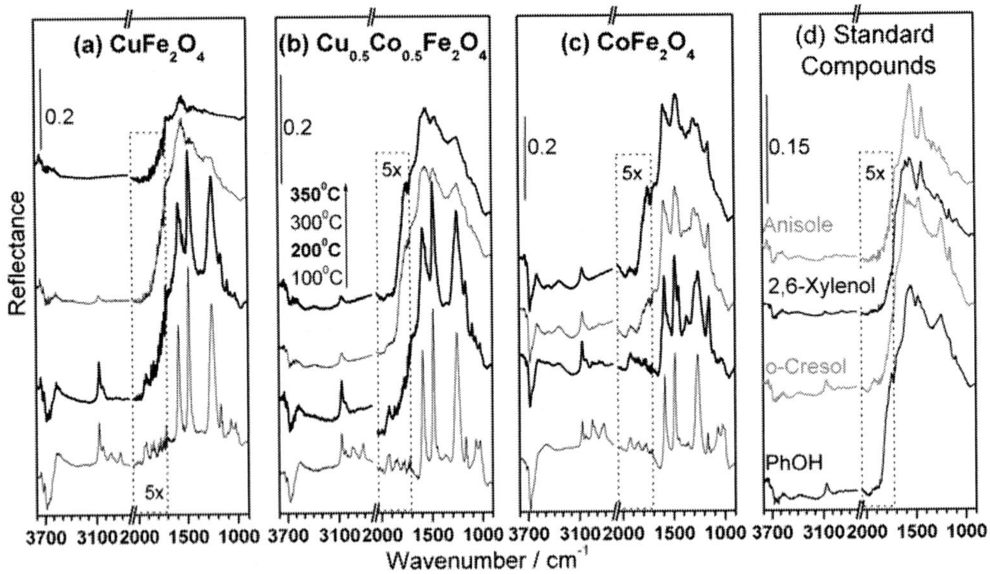

Figure 10. Temperature dependent FTIR spectra of methanol to phenol (5:1) mole ratio co-adsorbed on (a) CuFe2O4, (b) Cu0.5Co0.5Fe2O4, and (C) CoFe2O4 system. Out-of-plane C–H (overtone) vibrations between 2050 and 1700 cm−1 region is multiplied by a factor of 5 to show the features clearly. (d) Product compounds or phenol adsorbed on Cu0.5Co0.5Fe2O4 and IR recorded at 3000C. Note the similarity between o-cresol and/or 2,6-xylenol, and reaction mixture on x = 0.5 at 3000C suggesting the products formation between 200-3000C.

Another experiment in which sequential adsorption of phenol and pyridine then followed by methanol shows formation of pyridinium ion and phenolate anion; whereas no traces of methanol or electrophilic methyl species or formation of methylated products were identified on the catalysts surface. This result was supposedly confirmed from another experiment in which anisole and methanol were co-adsorbed on the catalyst. The spectra were referred to the molecular species of methanol and anisole without any significant interaction among them and above 200°C they simply desorbed from the catalyst.

Further, adsorption of methylated phenols (o-cresol and 2,6-xylenol) has given very weak and broad spectral features that were correlated to their weak interaction with the catalyst. Among the ortho-methylated phenols, 2,6-xylenol desorbs fast from the surface at 200°C than o-cresol. These methylated phenols, unlike phenol, desorbed from the catalyst above 200°C that was well below the actual reaction temperature (350°C). Hence, the desorption susceptible nature of the methylated phenols above 200°C facilitated the efficient methylation at 350°C [74]. In contrast to the Cu-containing ferrospinels, $CoFe_2O_4$ shows little interaction [74] of phenol with methanol, when they are co-adsorbed and this might be a limiting factor to the overall reaction.

3.2. Phenol Alkylation with Bulky Alkyl Groups

In contrast to phenol methylation, other alkyaltions are hard to carry out on spinel systems, as it is suggestive from very few available reports in the literature. Further, the ease of formation of the side products (alkyl phenols with small alkyl group) increases as the size of the alkyl group increases and hence poor selectivity for desired alkylated phenols is obtained. One possible way to avoid the decomposition of alkylating agents is to perform the reaction at relatively low temperatures; however, low reaction temperature at times may not effect good conversion of substrates. The catalytic study of phenol ethylation with ethanol on $Cu_{1-x}Co_xFe_2O_4$ ferrospinel shows 2-ethyl phenol (2EP) as the major product, irrespective of the reaction parameters and catalysts composition [80]. The main aspect is the low phenol conversion and relatively high optimum reaction temperature (375°C) for ethylation compared to methylation [74]. However, extent of diethylation is very low with significant side products (aromatics, alkyl phenols) formation in contrast to the very high selectivity observed for 2,6-xylenol in the case of methylation. This indicates the steric hindrance of alkylation species is an important factor for efficient ortho di-alkylation. Further, single ortho ethylated product is very much essential to reside sufficiently long time on catalyst surface for second ethylation to produce diethylated phenol. However, the steric hindrance of ethyl group might be unfavorable for the proper orientation of 2-ethyl phenol on $Cu_{1-x}Co_xFe_2O_4$ surface for diethylation, especially towards 2,6-diethylphenol. Further this also indicates the residence time of 2-ethylphenol is likely to be less and desorption might be preferred.

For detailed catalytic studies an optimum feed mixture using 5:1 molar ratio of EtOH:PhOH was chosen. Indeed the above ratio was selected by varying EtOH:PhOH ratio between 3 and 7, and the maximum PhOH conversion was observed at 5:1 [80]. The stability of the catalysts were studied from the TOS dependence at 375°C for 50 hours on three catalyst compositions namely, x = 0, 0.5 and 1 and shown in Figure 11. An increase in PhOH conversion is observed initially up to 5 h for x = 0.0 and 0.5; almost constant selectivity of 2EP is observed during the above time period with considerable amount of secondary products. Initial TOS up to 5 h is considered to be a transient state (TS). As the time progresses, PhOH conversion and 2EP selectivity reaches apparently to a steady state (SS) with a fluctuation of less than few percent up to 40 h and thereafter deactivation starts slowly. However, a continuous decline inactivity with increasing TOS is observed from the beginning on $CoFe_2O_4$. Progressive increase in 2EP selectivity for all x values is due to the retreat of secondary reactions, such as formation of aromatics, alkyl phenols etc. observed considerably at low TOS.

A better picture about the TS and SS can be seen in Figure 11b with 2EP yield as a function of TOS. Almost an invariable 2EP yield of 46±2% and 56±2% is seen in the SS for x = 0.0 and 0.5, respectively. No such SS is seen at x = 1.0 indicates a relatively poor activity of $CoFe_2O_4$ towards ethylation. Nonetheless, an equal bulk combination of Cu and Co at x = 0.5 displays a high and stable catalytic activity hints the Cu + Co combination is better than x = 0 or 1. Further, a product yield at 50 h is comparable or greater than that obtained at 1-5 h is observed for x = 0.0 and 0.5, demonstrates their superior stability. TS and SS observed in the above indicate a considerable modification of the surface from the initial state to the highly reactive state during SS.

It has been shown by several authors that a good correlation exists between activation energy (E_a) and ln A_0 for different reactions taking place over one catalyst [86-87]. Such a correlation is termed as ''compensation effect'' or ''isokinetic effect''. Indeed the higher catalytic activity found with intermediate catalyst compositions on $Cu_{1-x}Co_xFe_2O_4$ (x= 0.5) was attributed to the relatively low (E_a) for ethylation (Figure 12) on these compositions compared to the end compositions, x = 0 and 1.

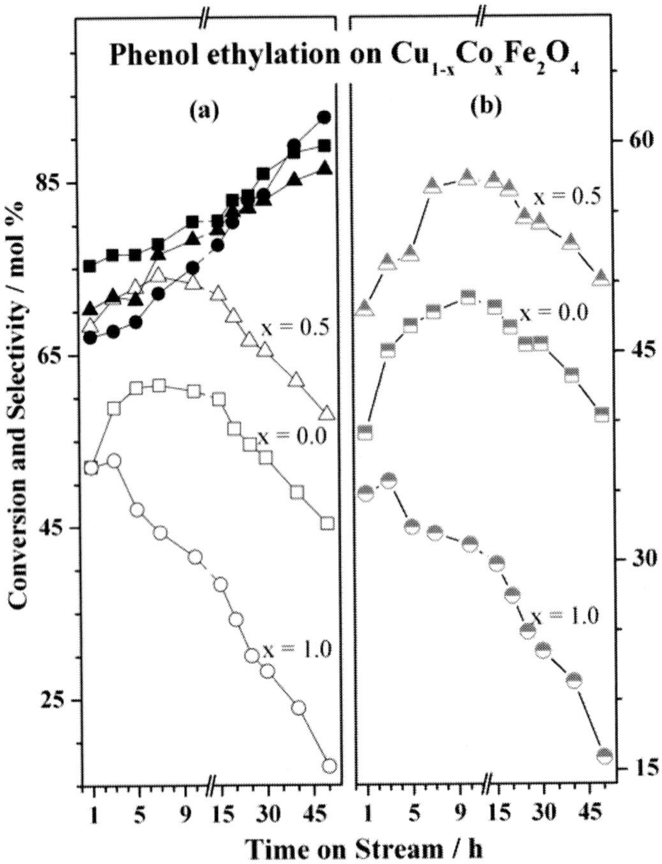

Figure 11. Time on stream dependence of (a) phenol conversion and 2-ethyl phenol selectivity and (b) 2-ethyl phenol yield at 3750C on Cu1-xCoxFe2O4 (x = 0.0, 0.5 and 1.0). Reactant feed ratio of EtOH:PhOH = 5:1 was used with a WHSV = 0.869 h-1. Conversion and selectivity is denoted by open and solid symbols, respectively. Note an increase in 2-ethyl phenol selectivity with increasing TOS on all catalyst compositions.

This is further confirmed from the calculation of compensation effect [80] by including all experimental values (ln A and E_a) within each group of related reactions in the linear regression analysis, and the compensation line was calculated (Figure 12).

Compensation effect on $Cu_{1-x}Co_xFe_2O_4$ can be explained by considering a range of acid sites with different acid strength present [78], and each one of them needing a different E_a for carrying out the reaction. In principle, it can be expected that increase in acidity decrease the increment of enthalpy between the reactants and the activated complex, and the entropy of such a complex. However, the nature of the active sites is not expected to change in an oxide series, but the acid strength and distribution and the resultant ''averaged-E_a'' changes and observed. A plot of E_a and ln A_o for $Cu_{1-x}Co_xFe_2O_4$ shows a straight line (Figure 12) with a very high linear regression coefficient of 0.9999 indicating the existence of a good compensation effect and hints the mechanism of PhOH ethylation remains the same at all x values.

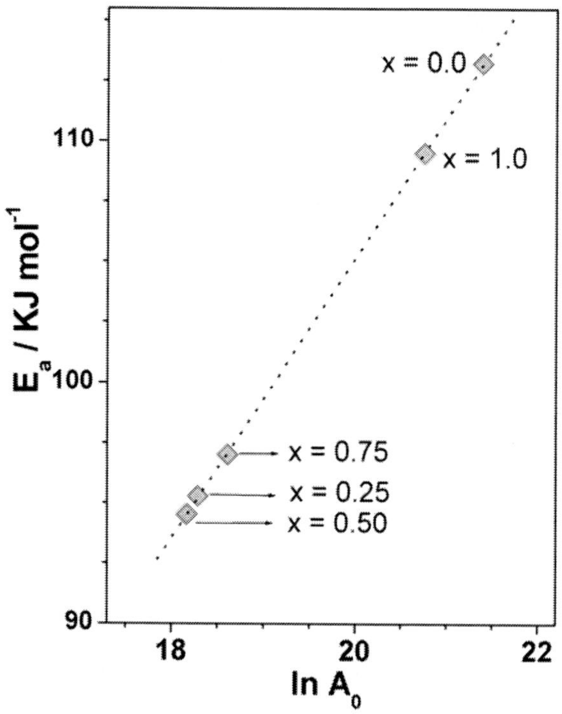

Figure 12. Activation energy dependence on catalyst composition for Cu1-xCoxFe2O4 catalysts. Note the relatively high Ea with end compositions than the intermediate compositions of catalyst.

A similar activity and selectivity trend was observed in the case of phenol isopropylation over $Cu_{1-x}Co_xFe_2O_4$ ferrospinel catalysts. Among the various products, 2-isopropyl phenol (2-IP) was obtained as the major product at 350°C, however, inevitable decomposition of isopropanol led to the formation of lower alkyl phenol like *o*-cresol and 2,6-xylenol as side products. The activation of the reactant molecules is associated with an acid-base pair site and this function plays a determining role in the catalytic activity towards phenol conversion and 2-IP selectivity. It was found that neither strong acid sites nor strong basic sites are suitable for high catalytic performance in phenol isopropylation (Table 3) with isopropanol to phenol molar ratio of 5 at 350°C. As demonstrated in the case of methylation (Figure 9), it is the optimum number and distribution of acid and basic sites makes the system efficient for phenol isopropylation and is achieved with x = 0.5 composition. This composition also displays promising activity for tertiary-butylation of phenol study on $Cu_{1-x}Co_xFe_2O_4$ catalysts using isobutene as alkylation agent [81]. 2-tert-butyl phenol (2-tBP), 4-tert-butyl phenol (4-tBP), and 2,4-di-tert-butyl phenol (2,4-tBP) are the major products obtained with trace amount of 2,6-di-tert-butyl phenol. A maximum phenol conversion (30 mol%) was observed at 200°C with isobutene to phenol molar ratio 3. Significant increase in 2,4-tBP formation with increase in phenol conversion and also at the expense of 4-tBP and 2-tBP selectivity indicates that at high phenol conversion more di-alkylation of phenol at 2 and 4 position has been effected. Further, one to one change observed between phenol conversion and 2,4-tBP selectivity indicates the first order dependence of phenol conversion for 2,4-tBP selectivity at 200°C.

Table 3. Metal content, relative acid-base property and isopropylation activity on $Cu_{1-x}Co_xFe_2O_4$

Catalysts Composition (x)	Metal content (Wt%)[a]			Relative Acid-base property		Phenol conversion (Wt%)	2-IP selectivity (Wt%)
	Co	Cu	Fe	Acidity	Basicity		
0.00	-	27.1	46.2	300	580	5.2	90.0
0.25	6.1	20.2	46.7	381	510	8.5	90.2
0.50	12.4	13.5	47.1	479	420	27.5	88.2
0.75	18.6	6.9	47.3	623	275	10.1	89.9
1.00	25.1	-	47.6	782	210	4.3	89.3

[a]Metal-ion content obtained from X-ray fluorescence

On the other hand 4-tBP formation was favored on catalyst having strong acid sites as evident from the catalytic results observed on x = 1 composition (Table 4) at 200°C, whereas selectivity of 2-tBP and 2,4-tBP is lowest on catalyst with strong acidity.

High activity associated with x = 0.5 composition demonstrates an optimum concentration of acid-base sites is needed for phenol adsorption and subsequent polarization of both phenol and isobutene as in the case of other alkylations. It was proposed that in the phenol t-butylation, t-butyl carbocation can attack phenol from the adsorbed as well as from the gaseous state resulted in the formation of para t-butylated products such as 4-tBP and 2,4-tBP. The steric hindrance of t-butyl group prevents the sequential attack of t-butyl cation at ortho position for dialkylation and that demonstrated the negligible formation of 2,6-di-t-butyl phenol.

Photoemission spectra from the Cu $2p_{3/2}$ core level for spent t-butylation $Cu_{1-x}Co_xFe_2O_4$ catalysts are shown in Figure 13a. It is already clear from Figure 5 that all fresh catalysts exhibit the Cu $2p_{3/2}$ main peak at a BE of 934.2 ± 0.2 eV with good satellite intensity around 942 eV, confirming the existence of Cu^{2+}. Two different features are observed on spent catalysts with lower satellite intensity. The first feature at 934.2± 0.2 eV corresponds to Cu^{2+} as in the calcined catalysts, and the second feature at 932.6 ± 0.2 eV corresponds to Cu^{+} (and/or Cu°) species. Cu^{2+} has a satellite feature due to its open-shell $3d^9$ configuration, while Cu°/Cu^{+} does not show a satellite due to its closed-shell configuration ($3d^{10}$). The intensity of main line also decreases from fresh to spent catalysts. Reducibility of Cu, calculated as the ratio of Cu^{+} to Cu^{2+}, given in Figure 13a increases linearly with increasing Co content. This trend is totally contrary to a decrease in Cu reducibility for phenol methylation and ethylation with increasing Co-content [74, 80]. Nonetheless, the results obtained from temperature programmed reduction (TPR) of $Cu_{1-x}Co_xFe_2O_4$ in H_2 atmosphere [13] is in good agreement with the Cu-reducibility observed in Figure 13a. Two important reduction processes involved are $Cu^{2+} \rightarrow Cu^{+}$ and $Cu^{+} \rightarrow Cu^{\circ}$, at increasing temperature between 180 and 280°C. $Cu^{2+} \rightarrow Cu^{+}$ occurs around 200°C and the amount of H_2 consumed increases with x. Since the tertiary butylation of phenol was carried out at 200°C, which is at the $Cu^{2+} \rightarrow Cu^{+}$ reduction temperature, no further reduction to Cu^{0} was observed under the experimental conditions employed [81]. Above TPR results entirely support increasing Cu reducibility (to Cu^{+}) with increasing x at 200°C as observed in XPS.

Lower-valent Cu species can be distinguished by examining the Cu–$L_3M_{45}M_{45}$ Auger peaks shown in Figure 13b. All the compositions exhibit a peak at a kinetic energy (KE) around 917.8 eV, characteristic of Cu^{2+} species. Additionally, a second component is also observed at about 916.2 eV, which is characteristic of Cu^+ species. It should be noted here that the KEs of the Cu–$L_3M_{45}M_{45}$ Auger peak values for Cu, CuO, and Cu_2O are 918.4, 917.6, and 916.5 eV, respectively [74, 81]. Present KE values of 916.2 and 917.8 eV suggest contributions of Cu^+ and Cu^{2+} species, respectively, on spent catalysts and no Cu-metal. Indeed the XRD pattern of spent catalysts obtained after reaction at 200°C shows hardly any Cu° [81]. This suggests the reaction temperature of 200°C is not capable of accelerating the reduction process is very much reflected in the XRD and XPS results.

Table 4. Tertiary butylation of phenol activity on $Cu_{1-x}Co_xFe_2O_4$ at 200°C

$Cu_{1-x}Co_xFe_2O_4$ Composition (x)	Phenol conversion (Wt%)	Selectivity (Wt%)		
		2-tBP	4-tBP	2,4-tBP
0.0	13.4	48.4	32.6	15.5
0.25	15.9	51.8	36.6	8.8
0.5	29.6	43	27.6	25.2
0.75	10.3	38.3	43.1	15.3
1	6.3	23.5	68.6	5.3

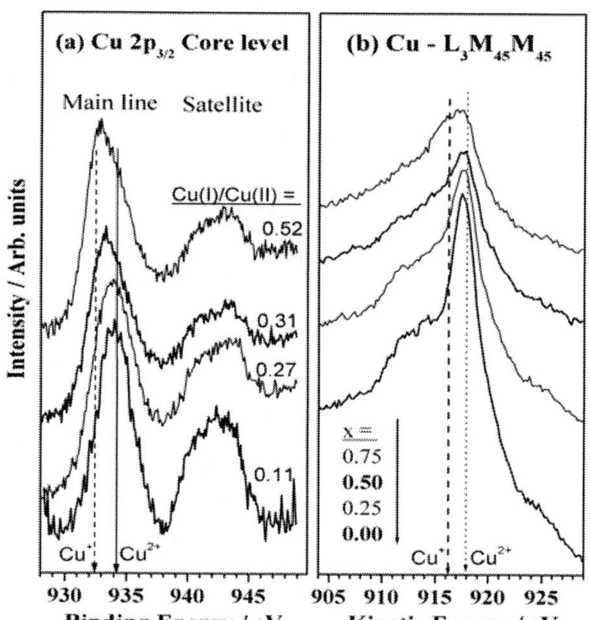

Figure 13. (a) Cu 2p3/2 photoemission spectra of spent Cu1−xCoxFe2O4 catalysts. Broadening on the lower-binding-energy side of main line demonstrates a partial reduction of Cu2+ on spent catalysts. Ratio of Cu+/Cu2+ on spent catalyst is given indicates the reducibility of Cu increases with Co content. (b) Cu L3M45M45 Auger electron spectra from spent catalysts.

A comparative study of various phenol alkylation reactions performed over $Cu_{1-x}Co_xFe_2O_4$ catalyst system with different alkylating agent. For simplicity, results exclusively from the active catalyst composition $Cu_{0.5}Co_{0.5}Fe_2O_4$ are compared in Figure 14. The above comparison clearly suggests that the total phenol conversion decreases in the following order: methanol~ dimethyl carbonate > ethanol > isopropanol > isobutene with regard to the alkylation agent [82]. The optimum reaction conditions for maximum phenol conversion and major product selectivity were different and indeed they vary, especially with the size of the alkylation agent. However, catalytic activity followed a similar trend with respect to the composition of the $Cu_{1-x}Co_xFe_2O_4$ catalyst systems for any type of alkylation, and $Cu_{0.5}Co_{0.5}Fe_2O_4$ always showed maximum phenol conversion and high yield of alkylated products. This suggests the influence of Cu+Co combination is an important factor towards phenol conversion, as larger conversion was obtained with intermediate compositions than end compositions, x = 0 and 1. Further, formation of 2,6-dialkylated phenols decreases as the alkylating agent chain length increases and indicating the big alkyl groups hinder the subsequent ortho- alkylation of 2-alkyl phenol.

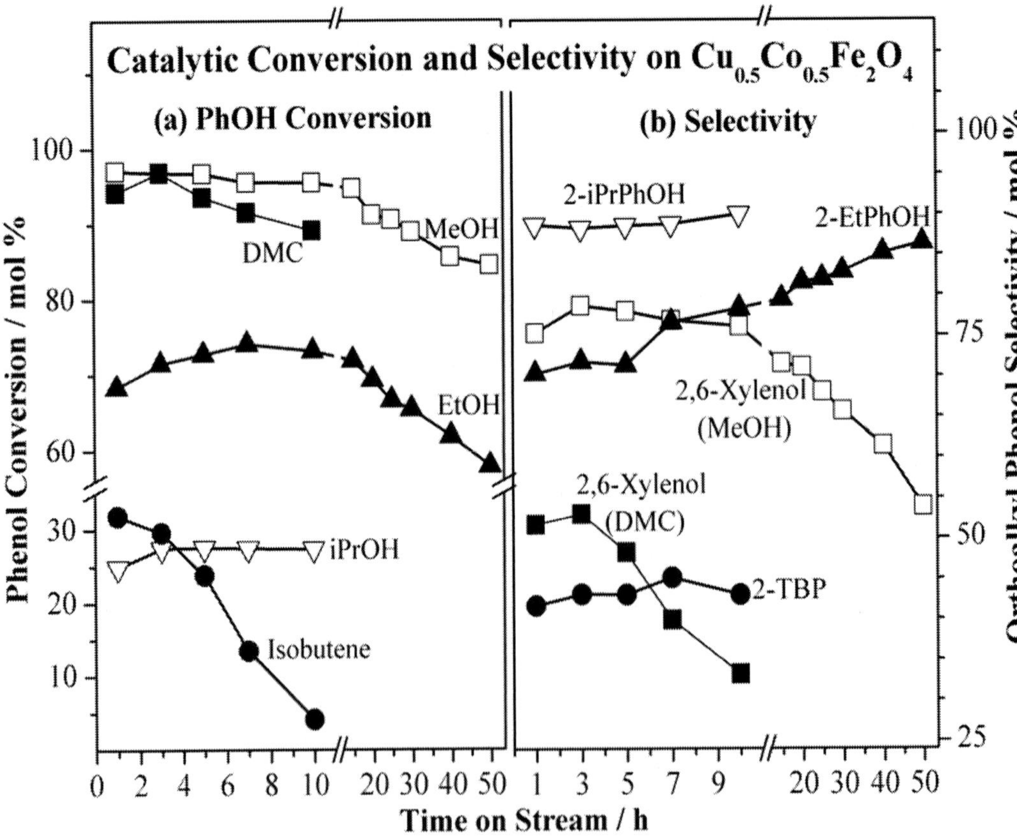

Figure 14. Time on stream dependence of (a) phenol conversion, and (b) orthoalkyl phenol selectivity on Cu0.5Co0.5Fe2O4 at optimum reaction conditions, described earlier. Different alkylation is indicated by the name of the alkylation agent in the left panel. 2,6-xylenol selectivity is given for methylation with MeOH/DMC.

3.3. Alkylation of M-Cresol to Thymol

Thymol is 2-isopropyl-5-methylphenol, an important pharmaceutical compound owing to its antibacterial activity on bacteria involved in upper respiratory tract infections, inhibitory activity against *Escherichia coli* and *Salmonella typhimurium* and on oral bacteria [88]. Further thymol exhibits the activity in protecting low-density lipoproteins and shows the high antioxidant effectiveness, for instance, during oxidation of triacylglycerols of sunflower oil. It is also an important intermediate used in perfumery and hydrogenation of thymol leads to menthol, a component of fragrances with a peppermint odour. Grabowska et al have studied the gas phase catalytic isopropylation of m-cresol on zinc aluminate spinel catalysts [89]. The reaction was carried out in a continuous process at atmospheric pressure in the temperature range 210-310°C. A maximum conversion of m-cresol (78%) and thymol selectivity up to 88% was obtained at 270°C over $ZnAl_2O_4$ that was prepared at hydrothermal conditions from basic aluminium nitrate and zinc acetate. Both C-alkylated (path 1 and 3 in scheme1) and O-alkylated (path 2) products were obtained at low reaction temperature in the above temperature range; however, thymol was the major product at high reaction temperature (270°C). At high temperatures, it was proposed that 3-methylphenyl-isopropyl ether undergoes rearrangement to yield thymol (path 4) and the possibility of isomerization of isothymol to thymol (path 5) was also considered.

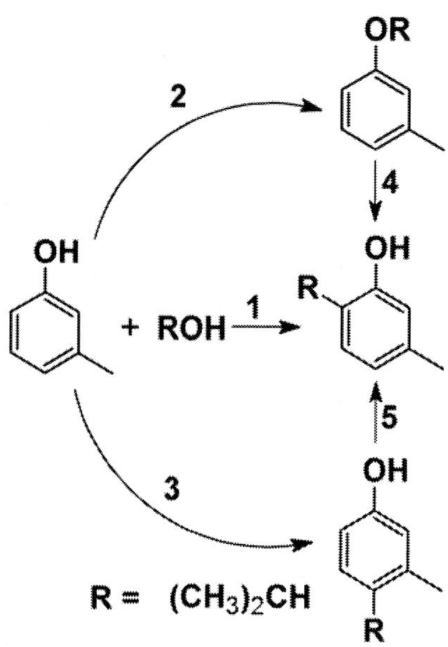

Scheme 1. Possible reaction pathways from m-cresol to thymol.

3. 4. N-Methylationof Aniline

N-methylation of aniline is an interesting reaction from the products selectivity point of view, especially N-methylaniline (NMA), due to the presence of chemically equivalent protons attached to the nitrogen atom. The high nucleophilicity of the amine nitrogen results in the formation of mixture of N-methyl and N, N-dimethylated products. Ko *et al.* [90]

reported the methylation of aniline with methanol at 425°C over alumina and formed both NMA and N,N-dimethylaniline (DMA) and the corresponding activation energies are 62.7±2.1 and 48.3±2.9 $KJmol^{-1}$, respectively. It is clear that NMA reacts faster than aniline and undergoes consecutive methylation to DMA that leads to poor product selectivity or highly selective DMA formation. Hence it is difficult to synthesize NMA selectively over DMA. However, by varying the reaction parameters, it is possible to control the formation of DMA [14]. Indeed industrial production of NMA being achieved by process control, rather than by catalyst control. N-methylaniline is a useful synthetic intermediate for the synthesis of several organic molecules [68].

Sreekumar and co-workers [91-94] used different ternary iron containing spinels for aniline N-methylation using methanol as methylating agent. The reaction was carried out over $Zn_{1-x}Ni_xFe_2O_4$ (x = 0, 0.2, 0.5, 0.8 and 1) in a fixed bed down-flow reactor [91]. It was observed that systems possessing low 'x' values are highly selective and active for NMA formation at 350°C. The electron acceptors like TCNQ (7,7,8,8-Tetracyanoquinodimethane and choloranil (2,3,5,6-Tetrachloro-4-benzoquinone) was used to study the surface acid base properties of these ferrospinels. It was found that substitution of Zn^{2+} by Ni^{2+} creates a decrease in acidity with concomitant increase in basicity. Thus, it was concluded that, the predominant Lewis acid sites found on Zn rich compositions are mainly responsible for the good catalytic performance [91]. Very similar catalytic aniline conversion and NMA selectivity was observed with $Zn_{1-x}Co_xFe_2O_4$ [92]. Zn rich compositions were found to be very active, whereas Co-rich compositions decreases the activity. This was attributed to the increasing catalysts acidity with Co^{2+} substitution was discerned from the electron acceptors studies [92].

In the above studies, $Zn_{1-x}Co_xFe_2O_4$ and Zn rich catalyst systems were found to be relatively active in the initial period, and exhibiting high aniline conversion than the Co-rich compositions. However, the high activity observed with $ZnFe_2O_4$ decreases rapidly with time on stream [92]. Further, significant amount of secondary products like DMA, benzene and toluene formed with all catalyst compositions. It is also to be mentioned that the feed ratio of $PhNH_2$:MeOH = 1:6 might lead to dimethylation due to the large excess amount of methanol. It is to be mentioned that, $ZnFe_2O_4$ is a normal spinel system containing Zn in T_d and Fe in O_h sites. As there is no Fe in T_d sites, the possibility of redox migration of Fe between the T_d and O_h sites is restricted. Hence, in the O_h sites, the reduced Fe^{3+} is unable to exchange electrons to regain their original states. Thus the initial high activity observed on this catalyst declines rapidly with time on stream. It is expected that the introduction of Fe, Cu, Ni and Co improves the stability of $ZnFe_2O_4$ through Fe migration between T_d and O_h sites; however, the addition of other metal ions affects the methylation activity for different reasons.

The merits and demerits of the choice of methylating agents (methanol or dimethyl carbonate) for aniline N-methylation to NMA were reported over $Zn_{1-x}Co_xFe_2O_4$ [93]. DMC acts as a better methylation agent at comparatively low reaction temperature (250°C), where methanol shows mild activity. However, employing DMC produces appreciable amount of DMA (12% selectivity) even at 250°C, whereas methanol gave nearly 97% NMA selectivity at 350°C. The product yield obtained during aniline methylation with methanol and DMC at reaction temperature 350 and 250°C, respectively, is given in Figure 15.

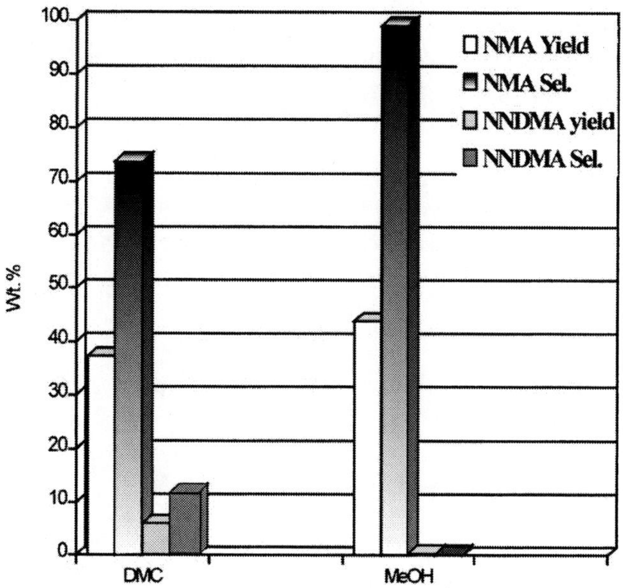

Figure 15. A comparison of aniline methylation using methanol at 350°C and DMC at 250°C on Zn0.8Co0.2Fe2O4. Reprinted from Applied Catalysis A: General, 159, Sreekumar K., et al., 2000, 327-334 with permission from Elsevier.

Though the observed NMA yield with DMC (at reaction temperature 250°C) was comparable to the yield obtained with methanol (at 350°C), the formation of NNDMA was significantly high in the case of DMC. Dimethylation rate was also increases with increasing reaction temperature and molar ratio (DMC/aniline).

The main advantage of the reaction is that alkylation is possible at sufficiently low temperature, where methanol shows mild activity. Nonetheless, the alkylation activity with MeOH and DMC followed a similar trend with respect to the catalyst composition indicating the participation of same active sites/species of catalysts irrespective of methylating agent.

Nishamol *et al* [95] observed similar catalytic aniline N-methylation activity and NMA selectivity with Cr-Mn ferrospinels ($Mn_{1-x}Cr_xFe_2O_4$) at reaction temperatures between 250 and 450°C. Surface acid-base property of these ferrospinels were studied by ammonia and 2,6-dimethylpyridine adsorption-desorption experiments. It was found that the progressive addition of Cr ions into the pure manganese ferrite decreases the total acidity, as well as an increase in weak and medium acidic sites at the cost of replacing strong acidic sties. Methylation activity increases with Cr substitution indicating the moderate acidity (weak and medium acid sites) favors aniline methylation, since remarkably high activity was found with intermediate compositions.

A thorough search for the selective catalytic production of NMA was reported [96] on different transition metals containing ferrospinels, $A_{0.5}A'_{0.5}Fe_2O_4$ (A = Fe, Cu, Zn and A' = Co, Ni, Cu and Zn) and the results are shown in the Figure 16. The high initial aniline conversion to NMA associated with Cu-containing spinels indicates that Cu is largely responsible for selective catalytic methylation activity. Addition of other metal ions to Cu improved the stability of the systems. Interestingly, $Cu_{0.5}Zn_{0.5}Fe_2O_4$ shows a stable and better catalytic activity than all other system. Thus $Cu_{1-x}Zn_xFe_2O_4$ where x = 0.05, 0.25, 0.5, 0.75

and 1 was pursued for systematic catalytic activity measurements. Individual oxides, including CuO, ZnO, Fe_2O_3 and a mixture of CuO and ZnO showed negligible catalytic activity toward aniline methylation suggesting the need for a suitable multi-component system [14].

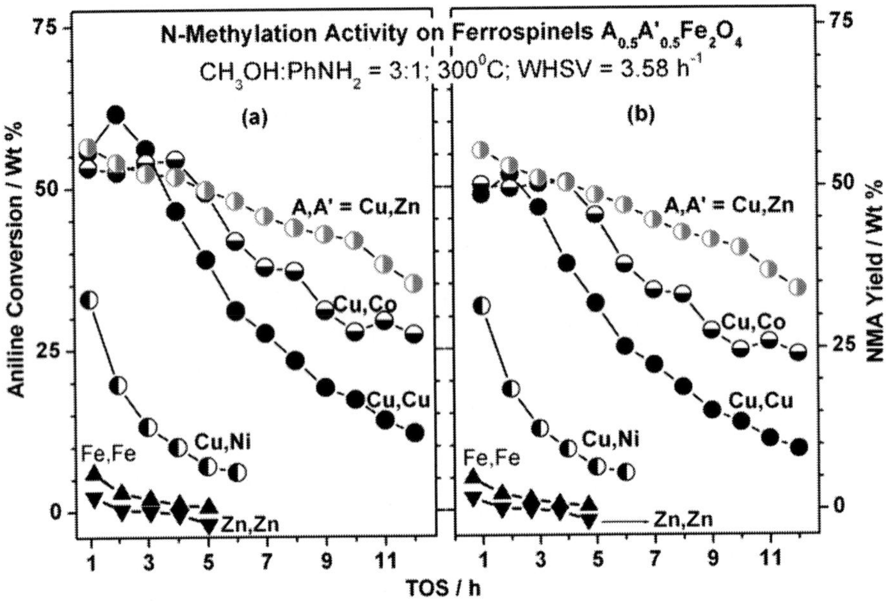

Figure 16. (a) Aniline conversion and (b) N-methylaniline yield obtained on A0.5A'0.5Fe2O4 (A = Fe, Zn, Cu and A' = Fe, Co, Ni, Cu and Zn) at 300°C with 3:1 ratio of CH3OH:PhNH2 at space velocity 3.58 h-1. Note a better initial catalytic activity was observed when Cu occupies the A-site in ferrites and the stability increases considerably when Zn also occupies the same site.

Due to promising activity trend, $Cu_{1-x}Zn_xFe_2O_4$ was subjected to systematic activity test. Figure 17 shows the (a) aniline conversion, and (b) NMA yield on $Cu_{1-x}Zn_xFe_2O_4$ (x = 0.05, 0.5, 0.75, and 1) at T = 300^{0}C, WHSV = 3.58 h^{-1}, with a feed composition of 3:1:1 molar ratio of CH_3OH:$PhNH_2$:H_2O. x = 0.25 composition showed similar initial high aniline conversion, as that of x = 0.05, that rapidly decreased with increasing TOS. The sample with x = 0.5 showed stable activity, and indeed, a marginal increase in activity occurred at TOS > 6 h. $ZnFe_2O_4$ showed negligible activity compared with other compositions, indicating a negligible role of direct methylation on Zn and Fe. Nevertheless, the catalyst stability associated with x = 0.5 under reaction conditions seemed to be enhanced, likely due to the presence of an optimum amount of Zn in it. Reasonable high activity reported by Sreekumar et al [94] on $ZnFe_2O_4$ is in contrast with the negligible activity displayed in Figures 16 and 17. Although the catalyst preparation methods are similar [94, 96] the reasons for different catalytic activity might be due to high methanol content in the feed and higher temperature (350^{0}C) employed in ref. 94.

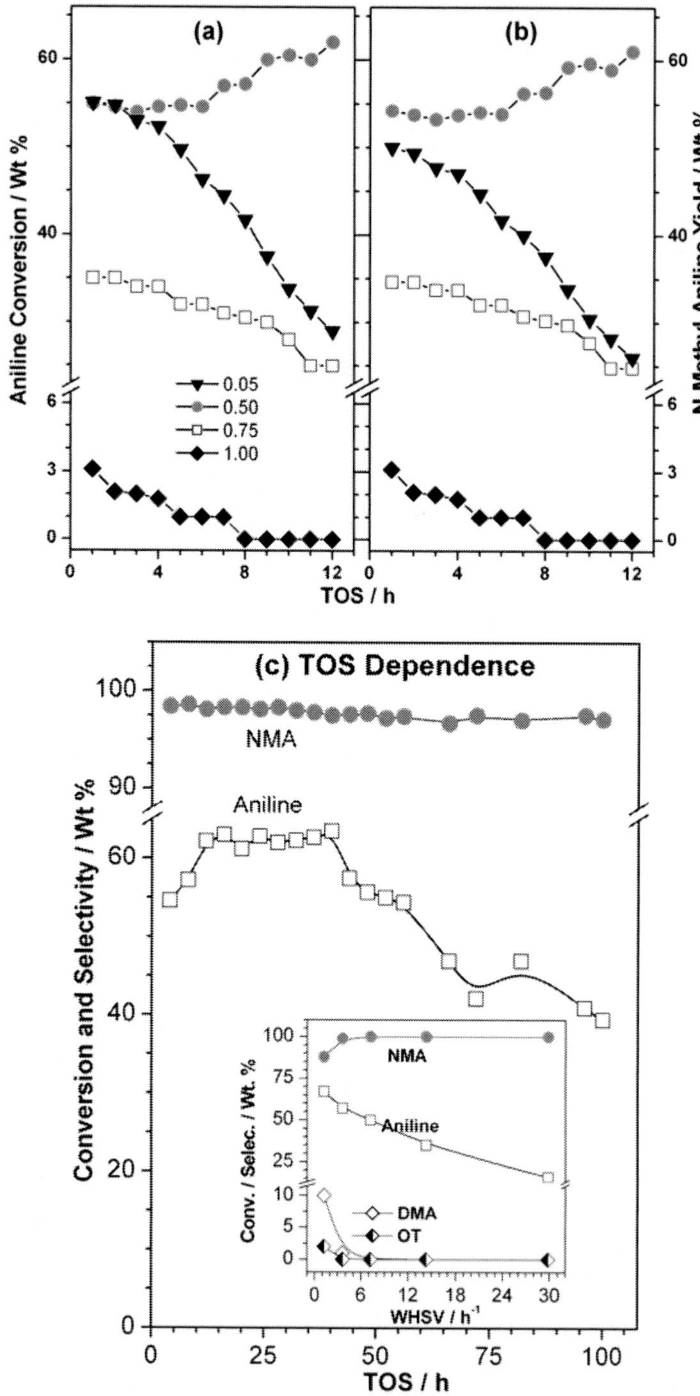

Figure 17. Time on stream dependence of (a) aniline conversion and (b) NMA yield on Cu1-xZnXFe2O4 at 3000C and WHSV = 3.58 h-1 with a CH3OH:PhNH2:H2O feed composition of 3:1:1 of. (c) Time on stream and weight hour space velocity (inset) dependence of aniline conversion and NMA selectivity on Cu0.5Zn0.5Fe2O4 at 300°C and WHSV = 3.58 h-1. Note the formation of secondary products at low WHSV.

A remarkable 100 h time on stream catalytic test was measured on potential composition $Cu_{0.5}Zn_{0.5}Fe_2O_4$ at optimum reaction conditions to show the stable activity and selectivity trend on ferrospinels (Figure 17c). Aniline conversion was stable around 60% for the first 60 h and thereafter it decreases slowly. A simple calcination of the spent catalyst in air restores its original activity. Space velocity study (Inset in Figure 17c) also indicated that it is possible to produce 100% NMA and totally subside the DMA formation applying relatively high space velocity (>3.6 h^{-1}). This suggests the high contact time is the prerequisite towards the formation of DMA and o-toluidine.

The results obtained from the catalytic study over wide range of reaction temperatures, with different catalyst compositions and different methanol to aniline molar ratio suggests that Cu^{2+} is largely responsible for the methylation activity than zinc and iron [96]. Detailed XPS and IR studies shed more light on the different roles of Cu and Zn in $Cu_{1-x}Zn_xFe_2O_4$ towards aniline methylation to selective NMA production. Although Cu^{2+} is the active site, it is assisted by Zn-site by the way of supplying methyl species as well as by preventing Cu-agglomeration as a spacer, so that long-term stability could be brought in.

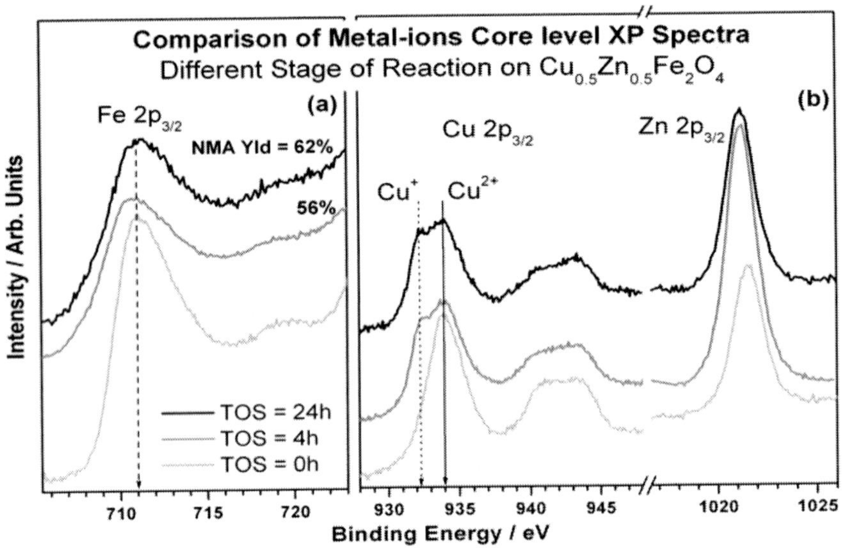

Figure 18. A comparison of (a) Fe 2p, and (b) Cu 2p and Zn 2p core level features on fresh and after two different stages of reaction on Cu0.5Zn0.5Fe2O4 catalyst. Note a drastic change in the transition metal composition accompanied with some reduction after different stages of reaction on the surface. Reprinted from Journal of Catalysis, 241, Vijayaraj M., et al., 2006, 83-95 with permission from Elsevier.

XPS and XAES investigations on fresh and spent $Cu_{1-x}Zn_xFe_2O_4$ catalysts were carried out in detail to understand the role of metal ions towards aniline N-methylation. The changes in the oxidation state of metal ions and the associated changes in electronic interaction, before and after the reaction were also investigated. Fresh catalysts clearly indicate the existence of Cu^{2+}, Zn^{2+} and Fe^{3+} species. However, the spent catalyst shows a composition dependent oxidation states, especially partially reduced Cu and Fe species [96]. Figure 18 shows XPS of metal ions Cu, Zn and Fe $2p_{3/2}$ core level before and after reaction (4h and 24 TOS) from $Cu_{0.5}Zn_{0.5}Fe_2O_4$ catalyst. The 2p level of Zn (Fe) did not change (changed to some extent) after reaction; whereas remarkable change was observed in Cu 2p level indicating the Cu^{2+}

reduction during reaction. However, Cu 2p core level remains similar at 4 and 24 h TOS and NMA yield increased significantly after 24 h. It is noteworthy that the redox migration of Fe between O_h and T_d sites in these spinel systems maintains the Fe^{3+} and Fe^{2+}concentration thereby retain the structural integrity.

In contrast to Sreekumar et al's observation [90,91], addition of Cu to $ZnFe_2O_4$ improved the methylation activity of $ZnFe_2O_4$. On the other hand addition of Zn improves the stability of the catalyst system as shown in Figure 16. These results show that Cu-Zn combination is better than Ni-Zn or Co-Zn combination with iron acting as the support for aniline methylation. The reduction of Cu^{2+} in $x \leq 0.25$ compositions is significant, whereas only a partial reduction of Cu^{2+} occurs in 0.5 and 0.75 systems after reaction. Wagner plot [97] was used to identify the nature of the reduced Cu species in a easy and direct manner. Wagner plot shown in Figure 19 is a plot of KE of Auger ($L_3M_{45}M_{45}$) transition on the *y*-axis and the Cu $2p_{3/2}$ BE on the *x*-axis for all of the catalyst compositions. The Auger parameters are described by a solid straight line with a slope of −1. Figure 19 clearly shows that Cu^{2+} in all of the fresh and spent catalysts were similar to that of CuO. Indeed, $x = 0.5$ and 0.75 systems resembles more closely to CuO than the other compositions. The nature of reduced species on the $x = 0.05$ and 0.25 systems corresponded to that of Cu^0 and further confirmed from the Cu reflections in XRD. However, the same corresponded to Cu_2O-like species on the $x = 0.5$ and 0.75 systems.

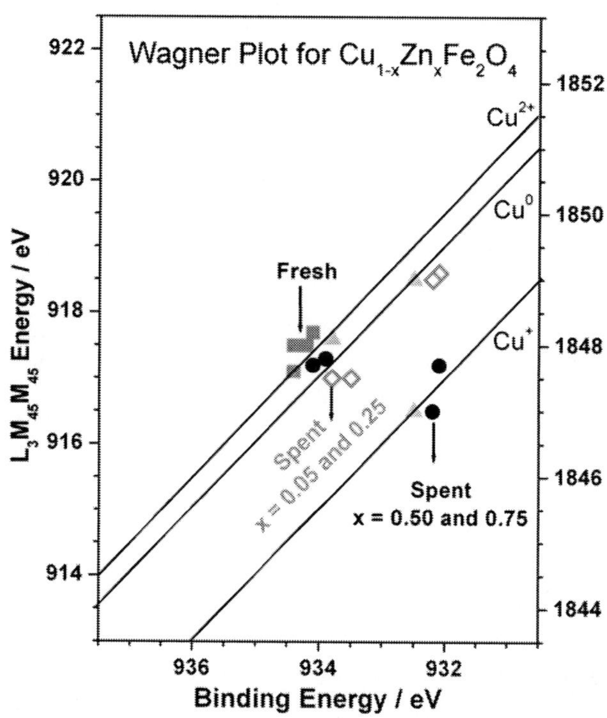

Figure 19. Wagner plot for Cu1-xZnxFe2O4 catalysts from the Cu 2p3/2 core level and Auger spectral data. Data points for standard compounds (Cu, Cu2O and CuO) are given by solid gray triangles and fresh catalyst by solid squares. Spent catalysts with x = 0.05 and 0.25 composition indicated by open diamond and x = 0.5 and 0.75 by solid circles, respectively. The Auger parameter is described by solid straight line with a slope of −1. Reprinted from Journal of Catalysis, 241, Vijayaraj M., et al., 2006, 83-95 with permission from Elsevier.

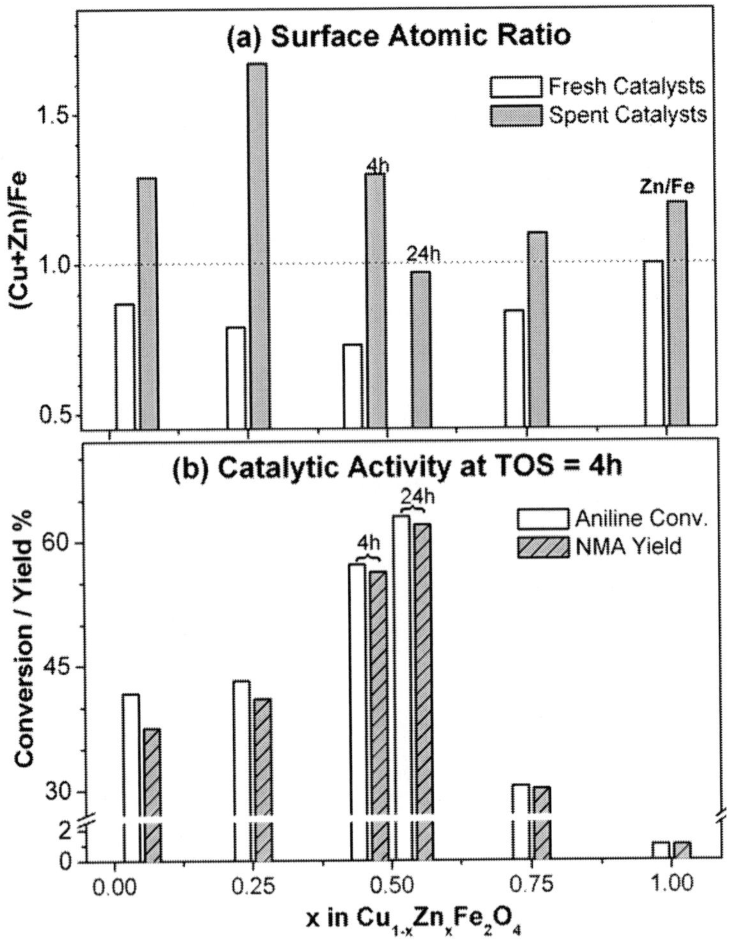

Figure 20. Comparison of (a) atomic ratio of (Cu+Zn)/Fe for all fresh and spent catalyst compositions and (b) aniline conversion and NMA selectivity on Cu1-xZnxFe2O4 catalysts. Note the Cu+Zn/Fe ~1 after 24 h of reaction on x = 0.5 and stable catalytic activity associated with it. Reprinted from Journal of Catalysis, 241, Vijayaraj M., et al., 2006, 83-95 with permission from Elsevier.

The surface redistribution of metal ions in the spinels during reaction conditions determines the course of the reaction (Figure 20). Catalytic activity measured was compared to the surface atomic ratio obtained on fresh and spent catalysts from XPS measurements. There is clearly a redistribution of metal ions on the surface, as the reaction progresses on all the compositions. Although, a same (Cu+Zn)/Fe ratio was observed for x = 0.05 and 0.5 (TOS = 4h), the nature of Cu-species dramatically changes from dominantly Cu^{2+} on x = 0.5 to Cu-metal on x = 0.05 composition. Further, an increase in the (Cu+Zn)/Fe ratio is almost entirely due to an increase in the Cu content on x = 0.05 and this leads to Cu agglomeration and consequent sintering. Although x = 0.75 composition shows the above ratio close to 1.1, the active Cu^{2+}-content on the surface is low and hence a relatively low catalytic activity.

These observations highlighted the necessity of sufficient concentration of oxidized Cu^{2+} for sustainable methylation activity. Only on x = 0.5 (after 24h reaction run) the (Cu+Zn)/Fe ratio was maintained at 1, and hence an extended catalytic activity was observed (Figure 17c). It is clear that not a single metal ion, but the presence of an equal amount of Cu and Zn on the surface is necessary for a longer methylation activity on these spinel systems, in which Cu^{2+} is the active species for methylation and Zn^{2+} acts as an "active spacer and stabilizer" to decrease the extent of Cu^{2+} reduction and Cu agglomeration on the surface. In fact, it is the complete heterogeneity of the surface metal ions composition that helps for the stable catalytic activity and retention of the structural integrity of the catalyst [96].

The best aniline N-methylation activity observed with $Cu_{0.5}Zn_{0.5}Fe_2O_4$ was reiterated with selective mono-N-methylation of substituted anilines (p-anisidine, p-aminoacetophenone, o-, m-, and p-toluidines, 2,6-xylidene) with methanol [98]. Nonetheless, the presence of ring directing groups does not have any significant influence on the N-methylanilines selectivity that remains very high at >95% and only traces of N,N-dimethylated and C-methylated products were observed (Figure 21). Electronic effects due to different groups in the above aniline substrates influence the reactivity of the substrates in terms of conversion and yield. Figure 21 displays the aniline substrate conversion increases in the following order: OT < PT~ PA < MT~PAA~Aniline. 2,6-xylidine shows no conversion at all. Toluidine reactivity varies with respect to the position (ortho, meta and para) of the methyl group on the phenyl ring. Para substituted anilines (p-toluidine and p-anisidine) exhibit comparable catalytic activity, while p-aminoacetophenone shows higher conversion. It is interesting to note that PT and PA, although they have different electronic effects +I and +R, respectively, at phenyl ring, show almost identical and stable activity. Nevertheless, the above reactivity trend indicates a complicated electronic effect due to the presence of substituents. This directly hints that the factors that helped stable NMA formation from aniline helps other aniline substrates too, in spite of different ring directing groups being present on the aromatic ring.

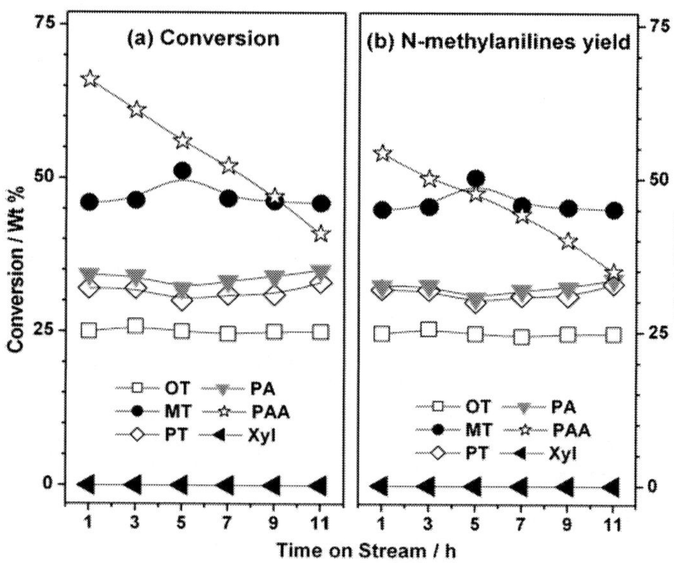

Figure 21. Conversion of substituted anilines (OT, MT, and PT are respective ortho-, meta-, and para-toluidine, PA = p-anisidine, and PAA = p-aminoacetophenone), corresponding mono-N-methylated anilines selectivity and yield obtained on Cu0.5Zn0.5Fe2O4 at 300oC with methanol. Only in the case of PAA, feed with CH3OH:PAA = 6 was employed to avoid solubility problem, and in all other cases a CH3OH:Sub. Aniline:H2O = 3:1:1 was maintained. Reprinted from Applied Catalysis A: General, 320, Vijayaraj M., et al., 2007, 64-68 with permission from Elsevier.

The perpendicular orientation of toluidine substrates on catalysts surfaces influences the conversion and N-methyltoluidine yield. Increased crowding at ortho-position and hence steric hindrance (aniline> o-toluidine> 2,6-xylidine) restrict the substrates interaction with the catalysts surface, and 2,6-xylidine shows no reactivity [95,98].

3.4a Active Spacer and Stabilizer Role of Zn in $Cu_{1-X}zn_xfe_2o_4$

XPS and XRD results [96] in the present study reveal that Cu-rich systems, $x = 0.05$ and 0.25, exhibited a significant Cu^{2+} reduction and agglomeration due to steam reforming of methanol and H_2 production and the same was confirmed from the gas products analysis. In contrast, for $x = 1$, no such dramatic changes were observed with Zn. However, at intermediate composition ($x = 0.5$), the catalytic activity was maintained for a longer period and Cu^{2+} reduction to metal was totally suppressed. The large energy gap observed between the Cu 3d (3-4 eV) or Fe 3d (4-5 eV) and Zn 3d (10 eV) valence bands [96] suggests that it had hardly any electronic interaction. Therefore the role of Zn might be non-electronic, unlike the role of Co in $Cu_{1-x}Co_xFe_2O_4$, and the following supports the role of Zn to be more of structural as well as a stabilizer to maintain the higher activity.

Surface atomic composition results (Figures 18 and 20) on fresh and spent $Cu_{1-x}Zn_xFe_2O_4$ catalysts clearly indicate that both Cu and Zn content on the surface changed significantly for the $x = 0.5$ system as the reaction proceeded in the first 10 h, after which an almost a 1:1 combination of Cu and Zn was maintained on the catalyst surface. The change in surface atomic composition, a transient change in catalytic activity up to 10 h before it reached stable steady state activity, and a comparison of deactivation trend and surface atomic composition of other catalyst compositions clearly indicate the role of Zn as an "active spacer" on $Cu_{0.5}Zn_{0.5}Fe_2O_4$ that helps prevent the reduction/agglomeration of Cu^{2+} under the present experimental conditions. A considerable increase in the lattice constant (a) for $x = 0.5$ from $a = 8.4052$ Å to 8.4133 Å [96] supports the view that the lattice can accommodate Cu^+ ions to certain degree without undergoing structural damage. Since the role of Zn-spacer increases the catalytic activity marginally and helps maintain the same activity in the present system, it is termed as "active spacer," in contrast to the lower activity observed, when any spacer is introduced in a catalytic system. The synergetic performance of Cu and Zn plays major roles in aniline methylation and catalyst stability.

A direct support to the stabilizer and active spacer role in preventing the reduction of Cu^{2+} is obtained from temperature programmed reduction (TPR) experiments, as shown in Figure 22a. All of the Cu-containing $Cu_{1-x}Zn_xFe_2O_4$ compositions clearly show a Cu^{2+} reduction peak at around 200^{0}C, and Fe and Zn reduction features in a broad peak between 400 and 650^{0}C. The Cu^{2+} reduction peak at around 200-225^{0}C for $x = 0.05$ and 0.75 shifts to lower temperatures (150–175^{0}C) for $x = 0.25$ and 0.5. Note that various Cu^{2+} reduction peaks have been observed in mixed Cu–Zn–Al oxide system, and that the Cu^{2+} reduction temperature is highly dependent on Zn content in this system [99]. Careful analysis revealed that the amount of H_2 uptake decreased nonlinearly with decreasing Cu content (Figure 22b). A quantitative analysis of the results for the $x = 0.5$ (0.05) systems hints that the amount of H_2 uptake in the temperature range of 140–225^{0}C (150–250^{0}C) corresponds to a Cu^{2+} to Cu^+ (Cu^{2+} to Cu^0) reduction. $x = 0.5$ composition also reveals a continuous H_2 uptake, at a significant level, between 270 and 380^{0}C.

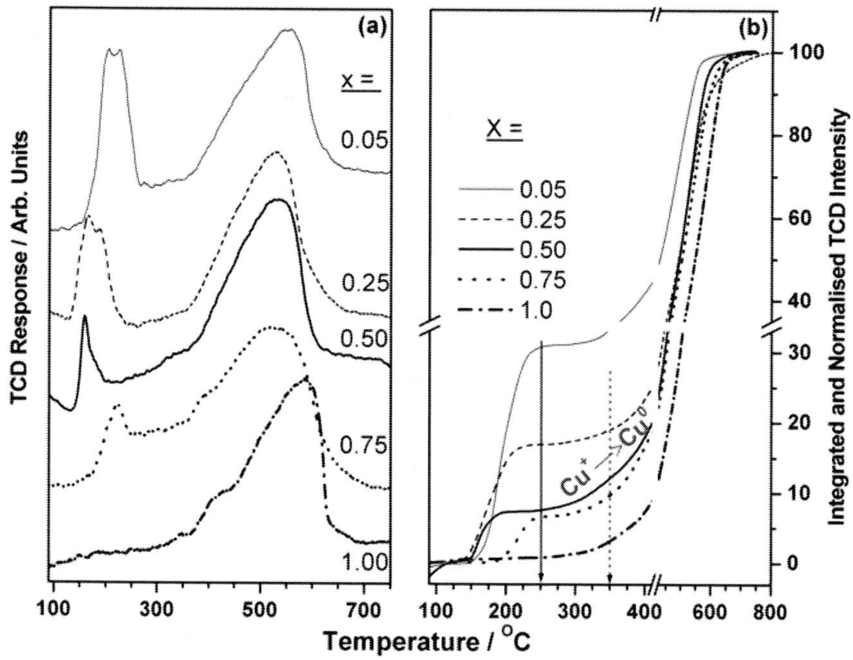

Figure 22. (a) TPR profiles of Cu1-xZnxFe2O4 catalysts measured with 5% H2 in Ar, under a flow rate of 30 ml/min at a heating rate of 5OC/min. (b) Integrated and normalized TCD intensity of TPR profiles. Complete reduction to Cu-metal was observed below 2500C for x = 0.05 and 0.25; however a continuous hydrogen consumption between 250 and 4000C to be noted on x = 0.5 and 0.75 indicating the role of Zn acting as spacer and preventing reduction below 3000C.

A similar H_2 uptake pattern can be seen for x = 0.25, at a still-lower intensity level; however, this feature was not observed on the Cu-rich composition (x = 0.05), indicating a complete reduction to Cu^0 <250°C. Fe^{3+} or Zn^{2+} reduction begins above 350°C on all compositions supporting the hydrogen uptake between 270 and 380°C is due to slow reduction of Cu^+ to Cu^0 on x = 0.25 and 0.5. This clearly supports the conclusion that total Cu^{2+} reduction occurred only above 300°C on the x = 0.5 composition due to the active spacer role of Zn.

3.4b Surface Acidity

It is important to understand the catalyst characteristics in detail, which in turn helps to understand the catalyst better and correlate the structure and composition of the catalysts with its performance, so that further improvement of the catalyst is possible. Acidity is an important property which influences the overall activity of the alkylation catalysts and the same was studied for $Cu_{1-x}Zn_xFe_2O_4$ by IR and TPD methods. The changes in acidity with respect to catalyst composition and temperature were studied through pyridine adsorption followed by IR measurements. In situ FTIR spectra of pyridine adsorbed on $Cu_{1-x}Zn_xFe_2O_4$ between 100 and 400°C (Figure 23) indicated Lewis acidity is the predominant active centers available on the surface [14].

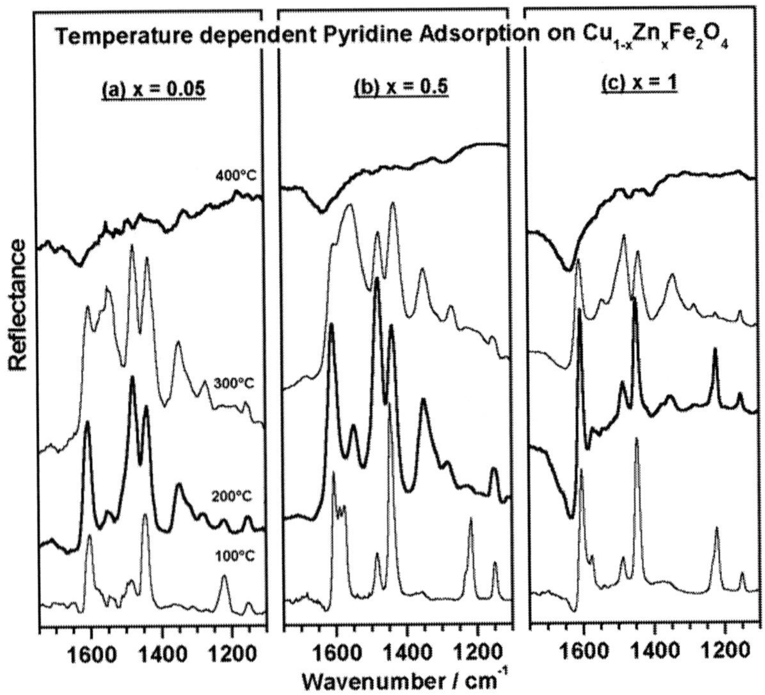

Figure 23. Temperature dependent DRIFT spectra of pyridine adsorbed on three selected compositions from Cu1-xZnxFe2O4 (x = 0.05, 0.5 and 1) between 100 and 400oC.

Important pyridine ring modes, namely, ν_{8a} (1606) and ν_{19b} (1445 cm^{-1}) were observed at 100°C on all of the catalyst compositions. In addition, auxiliary bands at 1580 and 1476 cm^{-1} were also observed that appeared to be strong only on x = 0.5 system, whereas on x = 0.05 these bands were very weak. On increasing the temperature from 100 to 200°C, the band at 1478 cm^{-1} (1580 cm^{-1}) became very strong (vanished) for all the compositions. Another band (present only for x = 0.05 at 100^{0}C) at 1550 cm^{-1} appeared on x = 0.5. This band became intense and broad on x = 0.05 and 0.5 with increasing temperature to 300°C, whereas very weak on x = 1. The peak intensity of the main stretching modes ν_{8a} and ν_{19b} remain hardly affected with increasing temperature. However, slight red shift in ν_{19b} band position is observed for all the compositions. This red shift in band position are 10, 10 and 5 cm^{-1}, respectively for x = 0.05, 0.5 and 1. A red shift for ν_{8a} band, which coupled with 1550 cm^{-1} band, is observed at 300°C only on Cu containing samples, that too significantly on x = 0.5. Further, a band at 1340 cm^{-1} (1220 cm^{-1}) newly formed (disappeared) on all of the catalyst compositions with increasing temperature. No distinct spectral features observed at 400°C indicate pyridine desorbed completely from the catalysts surface between 300 and 400°C.

It is clear from the above observations that pyridine molecule interacts on the catalyst surface in the following three modes: (1) interaction of the N lone pair electron and the H atom of the OH group, (2) transfer of a proton from surface OH group to the pyridine forming a pyridinium ion (Bronsted acidity), and (3) pyridine coordination to an electron deficient metal atom (Lewis acidity). Predominant IR bands, ν_{8a} and ν_{19b}, confirms that the major contribution of acidity is due to Lewis acid sites from all compositions. Between the above two modes of vibrations, ν_{8a} is very sensitive with respect to the oxidation state, coordination symmetry and cationic environment [100]. A broad feature for ν_{8a} band on Cu containing compositions indicates a large extent of disorder on these systems, whereas relatively sharp feature observed for $ZnFe_2O_4$ confirms that indeed it is a normal spinel. Upon pyridine adsorption, there is electron transfer from pyridine to the spinel surface resulting in the reduction of Cu^{2+} and the extent of reduction increases with temperature [78]. In consequent to that the charge density around Fe^{3+} is also modified, and hence the interaction of pyridine on the surface decreases and reflected by the broadening of ν_{8a} band. It is evident from the band at 1550 cm^{-1} that pyridinium ion is formed on all samples due to the proton transfer from the surface -OH groups. This band became very strong and intense at 300°C on Cu containing samples (significantly visible on x = 0.5) due to ν_{8a} band broadening. The stretching modes at 1580 and 1475 cm^{-1} are found to be very labile and their stability on the surface is temperature dependent. Further, α-pyridone [101] formation is evident from the observation of 1350 cm^{-1}, and other minor spectral features at high temperature are due to bipyridyl species formation [102,103] due to semiconducting nature of Fe^{3+}.

The observed FTIR results were correlated with NH_3 TPD studies. Figure 24 shows NH_3 TPD profiles from $Cu_{1-x}Zn_xFe_2O_4$, which indicate a broad distribution of binding strength of NH_3 and hence of acid strength of the catalysts [14]. A careful analysis of the results reveals that the high strength acid sites due to NH_3 desorption >300^{0}C, generally, increases with increasing Zn-content.

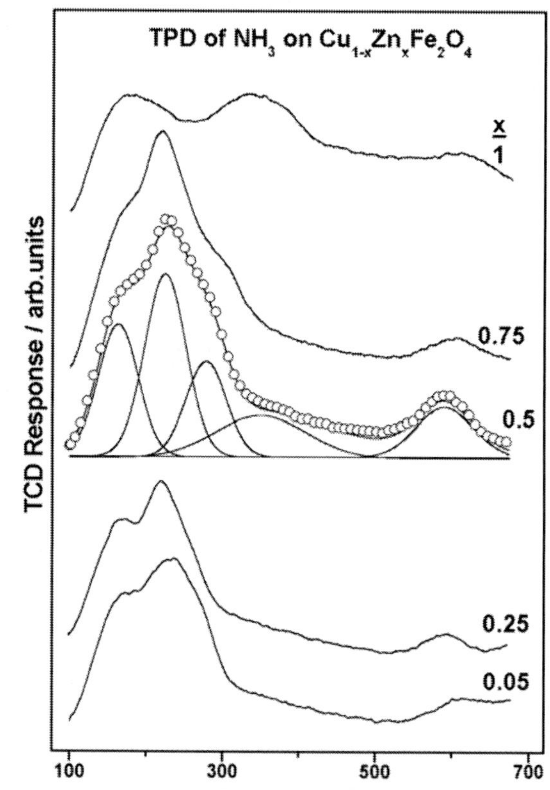

Figure 24. Temperature Programmed Desorption profile of NH3 on Cu1-xZnxFe2O4. Broad peak was deconvoluted for x = 0.5 composition.

Deconvolution of the profile from x = 0.5 indicates the acid sites distribution was divided into five distinct sites on Cu containing samples. Among the acid sites found, the one at 163°C (589°C) is weak (strong) whereas other three at 224, 277, and 357°C are moderate strength acid sites on Cu containing samples. Broad envelope observed around 370°C on $ZnFe_2O_4$ is attributed to a significantly enhanced strength acid site. Two broad peaks observed at 180 and 370^0C suggests that there might be contribution from new acid sites too. Nature of the acidity is attributed to Lewis type on all of the samples, as revealed from the results of FT-IR spectroscopy of adsorbed pyridine [104, 105]. However, contribution from Bronsted type acidity cannot be ruled out. It is evident from pyridine adsorption FT-IR study that $ZnFe_2O_4$ has very negligible Bronsted type acidity compared to Cu containing samples.

3.4c Aniline N-Methylation Mechanism on $Cu_{1-X}zn_xfe_2o_4$

A detailed investigation of aniline N-methylation on $Cu_{1-x}Zn_xFe_2O_4$ was carried out through in situ FTIR spectroscopy. The reactants (aniline and methanol) and possible products (NMA, DMA and o-toluidine) were adsorbed on the catalysts and analyzed [106,107]. Adsorption of methanol indicated a dissociative chemisorption as methoxy species on catalyst surface at 100°C. As the temperature increased, oxidation of methoxy species to formaldehyde to dioxymethylene to formate species was observed, and above 300°C complete oxidation takes place to CO, CO_2 and H_2. Indeed methanol alone on $Cu_{1-x}Zn_xFe_2O_4$ and $Cu_{1-x}Co_xFe_2O_4$ behaves in a similar way [79,107].

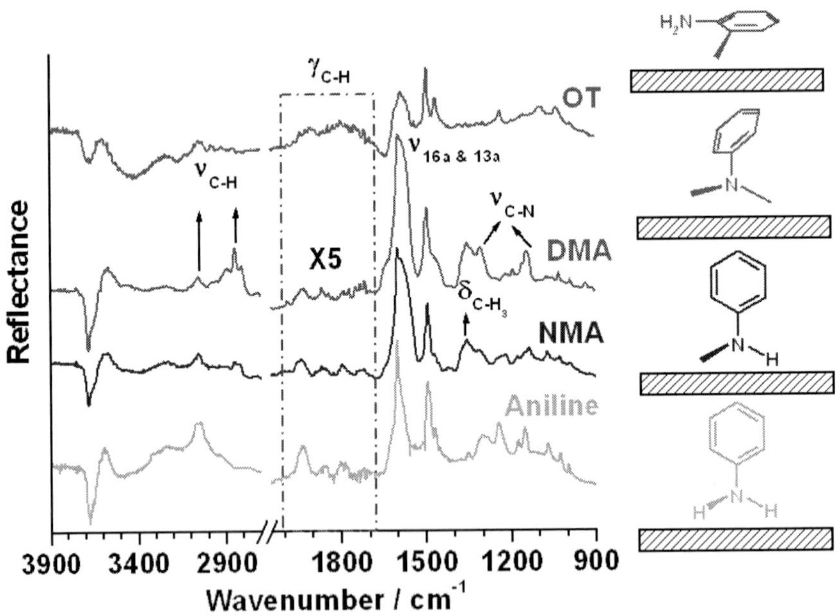

Figure 25. Various anilines adsorption on Cu0.5Zn0.5Fe2O4 at 100°C.

Adsorption of aniline on $Cu_{1-x}Zn_xFe_2O_4$ at $\leq 100^0$C indicates a simple molecular adsorption through N-atom on an acid-base pair site. However, above 100°C, N-H bond dissociates and aniline chemisorbed strongly on the catalyst surface. Chemisorbed aniline molecules are on the Lewis acid sites with phenyl ring perpendicular to the plane of the catalyst surface. Phenyl ring of aniline and NMA also interacts in a very similar manner and the out of plane C-H bending vibrations ($\gamma_{C\text{-}H}$) are observed between 2050-1700 cm^{-1} supporting this point. However, a very weak or poorly resolved $\gamma_{C\text{-}H}$ vibrations seen for DMA and o-toluidine indicating that the phenyl ring was fairly tilted or parallel to the plane of the oxide as shown in Figure 25.

A key unresolved problem in many catalytic studies is the reaction pathway and spectroscopic identification of intermediates and the possibility of examining in situ their occupation on catalyst surfaces under normal reaction conditions. To address this problem, it is necessary to employ the real-world catalyst and reaction conditions, or conditions that are close to the reaction conditions. Here, we present the results of selective mono-*N*-methylation of aniline through FTIR spectra of MeOH:$PhNH_2$ (3:1) mixtures co-adsorbed on $Cu_{1-x}Zn_xFe_2O_4$, (x = 0.05, 0.5, and 1) between 100 and 300^{0}C (Figure 26). NMA adsorbed on $Cu_{0.5}Zn_{0.5}Fe_2O_4$ at different temperature is given for comparison. The main features observed at 100^{0}C are: (a) IR bands observed for all reactant ratios suggest a simple co-adsorption of methoxy and aniline species with no major interaction among them (Figure 26a); (b) no sign of formate species indicate that the MeOH oxidation is hindered in the presence of aniline; (c) comparatively strong overtone bands for x = 0.5 indicate that the phenyl ring is highly perpendicular to the surface than in other cases.

Interaction among the adsorbates is evident from enormous changes observed in the spectra recorded at 200^{0}C (Figure 26b). Important points to be noted are: (1) x = 0.5 composition shows appreciable changes in band positions and intensity, including the appearance of N–C_{str} (dashed arrow at 1300 cm^{-1}) and C–H_{str} (dotted arrow at 2850 cm^{-1}) of N–CH_3 groups respectively, which corresponds to that of NMA and its formation around 200^{0}C. However, NMA desorption occurs above 250^{0}C which is the limiting factor to the reaction at lower temperatures, and a main reason for its spectroscopic observation. (2) On x = 0.05, no features correspond to NMA and no MeOH features were seen at 200^{0}C (Figure 26b); however, it resembles to that of adsorbed $PhNH_2$ at 200^{0}C [107], and suggests that the non-availability of methyl species is the limiting factor for the reaction. (3) OMe and N–H_{str} bands were observed and no N-C_{str} band at 1300 cm^{-1}, at 200^{0}C on x = 1 suggests that there is no interaction among the adsorbates; no overtone bands suggest that the phenyl ring is also not perpendicular to the surface, which is not a preferred orientation for *N*-methylation.

A very close similarity between NMA and MeOH:$PhNH_2$ (3:1) adsorbed separately on x = 0.5 at 300^{0}C (Figure 26c) further confirms the conversion of reactants to NMA. The low intensity of all the peaks at 300^{0}C compared to lower temperatures, suggests a desorption of NMA and aniline too. The methyl group seen at 2852 cm^{-1} suggests its great stability on $ZnFe_2O_4$ and, it is due to Zn–CH_3 species (Figure 26c, dotted arrow) and Zn acts as a methyl specie source for methylation reaction.

A summary of aniline N-methylation mechanistic features on $Cu_{1-x}Zn_xFe_2O_4$ ferrospinel catalysts is given in Figure 27. It was possible, due to in-situ IR studies, to observe a dissociative adsorption and possible orientation of reactants on the catalyst surface, their conversion to product at low temperatures, and desorption-limited kinetics, all under conditions that are close to the reaction conditions. Although Cu^{2+} is the active center for the aniline *N*-methylation reaction, and IR studies reveal that Zn^{2+} acts as the main methyl species source.

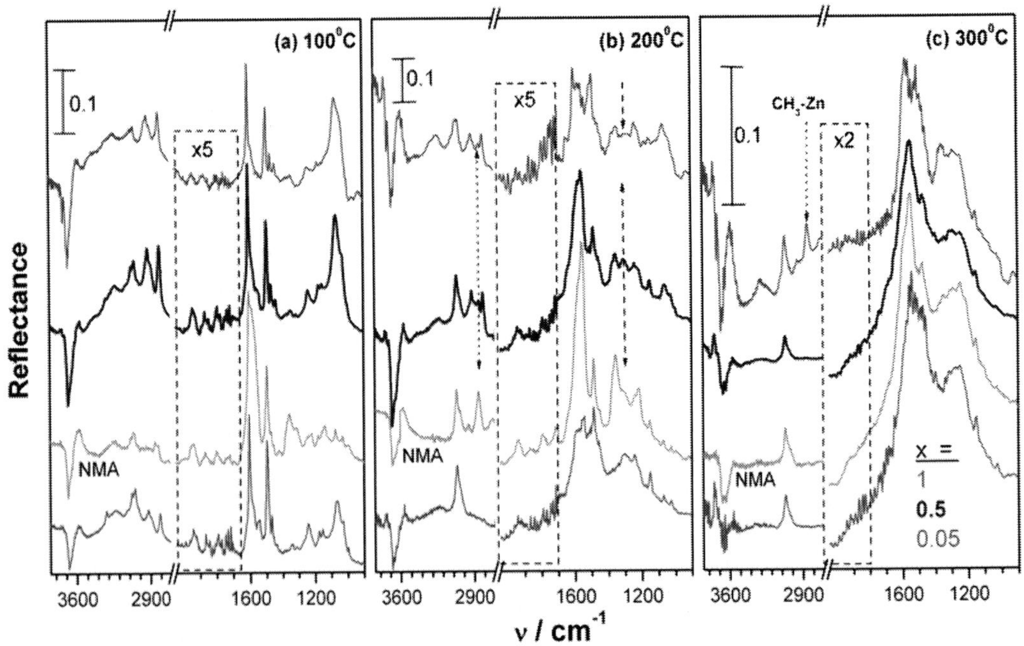

Figure 26. Representative FTIR spectra of MeOH:PhNH2 (3:1) mixture adsorbed on Cu1−xZnxFe2O4 (x = 0.05, 0.5, and 1) between 100 and 3000C. NMA adsorbed on x = 0.5 (gray traces) is shown for reference. On all panels 2100–1700 cm−1 range is multiplied on y-axis for clarity. Reprinted from Journal of Catalysis, 226, Vijayaraj M., et al., 2004, 230-234 with permission from Elsevier.

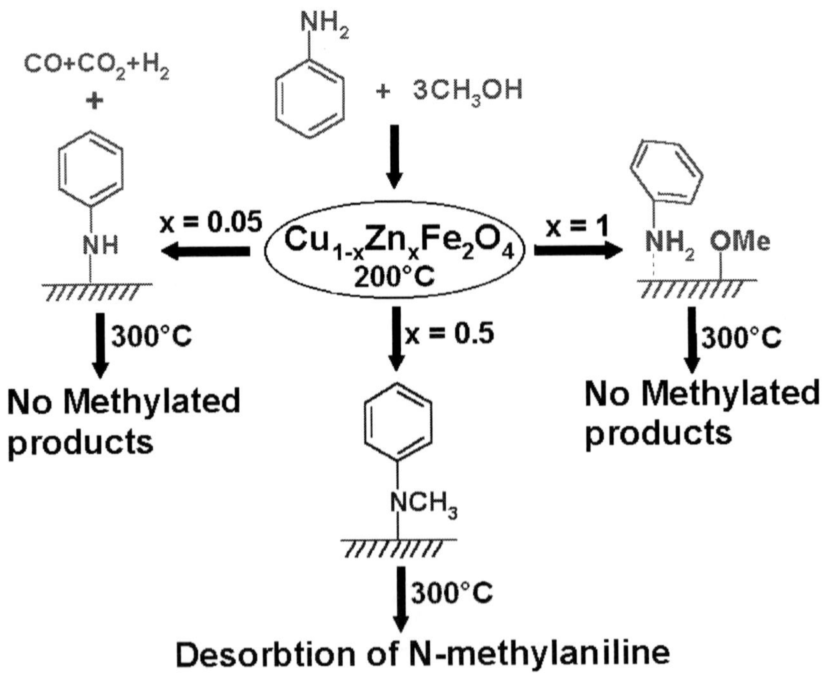

Figure 27. Aniline N-methylation trend on Cu1-xZnxFe2O4. Reprinted from Journal of Molecular Catalysis A, 231, Vijayaraj M., et al., 2005, 169-180 with permission from Elsevier.

A possible reaction mechanism is suggested from IR results. It is found that methylation of aniline is initiated by the protonation of methanol with aid of a Lewis acid-base pair. The protonation of methanol occurs on the oxides by accepting protons released from the adsorbed aniline and to form the N-methylated products. Thus the active center for the methylation is an acid-base pair site (Figure 28). This mechanism is further supported by the IR studies of competitive adsorption of reactants with acidity probe molecule pyridine [14, 107]. Aniline and methanol protonation is highly unlikely on the virgin catalyst surface since the surface –OH group was incapable of protonating even strong bases like pyridine (absence of pyridinium ion band or Bronsted acid sites) [107]. Hence, Lewis acid-base pair sites are the most probable sites for adsorption of aniline and methanol. Since aniline is more basic than methanol, Lewis acid-base pair sites are covered by aniline and that enhances aniline N-H bond scission. N-H bond scission on acid-base pair sites generates labile H^+ on the surface. This freshly formed surface H^+ has a tendency to protonate methanol molecules leading to $CH_3^{\delta+}$ cations. Methanol molecules are directed to adsorb on H^+ released from aniline and hence the active center for N-methylation is likely to be single acid-base pair site.

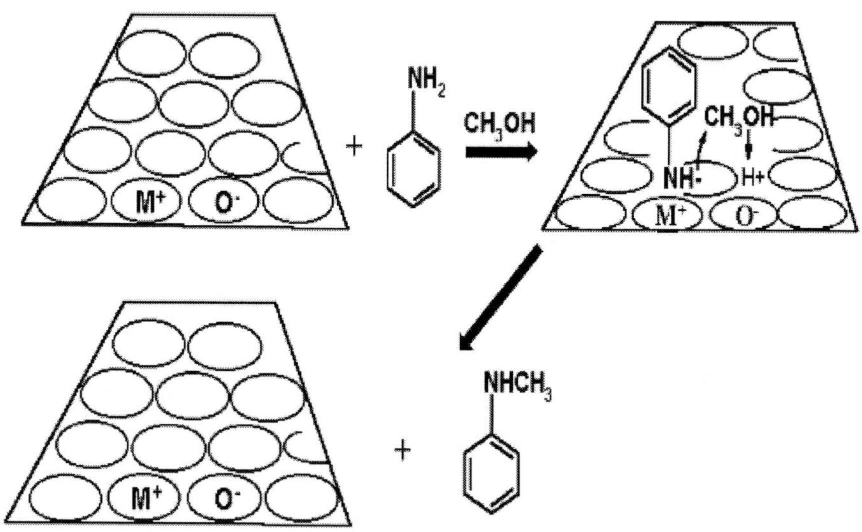

Figure 28. A pictorial representation of aniline N-methylation mechanism.

4. Alkylation of Heterocylces

Aromatic N-containing heterocycles (pyridine, imidazole etc.,) and their alkyl derivatives represent an important group of products and they have received considerable attention because of their various applications. For instance, methyl pyridines (picolines) and dimethyl pyridines (lutidines) are a class of industrially valuable compounds for the production of dyes and fine chemicals [108]. Sreekumar et al [109] reported pyridine methylation to 3-picolines over $Zn_{1-x}Co_xFe_2O_4$ spinel systems at reaction temperature between 325 and 425°C. The activity and selectivity were shown to be strongly dependent on the catalyst composition and reaction conditions. However, over all the catalyst compositions 3-picoline was formed as the major product. Pyridine conversion increases with increase in reaction temperature. Negligible conversion was observed below 375°C and the conversion occurs effectively (about 45%) around 400-425°C. 3-picoline was the major product formed in the range of temperature studied along with 3,5-lutidine, the only dialkylated side product. 3-picoline selectivity at 400°C was 97% that decreases to 83% at 425°C, whereas 3,5-lutidine selectivity increased from 2 to 14%. It is evident that 3,5-lutidine is formed by the alkylation of an initially formed 3-picoline, thus 3,5-lutidine selectivity increases at the expense of 3-picoline selectivity. Co substitution for Zn increases the surface acidity and activity of the catalyst. Hence pyridine conversion increases with increasing x values and reaches maximum conversion (49%) with x = 1 at 400°C. The combination of Co and Fe provides active sites, where methanol can be converted into HCHO that facilitates pyridine reduction to dihydropyridine, which in turn reacts with HCHO to yield 3-picoline. Trace amount of 2-picoline was also observed with Zn-rich catalysts [109].

Ni–Co ferrites with the general formula $Ni_{1-x}Co_xFe_2O_4$ were tested for the methylation of pyridine [110]. It was observed that the systems possessing x values ≥0.5 are selective for 3-picoline formation, whereas the ones with x values 0 and 0.2 give a mixture of 2- and 3-picolines. Pyridine conversion increased with the progressive substitution of Ni^{2+} ions by Co^{2+} ions. The cation distribution in the spinel lattice influences their acidic and basic properties, and these factors have been considered as helpful to evaluate the activity of the systems.

$Zn_{1-x}Cu_xFe_2O_4$ have been examined for pyridine methylation with methanol at vapor phase conditions [111]. The conversion of pyridine as well as yield to 3-methyl pyridine is found to be lowest in the case of $ZnFe_2O_4$, whereas conversion and selectivity increases as the copper content is increased and is maximum for x = 1 composition.

Grabowska *et al* [112] reported methylation of 2-hydroxypyridine with methanol over hydrothermally synthesized zinc aluminate spinels. N-methyl-2-pyridone was formed selectively (95%) and 100% 2-hydroxy pyridine conversion was achieved at 340°C. During the experiment the activity of the catalyst was shown to be stable and catalyst can be regenerated repeatedly without any significant loss of activity by passing air through the catalyst bed at 500°C.

Needle like nanoparticles of Mg-Al mixed spinel catalysts synthesized under hydrothermal conditions were used for the synthesis of 1-methylimidazole by the gas phase imidazole methylation with methanol performed at atmospheric pressure [113]. High yield and selectivity to desired product were obtained at a temperature range between 320 and 350°C. It was proposed that one of the nitrogen atoms participates in the bonding of the imidazole with the basic site of the catalyst, and the second nitrogen atom is accessible for the reaction with electrophilic methyl species formed from methanol on acid site of the catalysts.

5. CONCLUSION

Recent advances in the convenient synthesis of spinels has witnessed its application for a wide variety of catalytic reactions, such as oxidation, dehydrogenation, acylation, alkylation etc and contributed significantly to the progress of heterogeneous catalysis. An important aspect of spinels is that they can be easily prepared as a ternary (quarternary system is also possible) component system without altering the structure; however a careful selection of metal-ions is essential while designing a catalyst for specific reaction. Effectiveness of spinel materials towards catalytic alkylation reactions was demonstrated from high substrate conversion, selective product formation and long term stability for a number of reactions with different substrates in this chapter. Highly selective alkylation of organic substrates with alcohols in a fixed-bed reactor around 300^0C displays the possibility of employing the spinel catalysts in industrial applications. Particularly the ferrospinel with different metal caions exhibit the flexibility of accommodating the A-site cations, and hence a fine tuning of acid-base characteristics and energy overlap of valence band of metal ions could be achieved. Indeed the above factors contribute towards high yield with stable catalytic activity over a long period of reaction time. Further good ability to withstand some harsh reaction conditions, such as reduction atmosphere, by the redox behavior and hence structural integrity is maintained throughout the reaction period. Above characteristics of ferrospinels make them a class of structural catalytic materials, which is integrated chemically, and electronically.

An important attribute of these materials comes from the two distinct cationic sites (O_h and T_d), and the feasibility of cations migration among them due to the redox nature of metal-ions, while keeping the spinel structure intact. This particular aspect helps to avoid possible segregation/sintering of metal-ions and remain stable for longer period, compared to a mixed metal oxide catalyst. This unique property of spinels makes them an attractive candidate for number of catalytic reactions.

In-situ and ex situ analysis of active catalyst compositions show the complete heterogeneity of the surface, indicating the interaction among the same metal-ions is prevented and the structural integrity is maintained. This allows the different elementary reactions to occur on nearby active sites. Mechanism of phenol methylation to 2,6-xylenol and N-methyl aniline from aniline demonstrate the presence of acid-base pair site on ferrospinels. Further the role of Zn as an "*active spacer and stabilizer*" in the N-methylation of aniline on $Cu_{1-x}Zn_xFe_2O_4$ was brought out by a detailed characterization studies with XPS, in situ IR, TPR. Similarly, the presence of both Co and Cu are highly favorable for selective ortho alkylation and their synergetic interaction in Cu-Co-Fe ternary system was clearly elucidated in $Cu_{1-x}Co_xFe_2O_4$. Cu-Co synergism in this material is not only useful for achieving high phenol conversion and 2,6-xylenol selectivity, but also to avoid unwanted side reactions like methanol gasification and thereby stopping fast deactivation. The Co-Cu interaction is quite known in multi-component mixed oxides and/or supported mixed oxides, especially Fisher-Tropsch synthesis.

In spite of many positive points about the spinel system, not many industrial applications are available at present. Nonetheless, a continuing and thorough study would shed more light to improve the efficacy of the spinel system towards various industrial applications.

ACKNOWLEDGMENTS

We would like to thank the authors of the original article for providing the artwork and permission to use the figure in this review. Helpful discussions and fruitful collaborations are acknowledged with Drs. B.S. Rao, S.V. Chilukuri, and S.G. Hegde. Most part of the work reported in this chapter was carried out during the doctoral studies of MV and TM and financial support in the form of research fellowship to MV and TM from CSIR, New Delhi is gratefully acknowledged.

REFERENCES

[1] Author. Friedel-Crafts and Related Reaction, Interscience Publishers, New York, 1963, 1, pp 205.
[2] Narayanan, S.; Deshpande, K. *Appl. Catal.* A 2000, 199, 1-31.
[3] Tanabe, K; Misono, M; Ono, Y.; Hattori, H., *Stud. Surf. Sci. Catal.* 1989, 47, 51-56.
[4] Viswanathan, B. *Bull. Catal. Soc.* India. 2000, 9, 10-20.
[5] Clerici, M. G.; *Topics. Catal.* 2000, 13, 373-386.
[6] Viswanathan, B.; Jacob, B. *Catal. Rev.* 2005, 47, 1-82.
[7] Kotanigawa, T.; Yamamoto, M; Shimokawa, K.; Yoshida, Y. *Bull. Chem. Soc. Jpn.* 1971, 44, 1961-1964.
[8] Kotanigawa, T.; Yamamoto, M. *Bull. Chem. Soc. Jpn.* 1974, 47, 954-957.
[9] Kotanigawa, T. *Bull. Chem. Soc. Jpn.* 1974, 47, 950-953.
[10] Gregory, W. L. US Patent 4,933,509 (1990).
[11] Hill, A. G.; Shipps. J. H.; Hill, A. *J. Ind. Eng. Chem.* 1981, 43, 1579-1583.
[12] Parera, J. M.; Gonzalez, A.; Barral. M. M. *Ind. Eng. Chem. Prod. Res. Dev.* 1968, 7, 259-262.
[13] Mathew, T.; Synthesis and characterization of mixed oxides containing cobalt, copper and iron and study of their catalytic activity', Ph. D Thesis, University of Pune, 2002.
[14] Vijayaraj. M.; Heteroatom alkylation of aromatic compounds over metal oxides, Ph. D Thesis, University of Pune, 2006.
[15] Hill, R. J.; Craig, J. R.; Gibbs, G. V. *Phys. Chem. Minerals* 1979, 4, 317-339.
[16] Verwey, E. J. W.; Heilmaan, E. L. *J. Chem. Phys.* 1947, 15, 174-180.
[17] Bautiles, J. P.; Barboax, Y. *J. Appl. Cryst.* 1982, 15, 665-668.
[18] Jacobs, J. P.; Maltha, A.; Reintjes, J. G. H.; Drimal, J.; Ponec, V.; Brongersma, H. H. J. Catal. 1994, 147, 294-300.
[19] Ziolkowski, J.; Barbaun, Y. *J. Mol. Catal.* 1991, 67, 199-215.
[20] Goldstein, J. R.; Tseung, A. C. C. *J. Catal.* 1974, 32, 452-465.
[21] Onuchukwu, A. I.; J. Chem. Soc. Faraday Trans. I 1984, 80, 1447-1456.
[22] Cota, H. M.; Katan, J.; Chim. M.; Schoenweis, *J. Nature,* 1964, 203, 1281.
[23] Goldstein, J. R.; Tseung, A. C. C. *J. Phys. Chem.* 1972, 76, 3646-3656.
[24] Boreskov, G. K.; Proporski, V. V.; Sezonov, V, A. Proc. 4th Inter. Cong. Catal. Moscow, 1963.
[25] Omata, K.; Takada, T.; Kasahara, S.; Yamada, M. *Appl. Catal.* 1996, 146, 255-267.
[26] Ghose, J.; Murthy, K. S. R. C. *J. Catal.* 1996, 162, 359-360.
[27] Awe, A. A.; Miliades, G.; Vickerman, J. C. *J. Catal.* 1980, 62, 202-210.
[28] Murthy, K. S. R. C.; Ghose, J. *J. Catal. 1*994, 147, 171-176.
[29] Severino, F.; Brito, J.; Carias, O.; Lainc, J. *J. Catal.* 1986, 102, 172-179.
[30] Kehl, W. L.; Rennard, R. J. U. S. Patent, 1969, 3,450,787.

[31] Kung. H. H.; Kundalkar, B.; Kung, M. C.; Cheng, W. H. *J. Phys. Chem.* 1980, 84, 382-388.
[32] Sloczynski, J.; Ziolkowski, J.; Grzybowska, B.; Grabowski, R.; Jachewicz, D.; Wcislo, K.; Gengembre, L. *J. Catal.* 1999, 187, 410-418.
[33] Finocchio, E.; Busca, G.; Lorenzelli, V.; Willey, R. J. *J. Catal.* 1995, 151, 204-215.
[34] Cares, W. R.; Hightower, J. W. *J. Catal.* 1971, 23, 193-203.
[35] Rennard, R. J.; Kehl, W. L. *J. Catal.* 1971, 21, 282-293.
[36] Toledo-Antonio, J. A.; Nava, N.; Martinez, M.; Bokhimi, X. *Appl. Catal.* A 2002, 234, 137-144.
[37] Crivello, M.; Perez, C.; Herrero, E.; Ghione, G.; Casuscelli, S.; Rodriguez-Castellon, E. *Catal. Tod*ay 2005, 107-108, 215-222.
[38] Jenck, J.; Germain, E. *J. Catal.* 1980, 65, 133-140, 141-149.
[39] Evans, J. W.; Casey, P. S.; Wainwright, Trimm, D. L.; Cant, N. W. *Appl. Catal.* 1983, 7, 31-41.
[40] Shangguan, W. F.; Teraoka, Y.; Kagawa, S. *Appl. Catal.* B: Environmental 1998, 16, 149-154.
[41] Shangguan, W. F.; Teraoka, Y.; Kagawa, S. *Appl. Catal.* B: Environmental 1996, 8, 217-227.
[42] Narashimhan, C. S.; Swamy, C. S. *Appl. Catal.* 1982, 2, 315-328.
[43] Jebarathinam, N. J.; Eswaramoorthy, M.; Krishnaswamy, V. *Appl. Catal.* 1996, 145, 57-74.
[44] Chen. W. S.; Lee. M. D.; Lee, J. F. *Appl. Catal.* 1992, 83, 201-211.
[45] Dube, G. R.; Darshane, V. S.; *J. Mol. Catal.* 1993, 79, 285-296.
[46] Sreekumar, K.; Mathew, T.; Jyothi, T. M.; Bhiju, M. D.; Sugunan, S.; Rao, B. S. Pol. *J. Chem.* 2000, 74, 509-518.
[47] Roesky, R.; Weiguny, J.; Bestgen, H.; Dingerdissen, U. *Appl. Catal.* A 1999, 176, 213-220.
[48] Xanthopoulou, G. *Appl. Catal.* A 1999, 182, 285-295.
[49] Sloczynski, J.; Jan, J.; Machej, T.; Rynkowski, J.; Stoch, *J. Appl. Catal.* B 2000, 24, 45-60
[50] Fierro, G.; Morpurgo, S.; Lo Jacono, M.; Inversi, M.; Pettiti, I. *Appl. Catal.* A 1998 166, 407-417.
[51] Sato, S.; Iijima, M.; Nakayama, T.; Sodesawa, T.; Nozaki, F. *J. Catal.* 1997, 169, 447-454.
[52] Castiglioni, G. L.; Vaccari, A.; Fierro, G.; Inversi, M.; Lo Jacono, M.; Minclli, G.; Pettiti, I.; Porta, P.; Gazzano, M. *Appl. Catal.* A 1995, 123, 123-144.
[53] Miki, J.; Asanuma, M.; Tachibana, Y.; Shikada, T. *J. Catal.* 1995, 151, 323-329.
[54] Guin, D.; Baruwati, B.; Manorama, S. V. *J. Mol. Catal.* A 2005, 242, 26-31.
[55] Ramanathan, R.; Sugunan, S. *Catal. Comm.* 2007, 8, 1521-1526.
[56] Miya, B.; Hoshino, F.; Iwasa, I. *J. Catal*. 1966, 5, 401-411.
[57] Schoubye, P. *J. Catal.* 1969, 14, 238-246.
[58] Busca, G.; Trifiro, F.; Vaccari, A. Langmuir, 1990, 6, 1440-1447
[59] Kato, H.; Sano, T.; Wafa, Y.; Tamaura, M.; Tsuji, M.; Tsuji, T.; Miyazaki, S. J. Mater. Sci. 1995, 30. 6350-6354.
[60] Epling, W. S.; Hoflund, G. B.; Hartand, W. M.; Minahan, D. M. *J. Catal.* 1997, 169, 438-446.

[61] Kim, D.; IHM, S. Environ. *Sci. Technol.* 2001, 35, 222-226.
[62] Mei, Z.; Shen, Z.; Wang, W.; Zhang, Y. *Environ. Sci. Tehcnol.* 2008, 42, 590-595.
[63] Yoo, J. S.; Bhattacharyya, A. A.; Radlowski, C. A.; Karch, J. A.. *Ind. Eng. Chem. Res.* 1992, 31, 1252-1258.
[64] Bhattacharyya, A. A.; Woltermann, G. M.; Yoo, J. S.; Karch, J. A.; Cormier, W. E. Ind. *Eng. Chem. Res.* 1988, 27, 1356-1360.
[65] Li, W.; Cheng, H.; *Solid State Sciences* 2007, 9, 750-755.
[66] Anilkumar, P.; Shroti, J. J.; Kulkarni, S. D.; Deshpande, C. E.; Date, S. K . Mater. *Lett.* 1996, 27, 293-296.
[67] Platero, E. E.; Arean, C. O.; Parra, J. B. Res. *Chem. Intermed.* 1999, 25, 187-194.
[68] Industrial Organic Chemicals, Ullmann's Encyclopedia, Wiley- VCH, 1999, 6, 3713-3757.
[69] Ertl, G.; Knozinger, H.; Weitkamp, J. Handbook of Heterogeneous Catalysis, 1997, Vol. 5, Wiley- VCH, Weinheim, 2311-2324.
[70] Fiege, H.; Bayer, A. G.; Leverkusen. Federal Republic of Germany, in "*Ullmann's Encyclopedia of Industrial Chemistry*", 1991, A19, 313-369.
[71] Busca, G. Catal. Today 1996, 27, 457-496.
[72] Sreekumar, K.; Sugunan, S. *J. Mol. Catal.* A 2002, 185, 259-268.
[73] Sreekumar, K.; Sugunan, S.; *Appl. Catal.* A 2002, 230, 245-251.
[74] Mathew, T.; Shiju, N. R.; Sreekumar, K.; Rao, B. S.; Gopinath, C. S. *J. Catal.* 2002, 210, 405-417.
[75] Reddy, A. S.; Gopinath, C. S.; Satyanarayana, C. *J. Catal.* 2006, 243, 278-291.
[76] Ballarini, N.; Cavani, F.; Maselli, L.; Montaletti, A.; Passeri, S.; Scagliarini, D.; Flego, C.; Perego, C. *J. Catal.* 2007, 251, 423-436.
[77] Grabowska, H.; Mista, W.; Trawczynski, J.; Wrzyszcz, J.; Zawadski, M. Res. *Chem. Intermed.* 2001, 27, 305-313.
[78] Mathew, T.; Tope, B. B.; Shiju, N. R.; Hegde, S. G.; Rao. B. S.; Gopinath, C. S. *Phys. Chem. Chem. Phys.* 2002, 4, 4260-4267.
[79] Mathew, T.; Vijayaraj, M.; Pai, S.; Tope, B. B.; Hegde, S. G.; Rao, B. S.; Gopinath, C. S. *J. Catal.* 2004, 227, 175-185.
[80] Mathew, T.; Shiju, N. R.; Bokade, V. V.; Rao, B. S.; Gopinath, C. S. Cata. Lett. 2004, 94, 223-236.
[81] Mathew, T.; Rao, B. S.; Gopinath, C. S. *J. Catal.* 2004, 222, 107-116.
[82] Mathew, T.; Shylesh, S.; Devassy, B. M.; Vijayaraj, M.; Satyanarayana, C. V. V. S.; Rao, B. S.; Gopinath, C. S. *Appl. Catal.* A 2004, 273, 35-45.
[83] Van der Laan, G.; Westra, C.; Haas, C.; Sawatzky, G. A. *Phys. Rev.* B 1981, 23, 4369-4380.
[84] Sokoll, R. ; Hobert, H. *J. Catal.* 1990, 125, 285-291.
[85] Scire, S.; Crisafulli, C.; Maggiore, R.; Minico, S.; Galvagno, S. *Appl. Surf. Sci.* 1996, 93, 309-316.
[86] Bautista, F. M.; Campelo, J. M.; Garcia, A.; Luna, D.; Marinas, J. *M. J. Catal.* 1987, 107, 181-194.
[87] Corma, A.; Llopis, F.; Monton, J.B.; Weller, S. *J. Catal.* 1993, 142, 97-109.
[88] Helander, I. M.; Alakomi, H. L.; Latva-Kala, K.; Mattila-Sandholm, T.; Pol, I.; Smid, E. J.; Gorris, L. G. M.; von Wright, A. *J. Agric. Food Chem.* 1998, 46, 3590-3595.

[89] Grabowska, H.; Mista, W.; Trawczynski, J.; Wrzyszcz, J.; Zawadski, M. *Appl. Catal.* A 2001, 220, 207-213; and references therein.
[90] Ko, A. K.; Young, C. L.; Zhu, W. D.; Lin, H. E. *Appl. Catal.* A 1996, 134, 53-66.
[91] Sreekumar, K.; Raja, T.; Kiran, B. P, Sugunan, S.; Rao, B. S. *Appl. Catal.* A 1999, 182, 327-336.
[92] Sreekumar, K.; Jyothi, T. M.; Talawar, M. B.; Kiran, B. P.; Rao, B. S.; Sugunan, S. *J. Mol. Catal.* A 2000, 152, 225-236.
[93] Sreekumar, K.; Jyothi, T. M.; Mathew, T.; Talawar, M. B.; Sugunan, S.; Rao, B. S. *J. Mol. Catal.* A 2000, 159, 327-334.
[94] Sreekumar, K.; Mathew, T.; Mirajkar, S. P.; Sugunan, S.; Rao, B. S. *Appl. Catal.* A 2000, 201, L1-L8.
[95] Nishamol, K.; Rahna, K. S.; Sugunan, S. *J. Mol. Catal.* A 2004, 209, 89-96.
[96] Vijayaraj, M.; Gopinath, C. S. *J. Catal.* 2006, 241, 83-95.
[97] Moretti, J. J. Electron Spectrosc. *Relat. Phenom.* 1995, 76, 365-370.
[98] Vijayaraj, M.; Gopinath, C. S. *Appl. Catal.* A 2007, 320, 64-68.
[99] Velu, S.; Suzuki, K.; Okazaki, M.; Kapoor, M. P.; Osaki, T.; Ohashi, F. *J. Catal.* 2000, 194, 373-384.
[100] Busca, G. Catal. Today, 1998, 41, 191-206.
[101] Knözinger, H.; Krietenbrink, H.; Müller, H. D.; Schultz, W. Proceedings of Sixth International Congress on Catalysis, 1977, Vol. 1, p183-189.
[102] Busca, G.; Lorenzenlli, V. *Mater. Chem*, 1981, 6, 175-185.
[103] Harrison, P. G.; Thornton, E. W. *J. Chem. Soc. Faraday Trans* 1975, 71, 1013-1020.
[104] Rajagopal, S.; Grimm, T. L.; Collins, D. J.; Miranda, R. *J. Catal.* 1992, 137, 453-461.
[105] Khader, M. M. *J. Mol. Catal.* A 1995, 104, 87-94.
[106] Vijayaraj, M.; Gopinath, C. S. *J. Catal.* 2004, 226, 230-234.
[107] Vijayaraj, M.; Murugan, B.; Umbarkar, S.; Hegde, S. G.; Gopinath, C, S. *J. Mol. Catal.* A 2005, 231, 169-180.
[108] Kashiwagi, H.; Fujiki, Y.; Enomoto, S. *Chem. Pharm. Bull.* 1982, 30, 2575-2581.
[109] Sreekumar, K.; Mathew, T.; Devassy, B. M.; Rajagopal, R.; Vetrivel, R.; Rao, B. S. *Appl. Catal.* A 2001, 205, 11-18.
[110] Sreekumar, K.; Mathew, T.; Rajagopal, R.; Vetrivel, R.; Rao, B. S. *Cata. Lett.* 2000, 65, 99-105.
[111] Banerjee, M.; Verma, N.; Prasad, R. *J. Mater. Sci.* 2007, 42, 1833-1837.
[112] Grabowska, H.; Zawadski, M.; Syper, L. *Appl. Catal.* A 2006, 314, 226-232.
[113] Grabowska, H.; Zawadski, M.; Syper, L.; Mista, W. *Appl. Catal.* A 2005, 292, 208-214.

Reviewed by Dr. Subramani Velu BP Products North America, Inc.Refining Technology Naperville, IL, USA velu.subramani@bp.com.

In: Heterogeneous Catalysis Research Progress
Editor: Mathias B. Gunther, pp. 193-233
ISBN 978-1-60456-979-7

Chapter 5

HETEROGENEOUS FENTON CATALYSIS FOR THE DEGRADATION OF AZO TEXTILE DYES

Alberto Alvarez Gallegos[1,*], Susana Silva Martínez[1] and Roberto Flores Velazquez[2]

[1] Universidad Autónoma del Estado de Morelos, Centro de Investigación en Ingeniería y Ciencias Aplicadas. Av. Universidad 1001, Col. Chamilpa, Cuernavaca Morelos 62209, México

[2] Instituto de Investigaciones Eléctricas, Calle Reforma 113, Col. Palmira, Cuernavaca, Morelos 62490, México

ABSTRACT

The main experimental drawback raised in a homogeneous Fenton catalyst is analyzed in this chapter. In order to overcome its deficiency a heterogeneous catalyst is proposed.

In the homogeneous catalysis approach it was observed that hydrogen peroxide can be activated either by ions: Fe^{2+} or Fe^{3+}. The iron on its two oxidation states shows similar efficiencies on the degradation of a wide range of highly toxic organic molecules under mildly acid conditions. Amongst the organic compounds that can be mentioned are: synthetic textile dyes (amaranth, reactive black 5, acid green 25, basic blue 9, direct red 23 and acid orange 7), phenol, catechol, hydroquinone, p-benzoquinone, aniline, cresol and small open chain organic acids such as oxalic. Hence, strong oxidants such as $HO^{\bullet}$, $HO_2^{\bullet}$, FeO^{2+} among others are responsible of such degradation. Although homogeneous catalysts in the Fenton and Fenton-type reactions are working well, the catalytic activity ceases after a critical time because the iron ions undergo a chemical speciation during the reaction. It is not possible to restart it again either with the addition of more iron ions or with hydrogen peroxide. In order to maintain the catalytic activity after certain reaction time, a heterogeneous catalyst is proposed. This will improve the control over iron ions during the oxidation reaction. Iron (III) supported on fly ash and iron oxide supported on cerium oxide were used as heterogeneous Fenton catalysts and their preparations were performed by standardized thermal procedures. Chemical

* Tel/Fax: (777) 329 7084 and 329 7984.

oxidation of two azo dyes, reactive black 5 (RB5) and basic orange 2 (BO2), was conducted by using the heterogeneous Fenton catalysis. Experimental results show that it is possible to oxidize 0.061 mM RB5 (in a media 50 mM Na_2SO_4, ≈pH 2.8 adjusted with H_2SO_4) using a stoichiometric amount of H_2O_2. After 2 hours of treatment, reactive solutions were effectively colorless and the 80% of the original chemical oxygen demand was removed. A complete decoloration and 100% COD abatement was achieved for the degradation of 0.1 mM BO2 in 5 hours in the presence of 4 mM H_2O_2, 0.1 g/l catalyst (50 % iron oxide) at initial solution pH of 3.0. Low Fe(III) leaching in heterogeneous supported catalyst were observed.

INTRODUCTION

More than a century ago (*Fenton 1894*), it was reported that ferrous ion strongly catalyses the oxidation of maleic acid by hydrogen peroxide in mild conditions, thereafter the Fe^{2+}/H_2O_2 mixture was known as Fenton´s reagent. Soluble iron cations (normally Fe^{II}/Fe^{III}) interact with hydrogen peroxide as the primary oxidant. The conversion of the organic matter is achieved by using the oxidant that is liberated by a catalytic process, in which, a redox chain is maintained between Fe^{II}/Fe^{III}. The fraction of the organic conversion (or the fraction of the hydrogen peroxide consumed) provides a measure of the efficiency of the oxidation process. The chemistry behind the Fenton's reagent is quite complex and not always clearly established in all its details.

In this chapter, the main mechanisms of the homogeneous catalysis of the Fenton's reagent are discussed and special attention is paid in their main drawback as an effective oxidant. Indeed, the catalytic activity ceases after a critical time because iron ions undergoes a chemical speciation during the Fenton process. In order to maintain for a longer time the catalytic activity of the iron ions, a heterogeneous catalyst is proposed. This approach is better because the control over iron ions during the oxidation reaction is improved.

HOMOGENOUS CATALYSIS

Hydroxyl Radical Hypotheses

Hydrogen peroxide is an efficient oxidant and a high proportion of the peroxide's weight, 47%, is available as active oxygen. Hydrogen peroxide on its own can oxidize hydrogen sulphide, cyanides, sulphur and oxides of nitrogen (*Rajeshwar and Ibañes 1997; Hoffman 1977*). Although hydrogen peroxide alone can oxidize some pollutants, it cannot oxidize large organic pollutants. It needs to be activated in some way to become a powerful oxidant. After 38 years that Fenton's reagent was published, this process started receiving further attention, since then, a large amount of work has been carried out to understand the mechanism of catalysis. The first intermediate proposed, as the responsible of the organic oxidation, was FeO^{2+}, which is another way to denote Fe^{IV} (*Bray and Gorin 1932*). However, the mechanism proposed failed to describe the complexity of the Fenton chemistry. A few years later, Haber and Weiss (*Haber and Weiss 1934*) were the first researchers that suggested a hydroxyl radicals ($OH^{\bullet}$) production, from hydrogen peroxide and ion iron mixtures, as the responsible for organic oxidation. Based in such hypothesis, the first kinetics and stoichiometric mechanism were proposed (*Baxendale et al. 1946; Merz and Waters 1947; Uri 1952; Barb et al. 1951*) to understand the Fe^{2+}/Fe^{3+}-H_2O_2 interactions and hydroxyl radicals production (*Baxendale and Wilson 1957*) from hydrogen peroxide. However, a few years later the first experimental inconsistencies about the roll of hydroxyl radicals, during organic oxidations, were documented (*Shiga 1965*).

The hypothesis of hydroxyl radicals production from Fe^{2+}/H_2O_2 mixtures was retaken in the 1970´s with a fresh approach (*Walling and Weil 1974; Walling 1975*) and this hypothesis was repositioned because the presence of such radicals could explain organic hydroxylation: a controlled oxidation mechanism of aromatic compounds by $OH^{\bullet}$ radicals (*Tomat and Vecchi 1971; Tomat and Rigo 1976; Tomat and Rigo 1979; Zepp et al. 1992; Spadero et al. 1994*). Although the organic oxidation by Fenton´s reagent was not always considered clear (*Walling et al. 1970; Walling and Wei. 1974*) it was accepted that hydrogen peroxide is catalyzed by Fe^{2+} according to the following mechanism (*Walling 1975; Walling and Cleary 1977*):

$$Fe^{2+} + H_2O_2 \rightarrow Fe^{3+} + OH^{\bullet} + HO^{-} \quad k_1 = 76 \tag{1}$$

$$Fe^{2+} + OH^{\bullet} \rightarrow Fe^{3+} + OH^{-} \quad k_2 = 3x10^8 \tag{2}$$

$$OH^{\bullet} + H_2O_2 \rightarrow H_2O + HO_2^{\bullet} \quad k_3 = 1.2 - 4.5x10^7 \tag{3}$$

$$2HO^{\bullet} \rightarrow H_2O_2 \quad k_4 = 5.3x10^9 \tag{4}$$

$$Fe^{3+} + H_2O_2 \rightarrow Fe^{2+} + H^{+} + HO_2^{\bullet} \quad k_5 = 0.002 \tag{5}$$

$$Fe^{3+} + HO_2^{\bullet} \leftrightarrow Fe^{2+} + H^{+} + O_2 \quad k_6 \tag{6}$$

$$Fe^{2+} + HO_2^{\bullet} \rightarrow Fe^{3+} + HO_2^{-} \quad k_7 \tag{7}$$

Where k_1 to k_7 values are in M^{-1} s^{-1}, k_6 and k_7 values were only known approximately; as a result, different stages in the Fenton´s reagent were described as follows: reactions (1) and (2) will take place in a solution containing an excess of Fe^{2+} with respect to H_2O_2. With excess of hydrogen peroxide in acid solution, when reactions (5) and (6) can be neglected, reactions (1), (3) and (7) occur. Comparable concentrations of Fe^{2+} and H_2O_2 in acid media result in reactions (1), (2), (3) and (7).

Several research groups (*Pignatello 1992; Sun and Pignatello 1993; Legrini et al. 1993; Pignatello and Chapa 1994*) have reinterpreted the above mechanism (equations 1-7), accepting that hydroxyl radicals $OH^{\bullet}$ are the main oxidizing species in Fenton reaction, but the interactions of H_2O_2 with ferrous ions produce soluble ferric complex:

$$Fe^{2+} + H_2O_2 \rightarrow Fe(OH)^{2+} + OH^{\bullet} \tag{8}$$

Reaction (8) is still the main source of hydroxyl radicals $OH^{\bullet}$ but most of the soluble ferric ions are forming complexes: $Fe(OH)^{2+}$, $FeOOH^{2+}$. Ferrous and ferric ions can be regenerated supporting the Fenton process, according to the following equations:

$$Fe(OH)^{2+} + H_2O_2 \leftrightarrow Fe^{2+} + H_2O + HO_2^{\bullet} \tag{9}$$

$$Fe(OH)^{2+} + HO_2^{\bullet} \rightarrow Fe^{2+} + H_2O + O_2 \tag{10}$$

$$Fe^{3+}\ H_2O_2 \leftrightarrow Fe-OOH^{2+} + H^{+} \tag{11}$$

$$Fe-OOH^{2+} \rightarrow Fe^{2+} + HO_2^{\bullet} \tag{12}$$

$$Fe^{2+} + HO_2^{\bullet} + H^{+} \rightarrow Fe^{3+} + H_2O_2 \tag{13}$$

The fact that reaction (12) is much slower than reaction (8), implies that Fe^{2+} is faster depleted from the solution. As a result, Fenton process is halted because the redox chain cannot be supported itself. In addition, it is accepted that (*Pignatello 1992*; *Boye et al. 2003*) the hydroperoxyl radical ($HO_2^{\bullet}$) has a much lower oxidant power than $OH^{\bullet}$. In the presence of organics, Fenton chemistry is even more complex because hydroxyl radical, both iron cations and the oxidation products enter into a series of consecutive and parallel reactions. An example of the complexity of these reactions is discussed elsewhere (*Gozzo 2001*) but a brief description is given here. The initial step for an organic substrate (R-H) oxidation starts with the interaction of itself with $OH^{\bullet}$, according to (*Walling and Kato 1971*):

$$RH + OH^{\bullet} \rightarrow H_2O + \begin{Bmatrix} R^{\bullet}_{tertiary} \\ R^{\bullet}_{primary} \end{Bmatrix} \tag{14}$$

The tertiary and primary hydroxyalkyl radicals are produced in the ratio 7.2:1, respectively. The former is rapidly oxidized by Fe^{3+} (supporting the Fenton process) while a big proportion of the latter is accumulated in solution and terminates the redox chain by dimerization, according to the reactions (15) and (16), respectively:

$$Fe^{3+} + R^{\bullet}_{tertiary} \rightarrow R=O + H^{+} + Fe^{2+} \tag{15}$$

$$R^{\bullet}_{primary} + R^{\bullet}_{primary} \rightarrow R-R \tag{16}$$

In a more general situation, the redox chain depends, among other factors, of the type of $R^{\bullet}$ produced by the fast H-abstraction (reaction 14).

Ferryl Ion and Hydroxyl Radical Hypotheses

Since long time ago it was widely accepted that oxidation using Fenton reactions are initiated by free hydroxyl radical (*Haber and Weiss 1934; Walling and Cleary 1977; Lipczynska-Kochany 1991*) but some doubts have always raised about the nature of the real intermediate oxidant in the Fenton process. At the beginning, a sort of complex was proposed (*Shiga 1965*) as the main chemical species that initiate the oxidation rather than an $OH^{\bullet}$ radical. Thereafter, several studies (*Groves and Watanabe 1986; Kean et al. 1987; Rahhal and Richter 1988*) have retaken an old proposal and proposed high-valent iron-oxo intermediates (i.e. Fe^{IV}) as the responsible of the substrate oxidation. Under this approach, H_2O_2 interactions with ferrous ions produce soluble high-valent iron-oxo intermediates instead of soluble ferric or ferric complex (see reactions 1 and 8), according to the following consecutive reactions (*Bossman et al. 1998*):

$$Fe^{2+} + H_2O_2 \rightarrow Fe(H_2O_2)^{2+} \tag{17}$$

$$Fe(H_2O_2)^{2+} \rightarrow Fe^{4+}(OH^-)_2 \rightarrow \textit{initiate the oxidation} \tag{18}$$

$$Fe^{4+}(OH^-)_2 \rightarrow Fe^{3+}_{aq}(?) + OH^- + OH^{\bullet} \rightarrow \textit{continue the oxidation} \tag{19}$$

Under this approach, one of the biggest question is if the reaction (19) is not to slow to compete with the reaction (18). On the other hand, Fe^{IV} appears to be a weaker oxidant than $OH^{\bullet}$ radical (*Koppenol and Liebman 1984; Rahhal and Richter 1988).* Experimental evidences (using the electron paramagnetic resonance spin-trapping method) have shown the existence of several intermediated oxidant in the Fenton process, like $OH^{\bullet}$ bound $OH^{\bullet}$ and high-valence iron species (*Yamazaki and Piette 1991; De Laat and Gallard 1999; Gallard and De Laat 2001*).

Fe^{III} catalytically decomposes H_2O_2, unfortunately much less attention has been paid to this reaction and the possible mechanism varies according to the authors. Those that support the free radical $OH^{\bullet}$ approach, agree with the classical radical chain mechanism represented by equations (9, 10, 11 and 12). However, on the other side, high valent iron-oxo complex has also been proposed in Fe^{III}-H_2O_2 interactions (*Kremer 1985; Gozzo 2001*).

Ferryl Ion and Binuclear Species Intermediates Hypotheses

A detailed kinetic study in the Fenton chemistry, based on spectrophotometric measurements of evolved O_2 and the disappearance of Fe^{2+}, has shown the existence of a new intermediate (*Kremer 1999*). It was identified as the mixed valence binuclear species $(FeOFe)^{5+}$. The following mechanism is derived from experimental results:

$$Fe^{2+} + H_2O_2 \underset{k_2}{\overset{k_1}{\longleftrightarrow}} (Fe^{2+}H_2O_2) \underset{-H_2O}{\overset{k_3}{\longrightarrow}} FeO^{2+} \tag{20}$$

Fenton chemistry starts with the reversible formation of a complex, Fe^{2+} H_2O_2. From this primary complex a high-valent iron-oxo intermediates Fe^{IV} is formed by loss of H_2O. Although there is not a direct kinetic evidence for the existence of $Fe^{2+}H_2O_2$ it is included in equation (20) because it is more probable that FeO^{2+} is formed in more than one step. Under this assumption $k_2 << k_3$, in agreement with other experimental observations (*Bossman et al. 1998*). At this point, Fenton chemistry turns out very complex. The Fe^{IV} can react either with Fe^{2+} ions to produce Fe^{3+} or with H_2O_2 to produce O_2 and Fe^{2+} or even with Fe^{3+} to form a binuclear species or it can also decompose back into Fe^{IV} and Fe^{3+}, according to:

$$FeO^{2+} + \left\{ \begin{array}{l} H_2O_2 \xrightarrow{k_4} Fe^{2+} + O_2 + H_2O \\ Fe^{2+} + H_2O \xrightarrow{k_5} 2Fe^{3+} + 2OH^- \\ Fe^{3+} \underset{k_8}{\overset{k_6}{\leftrightarrow}} FeOFe^{5+} \end{array} \right\} \quad (21)$$

The binuclear species can react with H_2O_2 to form O_2 and a mixture of Fe^{2+}/Fe^{3+}, according to:

$$FeOFe^{5+} + H_2O_2 \xrightarrow{k_7} O_2 + Fe^{2+} + Fe^{3+} + H_2O \quad (22)$$

Based on the above chemical reactions, a qualitative description of the Fenton chemistry includes the following main stages. At the beginning of the reaction at high H_2O_2/Fe^{2+}, there is a competition between H_2O_2 and Fe^{2+} for FeO^{2+}, following the paths 4 and 5. As Fe^{3+} is produced a new intermediate is formed ($FeOFe^{5+}$) through the path 6 and it is further decomposed through the path 7. At low H_2O_2/Fe^{2+}, the abatement of FeO^{2+} is mainly carried out by steps 5 and 6, while step 4 is negligible. As the ratio Fe^{3+}/Fe^{2+} increases the concentration of the binuclear species increases as well and its decomposition is basically carried out through the step 8. According to the experimental results (*Kremer 1999; Kemer 2003*), it is difficult to support the existence of radical species ($OH^\bullet$, $HO_2^\bullet$) as the main responsible of Fe^{2+} and substrates oxidation in the Fenton chemistry. Under this approach, organic substrate (R-H) oxidation can be understood by the following reactions. The active intermediate may oxidize stoichiometrically organic substrate according to one-equivalent reducing agent:

$$FeO^{2+} + RH \rightarrow R + Fe^{3+} + OH^- \quad (23)$$

And/or two-equivalent reducing agent:

$$FeO^{2+} + H_2R \rightarrow R + Fe^{2+} + H_2O \quad (24)$$

It was needed more than one century to understand that the simple chemical interaction between H_2O_2 and Fe^{2+} turns out, surprisingly, in a very complex chemistry. The controversy is still open because we do not know the mechanism that liberates the enormous oxidant power for organic oxidation in mild conditions.

What is very clear is that a mixture of H_2O_2/Fe^{II} or H_2O_2/Fe^{III} produces a strong oxidant capable of oxidizing (large and small) organic pollutants in mild conditions and this approach can be applied in wastewater treatment. Fenton chemistry has been successfully applied for oxidizing a huge number of the strong oxidant from H_2O_2. For almost 25 years, Fenton chemistry has been recognized as a very good environmental method to treat toxic, pollutant or refractory wastewaters (*Bowers et al. 1989, Bigda 1995; Brillas et al. 1995; Brillas et al. 1998*; *Rodriguez et al. 2003; Brillas et al. 2003; Flox et al. 2006; Siminiceanu et al. 2006; Sires et al. 2007; Guinea et al. 2008*).

Catalytic Activation of Hydrogen Peroxide

The classical Fenton chemistry, in which diluted H_2O_2 is added to a mixture of Fe^{2+}/substrate solution, is the traditional laboratory approach used to assess its performance. Under this approach, homogeneous catalytic activation of H_2O_2 in 50 mM Na_2SO_4 acid solution ($\approx$ pH 2 and room temperature) can be studied, when an excess of commercial hydrogen peroxide (7 mM H_2O_2), with respect to either Fe^{2+} or Fe^{3+} (1 mM) is taken. For this simple aqueous system (*i.e* no complexing ligands other than water), the homogeneous catalytic activation of H_2O_2 (or the intermediated oxidant production) can be evaluated by analyzing the H_2O_2 remaining in solution, by means of the classical permanganate method (*Kolthoff and Belcher 1957*).

It can be seen from Figure 1 that the H_2O_2 activation rate is similar when either Fe^{2+} or Fe^{3+} is used. An interesting hypothesis is that both ions are producing the same quantity of the oxidant (Fe^{IV}, $OH^{\bullet}$ or $HO_2^{\bullet}$) and therefore an organic substrate should be oxidized at the same rate in both systems: H_2O_2/Fe^{2+} or H_2O_2/Fe^{3+}. Experimental results demonstrate that, although organic molecules are oxidized by both systems, as shown below, the degradation rate is slower in the presence of Fe^{III} (*Pignatello 1992*). The organic oxidation rates cannot be carried out faster than the H_2O_2 activation rate. From Figure 1 it can be seen that in 115 minutes almost the 50% of the initial hydrogen peroxide was consumed. Beyond this point, the catalytic activation of H_2O_2 is very slow. Under these experimental conditions H_2O_2 activation rate cannot be accelerated, either increasing the H_2O_2 concentration or the Fe^{2+}/Fe^{3+} concentration. A possible explanation is that, during the H_2O_2 activation the couple Fe^{2+}/Fe^{3+} regeneration is not efficient due to chemical speciation.

A similar procedure was used to degrade aqueous azo dyes solutions (Red M5B and Blue MR) by means of Fenton´s reagent in a stirred closed reactor. The effect of the catalyst amount (Fe^{2+}) was evaluated for an aqueous solution containing 250 mg l^{-1} of Red M5B and 500 mg l^{-1} of H_2O_2. It can be seen from Figure 2 (adapted from *Swaminathan et al. 2003*) that the minimal catalyst amount needed, for the maximum COD abatement for the dye solution, is about 10 mg l^{-1} of Fe^{2+}, shown in curve (a).

Adding more of Fe^{2+} does not increase the COD abatement. The effect of the H_2O_2 amount was evaluated for an aqueous solution containing 250 mg l^{-1} of Red M5B and 10 mg l^{-1} of Fe^{2+}. It can be seen from curve (b) (Figure 2) that the minimal H_2O_2 amount needed, for the maximum COD abatement of the dye solution, is about 400 mg l^{-1}.

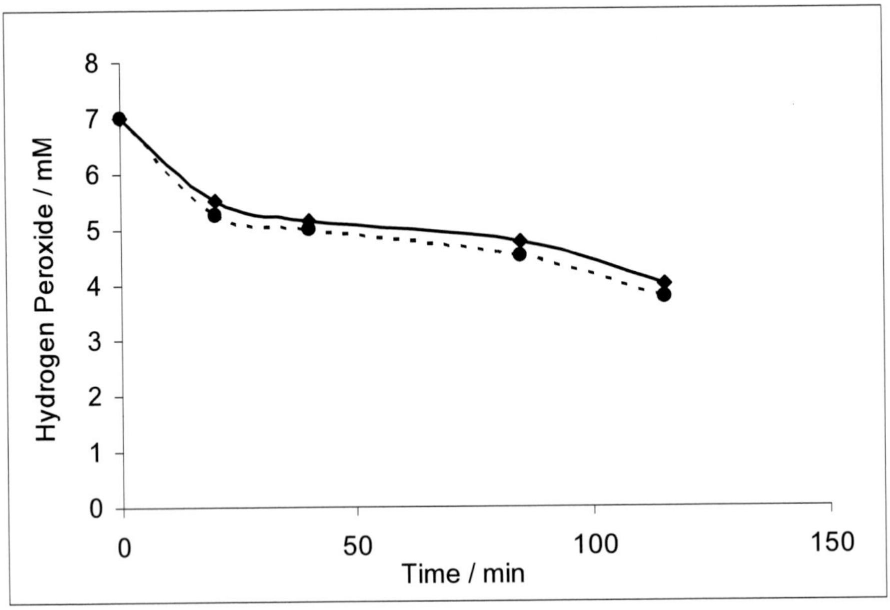

Figure 1. Homogeneous activation of commercial hydrogen peroxide (7 mM H_2O_2) in 50mM Na_2SO_4 acid solution (≈ pH 2 and room temperature), in the presence of: (●) 1 mM Fe^{3+}, (♦) 1mM Fe^{2+}.

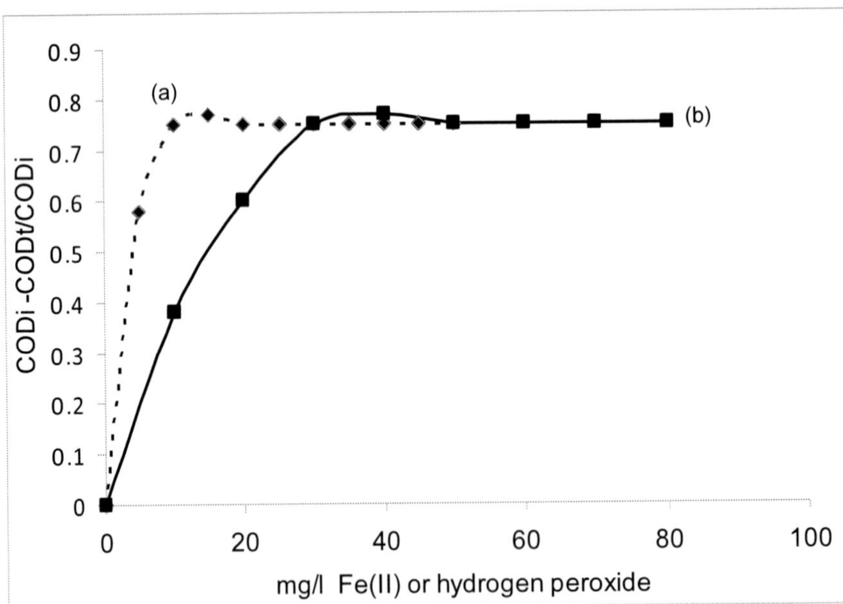

Figure 2. Effect of Fe^{2+} (a) and H_2O_2 (b) doses on COD removal (COD_i-COD_t/COD_i) during the degradation of 250 mg l^{-1} Red M5B in aqueous solution. In the plot the concentration of H_2O_2 has been divided by 10.

A further amount of H_2O_2 does not increase the COD abatement. The most common interpretation of similar situations (*Lucas and Peres 2006; Swaminathan et al. 2003*) is that, during the degradation, the big molecule (azo dye) breaks down into small ones that are very stable.

However, experimental evidence points towards another feasible interpretation: the couple Fe^{2+}/Fe^{3+} regeneration, during the Fenton chemistry is not efficient due to chemical speciation. Indeed, the efficiency of the Fenton chemistry can be enhanced (*Pignatello 1992; Huston and Pignatello 1999*) by UV light irradiation ($\lambda \approx 300$ nm) because the photoreduction of Fe^{III} (from equation 8) and H_2O_2 activation are another source of $OH^{\bullet}$ radicals and Fe^{2+} :

$$Fe(OH)^{2+} \xrightarrow{h\nu} OH^{\bullet} + Fe^{2+} \quad (25)$$

$$H_2O_2 \xrightarrow{h\nu} 2OH^{\bullet} \quad (26)$$

Although the approach discussed above is widely used in the laboratory (*Kuo 1992; Benitez et al. 2001; McGinnis 2001; Swaminathan et al. 2003*) its main drawbacks, for more realistic (full scale hazardous and industrial applications) applications, are the necessity to handle (transport and storage) large volumes of concentrated H_2O_2 and the design and construction of a proper chemical reactor. In recent years, academic activities are focused in an old electrochemical method for hydrogen peroxide production: reduction of dissolved oxygen on carbon cathodes surfaces, according to:

$$O_2 + 2H^+ + 2e^- \rightarrow H_2O_2 \quad (27)$$

Earlier works on this subject used gas diffusion O_2 cathodes (*Berl, 1939*). Subsequent works have improved the performance of several divided electrochemical reactors: trickle-bed reactor (*McIntyre 1995*), gas diffusion cathode (*Brillas et al. 1997; Harrington and Pletcher 1999*) and 3-dimensional carbon cathode electrode (*Ponce de Leon and Pletcher 1995; Alvarez-Gallegos and Pletcher 1998*). Under this approach, it is possible to electrogenerate Fenton´s reagent by bubbling O_2 directly in the catholyte (containing an organic acidic aqueous solution and a small amount of Fe^{2+} or Fe^{3+}) of a divided electrochemical reactor. As soon as H_2O_2 is produced it reacts with iron ions (Fe^{2+} or Fe^{3+}) to produce a strong oxidant (probably a mixture of $OH^{\bullet}$, $HO_2^{\bullet}$, FeO^{2+}). In theory, the cathode potential can simultaneously reduce both: Fe^{3+} and dissolved oxygen, as shown in equations (27) and (28), and this would allow the Fenton chemistry to be enhanced:

$$Fe^{3+} + \quad e^- \rightarrow Fe^{2+} \quad (28)$$

However, this is not possible, because equation (28) is just feasible for Fe^{3+} and not for high-valent iron-oxo intermediates (i.e. $FeH_2O_2^{2+}$, $Fe(OH)^{2+}$, $FeOOH^{2+}$, FeO^{2+}, Fe^{IV}) or other similar iron species, as discussed further.

Figure 3 shows the data from two electrolysis carried out in a divided flow cell, which is fully described elsewhere (*Ponce de Leon and Pletcher 1995; Alvarez-Gallegos and Pletcher 1998*), working at an averaged current of 0.220 Amperes and $\Delta E_{Cell} = 2.0\,V$ in a catholyte consisting of 50 mM Na_2SO_4 at pH close to 2 (adjusted with H_2SO_4).

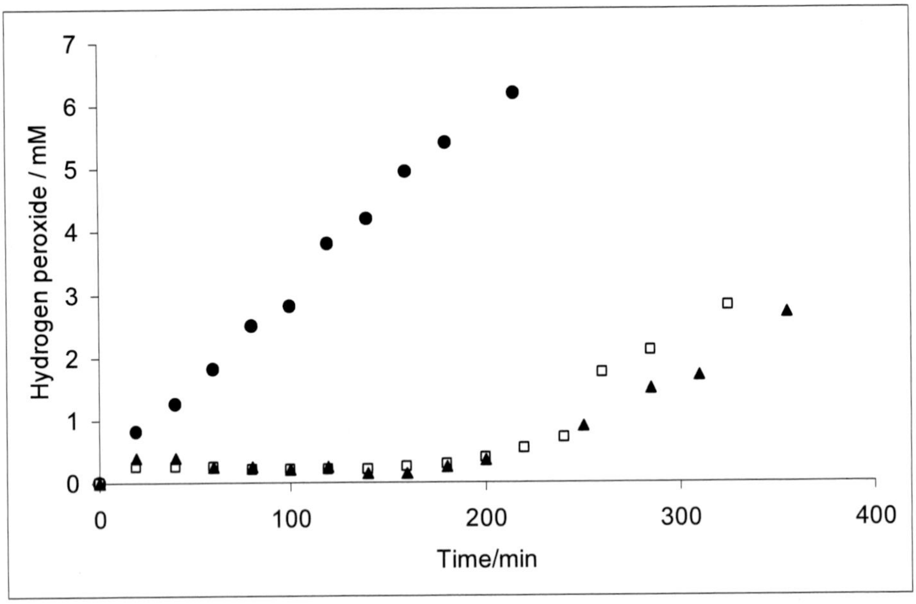

Figure 3. Electrolyses of 50 mM Na_2SO_4 adjusted to pH 2 continuously saturated with O_2 at a reticulated vitreous carbon cathode in a flow-cell. Plots of H_2O_2 formed versus electrolysis time for solution with: (●) 0 mM Fe^{2+}, (□)1mM Fe^{2+}, (▲) 1 mM Fe^{3+}.

In absence of Fe^{2+}, curve (●), hydrogen peroxide production can be accumulated (at ≈ 65% current efficiency) in the catholyte and is a linear function of the electrical charge passed during the oxygen reduction. In contrast, when iron ions (1mM Fe^{2+}, curve □ or 1 mM Fe^{3+}, curve ▲) are added to the catholyte, the current efficiency of hydrogen peroxide production is close to 0% during the passage of the first 2553 C of charge (≈ 200 minutes). This suggests that a strong oxidant (like: Fe^{IV}, $OH^{\bullet}$ or $HO_2^{\bullet}$) is formed when the hydrogen peroxide produced reacts with iron ions in the bulk solution. Thereafter, when the catalytic activity of iron ions is lost (the Fe^{2+} and Fe^{3+} are used up), the hydrogen peroxide concentration in the catholyte began to rise again, as shown in curves □ and ▲ for Fe^{2+} and Fe^{3+} respectively. These experimental results are consistent with those shown in Figure 1 and the implication is the same: both ions are producing a similar amount of the oxidant (i.e. Fe^{IV}, $OH^{\bullet}$ or $HO_2^{\bullet}$). By comparing Figure 1 and Figure 3, it can be seen that the simple mixture of H_2O_2 and Fe^{2+} or Fe^{3+} in a beaker, produce lower amount of strong oxidant than that produced by H_2O_2 electrogeneration in the presence of the same concentration of iron ions. In the former approach just the 50% (3.5 mM) of the H_2O_2 is used to produce the strong oxidant during 115 minutes. In contrast, in the last approach, almost the 100% (7 mM) of the H_2O_2 electroproduced is used to produce the strong oxidant for almost 200 minutes. Beyond this point, however, H_2O_2 is not longer activated and it starts to accumulate in the solution because Fe^{2+}/Fe^{3+} regeneration is not efficient due to chemical speciation.

In a similar electrolysis (*Alvarez-Gallegos and Pletcher 1998*), hydrogen peroxide was electrogenerated in the presence of 1 mM Fe^{2+} and during the electrolysis, aliquots of the catholyte were taken and voltammograms were recorded for each one at a rotating vitreous carbon disc electrode between the potential limits, + 1.100 V and –0.400 V (SCE). The oxidation and reduction waves observed at $E_{1/2}$ = + 0.780 V and + 0.015 V (SCE) were used to estimate the concentrations of electroactive Fe^{2+} and Fe^{3+} respectively in the catholyte as a function of the charge passed, as shown in Figure 4 (adapted from *Alvarez-Gallegos and Pletcher 1998*).

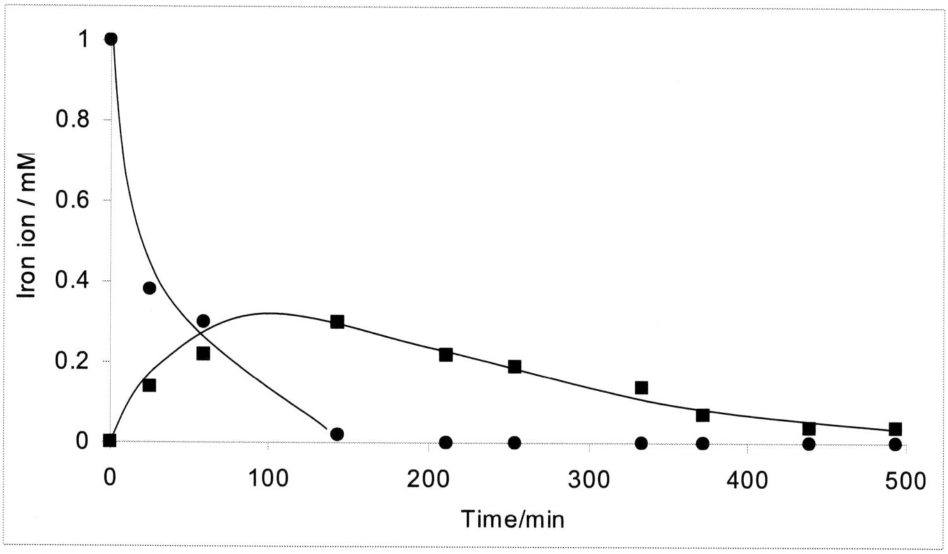

Figure 4. Electrolyses of 50 mM Na_2SO_4 adjusted to pH 2 continuously saturated with O_2 at areticulated vitreous carbon cathode in a flow-cell. Iron speciation versus electrolysis time: (●) Fe^{2+}, (■) Fe^{3+}.

It can be seen, from curve ● , that the concentration of Fe^{2+} decays steadily and has dropped to zero after about 4000 C (150 minutes). As soon as the Fe^{2+} concentration decreases, Fe^{3+} appears (curve ■) and it starts to accumulate in the solution. However, during the Fenton chemistry, mass balance for iron ions (mM Fe^{2+} + mM $Fe^{3+} \neq$ 1mM) is not conserved implying that both of them undergo individually chemical speciation.

Indeed, Fe^{3+} concentration increases over electrolysis time and passes through a broad maximum and then it decreases smoothly until it is almost zero after 13000 C (500 minutes), curve ■. The loss of Fe^{2+}/Fe^{3+} from the catholyte coincides with the formation of some stable hydrogen peroxide, as shown in Figure 3. Beyond this point, the catalytic activation of H_2O_2 is very slow because the catalytic activity of electroactive iron ions is lost. A feasible explanation is that, during the Fenton chemistry the ions Fe^{2+}/Fe^{3+} underwent a speciation (probably $FeH_2O_2^{2+}$, $Fe(OH)^{2+}$, $FeOOH^{2+}$, FeO^{2+}, among others), resulting in a loss of catalytic activity for the decomposition of hydrogen peroxide.

A similar situation was observed (*Kremer 2003*) during the homogeneous catalytic activation of H_2O_2 in the presence of Fe^{2+}: Fe^{2+} ions disappear from the bulk solution at the initial stage of the Fenton chemistry and they are not regenerated to maintain the redox chain.

Organic Degradation by Fenton´s Reagent

The aim of this section is to support the hypotheses that chemical speciation of iron ions is one of the main drawbacks in the performance of Fenton chemistry during the organic oxidation. In this context, speciation of iron ions is a critical step in the Fenton process efficiency rather than the formation of small and very stable organic molecules as it was documented elsewhere (*Lucas et al. 2006; Swaminathan et al. 2003*). Indeed, Fenton chemistry can work well as long as organic oxidation is achieved before catalytic activity stops. It is demonstrated that, although Fe^{III} can decolorize synthetic textile effluents, its catalytic activity is less efficient than Fe^{II}. A broad range of organic pollutants were oxidized in synthetic, acidic waste waters by Fenton's reagent electrogenerated at a reticulated vitreous carbon cathode using the flow-cell system described elsewhere (*Alvarez-Gallegos and Pletcher 1999*).

The complete oxidation of reactive black 5 (RB5) is a 152 electrons oxidation (the structure of this azo-dye is shown in Figure 5), if sulfur and nitrogen are transformed in sulfuric and nitric acids, respectively:

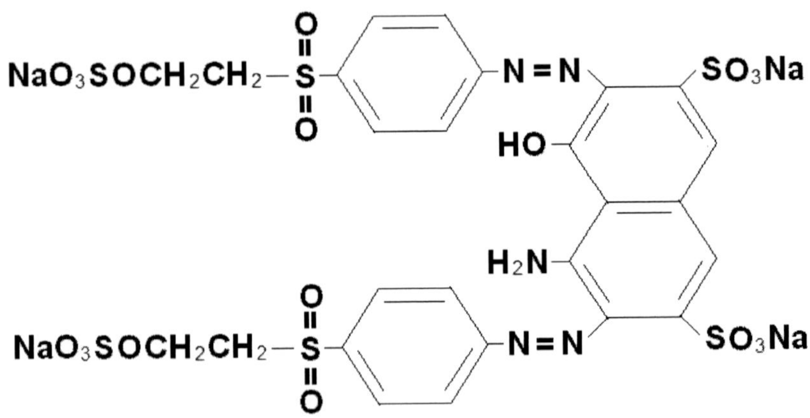

Figure 5. Molecular structure of reactive black 5.

$$C_{26}H_{21}O_{19}S_6N_5Na_4 + 72H_2O - 152e^- \rightarrow$$
$$6CO_2 + 152H^+ + 4H_2SO_4 + 5HNO_3 + 2Na_2SO_4 \quad (29)$$

It has been recognized (*Teel et al. 2001*) that one of the attractiveness of Fenton chemistry approach is its near-stoichiometric generation of a strong oxidant (probably a mixture of $OH^{\bullet}$, $HO_2^{\bullet}$, FeO^{2+}). If the ferryl ion hypothesis is taken into account, one molecule of H_2O_2 will give another FeO^{2+} (see equations 18 or 20), the ferryl ion in turn can accept two electrons in acid medium according to:

$$FeO^{2+} + 2H^+ + 2e^- \rightarrow Fe^{2+} + H_2O \quad (30)$$

Hence, the stoichiometric conversion of 1 mole of RB5 to CO_2 requires up to 76 mole of hydrogen peroxide (and therefore 76 mole ferryl ion, if equations 20, 24 and 30 are 100% efficient). The color of the dyestuff is a function of the conjugated length of the double bonds in the aromatic molecule. The ring opening process should lead to discoloration early in the oxidation process and requires only a small charge.

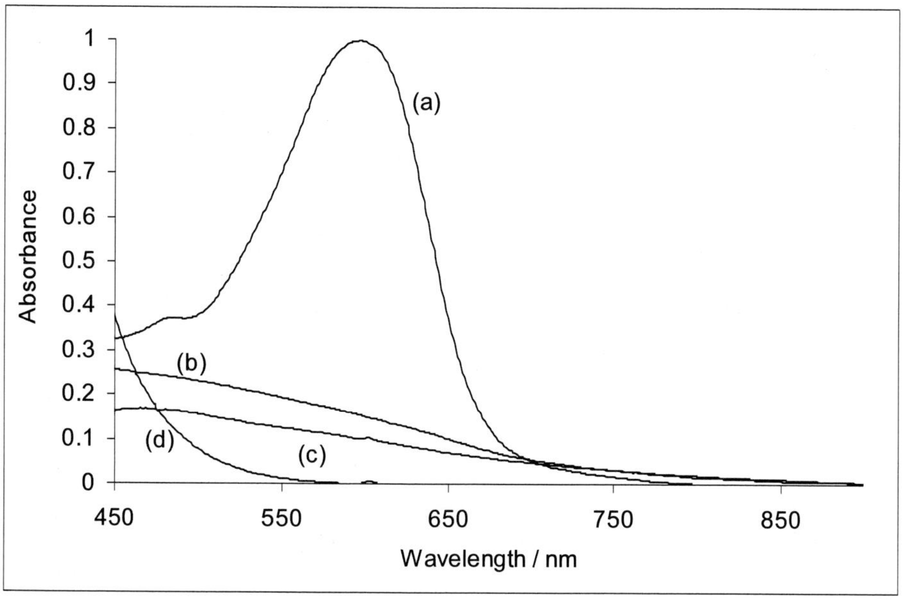

Figure 6. Spectra as a function of charge passed for a solution (1.5 liters) containing 0.1 mM reactive black 5 in 50 mM Na_2SO_4 + 1 mM Fe^{2+}; (a) initial solution diluted by a factor of ten, deep blue, 0 C; (b) diluted by a factor of ten, yellowish brown, 564 C; (c) diluted by a factor of ten, pale yellow, 1224 C; and (d) colorless, 5856 C. Theoretical charge for complete oxidation 3384 C. Applied ΔE_{Cell} = 2.0 V and I_{Cell} = 0.200 A.

The catalytic activity of Fe^{II} was tested in an electrolysis of 1.5 liters of an oxygen saturated solution of 50 mM Na_2SO_4 + 0.1 mM RB5 + 1 mM Fe^{2+} at pH ≈ 2. It was performed in a divided flow cell, (which is fully described elsewhere: *Ponce de Leon and Pletcher 1995; Alvarez-Gallegos and Pletcher 1998*), working at an averaged current of 0.220 amperes and ΔE_{Cell} = 2.0 V. Figure 6 reports the spectra recorded at various stages in the electrolysis.

The initial solution was deep blue and had to be diluted by a factor of ten before the spectrum was recorded, curve (a). It shows a peak in the spectrum at λ_{max} = 597 nm. After the passage of 564 C, the color changed to deep brown and it was diluted by a factor of ten before the spectrum was recorded, curve (b). The spectrum recorded after 1224 C was yellowish brown and the solution was diluted by a factor of ten, as shown in curve (c). On continuing the electrolysis by the passage of 2200 C, the solution has become pale yellow. After the passage of 5856 C the solution is effectively colorless and the spectrum (see curve d) shows only an absorption tail into the UV.

The theoretical charge and amount of H_2O_2 for complete oxidation of the RB5 can be evaluated by means of:

$$q = \frac{mnF}{\phi} \qquad (31)$$

Where q is the theoretical charge (in C) passed through the cell, m is the amount (in mol) of organic matter (or hydrogen peroxide electrogenerated), n is the number of electrons involved in the oxidation reaction, F is de Faraday constant (96 485 C mol^{-1}) and ϕ is the current efficiency for hydrogen peroxide production (≈ 65%). Substituting the appropriate values in equation (31) we obtain the theoretical charge for complete oxidation of 1.5 liters 0.1 mM RB5 (0.15 mmol RB5):

$$q = \frac{(0.00015\ mol)(152\ e^{-})(96485\ C\ mol^{-1})}{0.65} = 3384\ C \qquad (32)$$

Taking into account the theoretical charge, the theoretical amount of hydrogen peroxide produced is: 11.4 mmol in 1.5 liters of solution (7.6 mM H_2O_2 and therefore 7.6 mM FeO^{2+}, if equations 20, 24 and 30 are 100% efficient). The theoretical time needed for performing the electrolysis is 235 minutes (it can be estimated by dividing 3384 C by the experimental current, 0.24 A). However, before to perform the electrolysis, it is expected an incomplete RB5 oxidation because, by 200 minutes of electrolysis time the H_2O_2 electroproduced will not be longer activated by iron ions, as demonstrated in Figure 3.

The catalytic activity of Fe^{III} was tested in an electrolysis of 1.5 liters of an oxygen saturated solution of 50 mM Na_2SO_4 + 0.1 mM RB5 + 1 mM Fe^{3+} at pH ≈ 2. It was performed in a divided flow cell, working at an averaged current of 0.150 Amperes and ΔE_{Cell} = 2.0V. Figure 7 reports the spectra recorded at various stages in the electrolysis.

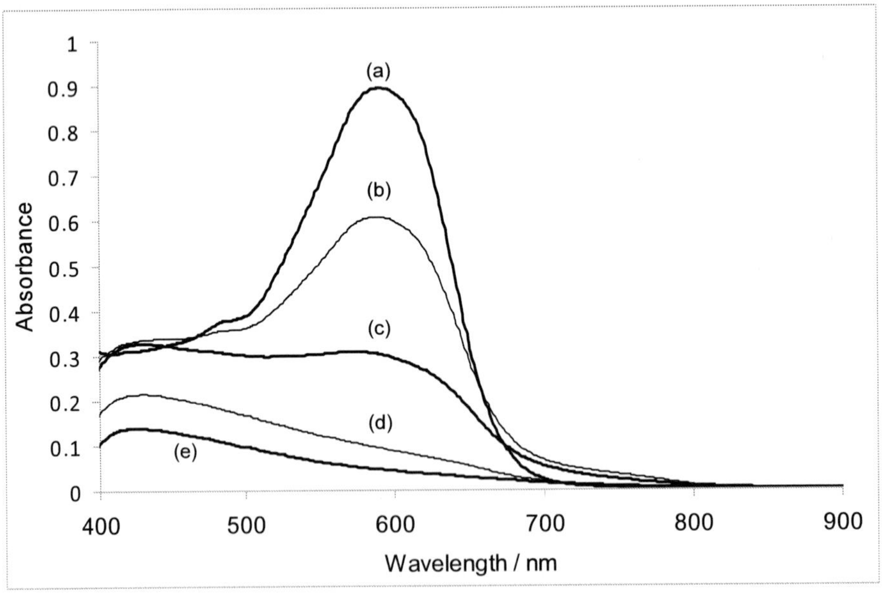

Figure 7. Spectra as a function of charge passed for a solution (1.5 liters) containing 0.1 mM reactive black 5 in 50 mM Na_2SO_4 + 1 mM Fe^{3+}. All spectra were recorded after diluting the solution by a factor of ten. (a) initial solution diluted, deep blue, 0 C; (b) deep blue, 348 C; (c) brown, 708 C; and (d) yellowish brown, 1416 C; (e) pale yellow. Theoretical charge for complete oxidation 3384 C. Applied ΔE_{Cell} = 2.0V and I_{Cell} = 0.150 A.

As it was expected from the results shown in Figure 1 and 3, both iron ions (Fe^{2+}/Fe^{3+}) can catalyze the H_2O_2 and produce a strong oxidant able to decolorize dyestuff. However, the efficiency of the oxidation process is boosted in the presence of Fe^{2+} rather than Fe^{3+}. Figure 8a, reports the COD as a function of the electrolysis time, for decolorizing 0.1 mM RB5 in the presence of Fe^{3+}. The COD was measured using a HACH 4000U spectrophotometer following standard procedures and standard vials (*Hach 1997*).

During the first 280 minutes of electrolysis the COD drops slowly and it is clear that RB5 molecule is broken and smaller and more oxidized organic molecules are being formed in solution. Beyond this time the decay in COD do not changes with time and it was not clear whether this decrease in rate results from the more difficulty in oxidizing the smaller molecules produced or it occurs because of a change in speciation of iron ions (Fe^{2+}/Fe^{3+}). Figure 8b, reports the chemical oxygen demand as a function of the electrolysis time (see as well Figure 6), for decolorizing 0.1 mM RB5 in the presence of Fe^{2+}. During the first 150 minutes of electrolysis the COD drops rapidly but beyond this time the decay in COD is slowly. It is clear that Fe^{2+} is a better catalyst than Fe^{3+}. The electrolysis was repeated under the same conditions and after 200 minutes of electrolysis an extra 1 mM Fe^{2+} was added but the degree of oxidation was essentially the same. It is concluded that extra iron ions (added at the beginning or during the Fenton chemistry) did not boost the catalytic activity. A similar behavior was documented elsewhere (*Swaminathan et al. 2003*), Figure 3.

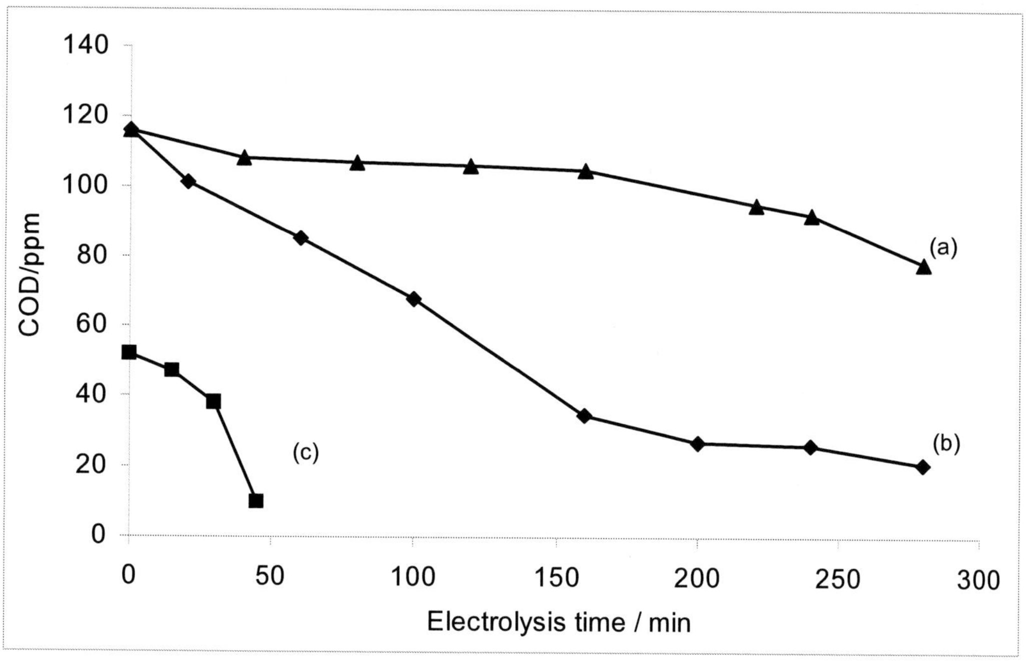

Figure 8. Chemical oxygen demand as a function of the electrolysis time, for decolorizing RB5 in 50 mM Na_2SO_4 : (a) 0.1 mM RB5 in the presence of 1mM Fe^{3+}, (b) 0.1 mM RB5 in the presence of 1 mM Fe^{2+} + and (c) 0.06 mM RB5 in the presence of 1 mM Fe^{2+}.

In order to test the hypothesis that, during the degradation the azo dye break down into small and very stable molecules, a further electrolysis was carried out, under the same experimental conditions, on a solution containing a smaller RB5 concentration (0.06 mM). It can be seen that, in this electrolysis, Figure 8c, the COD drops rapidly to below 10 ppm. Hence, the existence of very stable by-products cannot be correct and it must be concluded that the catalytic activity of iron ions (Fe^{2+}/Fe^{3+}) decreases due to a change in its speciation, as shown in Figure 4.

In order to test the hypothesis that, Fenton chemistry can work well, as long as organic oxidation was achieved before catalytic activity stops, several electrolyses for decolorizing synthetic wastewaters were performed in the same flow-cell. The dyestuff considered were: acid green 25 (AG25), basic blue 9 (BB9), direct red 23 (DR23), acid orange 7 (AO7) and amaranth (Am). The structure of these azo-dyes is shown in the Figure 9 and their initial concentrations, main experimental conditions, theoretical charge and time for complete oxidation are shown in table 1.

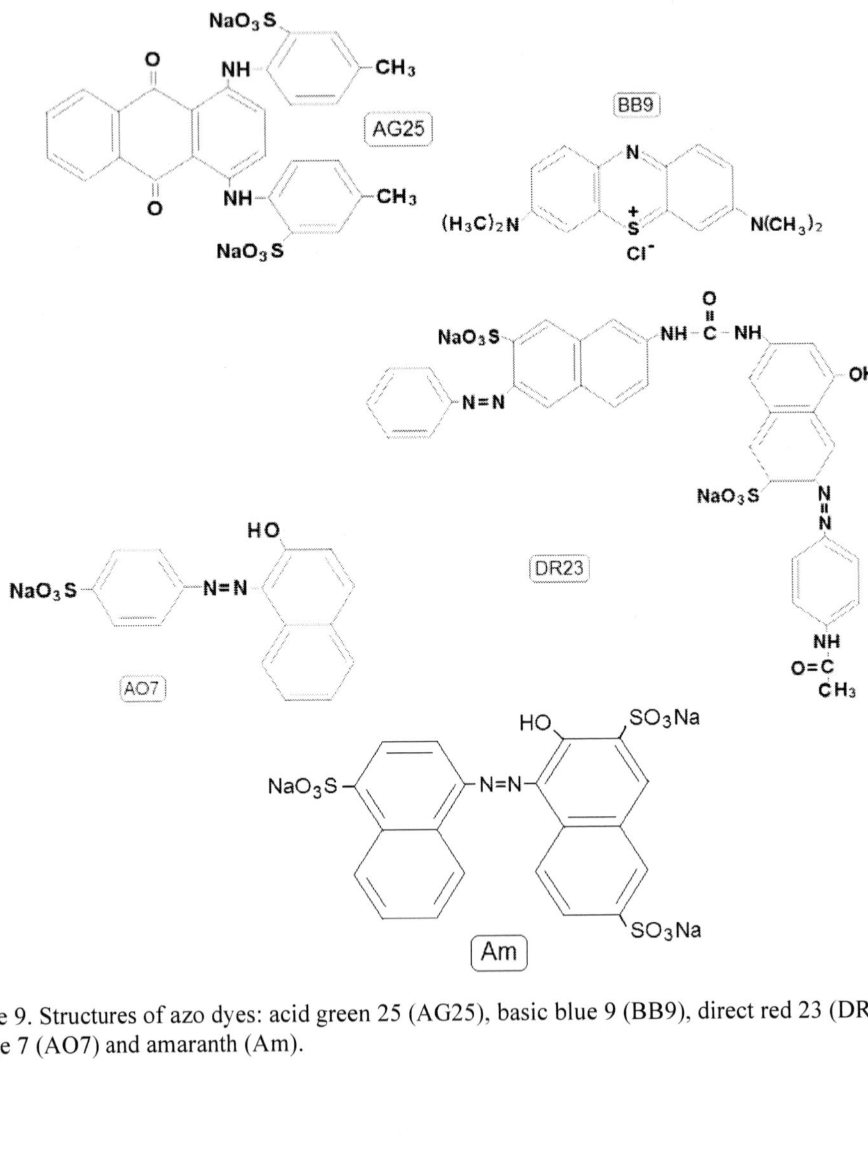

Figure 9. Structures of azo dyes: acid green 25 (AG25), basic blue 9 (BB9), direct red 23 (DR23), acid orange 7 (AO7) and amaranth (Am).

Table 1. Electrolyses performed in a divided flow-cell working at 65% current efficiency for hydrogen peroxide production

Dye	Catholyte volume litters	e^-	dye mM	$[FeO^{2+}]$ mM	I_{Cell} Amp	E_{Cell} Volts	q C	TET min.
RB5	1.3	152	0.06	4.6	0.170	2.00	1789	175
RB5	1.5	152	0.10	7.6	0.240	2.00	3384	235
AG25	1.5	124	0.10	10	0.120	1.83	3117	444
BB9	1.5	78	0.17	4.4	0.320	1.83	1968	102
DR23	1.5	140	0.08	8.0	0.320	1.83	3578	326
AO7	1.5	68	0.25	8.5	0.260	2.16	3785	243
Am	1.8	102	0.33	30.3	0.350	2.00	8994	428
Am	1.8	102	0.10	9.2	0.350	2.00	2725	130

Catholyte: 50 mM Na_2SO_4 + dye + 1 mM Fe(II) at pH ≈ 2. Reactive black 5 (RB5), acid green 25 (AG25), basic blue 9 (BB9), direct red 23 (DR23), acid orange 7 (AO), amaranth (Am); e^- = electrons to be lost for complete oxidation; initial dye concentration in mM (dye mM); ferryl ion required for complete oxidation ($[FeO^{2+}]$ mM); averaged cell current in amperes (I_{Cell}); applied voltage difference in volts (E_{Cell}); theoretical coulombs needed for complete dye oxidation (q); theoretical electrolysis time required for electrogenerate ferryl ion, in minutes (TET)

According to the theoretical calculations shown in Table 1, the COD will drop rapidly to below 10 ppm during the electrolysis time just for three catholytes containing the following colors: 0.06 mM RB5, 0.117 mM BB9 and 0.1 mM Am. Indeed, for these catholytes the flow-cell is able to produce the required amount of a strong oxidant (FeO^{2+}) before the critical time is reached (200 minutes). Experimental evidences shown in Figure 10 agree with the theoretical predictions, except for the catholyte 0.1 mM AG25, see Figure 10d. A possible explanation could be the oxidation path and the type of $R^\bullet$ produced by the fast H-abstraction (reaction 14).

The most difficult oxidation was the 0.08 mM DR23, one of the biggest molecules (MW 814), Figure 10a reports the COD as a function of the electrolysis time in the presence of 1 mM Fe^{2+}. Although the COD drops rapidly during electrolysis, the Fenton chemistry alts at 200 minutes of electrolysis. Further iron ion additions did not boost the oxidation. Similar results were found for 0.25 mM AO7 (Fig. 10b), although the COD drops constantly for the first 180 minutes, by 200 minutes COD abatement was similar to that of 180 minutes of reaction time. The oxidation of 0.10 mM Am, 0.17 mM BB9 and 0.06 mM RB5 was carried out without problem, see Figures 10e, 10f, and 8c, respectively, because the flow-cell can electroproduce enough FeO^{2+} in less than 200 minutes.

As it was commented above, during the degradation the azo dye breaks down into small molecules. Although the oxidation path during the electrolyses is unknown for the dyestuff shown in Figure 9, it can be taken, as an approximation, some paths proposed during biologic degradation of azo dyes. In this context, the first intermediates due to biological degradation for AO7 and BR5 are shown in Figures 11 (*Seshadri et al. 1994*) and 12 (*Storm. 2002*) respectively.

It is logic to expect naphthalenic and benzenic ring amines as one of the first intermediate species during the attack of FeO^{2+} on dyes because similar organic structures have been found during ozonation (*Lopez et al. 2004*) or combined sonolysis and ozonation of azo dyes (*He et al. 2007*).

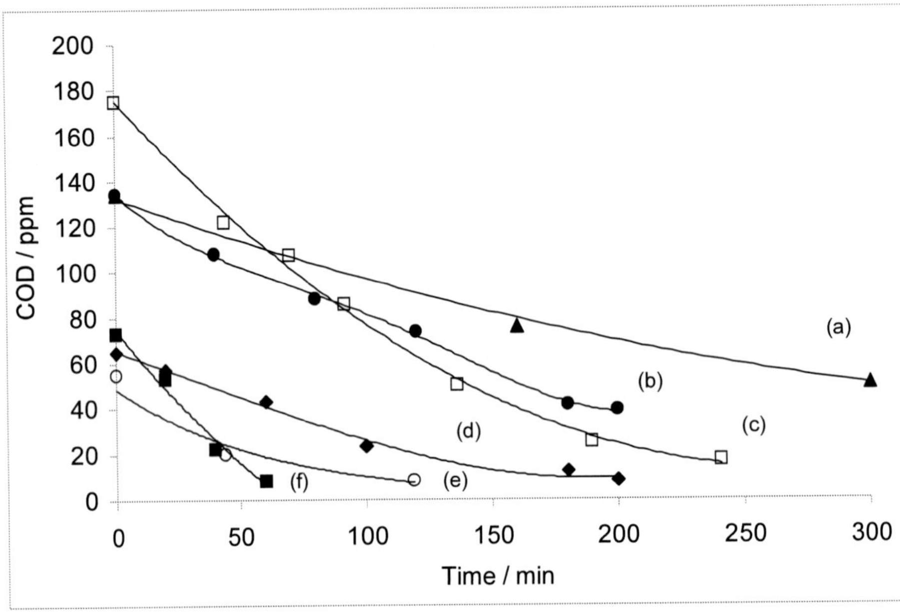

Figure 10. COD evolution as a function of electrolysis time for electrolysis of solutions containing 50 mM Na_2SO_4 + 1 mM Fe^{2+} + dye : (a) 0.082 mM DR23, (b) 0.25 mM AO7, (c) 0.33 mM Am, (d) 0.1 mM AG25, (e) 0.1 mM Am, (f) 0.17 mM BB9. Electrolyses were carried out in a membrane cell with a reticulated vitreous carbon cathode (5 cm x 5 cm x 1 cm).

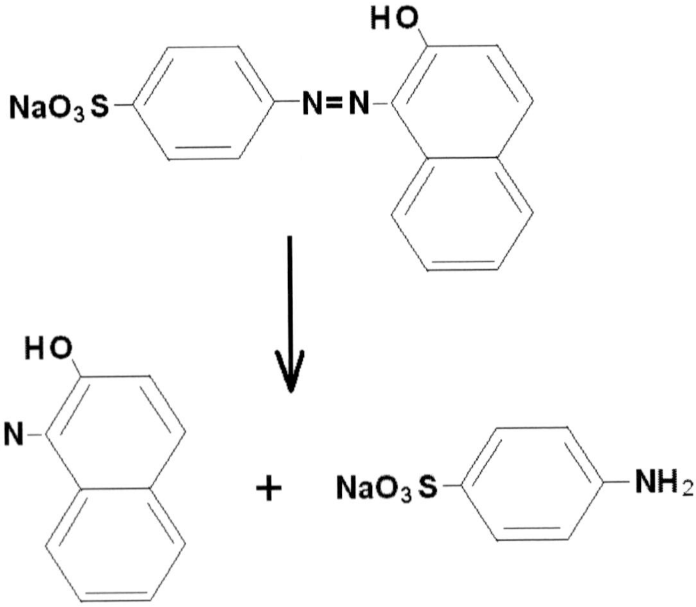

Figure 11. First intermediates due to biological degradation of AO7

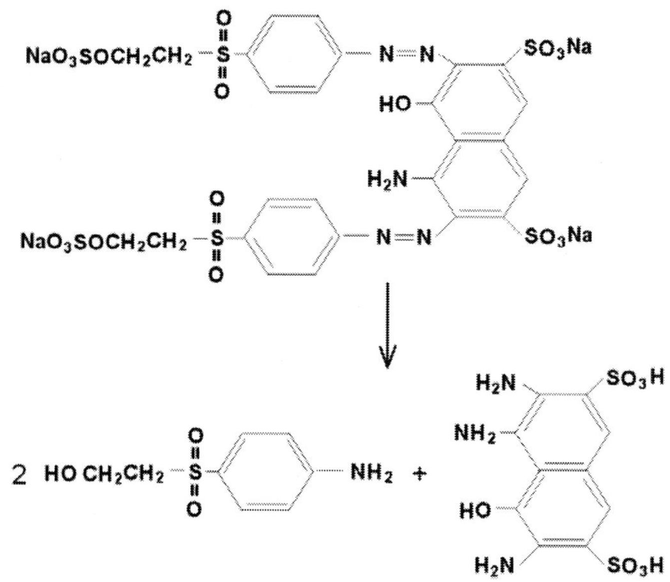

Figure 12. First intermediates due to biological degradation of RB5

Most of the naphthalenic ring amines produced during dye degradation are not commercially available and they could not be tested but, they should be oxidized by Fenton chemistry with the same degree of difficulty than the previous dyestuff studied.

Aniline, cresol and phenol can be taken as organic models for representing benzenic ring amines and benzenic ring structures, respectively. Phenol oxidation follows a general and defined degradation sequence (see equation 33) regardless of the oxidation method applied - e.g. anodic oxidation (*Sharifian and Kirk 1986; Gattrell and Kirk 1990*), indirect electrochemical oxidation (*Sudoh et al. 1986*), oxidative degradation using ultrasound (*Berlan et al. 1994*):

$$Phenol \xrightarrow{k_1} \begin{pmatrix} Aromatic \\ Intermediates \end{pmatrix} \xrightarrow{k_2} \begin{pmatrix} Aliphatic \\ Acids \end{pmatrix} \xrightarrow{k_3} CO_2 \qquad (33)$$

The aromatic intermediates are species such as hydroquinone, catechol and benzoquinone. Aliphatic acids include maleic, oxalic, fumaric acids, etc. There is a general agreement that the rate constant k_1 for phenol hydroxylation is very high in both chemical and electrochemical oxidation systems. Ring opening (rate constant k_2) and combustion (rate constant k_3) processes are relatively slow either in chemical (Fenton's reagent) or anodic oxidation. Therefore, aliphatic acids are often considered stable toward further oxidation at room temperature. Oxalic acid was taken as an aliphatic organic acid model for representing aliphatic acids in general.

The feasibility of benzenic ring amines, benzenic ring structures and aliphatic acids oxidation by means of Fenton chemistry was tested in synthetic, acidic wastewaters by Fenton's reagent electrogenerated at a reticulated vitreous carbon cathode using the flow-cell. The organic molecules considered were: phenol (Ph), cresol (Cr), aniline (An); hydroquinone (HQ), catechol (Cat), parabenzoquinone (pBQ) and oxalic acid (OxAc). Their initial concentrations, main experimental conditions, theoretical charge and time for complete oxidation are shown in table 2.

Table 2. Electrolysis performed in a divided flow-cell working at 65% current efficiency for hydrogen peroxide production

Organic	e^-	mM	$[FeO^{2+}]$ mM	I_{Cell} Amp	E_{Cell} Volts	q / C	TET min.
Ph	28	0.37	9.4	0.380	2.00	2785	122
Cr	34	0.33	10.1	0.430	2.00	2998	116
An	36	0.33	10.7	0.480	1.83	3174	110
HQ	26	0.33	7.7	0.355	1.83	2292	108
Cat	26	0.33	7.7	0.400	1.83	2292	96
pBQ	24	0.33	7.1	0.380	2.16	2116	93
OxAc	2	10	18.0	0.400	2.00	5344	223

1.8 liters of catholyte: 50 mM Na_2SO_4 + organic molecule + 1 mM Fe^{2+} at pH ≈ 2. Phenol (Ph); cresol (Cr); aniline (An); hydroquinone (HQ); catechol (Cat); parabenzoquinone (pBQ); oxalic acid (OxAc); electrons to be lost for complete oxidation (e^-); initial organic molecule concentration (mM); ferryl ion required for complete oxidation ($[FeO^{2+}]$ in mM); averaged cell current in amperes (I_{Cell}); applied voltage difference in volts (E_{Cell}); theoretical coulombs needed for complete organic molecule oxidation (q); theoretical electrolysis time required for electrogenerate ferryl ion, in minutes (TET)

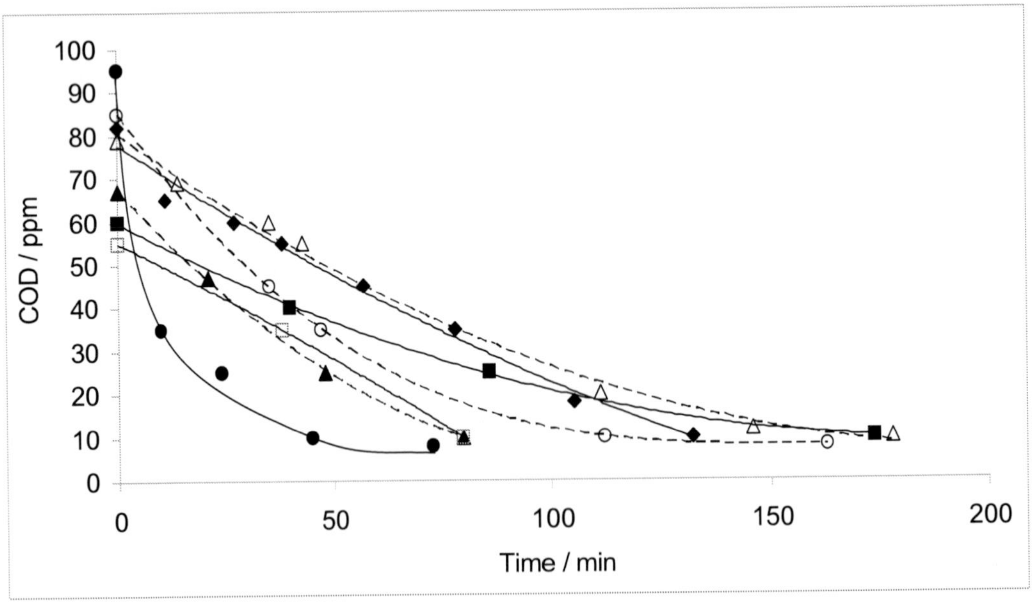

Figure 13. COD evolution as a function of electrolysis time for electrolysis of solutions (1.8 liters) containing 50 mM Na_2SO_4 + 1 mM Fe^{2+} + organic : (♦) 0.37 mM Ph, (■) 0.33 mM HQ, (▲) 0.33 mM Cat, (○) 0.33 mM Cr, (□) 0.33 mM pBQ, (Δ) 10 mM OxAc and (●) 0.33 mM An. Values corresponding to OxAc have been divided by 2. Electrolysis were carried out in a membrane cell with a reticulated vitreous carbon cathode (5 cm x 5 cm x 1 cm).

According to the theoretical calculations shown in Table 2, the COD will drop rapidly to below 10 ppm during the electrolysis time for the aniline, oxalic acid and all benzenic ring structures. Indeed, for these aqueous solutions, the flow-cell was able to produce the required amount of a strong oxidant (FeO^{2+}) before the critical time was reached (200 minutes). Experimental evidences shown in Figure 13 agree with theoretical predictions. The longest electrolysis was for the degradation 10 mM OxAc due to its concentration. Experimental electrolysis was stop by 355 minutes (≈ 9000 C) and the COD dropped to 12 ppm, see the curve (Δ) in Figure 13. As it can bee seen from Figure 13, the rest of benzenic ring structures were readily oxidized by hydrogen peroxide in the presence of Fe(II) at room temperature, provided that the catalyst remained in the active form.

According to the experimental evidences shown in Tables 1 and 2 and Figures 10 and 13 it can be concluded that Fenton chemistry can oxidize azo dyes, benzenic ring structures and aliphatic acids, provided that the catalyst remains in the active form.

Heterogeneous Catalysis

Nowadays, there is an imperative necessity to minimize the water pollution. The presence of different types of wastewaters containing several organic pollutants has an adverse effect in the environment. Therefore, the development of clean technologies compatible with the environment includes wastewater treatment and water recycling. There are several advanced oxidation processes, considered as non-waste generating technologies, which represent an alternative when common wastewater treatment technologies are insufficiently effective. Amongst these technologies, Fenton process offers a cost effective source of highly reactive oxidizing species that are responsible for the degradation of organic pollutants present in aqueous effluents. The Fenton process can be conducted homogeneously and heterogeneously. Even though, the homogeneous Fenton process is highly efficient in destroying the contaminants, its major drawbacks are: the difficulty for maintaining the redox chain and the necessity for the removal of iron ions to prevent from forming sludge at the end of the wastewater treatment.

These drawbacks can be overcome or at least minimized by the usage of a heterogeneous catalyst where the iron can be supported on diverse materials; in this manner, the heterogeneous iron catalyst can be removed by filtration. In the heterogeneous process the reaction occurs at the catalyst at active sites. An active site is a part of the surface which is particularly good at adsorbing molecules and helping them to react. One or more of the reactant molecules are adsorbed on to the surface of the catalyst at active sites. The product molecules are desorbed leaving the active site available for a new set of molecules to attach to the surface and react. One disadvantage of the heterogeneous catalyst is that its structure depends on the preparation method that may affect the reproducibility because it may have several types of active sites present in the form of surface defects. However, in the homogeneous reactions the chemical changes that take place are solely dependent on the nature of the interactions of the reacting substances.

In order to use a catalyst in the heterogeneous Fenton and Fenton-like processes, for industrial applications, there are several features that the catalyst needs to fulfill. Some of these characteristics are high activity, chemical and mechanical stability under several conditions, resistance to poisoning in prolonged use, low cost and minimum catalyst loss due to leaching (*Pirkanniemi and Sillanpää 2002*). A reaction believed heterogeneous might after all be homogeneous (*Pirkanniemi 2007*). According to Sheldon *et al.* (1998), many common heterogeneous liquid phase catalysts are not stable toward leaching under oxidizing conditions. Rigorous proof of heterogeneity can be obtained only by filtering the catalyst at the reaction temperature before completion of the reaction and testing the filtrate for activity (*Sheldon et al. 1998*).

The iron catalyst used in heterogeneous Fenton processes are supported on metal oxide surface such as TiO_2, Al_2O_3, and CeO_2 (*Pirkanniemi and Sillanpää 2002*; *Chan-Li et al. 2006; Lim et al. 2006*; *Tryba 2008*), synthetic zeolites (*Kušić et al. 2006*), pillared clays (*Ramirez et al. 2007a*), laponite and bentonite clay-based Fe-nanocomposites (*Feng et al. 2003*; *2006*), ceramic materials (*Pirkanniemi and Sillanpää 2002*), activated carbon (*Ramirez et al. 2007b*) and fly ash (*Flores et al. 2008*). The objective of this section is to study the oxidation of textile dye effluents by heterogeneous Fenton like process employing an iron oxide catalyst supported on fly ash and cerium oxides. It is expected that a heterogeneous supported catalyst approach will improve the control over iron ions during the oxidation reaction.

The mechanism of hydrogen peroxide decomposition on the homogeneous Fenton reaction is not well established because there is a controversy whether the chemical mechanism involves $OH^{\bullet}$ radical production or the ferryl ion (Fe(IV)) is produced as the active intermediate species in the Fenton reaction. For the heterogeneous Fenton reaction the chemical pathway is less understood. It has been suggested either the adsorption of the H_2O_2 molecule onto $\equiv Fe^{III}$ sites or the adsorption of the organics (*Dantas 2006; Ramirez et al. 2007a*). However, the following steps have been proposed where the Fe^{3+} is reduced with the generation of less oxidative $HO_2^{\bullet}$ radicals, followed by Fe^{3+} regeneration with the formation of $OH^{\bullet}$ radicals:

$$X - Fe^{3+} + H_2O_2 \rightarrow X - Fe^{2+} + HO_2^{\bullet} + H^+ \tag{34}$$

$$X - Fe^{2+} + H_2O_2 \rightarrow X - Fe^{3+} + OH^- + {}^{\bullet}OH \tag{35}$$

Where X represents the surface of the catalyst. It was also proposed (*Ramirez et al. 2007a*) that the radicals can also be generated in the surface of the solid so they are actually ''caged'' in the solid structure, subsequently reacting with the adsorbed reagents without radicals generation. Besides the indicated steps many other radical reactions can occur similar to those described above (see reactions 3 and 4 among others). In the presence of a substrate Fenton chemistry is very complex and can involve reactions like those represented by equations (14), (15) and (16).

Fly Ash

Fly ash from a coal thermal power plant is a good support due to its high surface area, which would provide more active sites to the catalyst. Moreover, some metals (i.e. Ca, K, Mg, Ti, among others) forming the fly ash, could be exchanged by Fe using proper procedures. The incorporation of Fe to the fly ash would provide more Fenton active sites to the catalytic structure. The oxidation of RB5 was investigated in batch test using a beaker containing the working solution (50 mM Na_2SO_4, pH = 2.8 adjusted with H_2SO_4) plus heterogeneous catalyst and H_2O_2.

Fly ash was used to prepare five types of heterogeneous catalyst following thermal procedures (*Geus and Von Veen 1999; Centi et al. 2000*), the full experimental description is given elsewhere (*Flores et al. 2008*). However, the main steps are the following: ferrous salt (i.e. $Fe(NO_3)_3{\cdot}9H_2O$) and fly ash are mix and heated together in an aqueous solution until evaporation. Then, the residue is dried at 100°C overnight and then calcined at 500°C during 4 h. Some variations on the procedure include mechanical stirring replacement by ultrasonic energy.

The catalyst characterization was performed basically by atomic absortion spectroscopy and X-ray diffraction techniques. The first technique indicates that the fly ash contains the following major elements: Si (26 wt%), Al (14 wt%), Ca (2 wt%), K (1.09 wt%) and Ti (1 wt%). Minors constituents, like Na(0.5 wt%) and Mg (0.5 wt%), were as well present. The second technique indicate that after preparation, fly ash is made of graphite, SiO_2 (quartz), $Al_6Si_2O_{13}$ (mullite), $CaAl_4Fe_8O_{19}$ and Al_2SiO_5 (Sillimanite). Hence, it is deduced that prepared catalysts (Fe^{3+}-containing ashes) mainly consist of Fe_2O_3 (Fe^{III}).

Leaching tests were carried out in the working solution (50 mM Na_2SO_4, pH = 2.8 adjusted with H_2SO_4) and applying the following procedure at room temperature: 0.1 g of any catalyst was added to a beaker containing 100 ml of the working solution. After stirring (120 rpm) for 2 h, the solution was allowed to settle down, filtrated and then, total iron and ferrous iron were analyzed. The catalyst was recuperated and added to a second beaker containing a fresh working solution, then, the previous methodology was applied. This methodology was applied three times. A resume of the main results are reported in Table 3. In all cases, leaching tests lasted 6 h and indicated that ion Fe^{3+} was lost in small amounts from the fly ash at constant rate. Therefore, the catalysts are very stable under the experimental studied condition.

Table 3. Amount of iron lost from prepared fly ash catalyst in 6 hours

Catalyst 0.1 g	Amount of immobilized iron		Leaching Test in 50 mM Na_2SO_4 (pH≈2.8 adjusted with H_2SO_4)	
	Fe^{3+} mg	Fe^{2+} mg	Fe^{3+} mg	Fe^{2+} mg
C1	8.2125	0.375	0.1487 (1.8%)	0.0128 (3.4%)
C2	0.600	0.12	0.0070 (1.2%)	0.0060 (5%)
C3	1.4025	0.06	0.022 (1.6%)	0.0065 (11%)
C4	0.7575	ND	0.0108 (1.4%)	ND
C5	0.885	0.495	0.008 (1.5%)	0.0047 (1%)

Organic degradation by Fenton-like chemistry. Taking into account the stoichiometric radical generation model represented by equations (34) and (35), one molecule of H_2O_2 will give either $HO_2^{\bullet}$ or $OH^{\bullet}$. Hence, the stoichiometric conversion of 1 mole of RB5 to CO_2 requires up to 76 mole of of hydrogen peroxide (if equations 3 and 4 are minimized and equations 34 and 35 are 100% efficient). According to this theoretical approach, the degradation of 100 ml 50 M Na_2SO_4 (pH≈ 2.8) containing 0.06 mM RB5 + 0.1 g catalyst C1 was investigated in batch tests, the iron ion content is equivalent to have 1.5 mM Fe^{3+} in solution. After a short mixing time a stoichiometric amount (4.56 mM) of H_2O_2 was added to the batch reactor and the RB5 oxidation started. During the degradation, samples were taken and the absorbance of the solution was immediately recorded in order to determine the concentration of RB5 in the reacting solution. Figure 14 shows RB5 oxidation with 4.56 mM H_2O_2 (stoichiometric amount, curve ●), 5.0 mM H_2O_2 (10% more than the stoichiometric amount, curve □) and 5.5 mM H_2O_2 (20% more than the stoichiometric amount, curve ▲).

As it was expected, the heterogeneous hydrogen peroxide activation can oxidize azo dye solutions and the rate of RB5 degradation increases with H_2O_2 concentration; however, is not advisable to increase further the H_2O_2 concentration, because the Fenton process is not longer profitable. After 165 min., of treatment and using a 20% in excess of H_2O_2 (curve ▲ in Figure 14), the solution was effectively colorless and the final COD was < 10 ppm O_2, representing an abatement of more than 80%.

Influence of catalyst load on the degradation of RB5 solution. In homogeneous catalysis, It is accepted that Fenton chemistry works well in the presence of small (~1 mM Fe^{2+}) quantities of Fe^{2+} (*Alvarez-Gallegos and Pletcher 1998; Meriç et al. 2004*). In heterogeneous catalysts Fenton chemistry behaves in a similar way, Figure 15 shows color removal as a function of the amount of catalyst C1 under the following experimental conditions: 100 ml 50 mM Na_2SO_4 + 0.06 mM RB5 + 5.5 mM H_2O_2, pH 2.8 and room temperature.

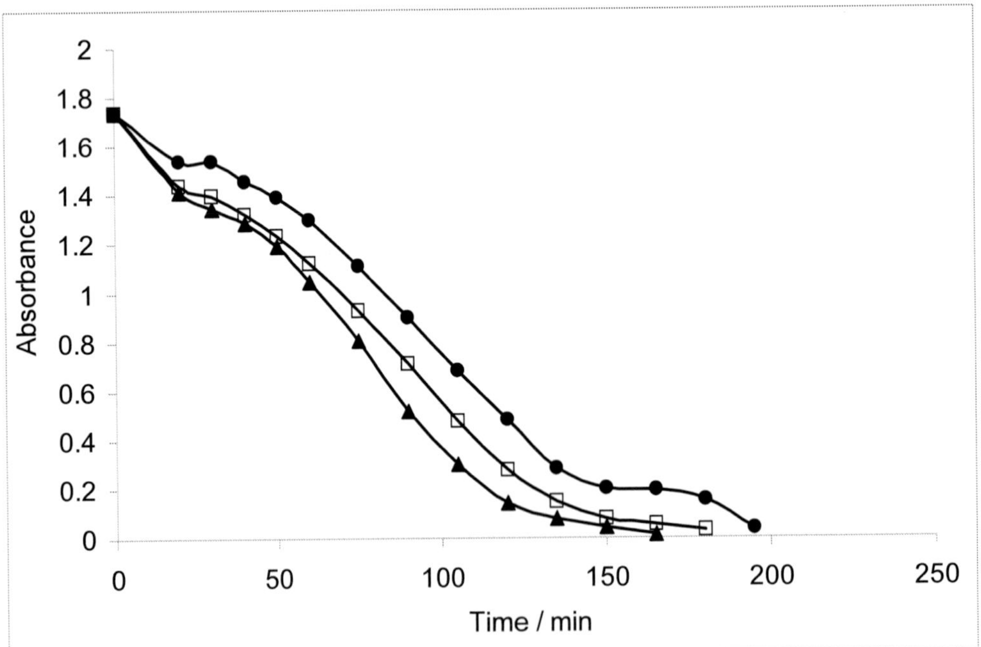

Figure 14. Degradation of RB5 with activated H_2O_2 by an heterogeneous catalyst (C1). Experimental conditions: 100 ml 50 mM Na_2SO_4 + RB5 + 0.100 g catalyst C1, pH 2.8 and room temperature. (●) 4.56 mM H_2O_2, (□) 5.0 mM H_2O_2, (▲) 5.5 mM H_2O_2.

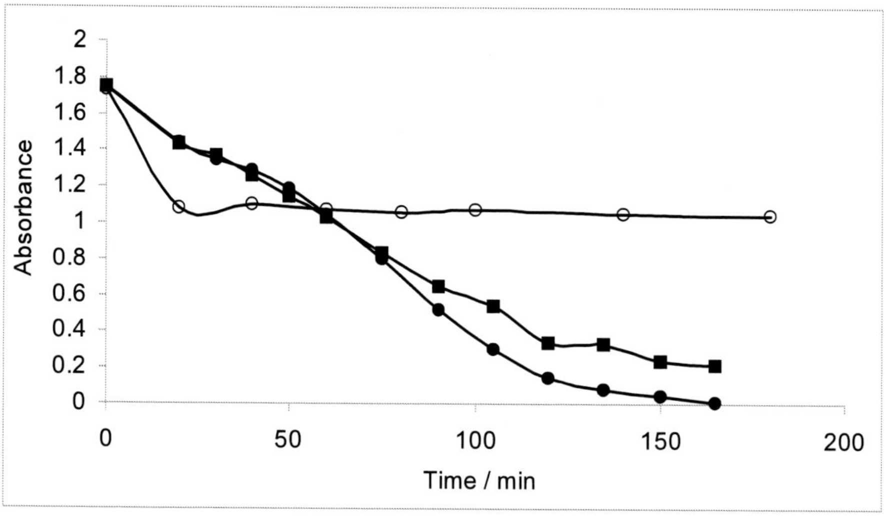

Figure 15. Color removal as a function of the amount of catalyst C1. Experimental conditions: 100 ml 50 mM Na_2SO_4 + 0.06 mM RB5 + 5.5 mM H_2O_2, pH 2.8 and room temperature. (○) 1.2 g original fly ash (0.1 mM Fe^{3+}), (■) 0.2 g catalyst C1 (3.0 mM Fe^{3+}, (●) 0.1 g catalyst C1 (1.5 mM Fe^{3+}).

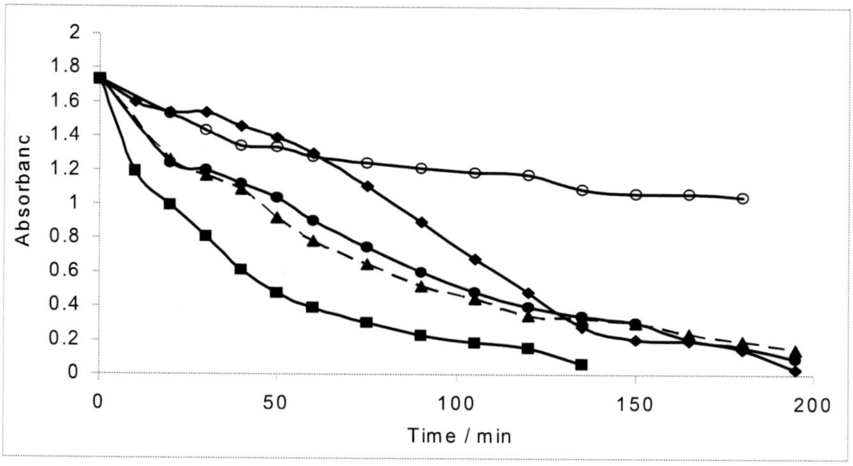

Figure 16. Degradation of RB5 as a function of heterogeneous catalyst performance. Experimental conditions: 100 ml 50 mM Na_2SO_4 + 0.06 mM RB5 + 4.56 mM H_2O_2 + required amount of catalyst, pH 2.8 and room temperature. (♦) 0.1 g catalyst C1; (●) 1.2 g catalyst C2; (▲) 0.59 g catalyst C3; (■) 1.13 g catalyst C4; (○) 0.90 g catalyst C5.Catalyst concentration: Curve (○): 1.2 g original fly ash (0.1 mM Fe^{3+}), curve (■): 0.2 g catalyst C1 (3.0 mM Fe^{3+}), curve (●): 0.1 g catalyst C1 (1.5 mM Fe^{3+}).

Catalytic behavior. The heterogeneous catalyst performance is shown in Figure 16. Color and COD abatement was assessed as a function of the prepared catalysts (C1 to C5) for an aqueous solution consisting of 100 ml 50 mM Na_2SO_4 + 0.06 mM RB5 + 4.56 mM H_2O_2 + required amount of catalyst, pH 2.8 and room temperature. Different amounts of catalyst were used to have always 1.5 mM Fe^{3+} in solution.

The poorest and the best catalytic performance correspond to catalyst C5 and C4, respectively. The rest of them (C1 to C3) are similar. Except for catalyst C5, the rest of them could abate both the COD (~80%) and the color from the solution.

In order to verify if the color and COD abatement from the solution was due to degradation by Fenton process rather than by an adsorption process, some experiments were repeated for an aqueous solution consisting of 100 ml 50 mM Na_2SO_4 + 0.06 mM RB5 + required amount of catalyst, pH 2.8 and room temperature. Hydrogen peroxide was not included to avoid the Fenton chemistry. Figure 17 shows the absorbance spectra for: (a) the initial solution, (b) after 180 minutes of contact time (catalyst C1+ BR5) without H_2O_2, and (c) in the presence of H_2O_2 after 180 minutes of Fenton process.

As it can be seen from Figure 17, adsorption process is not playing an important role in color and COD abatement from the solution. Similar results were found for catalyst C2, C3, C4 and C5. The exception was the original fly ash, Figure 18 shows a small amount of dye adsorbed on the original fly ash, see curve (b).

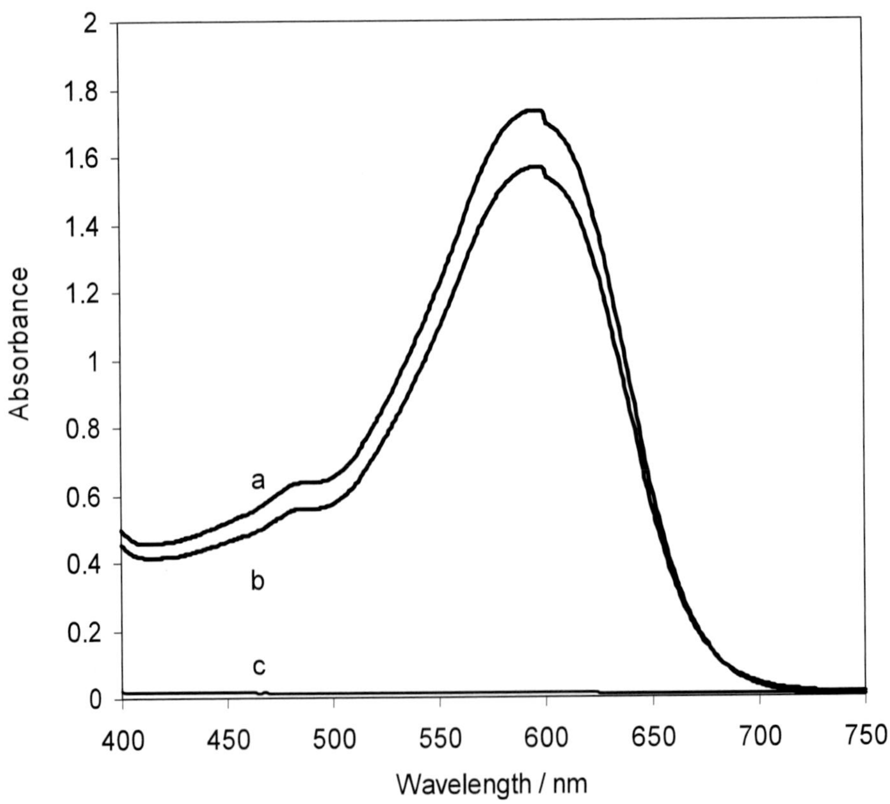

Figure 17. Absorbance spectra for: (a) the initial solution (100 ml 50 mM Na_2SO_4 + 0.06 mM RB5 + 0.100 g Catalyst 1, pH 2.8); (b) after 195 min of contact time (catalyst C1-RB5) but in the absence of H_2O_2; (c) after 195 min of Fenton process (100 ml 50 mM Na_2SO_4 + 0.06 mM RB5 + 0.10 g Catalyst C1, pH 2.8 + 4.56 mM H_2O_2).

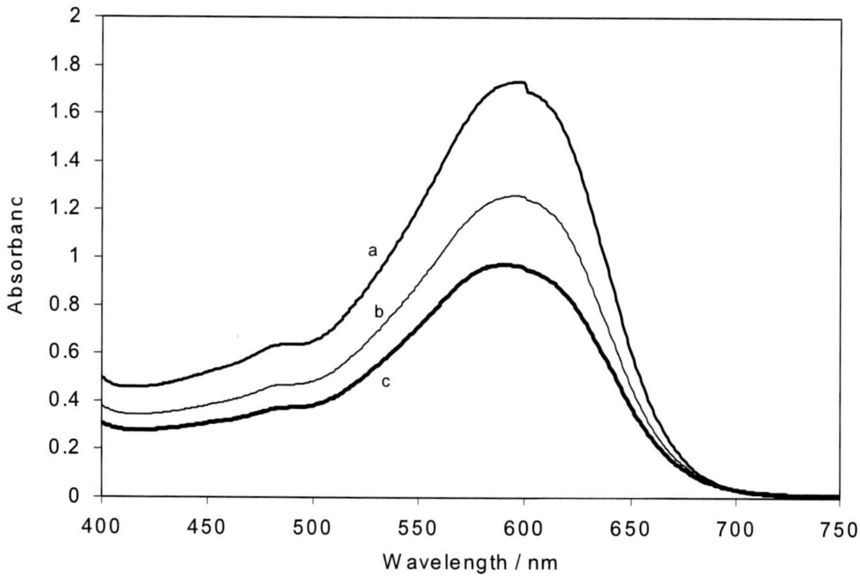

Figure 18. Absorbance spectra for: (a) the initial solution (100 ml 50 mM Na_2SO_4 + 0.06 mM RB5 + 0.50 g fly ash, pH 2.8); (b) after 180 min of contact time (fly ash-RB5) but in the absence of H_2O_2; (c) after 180 minutes of Fenton process (100 ml 50 mM Na_2SO_4 + 0.06 mM RB5 + 0.50 g fly ash 5, pH 2.8 + 4.56 mM H_2O_2).

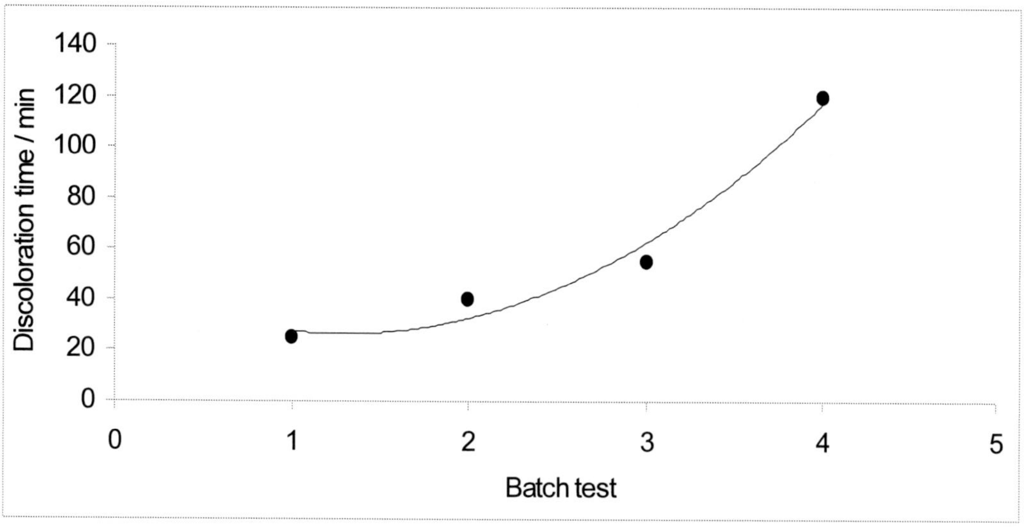

Figure 19. Consecutives discolorations of four different solutions consisting in 100 ml 50 mM Na_2SO_4 (pH≈ 2.8) + 0.06 mM RB5 + 10.7 mM H_2O_2. All degradations were performed with the same 0.1 g catalyst C1.

Catalyst recycling. It is demonstrated that H_2O_2 was activated by the heterogeneous catalyst and Fe^{3+}/H_2O_2 became a powerful oxidant, capable of the RB5 degradation, following a Fenton-type path. During this process most of the initial amount of immobilized ion Fe^{3+} in the support (fly ash) should stay, as it was suggested by experiments shown in

Table 3. In order to test the hypothesis that heterogeneous catalyst is better than the homogeneous one, because the former allows improving the control over iron ions during the Fenton chemistry, several consecutive degradations were performed with a fixed amount of catalyst. Figure 19 shows the time needed for four consecutive batch degradations of 100 ml 50 mM Na_2SO_4 (pH≈ 2.8) + 0.06 mM RB5 + 10.7 mM H_2O_2 (twice of the stoichiometric amount required) + 0.1 g catalyst C1. At the end of each discoloration the solution was allowed to settle down into the reactor and the solution was withdrawn carefully. Then, a fresh solution (100 ml 50 mM Na_2SO_4 + 0.06 mM RB5 + 10.7 mM H_2O_2) was reloaded over the previous 0.1 g catalyst C1 to start a new RB5 discoloration. The process was repeated three times.

It can be seen from Figure 19 that the catalyst C1 performance is good for three consecutive batch experiments and confirm the hypotheses that in the heterogeneous catalyst approach iron ions are in better control than in the homogeneous one.

Iron Oxide Supported on Cerium Oxides

Preparation of catalyst. Different amounts of iron oxide supported on cerium oxide were prepared under mechanical and ultrasonic stirring using an ultrasonic bath (450 kHz) followed by thermal treatment. The iron catalyst was prepared by mixing three different amounts of salts of ferric nitrate and cerium nitrate in an aqueous solution followed by an increase of the pH at 12 with a 2 mol l^{-1} sodium hydroxide solution. A mix of iron and cerium hydroxides (the catalyst) were precipitated and kept under vigorous mechanical stirring during four hours. After filtrating and washing trice with distilled water, the catalyst was dried at 110 °C for 16 hours. Previous to the calcination at 500 °C for 4 hours, the catalyst was finely ground. Three more catalysts were prepared in the presence of ultrasound following the same procedure describe above. The catalysts were shaken holding the ultrasonic irradiation for 20 minutes alternating the agitation with mechanical stirring for the same time during 4 hours. Table 4, shows the amount of catalytic agent and catalyst composition. The catalyst composition in iron content is between 25% to 75%.

Table 4. Catalyst type, amount of precursor salts and catalyst composition

Catalyst type ($Fe_2O_3 - CeO_2$)	Amount of the catalytic agent (g)		Catalyst composition	
	$Fe(NO_3)_3{\cdot}9H_2O$	$Ce(NO_3)_3{\cdot}6H_2O$	Fe_2O_3 (%)	CeO_2 (%)
C1-Ms	3.1614	9.4605	25	75
C1-Us/Ms	3.1614	9.4605		
C2-Ms	6.3248	6.3072	50	50
C2-Us/Ms	6.3248	6.3072		
C3-Ms	9.4873	3.1535	75	25
C3-Us/Ms	9.4873	3.1535		

Table 5. Iron content in the C2-Ms and C2-Us/Ms catalysts found after different amounts of each catalyst were subjected to the digestion process. It is observed that relatively higher iron content was found in the catalyst prepared by mechanical stirring (C2-Ms catalyst)

Iron content in the catalyst C2-Ms found after digestion in a volume of 100 ml.						
Catalyst (g)	Iron content in the catalyst C2-Ms (mg)			Iron content in the catalyst C2-Us/Ms (mg)		
	*Fe_{Total}	Fe^{2+}	Fe^{3+}	*Fe_{Total}	Fe^{2+}	Fe^{3+}
0.01	0.816	0.0158	0.80087	0.880	0.029	0.850
0.04	3.545	0.4136	3.1314	3.017	0.030	2.987
0.07	6.233	0.7592	5.4738	5.215	0.042	5.173
0.10	8.925	1.110	7.815	7.45	0.060	7.39

* $Fe_{Total} = Fe^{2+} + Fe^{3+}$

Organic degradation by Fenton-like chemistry. Chemical oxidation of basic orange 2 (industrial grade, provided by the local textile industry) was carried out using 0.1 l of dye solution, in a glass batch reactor under continuous stirring (using a magnetic stirrer) at room temperature. The powder catalyst was added and the beginning of the reaction (t = 0) was considered when H_2O_2 (30%, w/w, HACH) was added at the proper concentration. The influence of catalyst amount, initial dye concentration and pH of the solution was tested on the chemical oxidation of the textile dye.

A calibration curve was obtained by measuring the maxima absorbance of the solution at 456 nm at the pH under study. Measurements of absorbance were carried out as a function of time during the dye oxidation. The samples were subjected to filtration prior absorbance measurements and prior further analysis.

The chemical oxidation of the basic orange 2 (BO2) involves a complete discoloration and mineralization in order to achieve a conversion of the initial compound into CO_2 and H_2O. Therefore, it is important to evaluate the mineralization of organic dyes.

The BO2 contains an azo group, -N=N-, as part of its molecular structure connecting two aromatic ring compounds, as depicted by figure 20. The chemical name of the BO2 is 1,3-benzenediamine, 4-(phenylazo) with a molecular weight of 248.72 g mol^{-1} ($C_{12}H_{12}N_4HCl$). This dye is also known as Orcozine Chrysoidine 4.

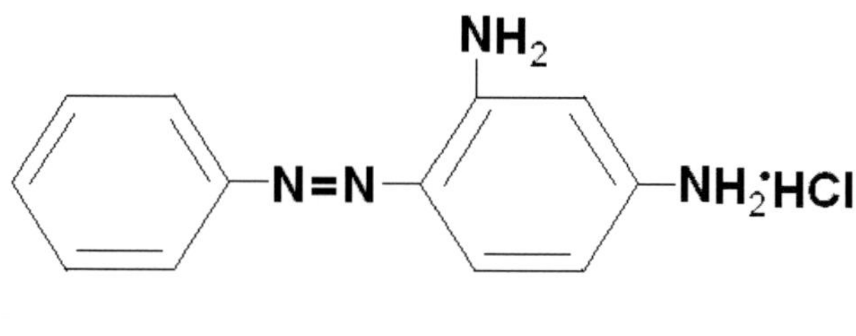

Figure 20. Molecular structure of basic orange 2.

The degradation of the BO2 was conducted by using iron oxide catalyst supported on cerium oxide as a heterogeneous Fenton catalyst in the presence of H_2O_2. The effect of the initial catalyst amount, initial pH of the solution, and initial dye concentration was studied as influencing factors on the BO2 degradation. Iron leaching from the catalyst was also tested.

Catalytic behavior. The catalytic experiments were performed using a 0.1 mM solution of BO2, pH 3 and room temperature. The concentrations of azo dyes found in industrial waste streams are usually around 0.1 mM. Initially, different amounts of the catalyst C2-Ms and C2-Us/Ms were employed inside the 0.01 g to 0.1 g range in the presence of H_2O_2. The mineralization of BO2 is *80e⁻* oxidation, as shown in reaction (36) with its transformation into carbon dioxide where the nitrogen atom undergoes a complete oxidation.

$$C_{12}H_{12}N_4HCl + 36H_2O \rightarrow 12CO_2 + 4HNO_3 + HCl + 80H^+ + 80e^- \qquad (36)$$

Thus the stoichiometric conversion of 1 mol of BO2 requires 40 moles of hydrogen peroxide, according to reaction (37):

$$C_{12}H_{12}N_4HCl + 40H_2O_2 \rightarrow 12CO_2 + 4HNO_3 + HCl + 44H_2O \qquad (37)$$

Therefore, a concentration of 4.0 mM of H_2O_2 was employed to carry out the chemical oxidation of 0.1 mM BO2.

Figure 21 shows the UV–Vis spectra recorded for the dye solution as a function of degradation time. The dye spectra shows one band in the ultraviolet region located at ~307 nm, attributed to the absorbance of benzene rings (*Thomas 1996*), and one band in the visible region with a maximum located at 456 nm due to the chromophore-containing azo group. The comparison of the original spectrum with that achieved after 340 min of oxidation demonstrates that the color removal of the treated dye solution is practically complete, as it did not show significant absorbance in the visible region. Therefore, the disappearance of the absorbance peak at 456 nm clearly shows the breakdown in the chromophoric group.

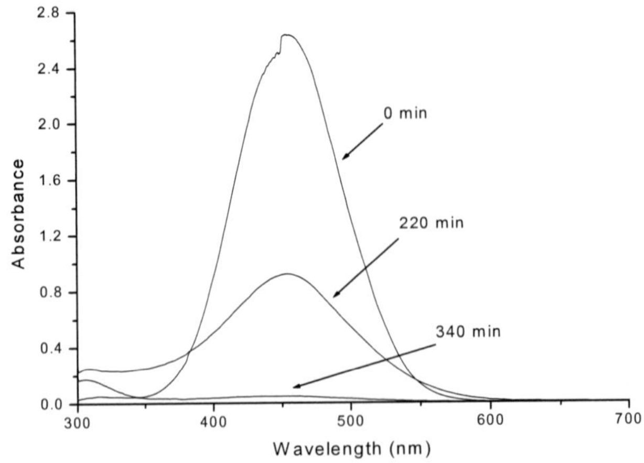

Figure 21. UV-Vis spectra evolution of a BO2 solution as a function of time using 0.01 g of C2-Ms catalyst at pH 3, 4.0 mM H_2O_2 and room temperature.

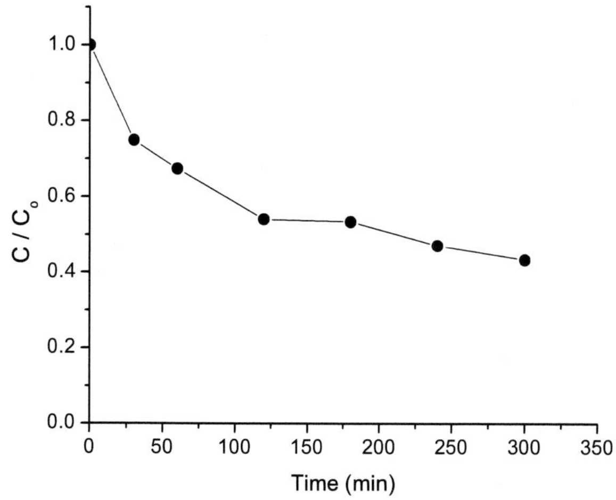

Figure 22. Discoloration by adsorption of 0.1mM NB2 solution in the presence of 0.01 g of C2-Ms at pH 3.

It is also observed, that the absorbance at ~307 nm of the spectrum recorded at 220 min has slightly increased with respect to the original one and then its intensity decreases almost completely after 340 min of oxidation. The decrease of the intensities of this band can be attributed to the formation of intermediates, resulting from the degradation of the azo dye, which still contain benzoic rings. Additionally, according to equation (37), there are other species that are generated by the azo dye oxidation; however, none of these compounds absorb in the visible region.

Influence of catalyst load on the degradation of BO2 solution. The profile for the color decay of 0.1 mM BO2 solution by 0.01 g of C2-Ms at pH 3, in the absence of H_2O_2, is shown in figure 22. It is observed a 57% of color removal by adsorption after 5 h (but no equilibration was yet reached) whereas the solution was effectively colorless when 4 mM H_2O_2 was present under similar experimental conditions, as depicted in figure 23 (plot A).

Figure 23, reports the normalized absorbance (at 456 nm) for the oxidation of a 0.1 mM dye solution as a function of reaction time under the reactor conditions of: pH 3, room temperature and different amounts of catalyst C2-Ms and C2-Us/Ms. The normalized absorbance was obtained by dividing the absorbance measured at 456 nm for the samples taken along the reaction time between the initial absorbance recorded at the same wavelength for the non degraded dye solution. It is observed that the normalized absorbance decreased with time for all the catalyst amounts tested following similar trend in all cases. There is an issue of interest at short oxidation time in which the normalized absorbance exhibits a sharp decrease that corresponds, for the C2-Ms (plot A), to a color removal between 30% and 46% from its initial value; while 58% to 73% color removal for the C2-Us/Ms (plot B) depending on the amount of catalyst used. Afterwards, the normalized absorbance increases and reaches a stage in which it varies slightly during a period of ~100 min or so, except for the catalyst amount of 0.07g where after the increase, it follows a slow decrease until the solution became colorless, in both catalysts. In all cases, except for that of 0.10g catalyst amount, the dye degradation achieved a complete color removal for the treated dye solution. According to these results, it is observed that the best amounts of catalysts are 0.01 g and 0.04 g for the catalyst C2-Ms and C2-Us/Ms respectively. The heterogeneous Fenton-like degradation profiles of BO2 exhibit an induction period (between ~30 to ~170 min) followed by an inflection point that represents a steady state decay.

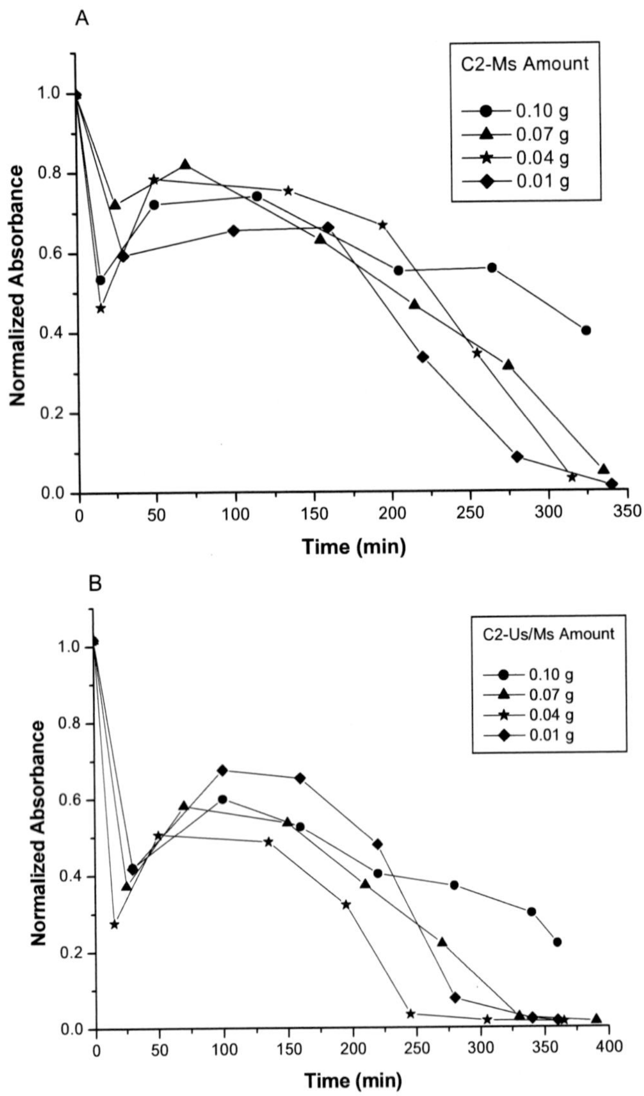

Figure 23. Heterogeneous Fenton degradation of 0.1mM BO2 as a function of different catalyst amounts (Plot A: C2-Ms, Plot B:C2-Us/Ms) in the presence of 4 mM H_2O_2 at pH 3.

Interestingly, the COD decay followed a similar trend of that exhibited by the absorbance during the dye oxidation. Figure 24, shows an increase in the COD at 10 min of reaction, achieving even a higher value than that recorded for the non degraded dye solution (initial COD). This reaction time corresponded to the moment in which the normalized absorbance sharply decreased. This COD increase can be explained by the production of intermediate compounds generated by the rapid oxidation of the BO2 that exhibited higher oxygen demand. Afterwards, the COD decreased ~20% from its initial COD value (at 100 min) and remains almost constant at ~21 mg l^{-1} with a very low variation during 125 min. After this reaction time, the COD started to decay progressively until it was not detected at 340 min of reaction. The estimated detection limit reported for the technique employed is 0.2 mg l^{-1} COD (*Hach 1997*). COD was completely abated for all the catalyst amounts employed except that for 0.1g of catalyst the COD was abated ~ 88%.

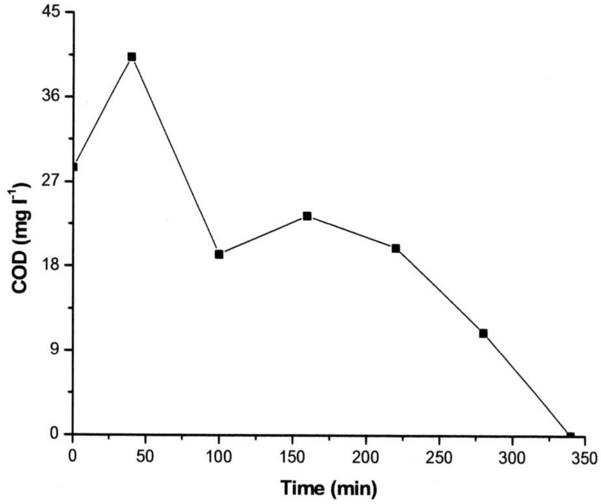

Figure 24. Chemical oxygen demand during the heterogeneous Fenton degradation of 0.1mM BO2 as a function of time employing 0.01 g de C2-Ms and 4 mM H_2O_2 at pH 3.

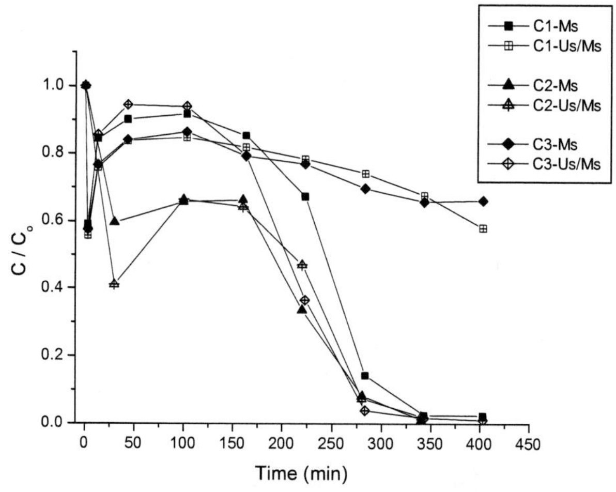

Figure 25. Heterogeneous Fenton degradation of 0.1mM BO2 as a function of the catalysts with an iron content increase (25% in C1, 50% in C2 and 75% in C3) using 0.01 g of catalyst, 6 mM H_2O_2 at pH 3.

The Heterogeneous Fenton degradation of 0.1 mM BO2 as a function of the catalysts with an iron content increase (25% in C1, 50% in C2 and 75% in C3) using 0.01 g of catalyst in the presence of 6 mM H_2O_2 at pH 3 is shown in figure 25. It is observed that dye degradation behavior is very similar in all the catalysts tested regarding of the increase in the iron content in the catalyst from 25% to 75%. These results have shown that faster dye degradation is achieved with the use of catalyst with less iron content, which represents an advantage from the economical point of view.

It is important to asses the loss of catalyst from the support in order to know the feasibility of the heterogeneous catalyst in the Fenton-like reaction for industrial applications. This was carried out by measuring the iron concentration in a stirring solution as a function of reaction time using a catalyst amount of 0.1 g (C2-Ms and C2-Us/Ms) at pH 3 in the absence of BO2 and H_2O_2. Samples were taken every 2 h up to a leaching time of 10 h. According to figure 26, the iron leaching is considerably low (<1 mg l^{-1}) for the C2-Us/Ms catalyst even after 10h. Higher iron leaching values were observed from the C2-Ms catalyst obtaining an iron concentration of 6.85 mg l^{-1} after 10 h of leaching time.

The low iron leaching from the C2-Us/Ms catalyst exhibits an interesting feature due to the possibility of using this catalyst for a longer operation time for industrial applications. The performance of this catalyst, that shows slower deactivation, seems to be mainly due to the heterogeneous Fenton-like process. Figure 27 shows that most of the iron leaching is as Fe^{3+} in both catalysts.

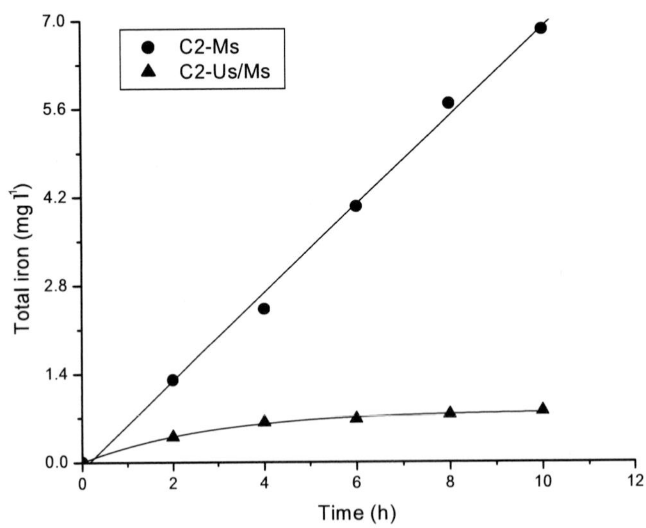

Figure 26. Total iron leaching from C2-Ms and C2-Us/Ms catalysts at pH 3, in the absence of BO2 and H_2O_2 and room temperature.

Initial H_2O_2 concentration effect. The hydrogen peroxide concentrations range was chosen by considering the stoichiometric conversion of BO2 into carbon dioxide according to equation (37) and varied between 1 to 7 mM. Figure 28 shows the results obtained using 0.01 g of C2-Ms catalyst in a 0.1 mM OB2 solution at pH 3 and room temperature. The dye degradation increases as the initial H_2O_2 concentrations is increased from 1 to 6 mM following all similar behavior, whereas the BO2 oxidation decreases when the H_2O_2 concentration is 7 mM. It is also observed that ~ 95% of discoloration is attained for all the H_2O_2 concentrations ≥ 2 mM after 340 min of dye oxidation. It seems that the optimum hydrogen peroxide concentration, typical and well-known in Fenton's oxidation, is 6.0 mM for the oxidation of 0.1 mM BO2 under the conditions stated. This optimal concentration corresponds to a H_2O_2 concentration 50% above of that predicted by equation (37). The increase of hydrogen peroxide concentration from 1 to 6 mM leads to an increase in the reaction rate as more radicals are formed according to the equations (34) and (35). However, when the hydrogen peroxide concentration is in excess ($\geq$ 7 mM) its performance decreases. In the presence of an excess of hydrogen peroxide the scavenging of $OH^{\bullet}$ radicals occurs, according to reaction (3).

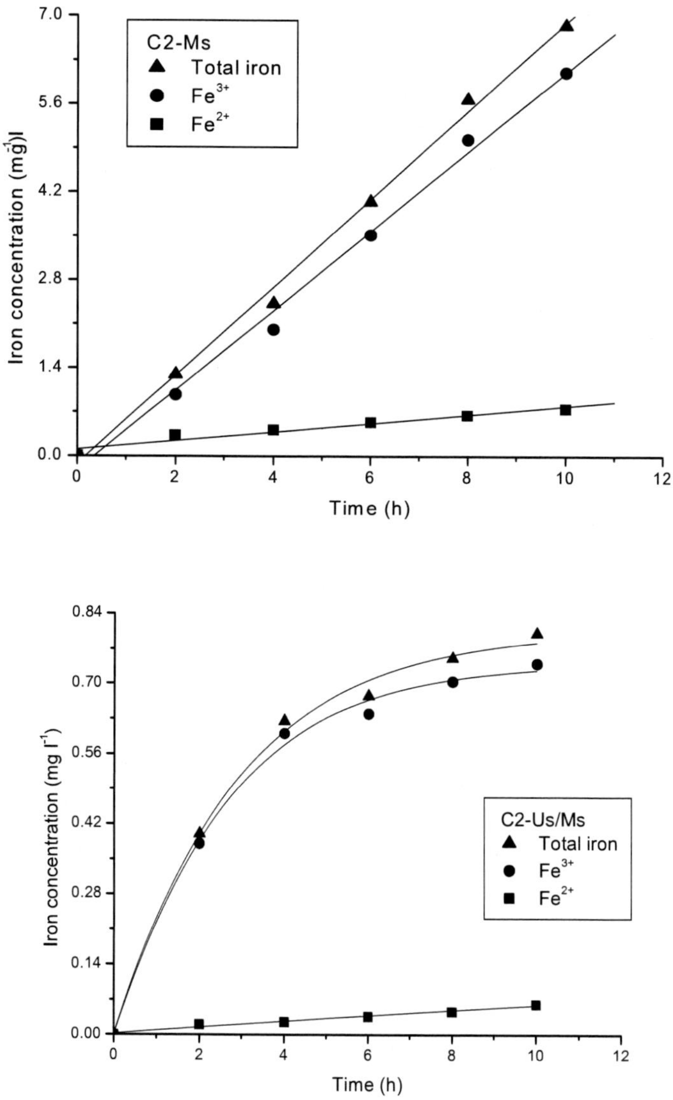

Figure 27. Iron leaching from C2-Ms and C2-Us/Ms catalysts at pH 3, in the absence of BO2 and H_2O_2 and room temperature.

Even though $HO_2^{\bullet}$ radicals are generated by the scavenging effect, they are much less reactive than $OH^{\bullet}$ radicals (*Bielski et al. 1985; Pignatello 1992*).

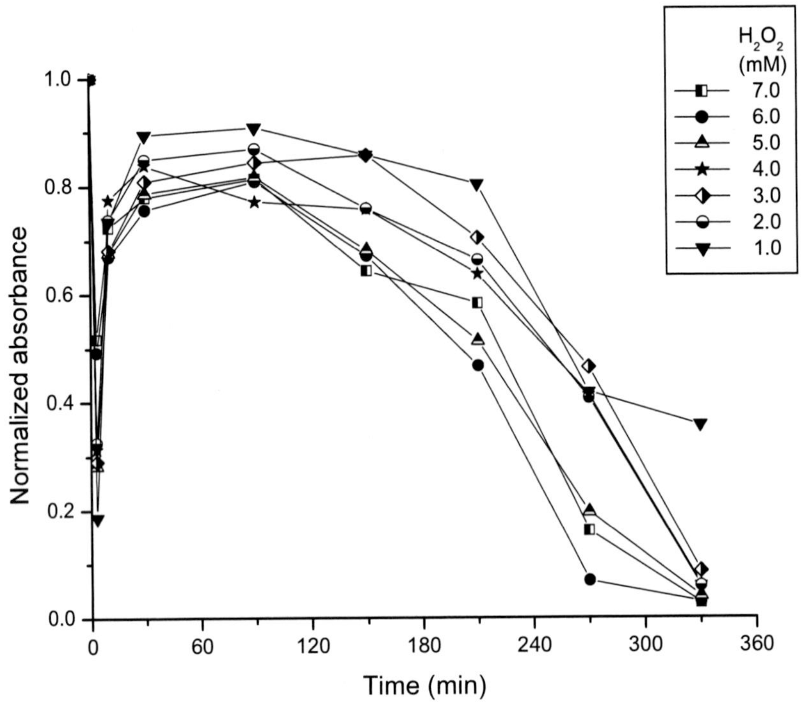

Figure 28. Degradation of 0.1 mM BO2 by the heterogeneous Fenton process as a function of H_2O_2 concentration with 0.01 g of C2-Ms at pH 3.

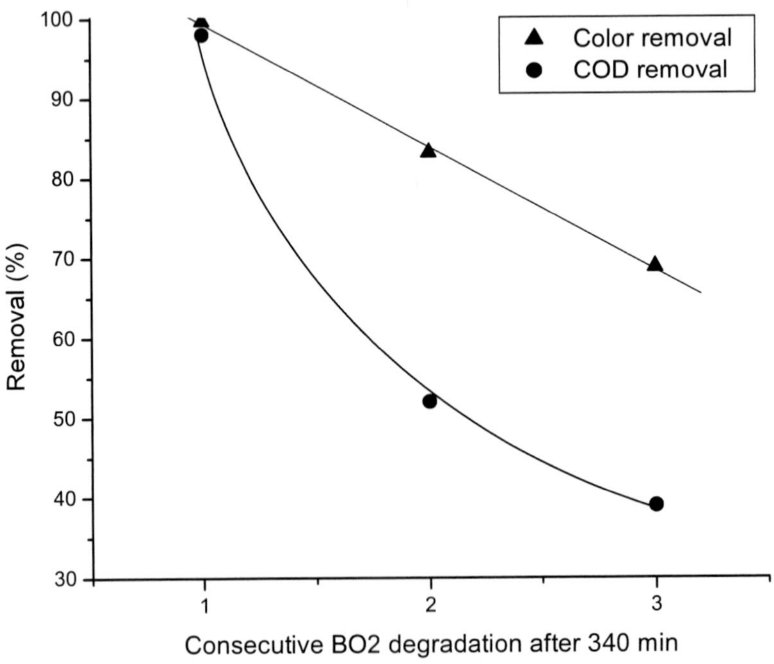

Figure 29. Effect of consecutive experiments with C2-Ms catalyst on the degradation of 0.1 mM BO2 solution in the presence of 4 mM H_2O_2. pH 3 and room temperature.

Catalyst recycling. Figure 29 shows the performance achieved for the BO2 degradation in three consecutive runs. In order to recover and reuse the C2-Ms catalyst, the catalyst was allowed to settle on the bottom of the reactor overnight after each oxidation cycle that lasted 340 min. Afterwards, the treated effluent was carefully decanted to minimize the loss of catalyst. After the first cycle, the dye solution became colorless and the measurement of the absorbance at 456 nm reported a color removal of 99%. For subsequent cycles, it was observed that both the color removal and COD abatement decreased revealing that the catalytic activity was diminishing. The BO2 discoloration decreased from 99 to 69%, whereas COD removal decreased from 98% to 39% in 3 cycles that corresponds to 17 h of operation. Iron loss from the catalyst per cycle corresponds to ~ 4.3% (0.388 mg l^{-1}).

Catalyst deactivation has been observed by other authors and it may occur due to several factors, amongst them: reduction of the catalyst specific surface area, poisoning of the catalytic agents by compounds formed during oxidation, surface deposition, etc. (*Catrinescu et al. 2003, Guo and Al-Dahhan 2006, Ramirez et al. 2007a*).

CONCLUSIONS

It was demonstrated that the couple Fe^{3+}/H_2O_2 is as efficient as the Fe^{2+}/H_2O_2 during the discoloration by Fenton and Fenton-type processes. The mixtures of Fe^{3+}/H_2O_2 and Fe^{2+}/H_2O_2 produce strong oxidants ($OH^{\bullet}$, $HO_2^{\bullet}$, FeO^{2+}) capable of oxidizing organic pollutants under mildly acidic conditions. This approach might be applied in wastewater treatment in more realistic experiments.

Fly ash, from a coal thermal power plant, is an economic material and it was used as iron support because of its high surface area, which would provide more active sites to the catalyst. Fe^{3+} was effectively immobilized in fly ash and cerium oxides using simple thermal procedures.

All the supported iron-fly ash and iron-cerium oxide catalysts have been prepared by means of salts of iron precursors at different iron loads. All the catalysts showed good performance in the Fenton-like oxidation of the textile dyes tested. The results showed a high degradation of the textile dyes (reactive black 5 and basic orange 2) and of the intermediary oxidized compounds. At optimal conditions, 100% discoloration and 80% of chemical oxygen demand abatement were achieved using the supported iron-fly ash catalyst after 2.5 h of reaction. Whereas, 99% discoloration and 88% of chemical oxygen demand abatement were achieved using the supported iron-cerium oxide catalyst after 5 h of reaction. All the catalysts exhibited not only good catalytic activity but also a reasonable small iron leaching (below the allowed iron content in the wastewater discharge by EPA regulations), demonstrating that the active phases are strongly fixed to the support. This feature makes possible the iron catalysts to have long-term stability, without generating iron hydroxide sludge. Consecutive oxidation cycles carried out with the same catalyst and untreated fresh dyestuff solution showed a low deactivation, which is possibly due to some iron leaching, thus evidencing the possibility of being used in continuous processes. Discoloration of 100% of reactive black 5 solution was achieved in all the consecutive oxidation cycles tested using the supported iron-fly ash catalyst. While, the discolorations of the basic orange 2 solutions were decreasing on each consecutive oxidation cycles employing the supported iron-cerium oxides (this catalyst was used up to 17 h).

Heterogeneous iron catalyst (Fe^{3+}-containing ashes) is stable for at least 6 hours in aqueous solutions (50 mM Na_2SO_4, pH ≈ 2.8) similar to those found in textile effluents. This implies that, under the heterogeneous catalyst approach, iron ion is better controlled making more efficient the heterogeneous Fenton-like process than in the homogeneous catalyst approach.

The extremely lightweight of the supported iron-fly ash and iron-cerium oxide catalysts generates a uniform suspension of the heterogeneous catalyst in the solution improving its efficiency in the heterogeneous Fenton-like process. However, this advantage could be a technological challenge to prevent from clogging an electrochemical flow-cell if this approach were applied in a more realistic experiment.

References

Alvarez-Gallegos, A.; Pletcher, D. *Electrochim Acta.* 1998, 44, 679–685.

Alvarez-Gallegos A.; Pletcher D. *Electrochimica Acta.* 1999, 44, 2483-2492.

Barb W.G.; Baxendale, J. H; George, P; Hargrave, K. R. *Trans. Faraday Soc.* 1951, 47, 462-469.

Baxendale, J.H.; Evans, M.G.; Park, C.S. *Trans. Faraday Soc.* 1946, 155-169.

Baxendale, J.H.; Wilson, J.A. *Trans. Faraday Soc..* 1957, 344-356.

Benitez, F.J.; Acero, J.L.; Real, F.J.; Rubio, F.J.; Leal, A.I., *Wat. Res.* 2001, 35, 1338-1343.

Berl, E. *Trans. Electrochem. Soc.* 1939, 76, 359-378.

Berlan, J.; Trabelsi, F.; Delmas, H.; Wilhelm, A.M.; Petrignani, J.F., Ultrasonics Sonochemistry. 1994, 1 (2), S97-S102.

Bielski, B.H.J.; Cabelli, D.E.; Arudi, R.L.; Ross, A.B. *J. Phys. Chem. Ref. Data.* 1985, 14, 1041-1100.

Bigda, R.J. *Chem .Eng. Prog.* 1995, 91, 62-66.

Bowers, A.R.; Gaddipati, P.; Eckenfelder, W.W.; Monsen, R.M. Water. *Sci. Technology.* 1989, 21, 477-486.

Bossmann, S.H.; Oliveros, E.; Göb, S.; Siegwart, S.; Dahlen, E.P.; Payawan L, Jr.; Straub, M.; Wörner, M.; Braun, A.M . *J. Phys. Chem.* A. 1998, 102, 5542-5550.

Boye, B.; Brillas, E.; Momar, M.D. *J. Electroanal. Chem.* 2003, 540, 25-34.

Bray, W.C.; Gorin, M.H. *J. Am. Chem. Soc.* 1932, 54, 2124-2125.

Brillas, E.; Banos, M.A.; Garrido, *J.A. Electrochim. Acta.* 2003, 48, 1697-1705.

Brillas, E.; Bastida, R.M.; Llosa, E.; Casado, J. *J. Electrochem. Soc.* 1995, 142, 1733-2379.

Brillas, E.; Mur, E.; Sauleda, R.; Sanchez, L.; Peral, J.; Domenech, X.; Casado, *J. Applied Catalysis B: Environmental.* 1998, 16, 31-42.

Brillas, E.; Sauleda, R.; Casado, J. *J. Electrochem. Soc.* 1997, 144, 2374-1741.

Catrinescu, C.; Teodosiu, C.; Macoveanu, M.; Miehe-Brendle, J.; Dred, R.L.;

Centi G.S.P.; Torre, T.; Verduna, M.G. Catal. Today. 2000, 55, 61-69.

Chan-Li, H.; Yao-Hui, H.; Cheng-Chien, W.; Chuh-Yung, *Chen. J. Mol. Catal.* A: Chemical. 2006, 245, 78–86.

Dantas, T.L.P.; Mendonça, V.P.; Jose, H.J.; Rodrigues, A.E.; Moreira, R.F.P.M. *Chem .Eng. J.* 2006, 118, 77–82.
De Laat, J.; Gallard, H. Environ. *Sci. Technol.* 1999, 33, 2726-2732.
Feng, J.; Hu, X.; Yue, P.L.; Zhu, H.Y.; Lu, G.Q. *Chem. Eng. Sci.* 2003, 58, 679 – 685.
Feng, J.; Hu, X.; Yue, P.L. *Water Res.* 2006, 40 641 – 646.
Fenton, H.J.H. J. Chem. Soc. 1894, 65, 899-901.
Flores, Y.; Flores, R.; Alvarez-Gallegos, A. Applied Catalysis B: *Environmental.* 2008, 281, 184-191.
Flox, C.; Ammar, S.; Arias, C.; Brillas, E.; Vargas-Zavala, A.V.; Abdelhedi, R. Applied Catalysis B: *Environmental.* 2006, 67, 93-104.
Gallard, H.; De Laat, *J. Chemosphere.* 2001, 42, 405-413.
Gattrell, M.; Kirk, D.W. *Can. J. Chem. Engineering.* 1990, 68, 997-1003.
Geus, J.W.; Von Veen, J.A.R., in: Catalysis: An Integrated Approach; Von Santer, R.A.; Von Leeumen, P.W.N.M.; Moulijn, A.J.; Averill, B.A.; Elsevier Science B.V., Amsterdam, 1999, p 459.
Gozzo, F. *J. Mol. Catal. A: Chem.* 2001, 171, 1-22.
Groves, J.T.; Watanabe, Y. *J. Am. Chem. Soc.* 1986, 108, 7834-7836.
Guinea, E.; Arias, C.; Cabot, P.L.; Garrido, J.A.; Rodriguez, R.M.; Centellas, F.; Brillas, E, *Water Reas. 2008, 42, 499-511.*
Guo, J.; Al-Dahhan, M. *Appl. Catal. A: Gen.* 2006, 299, 175–184.
Haber. F.; Weiss. J. Proc. R. Soc. London, Ser. A. 1934, 147, 332-351
Hach Company,1997-2003. Dr/4000 Spectrophotometer Procedures Manual. Printed in the U.S.A. 1997, pp. 1-6.
Harrington, T.; Pletcher, D. *J. Electrochem. Soc.* 1999, 146, 2983-2989.
He, Z.; Song, S.; Zhou, H.; Ying, H. *J. Chem. Ultrason. Sonochem.* 2007, 14, 298- 304.
Hoffmann, M.R. *Environmental Science and Technology Reports.* 1977, 11, 61-66
Huston, P.L.; Pignatello, J.J. *Wat. Res.* 1999, 33, 1238-1246.
Kean, R.T.; Oertling, A.; Babcock, G.T., *J. Am. Chem. Soc.* 1987, 109, 2185-2187.
Kolthoff, I.M.; Belcher, R., Volumetric Analysis. Titration. Methods: Oxidation Reduction Reaction. Interscience Publishers, Inc., NY 1957. Vol. III
Koppenol, W.H.; Liebman, J. *J. Phys. Chem.* 1984, 88, 99-101.
Kremer, M.L. *Int J. Chem. Kinetics.* 1985, 17, 1299-1314.
Kremer, M.L. *Phys. Chem. Chem. Phys.* 1999, 1, 3595-3605.
Kremer, M.L. *J. Phys. Chem. A.* 2003, 107, 1734-1741.
Kuo, W.G. *Wat. Res.* 1992, 26, 881-886.
Kušić, H.; Koprivanac, N.; Selanec, I. *Chemosphere.* 2006, 65, 65–73.
Legrini, O.; Oliveros, E.; Braun, A.M. *Chem. Rev.* 1993, 93, 671-698.
Lim, H.; Lee, J.; Jin, S.; Kim, J.; Yoon, J.; Hyeon, T. *Chem. Commun.* 2006, 463–465.
Lipczynska-Kochany, E. *Chemosphere.* 1991, 22, 529-536.
Lopez, A.; Benbelkacem, H.; Pic, J-S.; Debellefontaine, H. *Environmental Technology.* 2004, 25, 311-321.
Lucas M. S.; Peres, J.A. Dyes Pigments. 2006, 71, 236- 244.
McGinnis, B.D.; Adams, V.D.; Middlebrooks, E.*J. Chemosphere.* 2001, 45, 101-108.
McIntyre, L.A. Interface (The Electrochem Soc). 1995, 4, 29-34.
Meriç, S.; Kaptan, D.; Olmez, T. *Chemosphere.* 2004, 54, 435-441.
Merz, J.H; Waters, W. A. Discuss. *Faraday Soc.* 1947, 2, 179-188.

Pignatello, J.*J. Environ. Sci. Technol.* 1992, 26, 944-951.

Pignatello, J.J.; Chapa, G. *Environ. Toxicol. Chem.* 1994, 13, 423-427.

Pirkanniemi, K., Complexing Agents A Study of Short-term Toxicity, Catalytic Oxidative Degradation and Concentrations in Industrial Waste Waters. Kuopio University Publications C. Natural and Environmental Sciences, 209, 2007, Doctoral dissertation. Finland.

Pirkanniemi, K.; Sillanpää, M. *Chemosphere.* 2002, 48, 1047–1060.

Ponce de Leon, C.; Pletcher, D. *J. Applied Electrochem.* 1995, 25, 307-314.

Rahhal, S.; Richter, H.W. *J. Am. Chem. Soc.* 1988, 110, 3126-3133.

Rajeshwar K.; Ibanes J. Environmental Electrochemistry. Fundamentals and Applications in Pollution Abatement. Academic Press, Inc. San Diego, Ca. 1997.

Ramirez, J.H.; Costa, C. A.; Madeira, L. M.; Mata, G.; Vicente, M. A.; Rojas-Cervantes, M.L.; López-Peinado, A.J.; Martın-Aranda, R.M. *Applied Catalysis B: Environmental.* 2007a, 71, 44–56.

Ramirez, J.H.; Maldonado-Hódar, F.J.; Pérez-Cadenas, A.F.; Moreno-Castilla, C.; Costa, C.A.; Madeira, L.M. *Applied Catalysis B: Environmental.* 2007b, 75, 312–323.

Rodriguez, M.L.; Timokhin, V.I.; Contreras, S.; Chamarro, E.; Esplugas, S. *Advances in Environmental Research.* 2003, 7, 583-595.

Seshadri, S.; Bishop, P.L.; Agha, A.M. Waste Management. 1994, 14, 127-137.

Sharifian, H.; Kirk, D.W. *J. Electrochem. Soc.* 1986, 133(5), 921-924.

Sheldon, R.A.; Wallau, M.; Arends, I.W.C.E.; Schuchardt, U. *Acc. Chem. Res.* 1998, 31, 485–493.

Shiga, T. *J. Phys. Chem.* 1965; 69, 3805-3814.

Siminiceanu, I.; Alexandru, C.I.; Brillas, E. Revista de Chimie. 2006, 57, 1082-1085.

Sires, I.; Garrido, J.A.; Rodriguez, R.M.; Brillas, E.; Oturan, N.; Oturan, M.A. *Applied Catalysis B: Environmental.* 2007, 72, 382-394.

Spadero, J.; Isabelle L.; Renganathan, V. *Environmental Science and Technology.* 1994, 28, 1389-1393.

Storm, T., Aromatische Sulfonate: Untersuchungen zum Stoffverhalten in Industrieabwasser und aquatischer Umwelt mit HPLC-MS. Dissertation, Technischen Universität Berlin. 2002.

Sudoh, M.; Kodera, T.; Sakai, K.; Zhang, J.Q.; Koide, K. *J. Chem. Engineering of Japan,* 1986, 19, 513-518.

Sun, Y.; Pignatello,.*J. Environ. Sci. Technol.* 1993, 27, 304-310.

Swaminathan, K.; Sandhya, S.; Carmalin, S.A.; Pachhade, K.; Subrahmanyam, Y.V. *Chemosphere.* 2003, 50, 619-625.

Teel, A.L.; Warberg, C.R.; Atkinson, D.A.; Watts, R.J. *Wat. Res.* 2001, 35, 977- 984.

Thomas M., Ultraviolet and visible spectroscopy. Analytical chemistry by open lerning; Ando D.J. Editor. Second edition. 1996, John Wiley and Sons. NY p 173.

Tomat, R.; Vecchi, E. *J. Appl. Electrochem.* 1971, 1, 185-188.

Tomat, R.; Rigo, A. *J. Appl. Electrochem.* 1976, 6, 257-261.

Tomat, R.; Rigo, A. *J. Appl. Electrochem.* 1979, 9, 301-305.

Tryba, B. (2008). International Journal of Photoenergy. vol. 2008. Article ID 721824, 15 pages, 2008. doi:10.1155/2008/721824

Uri, N. *Chem. Rev.* 1952, 50, 375-454.

Walling, C.; Kurz, M.; Schugar H. *J. Inorg. Chem.* 1970, 9, 931-937.

Walling, C.; Weil, T. Int. *J.Chem. Kinetics.* 1974, 6, 507-516.
Walling, C.; El-Taliawi, G.M.; Johnson, R. A. *J. Am. Chem. Soc.* 1974, 96, 133-139.
Walling, C. Acc. *Chem. Res.* 1975, 8, 125-131.
Walling, C.; Cleary, M. *Int. J.Chem. Kinetics.* 1977, 9, 595-601.
Walling, C.; Kato, S.I. *J. Am. Chem. Soc.* 1971, 93, 4275-4281.
Yamazaki, I.; Piette, L.H. *J. Am. Chem. Soc.* 1991, 113, 7588-7593.
Zepp, R. G.; Faust, B. C.; Hoigne, *J. Environmental Science and Technology.* 1992, 26, 313-319.

In: Heterogeneous Catalysis Research Progress
Editor: Mathias B. Gunther, pp. 235-266
ISBN 978-1-60456-979-7

Chapter 6

REFLECTIONS ON APPLIED CATALYSIS AND FUNDAMENTAL MODEL STUDIES

Zhen Ma*
833 West Vanderbilt Drive, Oak Ridge, TN 37830, USA

ABSTRACT

In this chapter, our research on the decomposition of Freon-12 (CCl_2F_2) on solid acids, the oxidation of CO using functionalized gold nanocatalysts, the methodological pitfalls in the reaction testing of phenol hydroxylation, and the liquid-solid interfaces of catalysis relevance is delineated from the perspectives of applied catalysis and fundamental model studies. One strategy of our applied catalysis investigation is to identify key catalytic functionalities to be achieved, and then rationally design or functionalize catalytic materials according to the desired functionalities. On the other hand, one strategy of our fundamental model studies is to distill unanswered questions from applied catalysis research or the relevant literature, and then design simplified or idealized model systems in order to answer these questions and obtain general trends. Herein, these two strategies are reflected by selected examples from our research, and a few review articles are introduced for further reference.

1. INTRODUCTION

Catalysis refers to the phenomenon in which the rate of a chemical reaction is accelerated by a small quantity of substance (catalyst) that is not substantially consumed in the process [1]. Catalysis takes central stage in chemical industry via manufacturing chemicals and polymer materials, processing fossil fuels and biomass, as well as ablating environmental pollutants and developing cleaner processes [2-4]. Academically, catalysis research is tightly linked to the four main branches of chemistry (i.e., inorganic, organic, analytical, and physical chemistry) [5-7]. A typical research paper in the field deals with the synthesis, characterization, and catalytic performance of a few catalysts, with their synthesis parameters, characterization methods, and reaction conditions being varied systematically (see for example, refs. [8-11]). Although it is of practical merits to look into the deactivation and regeneration of catalysts [12-15], we believe that more interesting in catalyst development is to establish the promotional effects from putting a small amount of certain additives (promoters) in catalysts [16-19] or loading active components onto suitable supports [20-23]. To that end, an effective research strategy is to identify key functionalities (i.e., activity, selectivity, stability) to be realized [24], and then rationally design or functionalize catalytic materials according to the desired functionalities (Figure 1, panel A) [25-28]. This functionality-driven strategy, sometimes glorified as "molecular engineering of functional systems" [29, 30], is in contrast to a materials synthesis-oriented approach via which one first synthesizes novel materials with eye-catching morphologies, and then attempts to look for their potential applications in catalysis and other fields [31-34].

With the gradual evolution of the field of applied catalysis, it is more and more essential to obtain physicochemical insights into catalytic systems [35-37], because fundamental studies can not only help to shed new light on catalytic phenomena, but also accelerate the invention of high-performance catalysts [38-41]. In this regard, one salient research strategy is to first identify unanswered questions in catalysis research or the relevant literature, devise related and simplified model systems, scrutinize the factors that may perturb these systems, and then try to explain particular phenomena in applied catalysis based on the obtained results (Figure 1, panel B) [42-45]. For instance, surface chemistry studies [46-48] and first-principle calculations [49, 50] stemming from catalytic systems have brought new light to catalytic chemistry at the molecular level. In these model systems, model surfaces (e.g., metal single crystals, well-defined metal nanoclusters) and vacuum conditions (instead of the presence of gases and solvents commonly encountered in actual catalysis) are generally used to streamline the investigation.

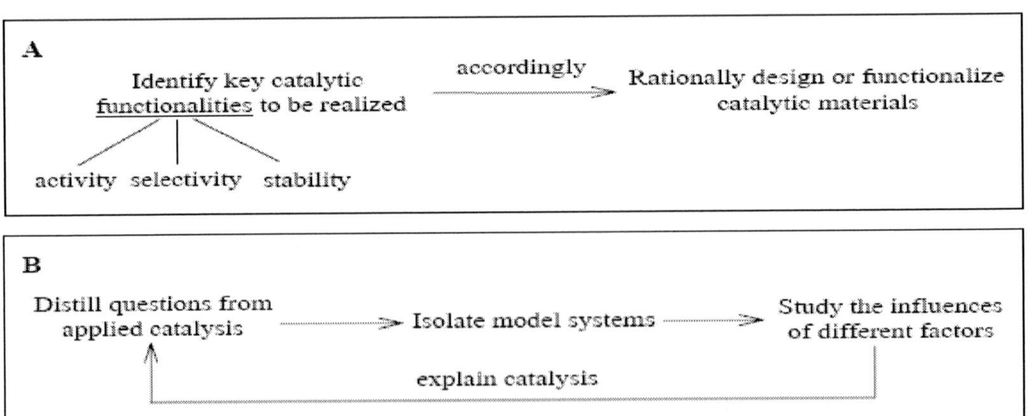

Figure 1. Schematic illustrations of a functionality-oriented strategy in applied catalysis research (panel A) and a catalytic insight-driven strategy in fundamental model studies (panel B).

However, some model systems constituted in surface chemistry studies are still far away from the realistic systems in applied catalysis [51, 52], so we reemphasize here the need for

* E-mail: zhen.ma@email.ucr.edu

designing fundamental studies aiming at answering specific questions in applied catalysis and getting some correlations.

We have been engaged in various catalysis-related projects, such as the decomposition of Freon-12 (CCl_2F_2) on solid acids [53-59], the development of solid acid catalysts for chemical transformation [60], the reaction testing of liquid-phase phenol hydroxylation [61], the in situ probing of liquid-solid interfaces of catalysis relevance [62-70], and the oxidation of CO on supported gold nanoparticles [71-80], in the last ten years. The research mentioned above has been conducted in the laboratories led by Gao [53-61], Zaera [62-70], and Dai [71-80]. In this chapter, we put some results from these projects in a new light by grouping our research into applied catalysis (Section 2) and fundamental model studies (Section 3) and highlighting the aforementioned research strategies. We then give a short discussion on the relation between these two research approaches in Section 4. Finally, in Section 5 we introduce a few review articles on heterogeneous catalysis and surface chemistry of catalysis relevance written by us for further reading [81-87]. This chapter is not intended to be comprehensive, because the progresses made in the areas of applied catalysis and fundamental model studies are certainly enormous. Rather, this chapter provides a personal reflection on our key research philosophies embedded in our work. Although these systems seem to be divergent and examples scattered, we hope that the commonality among these systems and examples could be appreciated.

2. Studies on Functionality-Oriented Applied Catalysis

2.1. Catalytic Decomposition of Freon-12 on Solid Acids

Freons (chlorofluorocarbons) have been widely used as coolants in refrigerators since the 1930s. However, they were determined to be the major cause of ozone layer depletion and were banned in many countries [88]. To properly dispose Freons in abandoned refrigerators, cylinders, and tanks, they may be either converted to alternative coolants via catalytic hydrodechlorination [89, 90] or decomposed to less harmful and easier-to-handle substances [59, 91]. Catalytic decomposition in the presence of water vapor (e.g., $CCl_2F_2 + 2H_2O \rightarrow CO_2 + 2HCl + 2HF$) followed by trapping CO_2, HCl, and HF by an aqueous base provides such a promising option, but the challenges encountered are low catalytic activity and acute catalyst deactivation during the course of the reaction. Therefore, the key functionalities identified in this case are to improve the catalytic activity and stability. It seems that active catalysts reported for the decomposition of Freons are often solid acids [92-97], and the catalysts containing SiO_2 or Al_2O_3 generally deactivate quickly due to the corrosion of catalysts by HF and HCl liberated during the reaction [92, 96]. Therefore, to realize the desired functionalities, one may make a point of using solid acids and avoiding the use of the SiO_2 and Al_2O_3-containing catalysts.

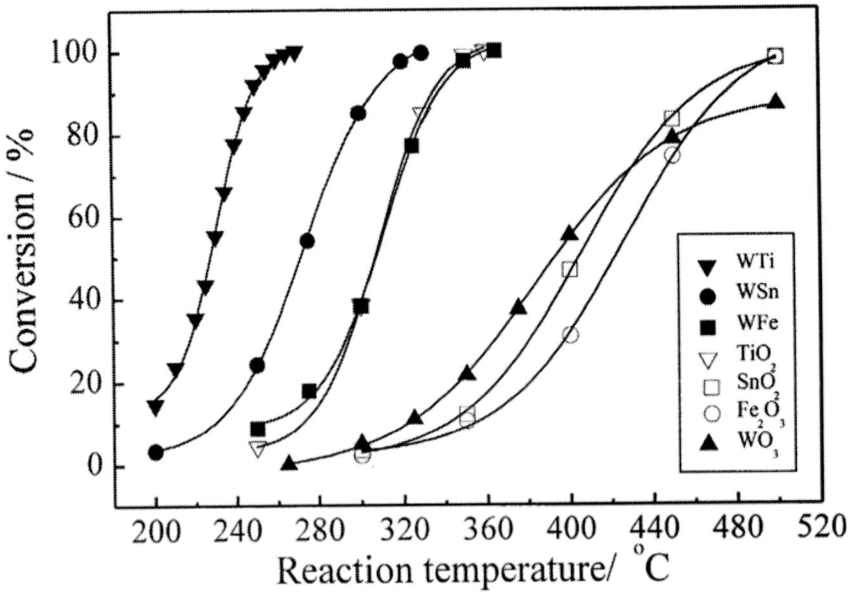

Figure 2. Activity of TiO2, SnO2, Fe2O3, WO3, WO3/TiO2 (WTi), WO3/SnO2 (WSn), and WO3/Fe2O3 (WFe) in Freon decomposition [55]. Reaction conditions: 1000 ppm CCl2F2; 6000 ppm water vapor; balance air; space velocity: 10000 h-1; 0.4 g catalyst. [Reproduced by permission of Elsevier from Ma, Z.; Hua, W. M.; Tang, Y.; Gao, Z. *J. Mol. Catal.* A 2000, 159, 335-345.].

With these criteria in mind, we searched the relevant reviews on solid acids [98-100], and efficiently located effective catalysts after trying some solid acids. As shown in Figure 2 [55], neat TiO_2, SnO_2, and Fe_2O_3 are able to catalyze the decomposition of Freon-12 (CCl_2F_2) at relatively high reaction temperatures, whereas the modification of TiO_2, SnO_2, and Fe_2O_3 surfaces by dispersed WO_3 significantly decreases the reaction temperature required to decompose CCl_2F_2. The superior activity of WO_3-modified metal oxides is owing to their enhanced acidity (in terms of the amount of medium-strong acid sites) [98-100], as evidenced by the NH_3-TPD data in Figure 3 [55]. The stability of these catalysts is quite good under our experimental conditions: there is no observable catalyst deactivation within 120 h on stream. The modification of TiO_2, SnO_2, and Fe_2O_3 surfaces by SO_4^{2-} leads to similar benefits [56], because these SO_4^{2-}-modified metal oxides are also solid acids with plenty of medium-strong acid sites [98-100] and neat TiO_2, SnO_2, and Fe_2O_3 supports themselves do not undergo deactivation on stream [56, 101]. This example shows that the rational functionalization of metal oxide surfaces can offer better catalysts for the decomposition of CCl_2F_2.

Takita and coworkers discovered that some metal phosphates, such as $AlPO_4$, can also catalyze the decomposition of CCl_2F_2 [102-104]. One rationale behind their choosing of catalysts is that according to their thermodynamic calculation, metal phosphates are less likely to react with HF to form metal fluorides than the corresponding metal oxides. However, the activities of metal phosphates in the decomposition of CCl_2F_2 are not particularly high, because their acid properties are not particularly superior [105]. It is known in the literature that some metal sulfates are also solid acids generally stronger than the corresponding metal phosphates [98]. In addition, metal sulfates are even less likely to react with HF than the corresponding metal phosphates according to our thermodynamic calculation.

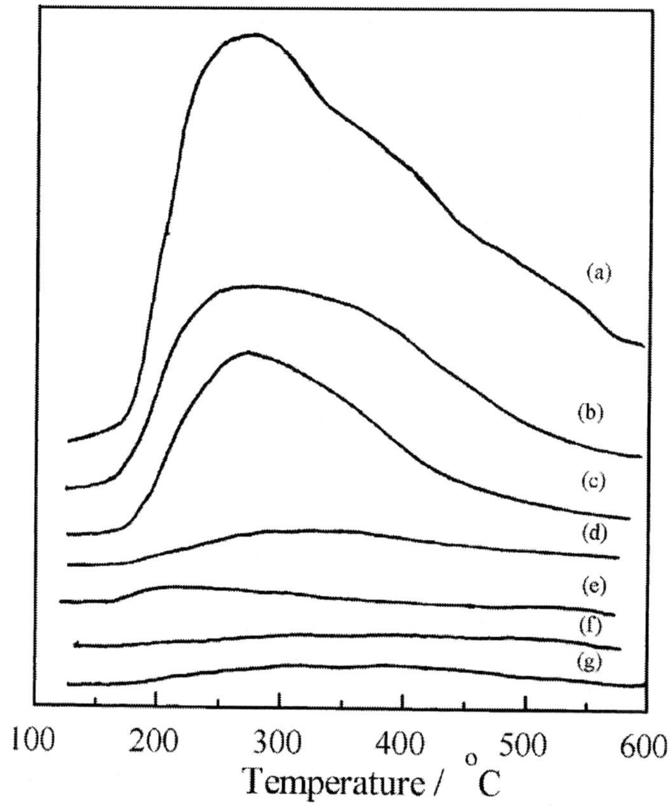

Figure 3. NH_3-TPD profiles of (a) WO_3/TiO_2, (b) WO_3/SnO_2, (c) WO_3/Fe_2O_3, (d) TiO_2, (e) SnO_2, (f) Fe_2O_3, and (g) WO_3 used for Freon decomposition [55]. Catalyst load: 0.2 g. [Reproduced by permission of Elsevier from Ma, Z.; Hua, W. M.; Tang, Y.; Gao, Z. *J. Mol. Catal. A* 2000, *159*, 335-345.].

Indeed, we demonstrated that titanium sulfate, zirconium sulfate, and tin sulfate are among the most active metal sulfates for the decomposition of CCl_2F_2, and they are reasonably stable on stream [57, 58]. Aluminum sulfate, iron sulfate, cerium sulfate, and nickel sulfate show moderate activity, whereas calcium sulfate and copper sulfate are almost inactive even when the reaction temperature is as high as 400°C. The reactivity trend observed in the decomposition of CCl_2F_2 on these metal sulfates is consistent with that observed in the dehydration of isopropanol to propene, an index reaction used to probe the acidity of solid catalysts [57]. After the publication of our papers [57, 58], Takita and coworkers also reported the decomposition of CCl_2F_2 on a series of metal sulfates [106, 107]. Our example again reveals that the rational choosing of catalysts considering the desirable catalytic functionalities and the database in the literature is an efficient way in catalysis research.

2.2. Designing Novel Tio$_2$-Based Gold Catalysts with Improved Performance

Since the pioneering finding by Haruta and coworkers that finely divided gold nanoparticles are very active for CO oxidation below room temperature, catalysis by gold nanoparticles has been one of the hot topics in catalysis research [108-111]. The majority of research in this area involves the use of gold nanoparticles supported on neat metal oxides, such as TiO_2, Al_2O_3, CeO_2, Fe_2O_3, and SiO_2 [108-111]. Neat metal oxide supports are frequently utilized because they are commercially available and it is relatively easier to isolate the structure-function correlations if neat supports (as opposed to mixed metal oxide supports) are used. Au/TiO_2 is very active for CO oxidation below room temperature and is so far the most studied gold catalyst, but it suffers from the sintering of gold nanoparticles and the significant decrease in activity after high-temperature treatment [112, 113]. Therefore, the functionalities to be realized here are to stabilize gold nanoparticles and to maintain their high activity after severe thermal treatment. These requirements are crucial for the applications of gold catalysts in car emission control and catalytic combustion of volatile organic compounds [114].

Supposing that the metal-support interaction may be tuned if the TiO_2 support is modified by another metal oxide, Yan et al. adopted a surface-sol-gel method to functionalize TiO_2 surface with one "monolayer" of Al_2O_3, and then loaded gold nanoparticles onto Al_2O_3-promoted TiO_2 via a deposition-precipitation method (Figure 4) [115, 116]. The surface-sol-gel method in that case consists of two half-reactions: (1) non-aqueous condensation of Al(sec-OC_4H_9)$_3$ with surface Ti-OH groups followed by non-aqueous washing to remove the unreacted aluminum alkoxide and (2) aqueous hydrolysis of the anchored aluminum alkoxide to regenerate surface hydroxyls (Al-OH groups). Yan et al. observed that after pretreating Au/Al_2O_3/TiO_2 in O_2-He at 500°C, the gold nanoparticles are still relatively small and the activity in CO oxidation is still high, namely, both as-synthesized and 500°C-pretreated Au/Al_2O_3/TiO_2 catalysts are quite active below room temperature. In contrast, the activity of Au/TiO_2 decreases dramatically because of significant sintering of gold nanoparticles after such thermal pretreatment [115, 116]. This example nicely reflects the spirit of the sol-called "molecular engineering of functional systems".

If what matters is that Al_2O_3 is homogeneously dispersed on TiO_2, then in principle, it should not matter whether the Al_2O_3 additive is derived from the surface-sol-gel method or another appropriate method. With this thought in mind, we then prepared Al_2O_3/TiO_2 via a conventional impregnation method using $Al(NO_3)_3$ as the precursor [72]. The impregnated $Al(NO_3)_3$ decomposes to amorphous Al_2O_3 on a commercial TiO_2 support (Degussa P25) upon calcination (Figure 5) [72]. We found that Au/Al_2O_3/TiO_2 catalysts prepared that way show catalytic performance comparable to that of Au/Al_2O_3/TiO_2 reported by Yan et al. [115, 116]. This discovery not only independently confirms the finding by Yan et al. [115, 116], but also argues that the surface-sol-gel method is not mandatory. Compared with the surface-sol-gel method, the preparation method employed by us involves simple steps, avoids non-aqueous environments, uses water instead of hazardous organic solvents (toluene and methanol), avoids the use of expensive metal alkaloids, and thus represents a step closer to the large-scale synthesis.

The next logical thinking is that, if the Al_2O_3 additive works, then some other metal oxides might also work. To check this prediction, we tried many metal oxide additives systematically, and found that the addition of some main-group, transition, and rare-earth metal oxides (M_xO_y, M = Ca, Ni, Zn, Ga, Y, Zr, La, Pr, Nd, Sm, Eu, Gd, Dy, Ho, Er, Yb) can indeed maintain the thermal stability of gold catalysts and thus their high activity after high-temperature treatment, whereas other metal oxide additives do not lead to obvious promotional effects.

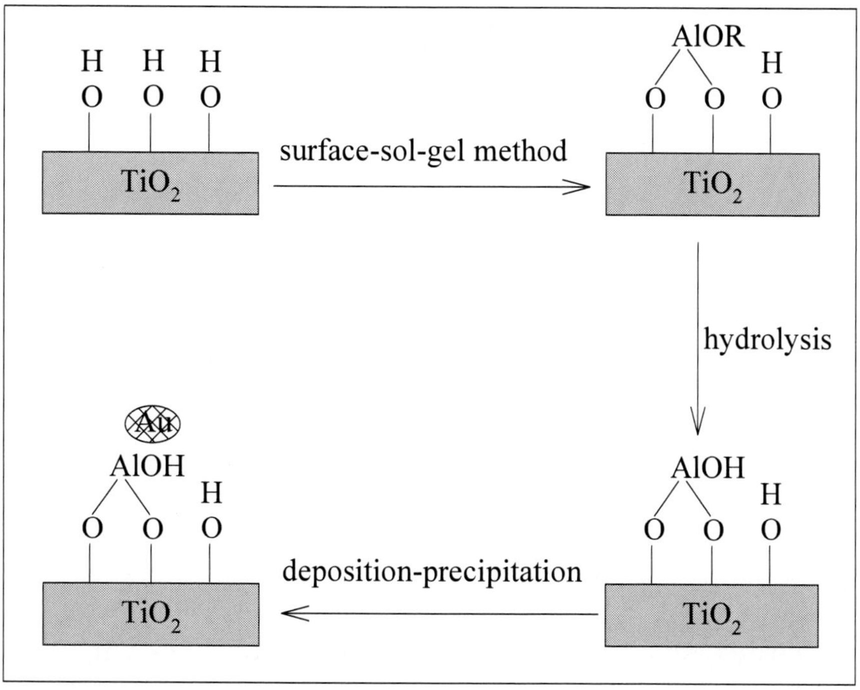

Figure 4. The modification of TiO2 support by Al2O3 using a surface-sol-gel method for loading gold nanoparticles via deposition-precipitation [115]. Al(sec-OC4H9)3 was used as the precursor. The scheme is drawn according to the description in Ref. [115].

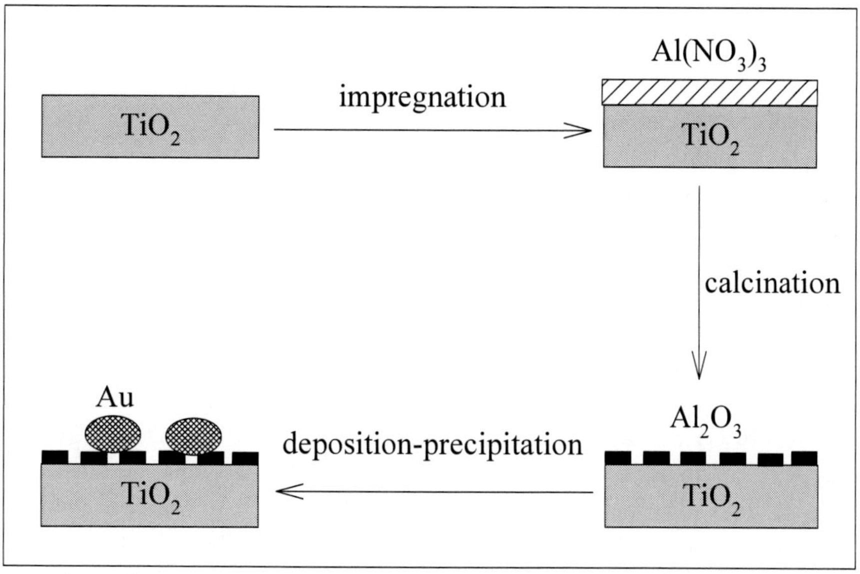

Figure 5. The modification of TiO2 support by Al2O3 using an impregnation method for loading gold nanoparticles via deposition-precipitation [72]. Al(NO3)3 was used as the precursor. The scheme is drawn according to the description in Ref. [72].

Figures 6 and 7 [72] show the CO conversion on different catalysts after catalyst pretreatment at 200 and 500°C, respectively. The intention of the catalytic experiments following catalyst pretreatment at 200°C is to see whether these catalysts are highly active below room temperature or not, whereas the objective of the reaction testing carried out after catalyst pretreatment at 500°C is to see whether these catalysts are still active after severe thermal aging. This example shows that the functionalization of TiO_2 support by suitable metal oxide additives can lead to gold catalysts more active for CO oxidation. The promotional effects established in this case may originate from the stabilization of gold nanoparticles [45], the supply for reactive oxygen by metal oxide additives, and the perturbation of the oxidation state of gold.

2.3. Designing Novel Carbon-Based Gold Catalysts with Improved Performance

Compared with metal oxides, carbon is less commonly used for loading gold [108-111]. It was reported that the activity of Au/C in CO oxidation is low [117-121], and gold particles on carbon supports are big [122]. We synthesized Au/C by loading gold onto graphite via deposition-precipitation. We found that its activity in CO oxidation is indeed low, and the X-ray diffraction (XRD) peaks of gold are sharp, indicating that the gold particles are big (Figure 8) [75]. According to calculation based on the sharpness of the XRD peak at $2\theta = 38^o$, the gold particle size of as-synthesized Au/C is 22 nm. However, the XRD results may be deceptive, because gold nanoparticles smaller than 3 nm can hardly be detected by XRD and the sharpness of the XRD peaks may be biased by a portion of bigger gold particles [123]. Indeed, our dark-field transmission electron microscopy (TEM) characterization of the catalyst collected after reaction revealed that in addition to occasionally observed gold chunks, there are numerous gold nanoclusters (bright dots in the TEM image) well dispersed on carbon (graphite) surfaces (Figure 9, panel A) [75]. Therefore, we speculated that maybe neat carbon cannot activate molecular oxygen for CO oxidation, and the key functionality to be realized here is to put something that can activate molecular oxygen and make the catalyst active for CO oxidation.

To that end, we modified a carbon support by MnO_x via an electroless deposition method introduced by electrochemistry workers to make high-performance capacitors [124-127]. The MnO_x additive is deposited onto the carbon surface when carbon is immersed into an aqueous $KMnO_4$ at room temperature, via a self-limiting surface reaction ($4KMnO_4 + 3C + 2H_2O \rightarrow 4MnO_2 + 3CO_2 + 4KOH$) [126]. The resulting Au/MnO_x/C catalyst is much more active than Au/C for CO oxidation (Figure 8, panel A) [75]. We performed scanning electron microscopy and energy-dispersive X-ray emission analysis (SEM-EDX) experiments to confirm the presence of Au-MnO_x interface (Figure 10) [75]. In addition, we found that MnO_x/Au/C made by soaking Au/C in an aqueous $KMnO_4$ is less active than Au/MnO_x/C, but is more active than Au/C and MnO_x/C (Figure 8, panel A) [75]. Furthermore, Au/MnO_x is even more active than Au/MnO_x/C (data not shown here). Therefore, the Au-MnO_x interface is important for improving the activity in CO oxidation. From this study we know that the carbon-based gold catalysts can also be active for gas-phase CO oxidation, given that the carbon surface is functionalized by suitable additives.

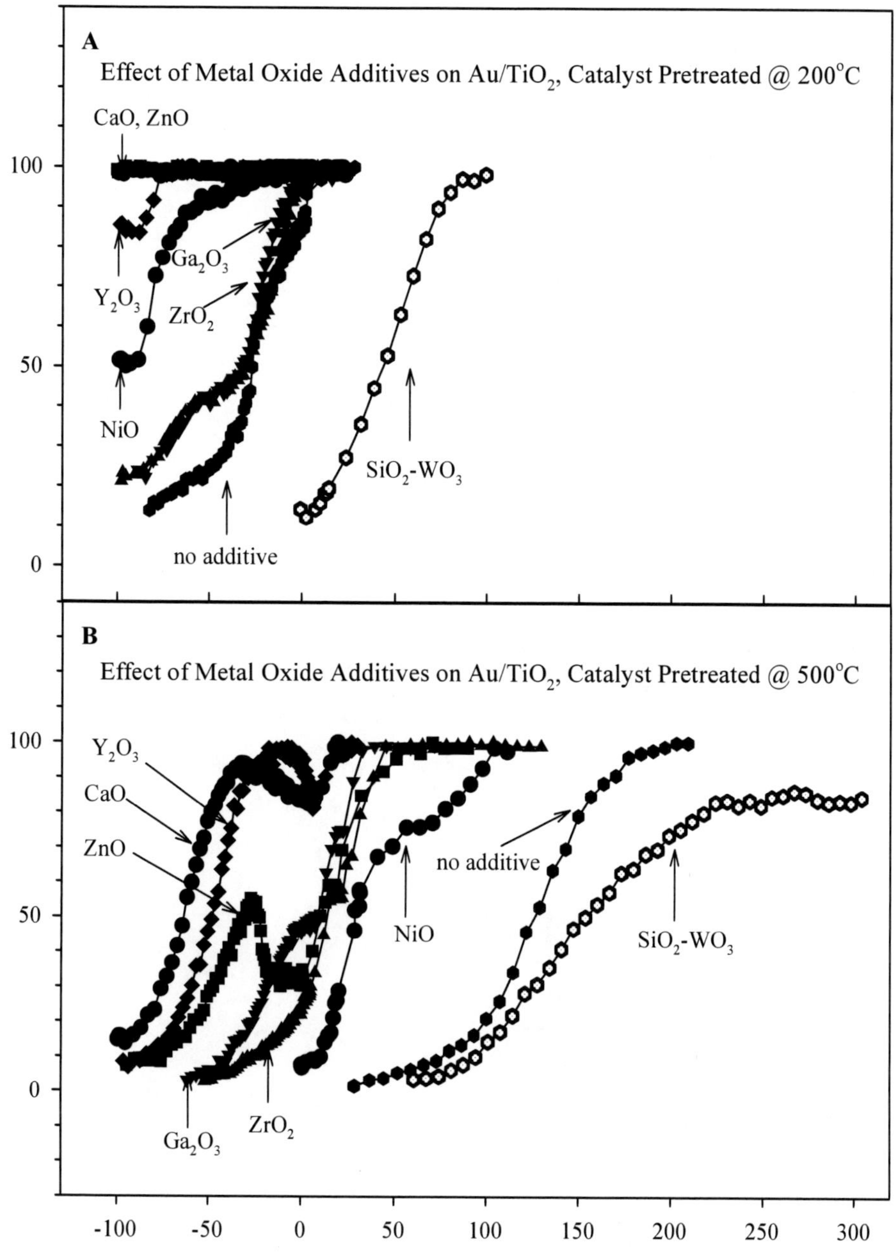

Figure 6. Au/Ga2O3/TiO2, Au/Y2O3/TiO2, Au/ZrO2/TiO2, and Au/SiO2-WO3/TiO2 in CO oxidation [72]. They were pretreated in O2-He at either 200oC (A) or 500oC (B) before the reaction. Reaction conditions: 1% CO in air; flow rate: 37 mL/min; 0.05 g catalyst. [Reproduced by permission of Elsevier from Ma, Z.; Overbury, S. H.; Dai, S. *J. Mol. Catal. A* 2007, 273, 186-197.]

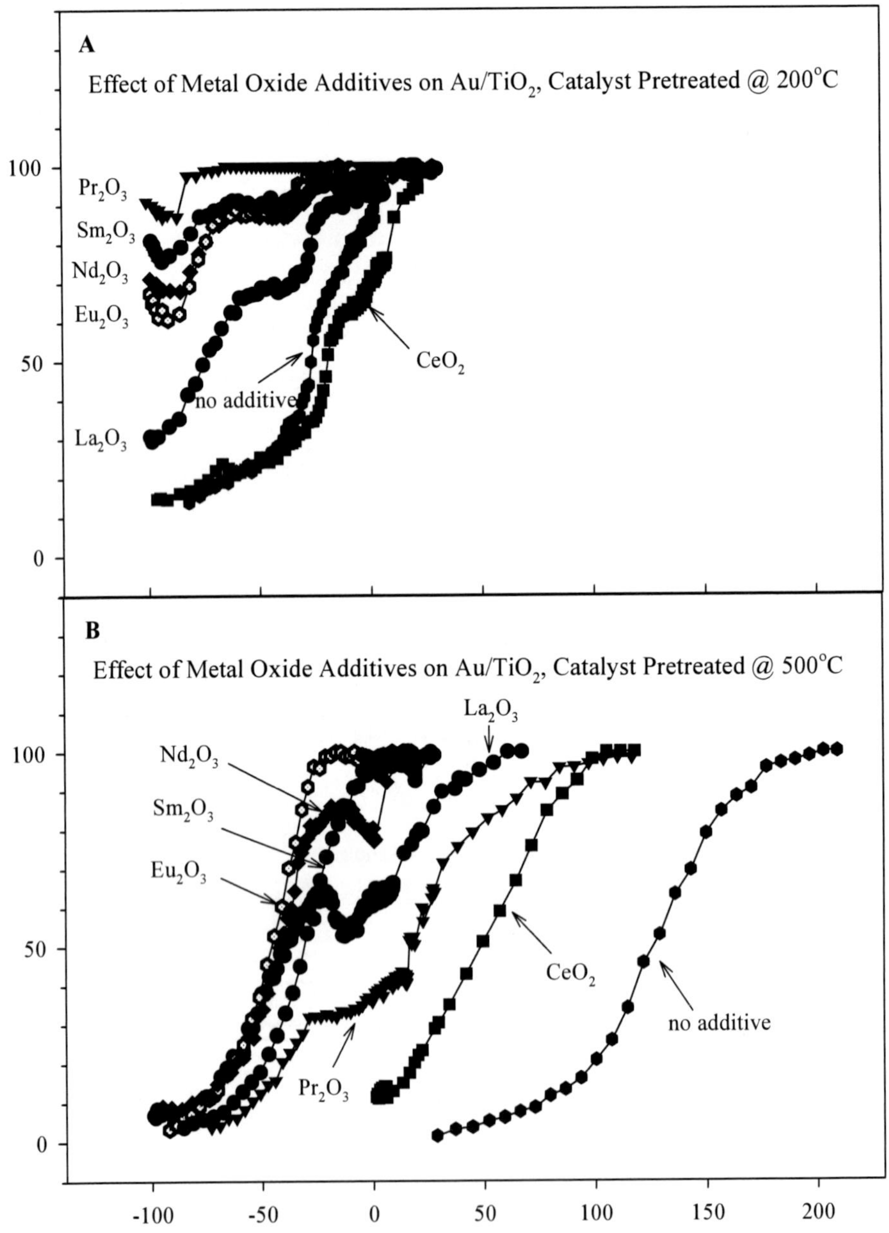

Figure 7. Activity of Au/TiO2, Au/La2O3/TiO2, Au/CeO2/TiO2, Au/Pr2O3/TiO2, Au/Nd2O3/TiO2, Au/Sm2O3/TiO2, and Au/Eu2O3/TiO2 in CO oxidation [72]. They were pretreated in O2-He at either 200oC (A) or 500oC (B) before the reaction. Reaction conditions: 1% CO in air; flow rate: 37 mL/min; 0.05 g catalyst. [Reproduced by permission of Elsevier from Ma, Z.; Overbury, S. H.; Dai, S. *J. Mol. Catal. A* 2007, 273, 186-197.]

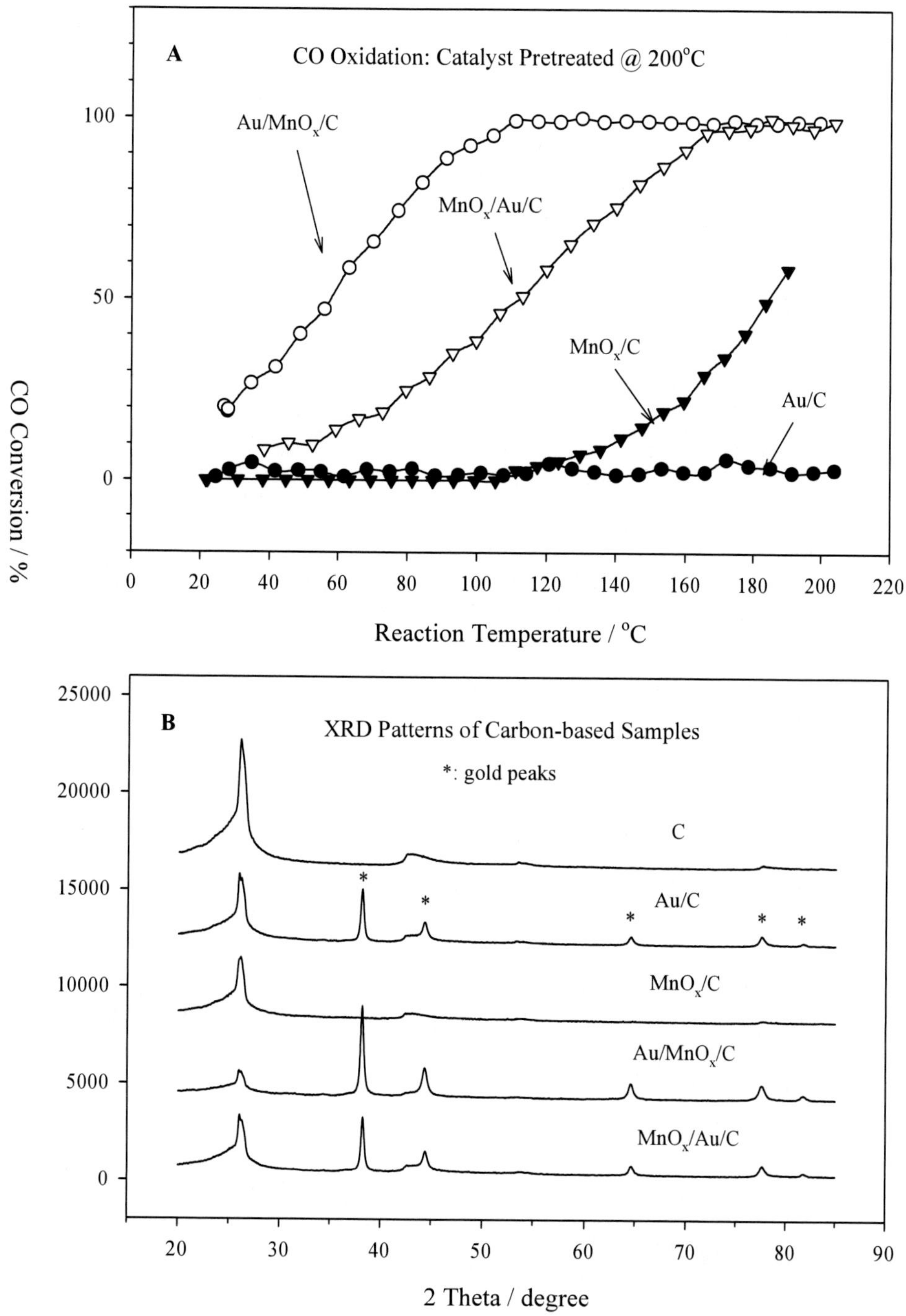

Figure 8. Activity of carbon-based gold catalysts in CO oxidation and their XRD results [75]. Panel A: the conversions on Au/C, Au/MnO$_x$/C, MnO$_x$/C, and MnO$_x$/Au/C pretreated in O_2-He at 200°C. Reaction conditions: 1% CO in air; flow rate: 37 mL/min; 0.05 g catalyst. Panel B: XRD patterns of as-synthesized catalysts. [Reproduced by permission of Elsevier from Ma, Z.; Liang, C. D.; Overbury, S. H.; Dai, S. *J. Catal.* 2007, *252*, 119-126.]

200°C-pretreated Au/C

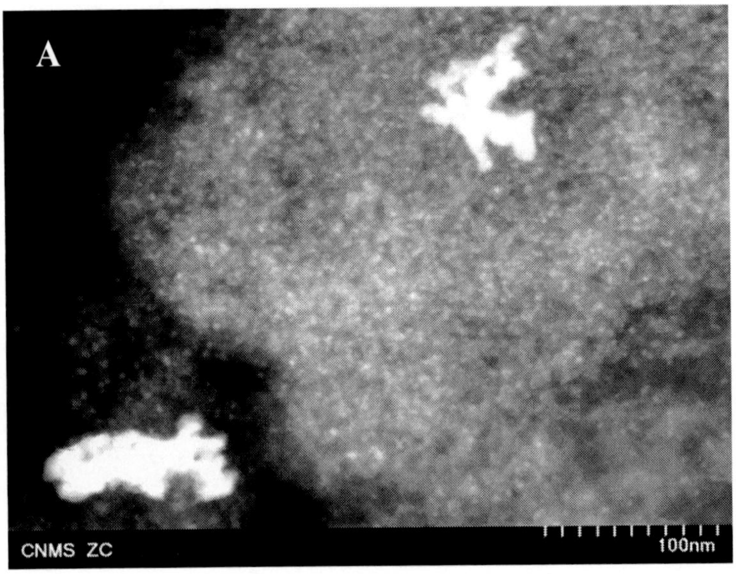

200°C-pretreated Au/MnO_x/C

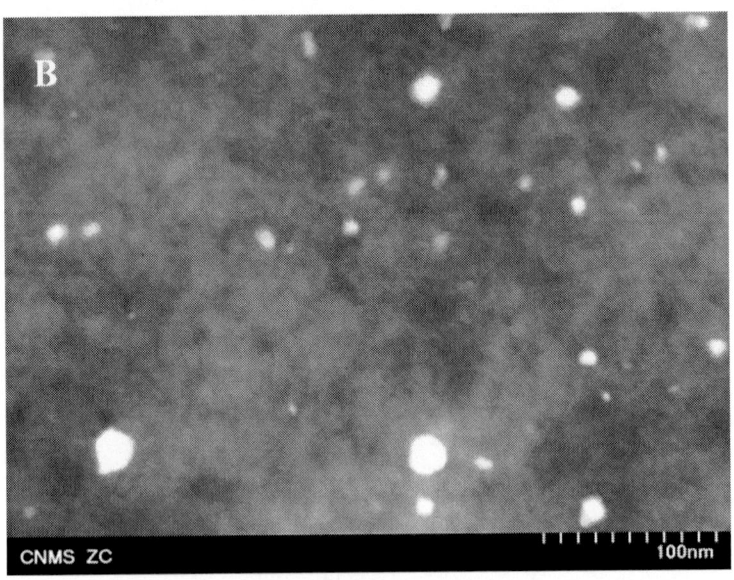

Figure 9. Dark-field TEM images of Au/C and Au/MnOx/C [75]. These catalysts were pretreated in O2-He at 200oC, tested in CO oxidation, and then collected for TEM analysis. [Reproduced by permission of Elsevier from Ma, Z.; Liang, C. D.; Overbury, S. H.; Dai, S. *J. Catal.* 2007, 252, 119-126.].

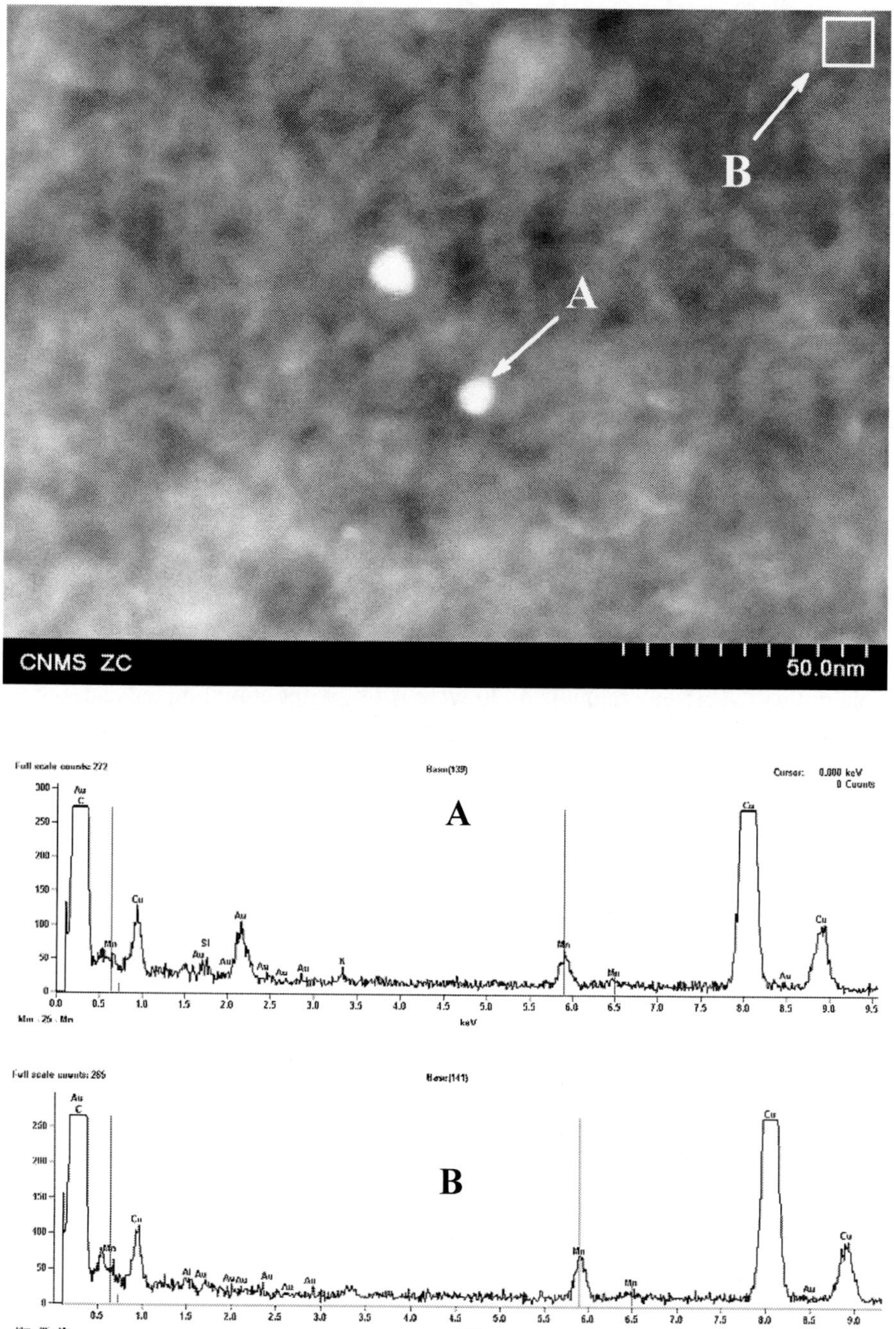

Figure 10. SEM-EDX results of selected areas of Au/MnOx/C [75]. The catalyst was pretreated in O2-He at 200oC, tested in CO oxidation, and then collected for SEM-EDX analysis. [Reproduced by permission of Elsevier from Ma, Z.; Liang, C. D.; Overbury, S. H.; Dai, S. *J. Catal.* 2007, 252, 119-126.].

3. Studies on Catalytic Insight-Driven Model Studies

3.1. Influence of Residual H_2O_2 on the Reaction Testing of Phenol Hydroxylation

Liquid-phase oxidation of phenol by H_2O_2 is an important reaction in industry to produce catechol and hydroquinone, and the most frequently used catalyst is titanium silicate-1 (TS-1) zeolite [128-131]. Many new catalysts have been developed in recent years [132-135]. For instance, Xiong et al. reported phenol hydroxylation in water solvent on $MgFe_2O_4/SiO_2$ [136]. This study promoted us to further investigate the influence of catalyst composition on the catalytic performance. However, when using gas chromatography (GC) to monitor this batch reaction as a function of reaction time, the initial conversion observed should have been fairly close to zero but in reality it is not zero (about 10%). As shown in Figure 11 [61], there are significant differences between the analysis results obtained by GC and high-performance liquid chromatography (HPLC), especially in the initial stage of the reaction when H_2O_2 is not sufficiently consumed (when the conversion of H_2O_2 is low in Figure 11, panel C) [61]. This unusual phenomenon led us to suspect that phenol could react with residual H_2O_2 in GC system, causing the analytical artifacts.

To verify the above-mentioned suspicion, we designed simplified model systems by mixing phenol with different amounts of 30 wt% H_2O_2, without putting any catalyst, and then injecting the solutions individually into GC system. The apparent conversion was indeed observed, and found to increase with the H_2O_2/phenol ratio (Table 1) [61]. Notably, when 4 g phenol is dissolved in 50 mL water and 1.4 g 30 wt% H_2O_2 is added, the apparently conversion in GC system is 10.1% (Table 1) [61], the same as the measured initial conversion in the presence of a catalyst (Figure 11, panel B) [61]. In addition, in our unpublished results, the apparent conversions amount to 40.5, 66.7, and 79.8%, respectively, if 4.2, 8.4, and 12.6 mL 30 wt% H_2O_2 is added into phenol solutions and subject to GC analysis. Therefore, we have every reason to argue that H_2O_2 causes artifacts in the analysis of phenol by GC. Interestingly, in Table 1 [61] the total yield of catechol (CAT), hydroquinone (HQ), and para-benzoquinone (BQ) products is always lower that the conversion of phenol (PHE) measured by using an inner stand (ethanol), implying that CAT, HQ, and BQ are further converted in GC system to products (e.g., CO_2 and tar) that fail to be detected by our GC.

We designed further experiments by preparing CAT + H_2O_2, HQ + H_2O_2, and BQ + H_2O_2 solutions with different concentrations of organic molecules and H_2O_2, and then injecting these solutions individually into GC system. For instance, (Figure 12) [61] shows the analysis data with CAT + H_2O_2 solutions. The conversion of CAT in GC system increases with the concentration of H_2O_2, and decreases with the concentration of CAT. To summarize, the apparent conversion increases with the H_2O_2/CAT ratio. Similar trends are seen with HQ + H_2O_2 and BQ + H_2O_2 mixtures. Unexpectedly, HQ converts in GC system even in the absence of H_2O_2, probably by reacting with trace O_2 coming from the not-perfectly-pure carrier gas (N_2), the solvent (water), and the syringe used to inject the liquid samples [75].

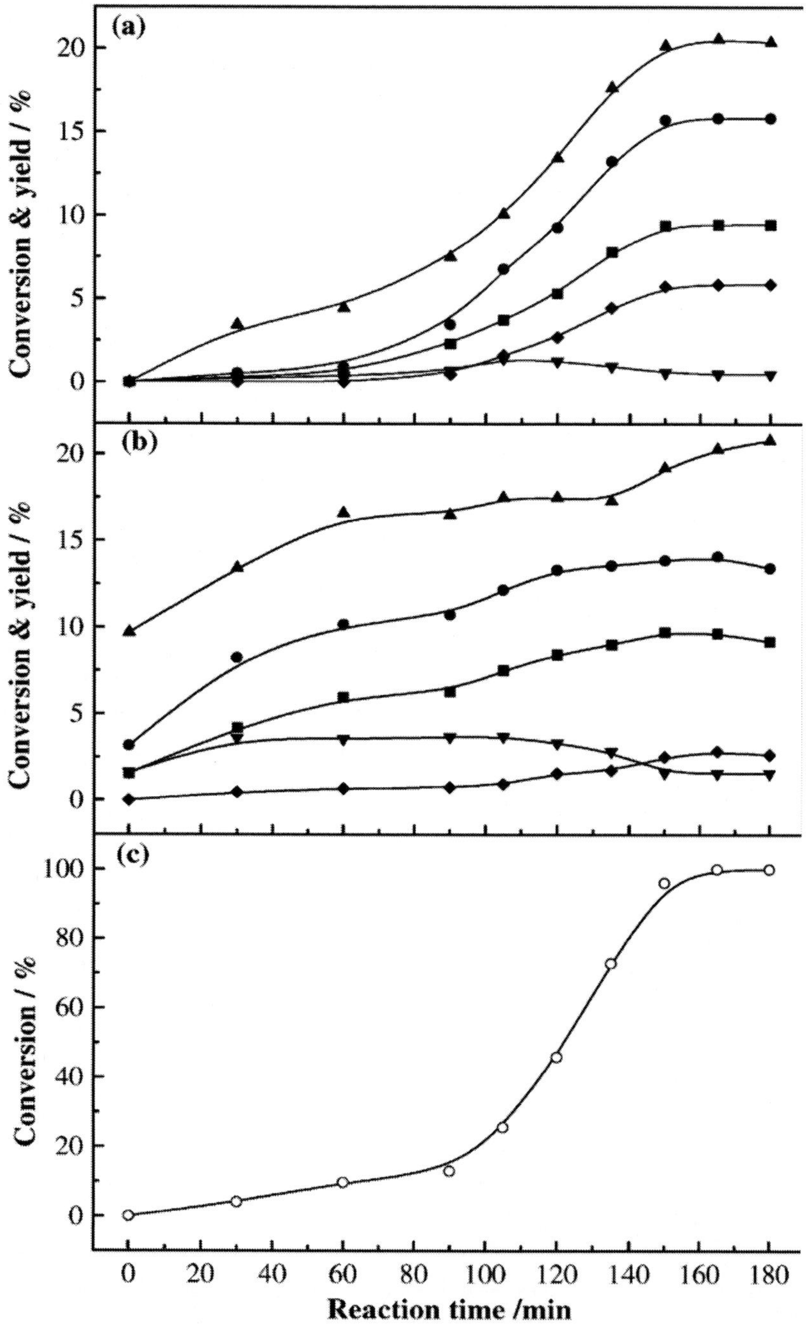

Figure 11. The influence of residual H_2O_2 on the reaction testing of phenol hydroxylation to catechol (CAT), hydroquinone (HQ), and para-benzoquinone (BQ) [61]. Reaction conditions: 4 g phenol; 50 mL water solvent; 0.2 g α-Fe_2O_3 catalyst; inner standard: ethanol; reaction temperature: 70°C. Aliquots were sampled at different times and analyzed by (a) HPLC and (b) GC to determine the conversions of PHE (▲) and yields of CAT + HQ + BQ (●), CAT (■), and BQ (▼). Aliquots were also analyzed by (c) iodometric titration to determine the conversion of H_2O_2 (○). [Reproduced by permission of Elsevier from Ma, N.; Ma, Z.; Yue, Y. H.; Gao, Z. *J. Mol. Catal. A* 2002, *184*, 361-370.]

Table 1. Influence of H_2O_2 on the analysis of phenol by gas chromatography (GC) [61]

H_2O_2 added / mL	X_{PHE} / %	Product yield / %			
		Total	CAT	HQ	BQ
0	0	0	0	0	0
0.05	0	0	0	0	0
0.1	0	0	0	0	0
0.2	0	0	0	0	0
0.5	4.6	0	0	0	0
0.8	5.0	0	0	0	0
1.1	6.5	0.8	0	0	0.8
1.4	10.1	2.8	0.6	0	2.2
1.7	13.3	4.0	1.2	0.1	2.7
2.0	19.3	6.9	2.6	0.3	4.0

* In independent experiments, 4 g phenol (PHE) was dissolved in 50 mL water, the solution mixed with 0-2.0 mL H_2O_2, and injected immediately into GC system to measure the apparent conversion of PHE as well as yields of catechol (CAT), hydroquinone (HQ), and para-benzoquinone (BQ). Ethanol was used as the inner standard. [Reproduced by permission of Elsevier from Ma, N.; Ma, Z.; Yue, Y. H.; Gao, Z. *J. Mol. Catal. A* 2002, *184*, 361-370.]

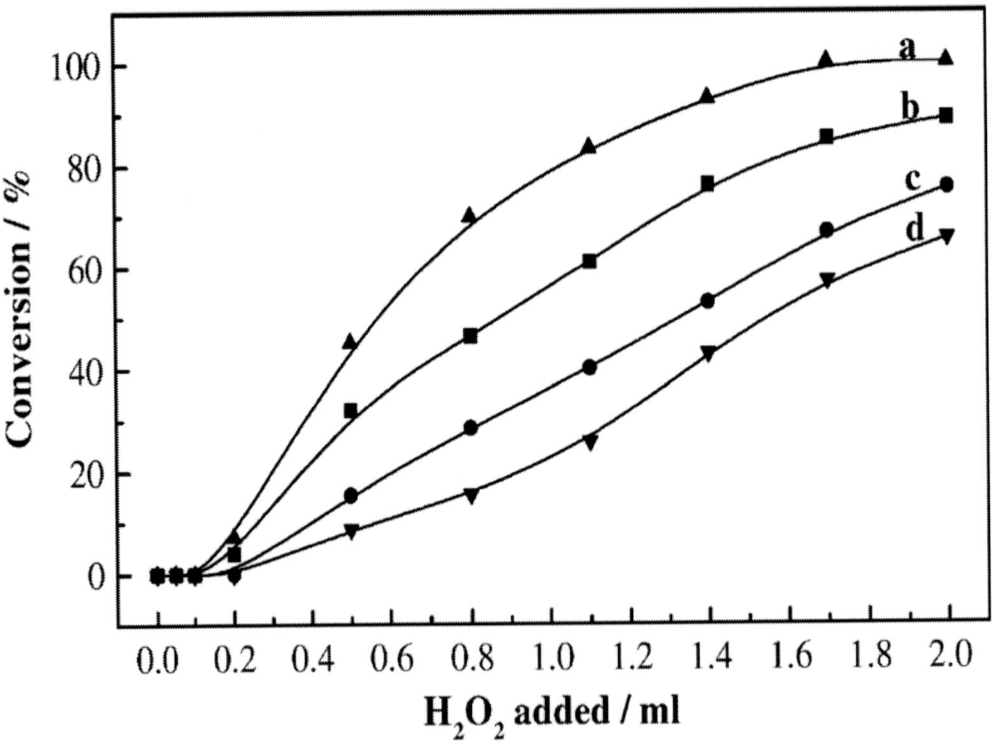

Figure 12. The influence of H2O2 on the analysis of catechol (CAT) by GC [61]. Aqueous CAT solutions (50 mL in each of 40 flasks) with concentrations of (a) 0.04, (b) 0.08, (c) 0.12, and (d) 0.16 mol/L were mixed with different amount of 30 wt% H2O2 (0-2.0 mL), and immediately injected into GC system. Ethanol was used as the inner standard. [Reproduced by permission of Elsevier from Ma, N.; Ma, Z.; Yue, Y. H.; Gao, Z. *J. Mol. Catal.* A 2002, 184, 361-370.]

The above experiments were conducted using water as the solvent. For comparison, if acetone is used as the solvent in the model studies, the apparent conversions of PHE and HQ in GC system are lower than those observed when using water as the solvent, but the apparent conversions of CAT and HQ are still high [61]. The differences with different solvents are not completely understood, but are vaguely called "solvent effects". It should be mentioned that van der Pol et al. previously communicated that at the high temperature of the injection port and the GC column, reactions between CAT and HQ with H_2O_2 could be possible [137]. Therefore, GC is not appropriate for reaction testing of phenol hydroxylation as long as H_2O_2 is present, and the reaction testing results in many publications on phenol hydroxylation deserve reconsideration. Our current example illustrates the isolation of simplified model systems based on unexpected observations in applied catalysis.

3.2. Decarbonylation of Organic Molecules on Platinum Surface

In catalysis, adsorbed CO may retard some reactions such as olefin hydrogenation, fuel cell conversion, and enantioselective hydrogenation. For instance, Lercher and coworkers observed the deactivation of Pt/SiO_2 in the liquid-phase hydrogenation of crotonaldehyde, and ascribed this deactivation to the decomposition of crotonaldehyde on platinum surface to adsorbed CO [138]. Blaser and coworkers found that the addition of a small amount of formic acid decreases the rate of liquid-phase hydrogenation of ethyl pyruvate on cinchonidine-modified Pt/Al_2O_3 catalyst, which they explained as the decomposition of formic acid on the catalyst to adsorbed CO. Interestingly, the addition of acetic acid does not decrease the reaction rate, but whether acetic acid decomposes on the catalyst as formic acid does was not mentioned [139].

To grasp the trend of the decarbonylation of different organic solvents and reactants used in liquid-phase heterogeneous catalysis, we designed model studies by injecting different organic molecules (sometimes with a series of concentrations) individually into an in situ infrared cell with a polycrystalline platinum disk being mounted on a supporting rod inside, and then recording the CO peak intensities after such exposure and H_2-treatment used to clean the platinum surface for reaction and adsorption. Figure 13 [63] depicts a schematic diagram of the experimental setup. Sometimes the infrared cell was flushed by small batches of water to minimize the spectroscopic interference from organic molecules in the solution phase, i.e., to flush away the organic molecules in the solution phase while leaving behind the CO adsorbed on platinum surface. Typical experiments with formic acid, methyl formate, ethyl formate, and ethyl acetate are shown in Figure 14 [64].

A few trends can be generalized from our model studies (Figure 15) [64]. First, organic molecules containing C=O groups (such as aldehydes, carboxylic acids, and esters) more easily decompose to adsorbed CO than those with C-O single bonds (such as alcohols and ethers), because the C=O group, with its higher bond order than C-O, may be a direct precursor to adsorbed CO with the C≡O triple bond [140]. Williams and coworkers, applying attenuated total reflection infrared spectroscopy (AIR-IRS), also found that formaldehyde decomposes much more easily to adsorbed CO on Pt/Al_2O_3 than ethanol does [141]. Second, the substituent groups linked to the carbonyl group in the organics influences the propensity for decarbonylation.

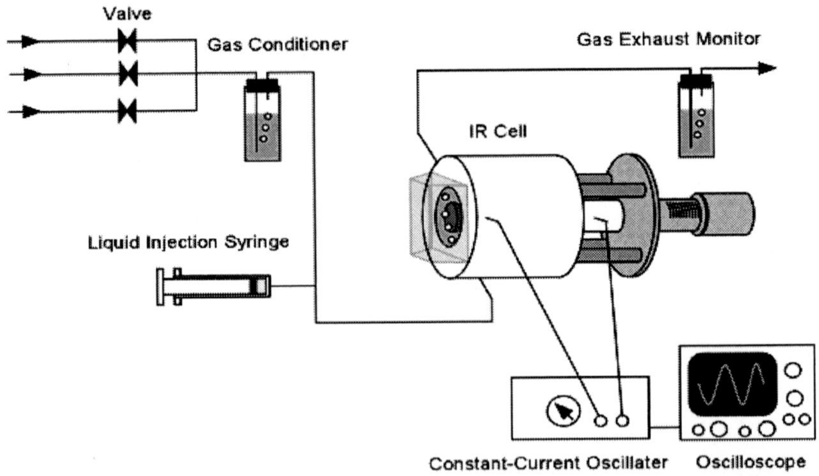

Figure 13. Schematic representation of the setup used for the infrared characterization of liquid-solid interfaces [63]. The main cell consists of a platinum disk used for adsorption and reaction, a CaF_2 prism for guidance of the infrared beam, and a liquid solution trapped between those two elements. The overall arrangement includes gas and liquid sample introduction stages as well as the electronics used for the electrochemical oxidation-reduction cycles needed to preclean the platinum surface. [Reproduced by permission of Elsevier from Ma, Z.; Kubota, J.; Zaera, F. *J. Catal.* 2003, *219*, 404-416.]

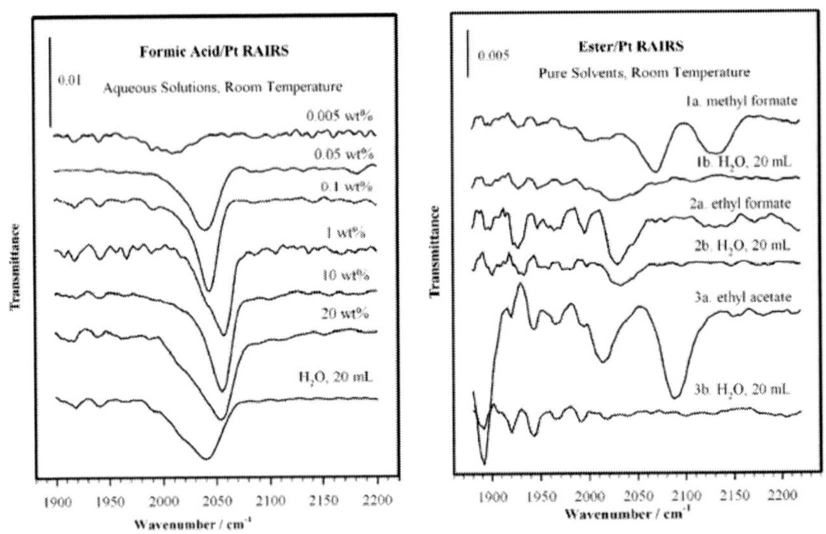

Figure 14. Infrared data summarizing the decomposition of formic acid, methyl formate, ethyl formate, and ethyl acetate on platinum [64]. Panel A: in independent experiments, the platinum disk was exposed to formic acid solutions with different concentrations (0.005-20 wt%). For the last experiment (20 wt% formic acid), the CO-adsorbed platinum disk was further flushed by water. Panel B: the platinum disk was exposed to methyl formate, ethyl formate, and ethyl acetate, respectively, and the CO-adsorbed platinum disk was rinsed by water to remove the spectroscopic interference from the organics. The positions of CO peaks are in the range of 1950-2100 cm-1. [Reproduced by permission of Springer from Ma, Z.; Zaera, F. *Catal. Lett.* 2004, 96, 5-12.]

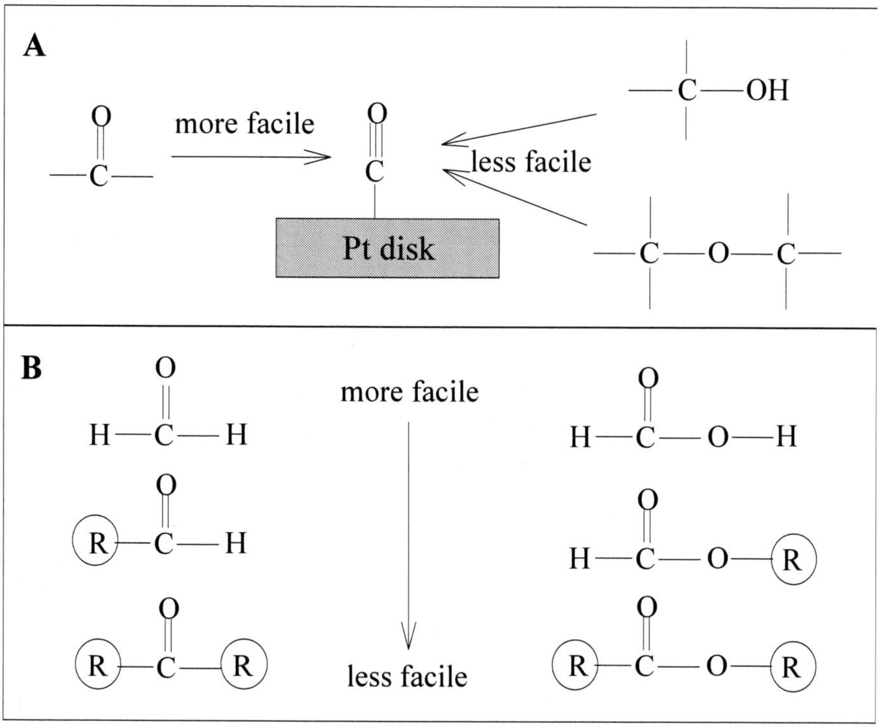

Figure 15. Schematic generalizations of the easiness of decarbonylation of organic molecules on platinum surfaces [64]. The schemes is drawn according to the conclusions in Ref. [64].

Carbonyl compounds with smaller substituents (such as –H and –OH) are more likely to decompose, whereas the propensity for decarbonylation is compromised when increasing the bulkiness of the substituent groups, possibly due to steric effects [142]. For instance, for carboxylic acids and esters, the propensity for decarbonylation follows a formic acid (HCOOH) >> methyl formate ($HCOOCH_3$), ethyl formate ($HCOOC_2H_5$) > acetic acid (CH_3COOH), propionic acid (C_2H_5COOH), acrylic acid (CH_2=CHCOOH), ethyl acetate ($CH_3COOC_2H_5$) sequence. For aldehydes and acetone, the observed trend is formaldehyde (HCHO) >> acetaldehyde (CH_3CHO) > acrolein (CH_2=CHCHO), crotonaldehyde (CH_3CH=CHCHO) > propionaldehyde (C_2H_5CHO), acetone (CH_3COCH_3) [64]. These results not only reasonably justify some empirical observations in the previous catalysis work [138, 139], but also go beyond that in the sense that they involve more molecules and show the trends of decarbonylation with respect to molecular structures.

3.3. Solvent Effect on the Desorption of Chiral Modifier Used in Chiral Hydrogenation

Enantioselective hydrogenation of α-ketoesters on cinchona alkaloid-modified Pt/Al_2O_3 is an interesting system in heterogeneous catalysis [143-146]. The key feature is that on cinchonidine-modified platinum, ethyl pyruvate is selectively hydrogenated to R-ethyl lactate, whereas on cinchonine-modified platinum, S-ethyl pyruvate is the dominant product (Figure 16) [143].

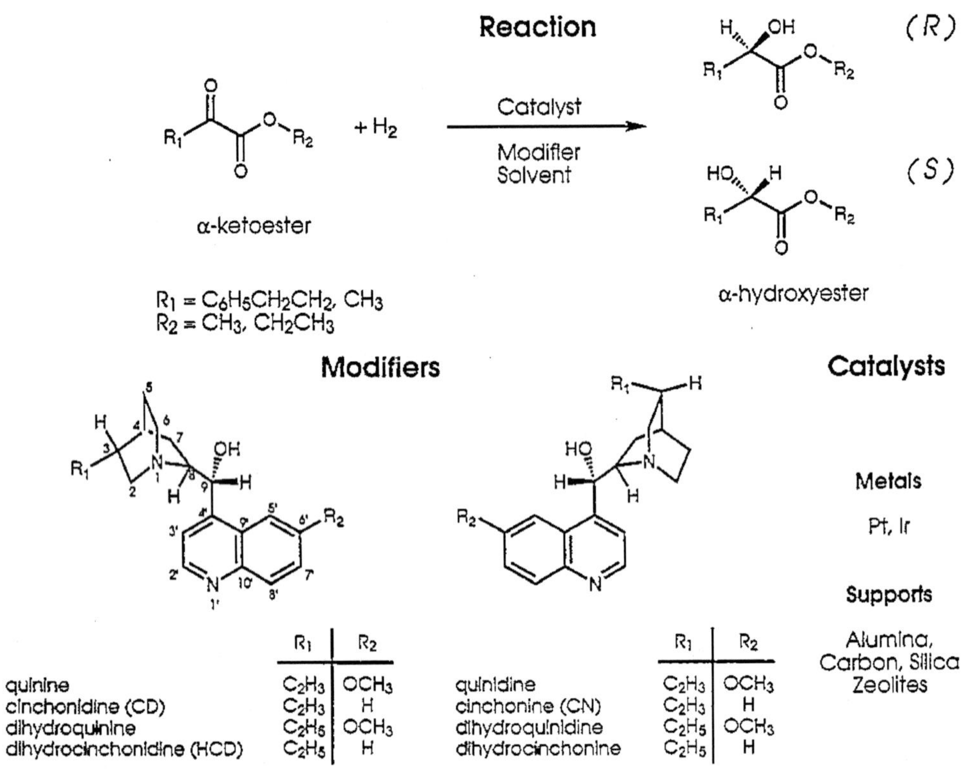

Figure 16. Main features in enantioselective hydrogenation of α-ketoesters on cinchona alkaloid-modified metal catalysts [143]. [Reproduced with permission of Elsevier from Baiker, A. *J. Mol. Catal. A* 1997, *115*, 473-493.]

Theoretical calculations proved that the reaction intermediate leading to R-ethyl lactate on cinchonidine-modified Pt(111) is energetically more stable than the intermediate leading to the S-ethyl lactate [147]. However, the catalytic system is complex and the formation and breaking of intermediates are transient, so it is certainly difficult to obtain direct information spectroscopically. It is therefore advisable to use simplified model systems and investigate each possible pairwise interaction among reactants, products, catalyst, chiral modifier, and solvent separately [147, 148]. In order to constitute these model systems, it is important to get initial inputs from specific catalytic phenomena.

One group of such phenomena is related to the solvent effect in enantioselective hydrogenation. Bartók and coworkers reported that the loss of cinchonidine preimmobilized on clay-supported platinum catalysts can be reduced if toluene is used instead of acetic acid as the solvent [149]. Li and coworkers studied the enantioselective hydrogenation of ethyl pyruvate in a fixed bed reactor [150]. They observed that if the Pt/Al_2O_3 catalyst is pre-modified by cinchonidine and no extra cinchonidine is added during the course of reaction, the enantioselectivity decreases the most obviously when using acetic acid as the solvent, less obviously with ethanol, and the least obviously with toluene [150]. When examining the reaction with different solvents, Blaser and coworkers found that the coverage of cinchonidine adsorbed from a methanol solution onto Pt/Al_2O_3 amounts to only one-eighth of the obtained with toluene as the solvent, although the initial concentration of cinchonidine in methanol is twice that in toluene [151]. It was speculated that a difference in the solvation of cinchonidine in polar versus non-polar solvents may influence the adsorption equilibrium of cinchonidine [152, 153], but neither quantitative solubility data nor spectroscopic evidence was furnished to support their claim.

We then designed model studies by adsorbing cinchonidine from CCl_4 solution onto a polycrystalline platinum disk, and then rinsing the platinum surface with a solvent. The fate of the adsorbed cinchonidine was monitored by reflection-absorption infrared spectroscopy (RAIRS) that probes the adsorbed cinchonidine on the surface. By trying 54 different solvents, we are able to identify two broad trends (Figure 17) [66]. For the first trend, the cinchonidine initially adsorbed at the CCl_4-Pt interface is not easily removed by the second solvent such as cyclohexane, n-pentane, n-hexane, carbon tetrachloride, carbon disulfide, toluene, benzene, ethyl ether, chlorobenzene, and formamide. For the second trend, the initially established adsorption-desorption equilibrium at the CCl_4-Pt interface is obviously perturbed by flushing the system with another solvent such as dichloromethane, ethyl acetate, methanol, ethanol, and acetic acid. These trends can already explain the above-mentioned observations made by catalysis researchers, in the sense that the perturbation of initially established adsorption-desorption equilibrium is related to the nature of the solvent.

The next question is, what physicochemical parameters may influence the adsorption-desorption equilibrium? We suspected that the difference with different solvents may be due to the fact that the solubilities of cinchonidine in different solvents are different, so we tested the solubilities of cinchonidine in 54 solvents, and found that if the initially established adsorption-desorption equilibrium is perturbed, that is because the solubility of cinchonidine in that flushing solvent is relatively big (e.g., 12 g/L in dichloromethane). On the other hand, the adsorption-desorption equilibrium is not perturbed by cyclohexane, because the solubility of cinchonidine in cyclohexane is quite small (0.46 g/L). By plotting the measured cinchonidine solubility versus solvent polarity reported in the literature, nice volcano-like correlations can be identified (Figure 18) [66]. This example shows that some empirical observations in enantioselective hydrogenation may be traced back to basic physicochemical properties such as the solubility of cinchonidine and the polarity of the solvent.

4. Discussion and Perspective

We have highlighted some of our research progresses from the vantage points of applied catalysis and fundamental model studies. These two fields seem to be divergent: whereas the former is aimed at looking for better catalysts with improved activity, selectivity, and stability, the latter inquires into how things work or don't work. However, they are interrelated because applied catalysis research can provide grounds for further model studies, whereas model studies can furnish new insights for understanding catalysis and improving the performance of catalysts. In addition, a common feature is to design experiments with objectives (i.e., to realize catalytic functionalities or to gain fundamental insights) in mind, and then take effective actions to meet these objectives. In more general terms, these goal-oriented strategies are of philosophical meanings, because "goal-oriented" is an important concept in engineering and business as well.

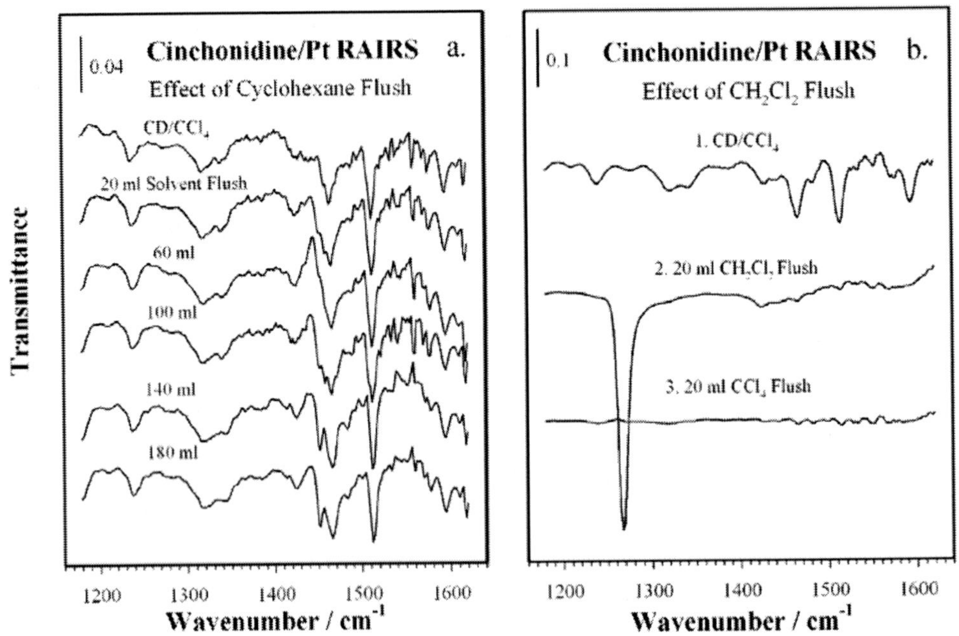

Figure 17. The effect of cyclohexane (A) and dichloromethane (B) solvents on the desorption of cinchonidine (abbreviated as CD) from platinum [66]. In both cases, a clean platinum surface was first exposed to a cinchonidine solution in CCl4 to allow for the adsorption of cinchonidine, and the platinum disk was then exposed to either cyclohexane or dichloromethane. In the case of cyclohexane, a total rinsing with 180 mL in several sequential flushings did not lead to significant change of the infrared spectra. On the other hand, with dichloromethane (B), one flush was sufficient to remove most of the adsorbate. [Reproduced by permission of the American Chemical Society from Ma, Z.; Zaera, F. *J. Phys. Chem. B* 2005, *109*, 406-414.]

Although we emphasized our favorite strategies from personal perspectives, catalysis and fundamental research should not be simply viewed as functionality-driven and catalytic insight-driven, respectively, just as the world should not be perceived as either black or white. Many publications in the field of catalysis are devoted to the scientific understanding of catalyst preparation [154, 155], catalyst characterization [123], reaction kinetics [156], or development of new catalytic processes [157-159], but not necessarily to improving catalytic performance. In addition, for some of those who are working on inorganic materials, the key focus there may be to synthesize novel materials with interesting structures, but not necessarily to improve the catalytic performance in the first place [31-34]. For instance, Wang et al. at Pacific Northwest National Laboratory invented mesoporous crystalline metal oxides via a surfactant-templating methodology, and we then collaborated with each other to see whether gold nanoparticles supported on these mesoporous materials are catalytic active or not [160]. It is certainly difficult to predict whether these new materials are better catalysts or not without putting them into the reactor, and there is always a concern that whether novel catalysts synthesized by demanding and often tedious synthetic methods or using unique nanostructured supports (e.g., nanotubes and nanobelts) are better than those synthesized via conventional methods or using commercialized supports [24, 85].

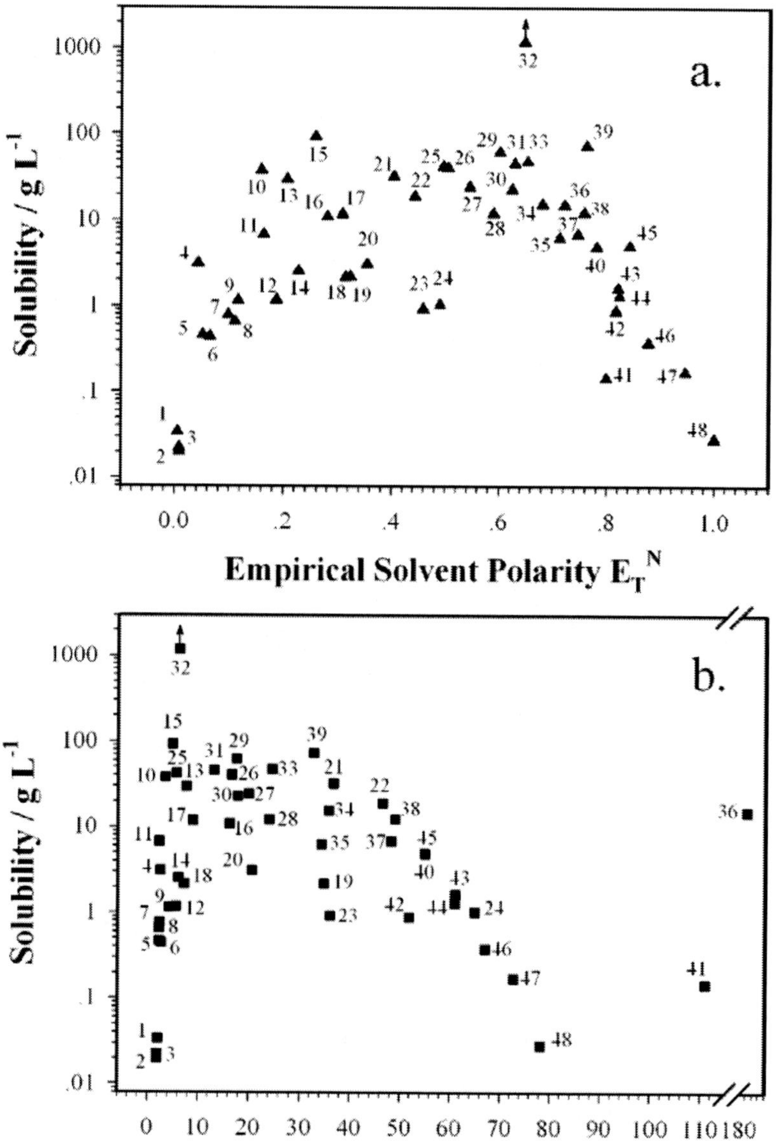

Figure 18. Correlations between the solubility of cinchonidine and the reported empirical polarity (A) and dielectric constants (B) of 48 solvents [66]. Those solvents are indicated by the numbers in the figures: 1: cyclohexane; 2: n-pentane; 3: n-hexane; 4: triethylamine; 5: carbon tetrachloride; 6: carbon disulfide; 7: toluene; 8: benzene; 9: ethyl ether; 10: trichloroethylene; 11: 1,4-dioxane; 12: chlorobenzene; 13: tetrahydrofuran; 14: ethyl acetate; 15: chloroform; 16: cyclohexanone; 17: dichloromethane; 18: ethyl formate; 19: nitrobenzene; 20: acetone; 21: N,N-dimethyl formamide; 22: dimethyl sulfoxide; 23: acetonitrile; 24: propylene carbonate; 25: dioxane (90 wt%)-water; 26: 2-butanol; 27: 2-propanol; 28: acetone (90 wt%)-water; 29: 1-butanol; 30: dioxane (70 wt%)-water; 31: ethyl lactate; 32: acetic acid; 33: ethanol; 34: acetone (70 wt%)-water; 35: dioxane (50 wt%)-water; 36: N-methylformamide; 37: acetone (50 wt%)-water; 38: ethanol (50 wt%)-water; 39: methanol; 40: ethanol (40 wt%-water); 41: formamide; 42: dioxane (30 wt%)-water; 43: ethanol (30 wt%)-water; 44: acetone (30 wt%)-water; 45: methanol (50 wt%)-water; 46: ethanol (20 wt%)-water; 47: ethanol (10 wt%)-water; 48: water. [Reproduced by permission of the American Chemical Society from Ma, Z.; Zaera, F. *J. Phys. Chem. B* 2005, 109, 406-414.]

As for fundamental research, sometimes it is not driven by the existing questions in applied catalysis, but is driven by the fundamental curiosity in materials chemistry [51, 52] or inspired by the results from basic research reported previously. For example, after studying the decarbonylation of organic molecules on platinum surface [64], we wondered that if adsorbed CO is produced on platinum surface, then how to remove it.

Williams and coworkers preliminarily reported that CO oxidation on Pt/Al_2O_3 is faster in the presence of water solvent than in the presence of ethanol [141]. We then studied CO oxidation on platinum surface in the presence of different solvents, and identified obvious solvent effects, namely, CO oxidation takes place the most easily with water solvent, the least easily with carbon tetrachloride solvent, and follows the overall trend of water > ethanol > methanol > cyclohexane > benzene ~ carbon tetrachloride [67]. We subsequently took advantage of the solvent effect to design a diagnosing tool to pin down low-coverage CO at the liquid-solid interface, by flushing the liquid-solid interface with water and carbon tetrachloride individually [67].

Traditionally, the fields of applied catalysis and fundamental research have been advanced by two schools of researchers individually, with limited input from each other. As described by a recent review on heterogeneous enantioselective catalysis [161], "Presently, there are two main directions in catalysis at chiral metal surfaces. A major group of scientists are engaged in the discovery of new synthetic applications, apparently with only marginal interest in the surface science aspects of the reactions; in some cases, it is not clear at all what the role of the solid surface is. One the other hand, surface scientists focus on the fundamental aspects of idealized chiral surfaces, and the catalytic action under practical conditions remains in the background." As we look to the future, it is essential to bridge the gap between these two fields by conducting reaction testing, spectroscopic characterization, and theoretical calculations collectively [162-166]. However, due caution still needs to be exercised when attempting to explain applied catalysis by fundamental data because the parameters used in constructing fundamental model studies may not the same as those in applied catalysis.

5. Further Reading

5.1. Our Reviews on Heterogeneous Catalysis

The chapter "Heterogeneous Catalysis by Metals" [81] in Encyclopedia of Inorganic Chemistry (Second Edition) presents an overview on the basic concepts of catalysis, on the applications of heterogeneous metal catalysis in energy production, chemical manufacturing, materials synthesis, and environmental control, and on several promising directions (i.e., studies on novel catalytic materials, surface science of catalysis, high throughput catalyst testing, and green chemistry). The recent progress in the synthesis of novel advanced structured metal catalysts via encapsulation of metal nanoparticles into oxide shells, immobilization of metal-oxide core-shell structures on solid supports, post-modification of supported metal nanoparticles by surface coating, and pre-modification of supports before loading metal nanoparticles is published elsewhere [87].

The chapter "Characterization of Heterogeneous Catalysts" [82] in Surface and Nanomolecular Catalysis gives an overview on the methods (i.e., structural techniques, adsorption-desorption and thermal techniques, optical spectroscopies, and surface-sensitive spectroscopies) used for the elucidation of the structures, compositions, and chemical properties of both solid catalysts and the adsorbates and reaction intermediates on catalyst surfaces. Figures are selected from the recent literature to show the applications of these techniques.

The chapter "Gold Nanoparticles as Chemical Catalysts" [85] in Inorganic Materials: Inorganic and Bioinorganic Perspectives reviews the synthesis of Au/SiO_2 catalysts by different inorganic synthesis approaches, the pre-modification of supports by metal oxides before loading gold, the post-modification of supported gold catalysts, gold on nanostructured and nanosized supports, and the synthesis of supported gold catalysts using other methods from the perspective of inorganic materials synthesis. A more specific account on the synthesis and catalytic performance of mesoporous materials-supported gold catalysts is given elsewhere [86].

5.2. Our Reviews on Surface Chemistry of Catalysis Relevance

The review "Organic Chemistry on Solid Surfaces" [83] in Surface Science Reports broadly summarizes the chemical reactions of organic molecules on solid surfaces, including metals, metal oxides, carbides, semiconductors, bimetallics, and supported metal particle surfaces, with an emphasis on the reactivity of organic molecules instead of on the structures of the surfaces and adsorbates. Attempts are made to provide representative examples for each family of reactions, to highlight important mechanistic evidence and insights into each surface step, to identify general trends concerning reactivity, and to contrast reactivity on different solids.

The chapter "Chiral Modification of Catalytic Surfaces" [84] in Design of Heterogeneous Catalysts: New Approaches based on Synthesis, Characterization and Modelling summarizes the fundamental research related to the chiral hydrogenation of α-ketoesters on cinchona-modified platinum catalysts and that of β-ketoesters on tartaric acid-modified nickel catalysts. Emphasis is placed on the adsorption of chiral modifiers as well as on the interaction of the modifier and the organic reactant on catalytic surfaces.

6. Conclusion

Applied catalysis and fundamental model studies are important for developing practical catalysts and obtaining basic understanding. One strategy in applied catalysis research is to identify key catalytic functionalities to be actualized or improved, and then rationally design or functionalize catalytic materials. One strategy in fundamental investigation is to distill unanswered questions from applied catalysis, and then isolate simplified model systems in order to answer these questions and even go beyond that. In this chapter, we have briefly outlined our research on the catalytic decomposition of Freon-12, catalysis by gold nanoparticles, reaction testing of phenol hydroxylation, and liquid-solid interfaces of catalysis relevance within the frameworks of these fields and strategies. Although the actual scopes of applied catalysis and fundamental model studies are very wide whereas the current review is mainly limited to our own research, we hope that the reader would find many applications of these highlighted research strategies in the literature. In the future, it is worthwhile to bring the fields of applied catalysis and fundamental studies closer in order to gain comprehensive understanding of catalysis, and there certainly will be more interesting topics to be explored.

Acknowledgments

This invited account is put together mainly based on our published work carried out in the laboratories led by Prof. Zi Gao (Department of Chemistry, Fudan University), Prof. Francisco Zaera (Department of Chemistry, University of California, Riverside), and Dr. Sheng Dai (Chemical Sciences Division, Oak Ridge National Laboratory). I would like to thank them for their dedicated mentoring and continuous encouragement throughout the years. I would also like to thank the research institutions mentioned above for being part of my unfading memory. Many thanks also go to all of our previous coworkers and the authors whose names appear in the references for their collaboration and inspiration, to my spouse, Dr. Lei Shu, for her everlasting support, as well as to the publishers (Elsevier, the American Chemical Society, and Springer) who kindly permitted me to reproduce some published figures. For those of you who wish to contact me in the future, my "permanent" e-mail address is zhen.ma@email.ucr.edu.

References

[1] Cornils, B.; Herrmann, W. A.; Schlögl, R.; Wang, C.-H.; Ed.; *Catalysis from A to Z: A Concise Encyclopedia*; Wiley-VCH: Weinheim, 2000.

[2] Satterfield, C. N. *Heterogeneous Catalysis in Industrial Practice (Second Edition)*; Krieger Publishing: Malabar, FL, 1996.

[3] Hagen, J. *Industrial Catalysis: A Practical Approach (Second Edition)*; Wiley-VCH: Weinheim, 2006.

[4] Ertl, G.; Knözinger, H.; Schüth, F.; Weitkamp, J.; Ed.; *Handbook of Heterogeneous Catalysis (Second Edition)*; Wiley-VCH: Weinheim, 2008.

[5] Bond, G. C. *Heterogeneous Catalysis: Principles and Applications*; Clarendon Press: Oxford, 1974.

[6] Thomas, J. M.; Thomas, W. J. *Principles and Practice and Heterogeneous Catalysis*; VCH: Weinheim, 1996.

[7] Van Santen, R. A.; Van Leeuwen, P. W. N. W.; Moulijin, J. A.; Averill, B. A.; Ed.; *Catalysis: An Integrated Approach*; Elsevier: Amsterdam, 1999.

[8] Chen, A. M.; Xu, H. L.; Yue, Y. H.; Hua, W. M.; Shen, W.; Gao, Z. *Appl. Catal. A* 2004, *274*, 101-109.

[9] Natesakhawat, S.; Oktar, O.; Ozkan, U. S. *J. Mol. Catal. A* 2005, *241*, 133-146.

[10] Kónya, Z.; Molnar, E.; Tsai, K.; Niesz, K.; Somorjai, G. A.; Kiricsi, I. *Catal. Lett.* 2007, *113*, 19-28.

[11] Sato, S.; Takahashi, R.; Yamamoto, N.; Kaneko, E.; Inoue, H. *Appl. Catal. A* 2008, *334*, 84-91.

[12] Wang, W.-J.; Li, H.; Li, H.-X.; Li, Y.-J.; Deng, J.-F. *Appl. Catal. A* 2000, *203*, 301-306.
[13] Xie, S. X.; Wang, J.; He, H. *J. Mol. Catal. A* 2007, *266*, 166-172.
[14] Wang, Q. F.; Wang, L.; Chen, J. X.; Wu, Y. L.; Mi, Z. T. *J. Mol. Catal. A* 2007, *273*, 73-80.
[15] Choudhary, V. R.; Mondal, K. C.; Choudhary, T. V. *Catal. Commun.* 2007, *8*, 561-564.
[16] Hua, W. M.; Xia, Y. D.; Yue, Y. H.; Gao, Z. *J. Catal.* 2000, *196*, 104-114.
[17] Li, H. X.; Luo, H. S.; Zhuang, L.; Dai, W. L.; Qiao, M. H. *J. Mol. Catal. A* 2003, *203*, 267-275.
[18] Wu, J. J.; Zhang, H. L.; Qin, S.; Hu, C. W. *Appl. Catal. A* 2007, *323*, 126-134.
[19] Yao, W.; Lu, G. Z.; Guo, Y. L.; Guo, Y.; Wang, Y. Q.; Zhang, Z. G. *J. Mol. Catal. A* 2007, *276*, 162-167.
[20] Li, H. X.; Wang, W. J.; Li, H.; Deng, J. F. *J. Catal.* 2000, *194*, 211-221.
[21] Chen, A. M.; Xu, H. L.; Yue, Y. H.; Hua, W. M.; Shen, W.; Gao, Z. *J. Mol. Catal. A* 2003, *203*, 299-306.
[22] Xu, B. J.; Zheng, B.; Hua, W. M.; Yue, Y. H.; Gao, Z. *J. Catal.* 2006, *239*, 470-477.
[23] Sandoval, A.; Gomez-Cortes, A.; Zanella, R.; Diaz, G.; Saniger, J. M. *J. Mol. Catal. A* 2007, *278*, 200-208.
[24] Armor, J. N. *Appl. Catal. A* 2005, *282*, 1-4.
[25] Xiao, F.-S.; Han, Y.; Yu, Y.; Meng, X. J.; Yang, M.; Wu, S. *J. Am. Chem. Soc.* 2002, *124*, 888-889.
[26] Collier, P.; Golunski, S.; Malde, C.; Breen, J.; Burch, R. *J. Am. Chem. Soc.* 2003, *125*, 12414-12415.
[27] Yan, W. F.; Chen, B.; Mahurin, S. M.; Hagaman, E. W.; Dai, S.; Overbury, S. H. *J. Phys. Chem. B* 2004, *108*, 2793-2796.
[28] In, S.; Orlov, A.; Berg, R.; Garcia, F.; Pedrosa-Jimenez, S.; Tikhov, M. S.; Wright, D. S.; Lambert, R. M. *J. Am. Chem. Soc.* 2007, *129*, 13790-13791.
[29] Tang, Y. Q.; Gui, L. L.; Ed.; *Going in the Opposite Direction: Studies on Molecular Engineering of Functional Systems*; Hunan Science and Technology Press: Changsha, 1997.
[30] Tang, Y. G.; Xu, R. R. *Top. Catal.* 2005, *35*, 1-2.
[31] Wang, D. S.; Xu, R.; Wang, X.; Li, Y. D. *Nanotechnol.* 2006, *17*, 979-983.
[32] Zhong, L.-S.; Hu, J.-S.; Cao, A. M.; Liu, Q.; Song, W. G.; Wan, L.-J. *Chem. Mater.* 2007, *19*, 1648-1655.
[33] Chen, H. M.; He, J. H.; Zhang, C. B.; He, H. *J. Phys. Chem. C* 2007, *111*, 18033-18038.
[34] Wang, D. S.; Wang, X.; Xu, R.; Li, Y. D. *J. Nanosci. Nanotechnol.* 2007, *7*, 3602-3606.
[35] Gates, B. C.; Knözinger, H.; Ed.; *Impact of Surface Science on Catalysis*; Academic Press: San Diego, CA, 2000.
[36] Zaera, F. *Surf. Sci.* 2002, *500*, 947-965.
[37] Chorkendorff, I.; Niemantsverdriet, J. W. *Concepts of Modern Catalysis and Kinetics*; Wiley-VCH: Weinheim, 2003.
[38] Besenbacher, F.; Chorkendorff, I.; Clausen, B. S.; Hammer, B.; Molenbroek, A. M.; Nørskov, J. K.; Stensgaard, I. *Science* 1998, *279*, 1913-1915.
[39] Linic, S.; Jankowiak, J.; Barteau, M. A. *J. Catal.* 2004, *224*, 489-493.
[40] Nikolla, E.; Holewinski, A.; Schwank, J.; Linic, S. *J. Am. Chem. Soc.* 2006, *128*, 11354-11355.

[41] Nikolla, E.; Schwank, J.; Linic, S. *J. Catal.* 2007, *250*, 85-93.
[42] Ferri, D.; Bürgi, T. *J. Am. Chem. Soc.* 2001, *123*, 12074-12084.
[43] Linic, S.; Barteau, M. A. *J. Am. Chem. Soc.* 2004, *126*, 8086-8087.
[44] Chu, W.; Leblanc, R. J.; Williams, C. T.; Kubota, J.; Zaera, F. *J. Phys. Chem. B* 2003, *107*, 14365-14373.
[45] Wang, C.-M.; Fan, K.-N.; Liu, Z.-P. *J. Phys. Chem. C* 2007, *111*, 13539-13546.
[46] Somorjai, G. A. *Introduction of Surface Chemistry and Catalysis*; John Wiley and Sons: New York, NY, 1994.
[47] Zaera, F. *Prog. Surf. Sci.* 2001, *69*, 1-98.
[48] Wallace, W. T.; Goodman, D. W. In *Surface and Nanomolecular Catalysis*; Richards, R.; Ed.; Taylor and Francis (CRC Press): Boca Raton, FL, 2006; pp 337-371.
[49] Hammer, B.; Nørskov, J. K. *Adv. Catal.* 2000, *45*, 71-129.
[50] Van Santen, R. A.; Neurok, M. *Molecular Heterogeneous Catalysis: A Conceptual and Computational Approach*; Wiley-VCH: Weinheim, 2006.
[51] De Jesús, J. C.; Pereira, P.; Carrazza, J.; Zaera, F. *Surf. Sci.* 1996, *369*, 217-230.
[52] De Jesús, J. C.; Carrazza, J.; Pereira, P.; Zaera, F. *Surf. Sci.* 1998, *397*, 34-47.
[53] Ma, Z.; Hua, W. M.; Tang, Y.; Gao, Z. *Chem. Lett.* 1999, 1215-1216.
[54] Hua, W. M.; Zhang, F.; Ma, Z.; Tang, Y.; Gao, Z. *Catal. Lett.* 2000, *65*, 85-89.
[55] Ma, Z.; Hua, W. M.; Tang, Y.; Gao, Z. *J. Mol. Catal. A* 2000, *159*, 335-345.
[56] Ma, Z.; Hua, W. M.; Tang, Y.; Gao, Z. *Chin. J. Chem.* 2000, *18*, 341-345.
[57] Ma, Z.; Hua, W. M.; Tang, Y.; Gao, Z. *Chin. Chem. Lett.* 2000, *11*, 311-314.
[58] Deng, X. Y.; Ma, Z.; Yue, Y. H.; Gao, Z. *J. Catal.* 2001, *204*, 200-208.
[59] Ma, Z.; Hua, W. M.; Gao, Z. *Chin. Bull. Chem.* 2001, *64*, 339-342.
[60] Ma, Z.; Yue, Y. H.; Deng, X. Y.; Gao, Z. *J. Mol. Catal. A* 2002, *178*, 97-104.
[61] Ma, N.; Ma, Z.; Yue, Y. H.; Gao, Z. *J. Mol. Catal. A* 2002, *184*, 361-370.
[62] Kubota, J.; Ma, Z.; Zaera, F. *Langmuir* 2003, *19*, 3371-3376.
[63] Ma, Z.; Kubota, J.; Zaera, F. *J. Catal.* 2003, *219*, 404-416.
[64] Ma, Z.; Zaera, F. *Catal. Lett.* 2004, *96*, 5-12.
[65] Ma, Z.; Lee, I.; Kubota, J.; Zaera, F. *J. Mol. Catal. A* 2004, *216*, 199-207.
[66] Ma, Z.; Zaera, F. *J. Phys. Chem. B* 2005, *109*, 406-414.
[67] Ma, Z. *J. Colloid Interface Sci.* 2006, *304*, 419-430.
[68] Ma, Z.; Zaera, F. *J. Am. Chem. Soc.* 2006, *128*, 16414-16415.
[69] Ma, Z.; Lee, I.; Zaera, F. *J. Am. Chem. Soc.* 2007, *129*, 16083-16090.
[70] Mink, L.; Ma, Z.; Olsen, R. A.; James, J. N.; Sholl, D. S.; Mueller, L. J.; Zaera, F. *Top. Catal.* 2008, in press.
[71] Zhu, H. G.; Ma, Z.; Clark, J. C.; Pan, Z. W.; Overbury, S. H.; Dai, S. *Appl. Catal. A* 2007, *326*, 89-99.
[72] Ma, Z.; Overbury, S. H.; Dai, S. *J. Mol. Catal. A* 2007, *273*, 186-197.
[73] Ma, Z.; Brown, S.; Overbury, S. H.; Dai, S. *Appl. Catal. A* 2007, *327*, 226-237.
[74] Zhu, H. G.; Ma, Z.; Overbury, S. H.; Dai, S. *Catal. Lett.* 2007, *116*, 128-135.
[75] Ma, Z.; Liang, C. D.; Overbury, S. H.; Dai, S. *J. Catal.* 2007, *252*, 119-126.
[76] Yan, W. F.; Ma, Z.; Mahurin, S. M.; Jiao, J.; Hagaman, E. W.; Overbury, S. H.; Dai, S. *Catal. Lett.* 2008, *121*, 209-218.
[77] Yin, H. F.; Ma, Z.; Overbury, S. H.; Dai, S. *J. Phys. Chem. C* 2008, submitted.
[78] Ma, Z.; Brown, S.; Howe, J. Y.; Overbury, S. H.; Dai, S. *J. Phys. Chem. C* 2008, submitted.

[79] Ma, Z.; Yin, H. F.; Overbury, S. H.; Dai, S. *Abstr. Pap. Am. Chem. Soc.* 2008, in press.
[80] Ma, Z.; Yin, H. F.; Overbury, S. H.; Dai, S. *Abstr. Pap. Am. Chem. Soc.* 2008, in press.
[81] Ma, Z.; Zaera, F. In *Encyclopedia of Inorganic Chemistry (Second Edition)*; King, R. B.; Ed.; John Wiley and Sons: Chichester, 2005; pp 1768-1784.
[82] Ma, Z.; Zaera, F. In *Surface and Nanomolecular Catalysis*; Richards, R. M.; Ed.; Taylor and Francis (CRC Press): Boca Raton, FL, 2006; pp 1-37.
[83] Ma, Z.; Zaera, F. *Surf. Sci. Rep.* 2006, *61*, 229-281.
[84] Ma, Z.; Zaera, F. In *Design of Heterogeneous Catalysts: New Approaches based on Synthesis, Characterization and Modelling*; Ozkan, U. S.; Ed.; Wiley-VCH: Weinheim, 2008; pp in press.
[85] Ma, Z.; Overbury, S. H.; Dai, S. In *Nanomaterials: Inorganic and Bioinorganic Perspectives*; Lukehart, C. M., Scott, R. A.; Ed.; John Wiley and Sons: Chichester, 2008; pp in press.
[86] Ma, Z.; Zhu, H. G.; Yan, W. F.; Overbury, S. H.; Dai, S. In *Nanoporous Materials: Proceedings of the 5th International Symposium*; Sayari, A., Jaroniec, M.; Ed.; World Scientific Publishing: Singapore, 2008; pp in press.
[87] Ma, Z.; Dai, S. *Mater. Technol.* 2008, in press.
[88] Brune, W. *Nature* 1996, *379*, 486-487.
[89] Manzer, L. E.; Rao, V. N. M. *Adv. Catal.* 1993, *39*, 329-350.
[90] Yu, H.; Kennedy, E. M.; Adesina, A. A.; Dlugogorski, B. Z. *Catal. Surv. Asia* 2006, *10*, 40-54.
[91] Takita, Y.; Ishihara, T. *Catal. Surv. Jpn.* 1998, *2*, 165-173.
[92] Tajima, M.; Niwa, M.; Fujii, Y.; Koinuma, Y.; Aizawa, R.; Kushiyama, S.; Kobayashi, S.; Mizuno, K.; Ohûchi, H. *Appl. Catal. B* 1996, *9*, 167-177.
[93] Fu, X. Z.; Zeltner, W. A.; Yang, Q.; Anderson, M. A. *J. Catal.* 1997, *168*, 482-490.
[94] Tajima, M.; Niwa, M.; Fujii, Y.; Koinuma, Y.; Aizawa, R.; Kushiyama, S.; Kobayashi, S.; Mizuno, K.; Ohuchi, H. *Appl. Catal. B* 1997, *12*, 263-276.
[95] Tajima, M.; Niwa, M.; Fujii, Y.; Koinuma, Y.; Aizawa, R.; Kushiyama, S.; Kobayashi, S.; Mizuno, K.; Ohuchi, H. *Appl. Catal. B* 1997, *14*, 97-103.
[96] Ng, C. F.; Shan, S. C.; Lai, S. Y. *Appl. Catal. B* 1998, *16*, 209-217.
[97] Lai, S. Y.; Pan, W. X.; Ng, C. F. *Appl. Catal. B* 2000, *24*, 207-217.
[98] Tanabe, K.; Misono, M.; Ono, Y.; Hattori, H. *New Acids and Bases*; Elsevier: Amsterdam, 1989.
[99] Arata, K. *Adv. Catal.* 1990, *37*, 165-211.
[100] Arata, K. *Appl. Catal. A* 1996, *146*, 3-32.
[101] Karmakar, S.; Greene, H. L. *J. Catal.* 1995, *151*, 394-406.
[102] Takita, Y.; Ninomiya, M.; Matsuzaki, R.; Wakamatsu, H.; Nishiguchi, H.; Ishihara, T. *Phys. Chem. Chem. Phys.* 1999, *1*, 2367-2372.
[103] Takita, Y.; Wakamatsu, H.; Li, G.-L.; Moro-Oka, Y.; Nishiguchi, H.; Ishihara, T. *J. Mol. Catal. A* 2000, *155*, 111-119.
[104] Takita, Y.; Wakamatsu, H.; Tokumaru, M.; Nishiguchi, H.; Ito, M.; Ishihara, T. *Appl. Catal. A* 2000, *194-195*, 55-61.
[105] Zhao, J.; Tian, B. Z.; Yue, Y. H.; Hua, W. M.; Zhao, D. Y.; Gao, Z. *J. Mol. Catal. A* 2005, *242*, 218-223.
[106] Moriyama, J.; Nishiguchi, H.; Ishihara, T.; Takita, Y. *Ind. Eng. Chem. Res.* 2002, *41*, 32-36.

[107] Takita, Y.; Moriyama, J.; Nishiguchi, H.; Ishihara, T.; Hayano, F.; Nakajo, T. *Catal. Today* 2004, *88*, 103-109.
[108] Haruta, M.; Daté, M. *Appl. Catal. A* 2001, *222*, 427-437.
[109] Hashmi, A. S. K.; Hutchings, G. J. *Angew. Chem. Int. Ed.* 2006, *45*, 7896-7936.
[110] Bond, G. C.; Louis, C.; Thompson, D. T. *Catalysis by Gold*; Imperial College Press: London, 2006.
[111] Kung, M. C.; Davis, R. J.; Kung, H. H. *J. Phys. Chem. C* 2007, *111*, 11767-11775.
[112] Wu, S.-H.; Zheng, X.-C.; Wang, S.-R.; Han, D.-Z.; Huang, W.-P.; Zhang, S.-M. *Catal. Lett.* 2004, *96*, 49-55.
[113] Yan, W. F.; Chen, B.; Mahurin, S. M.; Dai, S.; Overbury, S. H. *Chem. Commun.* 2004, 1918-1919.
[114] Thompson, D. T. *Top. Catal.* 2006, *38*, 231-240.
[115] Yan, W. F.; Mahurin, S. M.; Pan, Z. W.; Overbury, S. H.; Dai, S. *J. Am. Chem. Soc.* 2005, *127*, 10480-10481.
[116] Yan, W. F.; Mahurin, S. M.; Overbury, S. H.; Dai, S. *Top. Catal.* 2006, *39*, 199-212.
[117] Okumura, M.; Tsubota, S.; Haruta, M. *J. Mol. Catal. A* 2003, *199*, 73-84.
[118] Bulushev, D. A.; Yuranov, I.; Suvorova, E. I.; Buffat, P. A.; Kiwi-Minsker, L. *J. Catal.* 2004, *224*, 8-17.
[119] Wang, F.; Lu, G. X. *Catal. Lett.* 2007, *115*, 46-51.
[120] Ketchie, W. C.; Murayama, M.; Davis, R. J. *Top. Catal.* 2007, *44*, 307-317.
[121] Ketchie, W. C.; Fang, Y.-L.; Wong, M. S.; Murayama, M.; Davis, R. J. *J. Catal.* 2007, *250*, 95-102.
[122] Prati, L.; Martra, G. *Gold Bull.* 1999, *32*, 96-101.
[123] Niemantsverdriet, J. W. *Spectroscopy in Catalysis (Second Edition)*; Wiley-VCH: Weinheim, 2000.
[124] Lee, H. Y.; Kim, S. W.; Lee, H. Y. *Electrochem. Solid St.* 2001, *4*, A19-A22.
[125] Wu, M. Q.; Snook, G. A.; Chen, G. Z.; Fray, D. J. *Electrochem. Commun.* 2004, *6*, 499-504.
[126] Huang, X. K.; Yue, H. J.; Attia, A.; Yang, Y. *J. Electrochem. Soc.* 2007, *154*, A26-A33.
[127] Fischer, A. E.; Pettigrew, K. A.; Rolison, D. R.; Stroud, R. M.; Long, J. W. *Nano Lett* 2007, *7*, 281-286.
[128] Tuel, A.; Moussa-Khouzami, S.; Taarit, Y. B.; Naccache, C. *J. Mol. Catal.* 1991, *68*, 45-52.
[129] Thangaraj, A.; Kumar, R.; Ratnasamy, P. *J. Catal.* 1991, *131*, 294-297.
[130] Van Der Pol, A. J. H. P.; Verduyn, A. J.; Van Hooff, J. H. C. *Appl. Catal. A* 1992, *92*, 113-130.
[131] Notari, B. *Catal. Today* 1993, *18*, 163-172.
[132] Wang, D. A.; Liu, Z. Q.; Liu, F. Q.; Ai, X.; Zhang, X. T.; Cao, Y.; Yu, J. F.; Wu, T. H.; Bai, Y. B.; Li, T. J.; Tang, X. Y. *Appl. Catal. A* 1998, *174*, 25-32.
[133] Yu, R. B.; Xiao, F. S.; Wang, D.; Sun, J. M.; Liu, Y.; Pang, G. S.; Feng, S. H.; Qiu, S. L.; Xu, R. R.; Fang, C. G. *Catal. Today* 1999, *51*, 39-46.
[134] Sun, J. M.; Meng, X. J.; Shi, Y. H.; Wang, R. W.; Feng, S. H.; Jiang, D. H.; Xu, R. R.; Xiao, F. S. *J. Catal.* 2000, *193*, 199-206.
[135] Xiao, F. S.; Sun, J. M.; Meng, X. J.; Yu, R. B.; Yuan, H. M.; Xu, J. N.; Song, T. Y.; Jiang, D. Z.; Xu, R. R. *J. Catal.* 2001, *199*, 273-281.

[136] Xiong, C. R.; Chen, Q. L.; Lu, W. R.; Gao, H. X.; Lu, W. K.; Gao, Z. *Catal. Lett.* 2000, *69*, 231-236.
[137] Van De Pol, A. J. H. P.; Verduyn, A. J.; Van Hooff, J. H. C. *Appl. Catal. A* 1993, *96*, L13-L20.
[138] Englisch, M.; Ranade, V. S.; Lercher, J. A. *Appl. Catal. A* 1997, *163*, 111-122.
[139] Blaser, H. U.; Jalett, H. P.; Wiehl, J. *J. Mol. Catal.* 1991, *68*, 215-222.
[140] Masel, R. I. *Chemical Kinetics and Catalysis*; Wiley-Interscience: New York, NY, 2001.
[141] Ortiz-Hernandez, I.; Williams, C. T. *Langmuir* 2003, *19*, 2956-2962.
[142] Davis, J. L.; Barteau, M. A. *J. Am. Chem. Soc.* 1989, *111*, 1782-1792.
[143] Baiker, A. *J. Mol. Catal. A* 1997, *115*, 473-493.
[144] Studer, M.; Blaser, H. U.; Exner, C. *Adv. Synth. Catal.* 2003, *345*, 45-65.
[145] Murzin, D. Y.; Maki-Arvela, P.; Toukoniitty, E.; Salmi, T. *Catal. Rev.-Sci. Eng.* 2005, *47*, 175-256.
[146] Bartók, M. *Curr. Org. Chem.* 2006, *10*, 1533-1567.
[147] Baiker, A. *J. Mol. Catal. A* 2000, *163*, 205-220.
[148] Baiker, A. In *Chiral Catalysts Immobilization and Recycling*; De Vos, D. E., Vankelecom, I. F. J., Jacobs, P. A.; Ed.; Wiley-VCH: Weinheim, 2000.
[149] Török, B.; Balázsik, K.; Kun, I.; Szöllösi, G.; Szakonyi, G.; Bartók, M. *Stud. Surf. Sci. Catal.* 1999, *125*, 515-522.
[150] Li, X. H.; Li, C. *Catal. Lett.* 2001, *77*, 251-254.
[151] Blaser, H. U.; Jalett, H. P.; Montini, D. M.; Reber, J. F.; Wehrli, J. T. *Stud. Surf. Sci. Catal.* 1988, *41*, 153-163.
[152] Gamez, A.; Köhler, J.; Bradley, J. *Catal. Lett.* 1998, *55*, 73-77.
[153] Nitta, Y. *Top. Catal.* 2000, *13*, 179-185.
[154] Ertl, G.; Knözinger, H.; J., W.; Ed.; *Preparation of Solid Catalysts*; Wiley-VCH: Weinheim, 1999.
[155] Regalbuto, J.; Ed.; *Catalyst Preparation: Science and Technology*; Taylor and Francis (CRC Press): Boca Raton, FL, 2007.
[156] Vannice, M. A. *Kinetics of Catalytic Reactions*; Springer: Heidelberg, 2005.
[157] Thomas, J. M.; Raja, R. *Proc. Natl. Acad. Sci. U.S.A.* 2005, *102*, 13732-13736.
[158] Huber, G. W.; Chheda, J. N.; Barrett, C. J.; Dumesic, J. A. *Science* 2005, *308*, 1446-1450.
[159] Corma, A.; Serna, P. *Science* 2006, *313*, 332-334.
[160] Wang, D. H.; Ma, Z.; Dai, S.; Liu, J.; Nie, Z. M. *AIChE Annual Meeting, Salt Lake Ciry, November* 2007.
[161] Mallat, T.; Orglmeister, E.; Baiker, A. *Chem. Rev.* 2007, *107*, 4863-4890.
[162] Podkolzin, S. G.; Stangland, E. E.; Jones, M. E.; Peringer, E.; Lercher, J. A. *J. Am. Chem. Soc.* 2007, *129*, 2569-2576.
[163] Hoxha, F.; Konigsmann, L.; Vargas, A.; Ferri, D.; Mallat, T.; Baiker, A. *J. Am. Chem. Soc.* 2007, *129*, 10582-10590.
[164] Bonalumi, N.; Vargas, A.; Ferri, D.; Baiker, A. *J. Phys. Chem. C* 2007, *111*, 9349-9358.
[165] Ishikawa, A.; Neurock, M.; Iglesia, E. *J. Am. Chem. Soc.* 2007, *129*, 13201-13212.
[166] Boronat, M.; Concepción, P.; Corma, A.; González, L.; Illas, F.; Serna, P. *J. Am. Chem. Soc.* 2007, *129*, 16230-16237.

Reviewed by Ru Ren, School of Chemistry, University of St Andrews, U.K; Min Shen, School of Chemical, Biological and Materials Engineering, University of Oklahoma, USA.

In: Heterogeneous Catalysis Research Progress
Editor: Mathias B. Gunther, pp. 267-297
ISBN 978-1-60456-979-7

Chapter 7

SUPPORTED POLYOXOMETALATE-BASED HETEROGENEOUS CATALYSIS FOR LIQUID PHASE SELECTIVE OXIDATIONS

Oxana A. Kholdeeva[1]
Boreskov Institute of Catalysis, Prospekt Ak. Lavrentieva 5,
Novosibirsk, 630090, Russia

ABSTRACT

The design of active, selective, stable and easily recyclable heterogeneous catalysts is a challenging goal of liquid phase oxidation catalysis. Early transition-metal oxygen-anion clusters or polyoxometalates (POMs) have received increasing attention as oxidation catalysts because of their unique ensemble of properties, such as inorganic nature, metal oxide-like structure, thermodynamic stability to oxidation, thermal and hydrolytic stability, tunable acidities, redox potentials and solubility, etc. The apparent structural analogy of POMs and metal oxide surfaces allows considering POMs as discrete, soluble fragments of extended metal oxide lattices, which can be comprehensively investigated at the atomic level. In the structure of transition-metal-monosubstituted POMs (M-POMs), the active center M is isolated in and strongly bound to an inert metal oxide matrix and is thus prevented from hydrolysis and oligomerization, which makes M-POMs perspective building blocks for designing single site heterogeneous catalysts. The development of strategies for converting homogeneous POMs to solid, true heterogeneous catalysts is of primary importance. A simple wet impregnation usually did not allow obtaining catalysts which would be stable to leaching in polar reaction media. In the past decade, a great deal of work was directed to immobilize POMs on specially modified supports *via* the formation of a chemical bond (ionic, covalent or dative). Embedding POM into an inert matrix using sol-gel method may also lead to stable heterogeneous catalysts. Importantly, the molecular structure of POM can be maintained after immobilization, which ensures uniformity of spatially separated active centers on the catalyst surface or inside pores. In this chapter, we survey recent achievements in the synthesis of POM-based heterogeneous catalysts using different immobilization techniques, their characterization, and application for liquid

[1] email: khold@catalysis.ru.

phase selective oxidation of organic compounds. Special attention is paid to analyze the frequently success-limiting issues of catalyst activity and selectivity after immobilization and, especially, after recycling, catalyst stability to leaching and nature of catalysis.

"Heterogeneous Catalysts for Liquid-Phase Oxidations: Philosophers' Stones or Trojan Horses?"
R.A. Sheldon, et al. *Acc. Chem. Res.* 1998, 31, 485.

1. Introduction

The selective catalytic oxidation of organic compounds with environmentally benign, cheap and readily available oxidants is the most economic and ecological route to a wide variety of valuable oxygen-containing products and intermediates [1-5]. Although heterogeneous catalysis offers clear advantages of an easier catalyst separation and recycling and thus better meets the requirements of sustainable chemistry, many bulk chemicals (as well as some fine and specialty chemicals) are produced using homogeneous catalysts [1,5-7]. The major part of manufactures in fine chemicals industry still employs conventional stoichiometric oxidations thus producing a huge amount of inorganic waste [2]. In the past decades, replacement of stoichiometric and homogeneous processes by environmentally friendly ones, which would employ heterogeneous catalysts and clean oxidants – O_2 and H_2O_2, has become a widely accepted strategy; however, progress in this direction is obstructed by the limited availability of efficient heterogeneous catalysts for selective oxidations in liquid phase [2,3,8-10].

In the last years, much attention was focused on elaboration of different approaches, which allow constructing on surface or within pores of solids spatially well separated active sites, uniform in composition and distribution. Such catalysts are usually called "single-site heterogeneous catalysts" (SSHC) [11]; they are expected to combine the main advantages of both homogeneous (activity and selectivity) and heterogeneous catalysts (simplicity of separation and recycling). The key strategies to their design involve introduction of an active center into an inert matrix by framework substitution, grafting, tethering, encapsulation, intercalation, and some other techniques [11-18]. Many research efforts have been directed to synthesize mesoporous redox active molecular sieves [19-24], thus solving the problem of diffusion limitation in liquid phase, which takes place for microporous catalysts and bulky reagents. However, numerous catalytic studies have shown that a great majority of known SSHC undergo irreversible deactivation in liquid phase oxidation reactions (especially in polar solvents) and, in effect, operate as homogeneous rather than true heterogeneous catalysts [25-30]. Another problem of such catalysts is frequently occurring irreversible deactivation in the presence of water/hydrogen peroxide due to hydrolysis of M-O-Si bonds and oligomerization of active metal on the catalyst surface [28-32]. Very few catalytic materials are exception of this rule. A rare example of a highly efficient, stable and true heterogeneous liquid phase selective oxidation catalyst is the microporous titanium silicalite TS-1 developed by Enichem at the early 1980th [33], which is currently employed in three industrial liquid phase oxidation processes [3,12,34].

It is well-known that conventional transition metal complexes, homogeneous or heterogenized by means of different techniques, are disposed to destruction owing to thermodynamically favorable oxidation of organic ligands. Although several techniques have been developed to improve the stability of such catalysts, it is believed that fully inorganic systems have more scope for practical applications than those containing organic ligands or supports [35]. This stipulated for a growing interest to a big family of early-transition-metal oxygen-anion nano-size clusters or polyoxometalates (POMs), which have completely inorganic, metal oxide-like structure and are, therefore, thermodynamically stable to oxidation [36-50]. Yet, this class of compounds possesses a unique combination of properties, such as good thermal and hydrolytic stability, a wide variety of structures and compositions, tunable acidity, redox potentials and solubility, etc. The apparent structural analogy of POMs and metal oxide surfaces allows viewing POMs as discrete, soluble fragments of extended metal oxide lattices [51-55]. Significantly, POMs can be comprehensively investigated at the atomic level, both structurally and mechanistically, which makes them good objects for studying structure-activity relationships and mechanisms of catalysis. Transition-metal-monosubstituted M-POMs are perspective building blocks for designing and studying SSHC [11, 56, 57]. The most widely used M-POM of the Keggin structure $[XW_{11}M(L)O_{39}]^{n-}$ is shown in Figure 1. The active transition metal M is isolated in and strongly bound to a tungsten oxide matrix and is thus prevented substantially from hydrolysis, olygomerization and leaching.

In view of the numerous advantages of POMs the development of strategies for converting them to solid catalysts is of primary interest. First, catalytically active POMs can be heterogenized in the form of insoluble salts using Cs^+, Ag^+, K^+, NH_4^+ and some organic cations [37,49, 58-64]. Such salts possess micro/mesoporous structure and their surface area is typically in the range of 10-150 m^2/g.

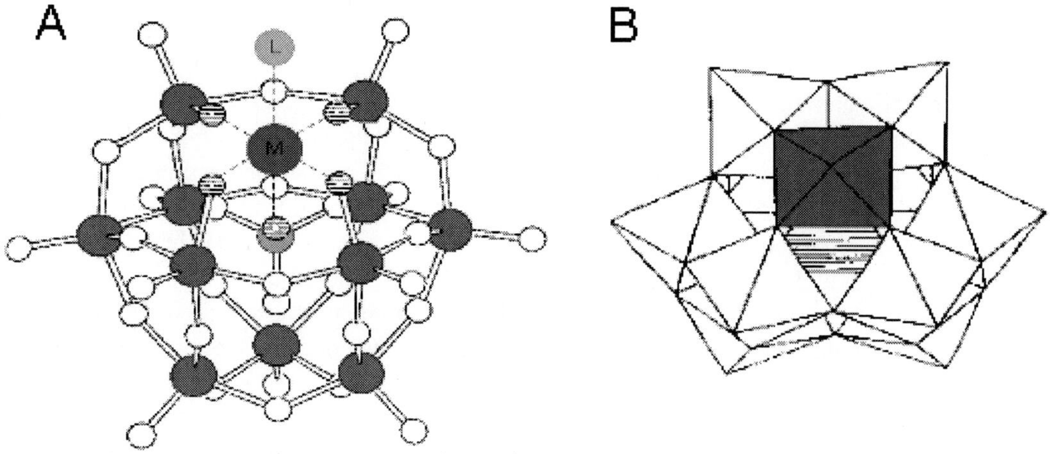

Figure 1. Ball and stick (A) and polyhedral (B) representation of the monosubstituted Keggin-type polyanion [XW11M(L)O39]n-. Reprinted from J. Mol. Catal. A: Chem. 262, Kholdeeva, O. A.; Maksimovskaya, R. I., 7-24, Copyright (2007), with permission from Elsevier.

The second general approach involves immobilization of POMs on a suitable carrier material. Importantly, the POM molecular structure can be maintained after immobilization, which ensures uniformity of spatially separated active centers on the catalyst surface. So far, the main areas of application of insoluble POM salts and solid ionic materials remain acid catalysis and oxidation gas-phase catalysis as well as material science. These topics have been extensively reviewed [37,39,41-44,58-61] and are outside the scope of this paper, which deals with the preparation of immobilized POM catalysts and their application in liquid phase oxidation reactions. A recent review of Mizuno and co-workers [49] covers a considerable part of the literature in this specific area until 2005, but comparison with the state-of-the-art at the beginning of 2008 shows that this field has rapidly expanded over the past few years.

In the beginning, simple wet impregnation or nonspecific adsorption techniques were mainly employed to produce supported POM catalysts. A great deal of work was aimed at depositing POMs on various supports, including silica, active carbon, alumina, metal oxides, MgF_2, SiC, etc., and to employ them as catalysts in liquid phase processes. Several monographs and review papers summarized the relevant literature in this area [37,44,49,58-61]. Silica and active carbon were the most frequently used carriers because of their acidic or neutral surface properties which allow avoiding POM destruction to a considerable degree. Typically, catalysts with POM loading below 14% are obtained on carbon supports. The carbon textural properties strongly affect its entrapping ability, microporous carbons being preferable. Using mesoporous silica supports, one can prepare materials containing up to 50 wt.% of POM. However, many researchers established that the catalysts obtained by wet impregnation undergo POM leaching and rapid deactivation in a polar reaction medium (note that the polarity can be increased during the oxidation process due to formation of polar products).

In the past two decades, several concepts of attaching POMs to specially modified supports through chemical (ionic, dative or covalent) bonds have been elaborated to design heterogeneous POM catalysts [42,43,49,50,65]. In this short review paper, we would like to survey recent achievements in the field of chemical immobilization of POMs by means of different techniques, characterization of the POM-based heterogeneous catalysts, and their application for liquid phase selective oxidation of organic substrates. Some other methodologies based on embedding (occluding, entrapping) POMs within a silica matrix or other carrier materials will be briefly discussed with a focus on their relevance to liquid-phase oxidation processes. Composite heterogeneous photocatalysts based on POMs have been comprehensively reviewed by Guo and Hu [66,67] and will not be considering here.

In this paper, special attention is paid to analyze the frequently success-limiting issues of the effect of support on the catalytic activity and reaction selectivity, POM leaching under the conditions of liquid phase oxidation, catalyst reusability, and nature of catalysis. The latter problem is addressed in accordance with the methodology suggested by R.A. Sheldon [25,26], who demanded from researchers to obtain rigorous proof of heterogeneity *by filtering the catalysts at the reaction temperature before completion of the reaction and testing the filtrate for activity.* Representative examples are given from our recent work, and conclusions are based on both the analysis of the literature and our own research experience in the field.

2. Chemical Binding of POMs to Supports

2.1. Electrostatic Attachment

The first examples of an electrostatic attachment of POMs on solid supports are dated to the mid 1980th. Baba *et al.* reported immobilization of heteropolyacids on the anion-exchange resin Amberlist-15 [68]. Later on, this type of the support was successfully used by Jacob's group to immobilize the Venturello complex $\{PO_4[WO(O_2)_2]_4\}^{3-}$, well known as highly selective homogeneous catalyst for H_2O_2-based epoxidation of alkenes [35,38,40,45,46,49,69-72]. The catalyst obtained *via* anion exchange with the commercial resin Amberlite IRA-900 showed excellent selectivity in epoxidation of acid-sensitive natural terpenes and allylic alcohols [73-75]. The selectivity of 92% at 83% conversion was attained in epoxidation of α-pinene and limonene. The catalytic activity of the reused catalyst was completely maintained after several catalytic cycles, and the filtrate was catalytically inactive [75].

Transition metal polyoxoanions were intercalated in hydrotalcite-like anionic clays or layered double hydroxides (LDHs, $[M(II)_{1-x}M(III)_x(OH)_2](A^{n-})_{x/n}$ z H_2O, where M(II) = Mg, Zn, Co, Ni, Cu…, M(III) = Al, Cr, Fe… and A^{n-} = CO_3, SO_4, NO_3, OH, Cl…) by anion exchange and some direct synthesis methods [76,77,78,79,80,81,82,83]. The POM-LDHs materials typically possess a surface area of about 90-200 m^2g^{-1}. The structural, thermal and textural properties of the LDH-POM intercalates were comprehensively studied by XRD, FTIR, TEM, EDS, and N_2 adsorption techniques. A gallery height of about 10 Å was observed for the LDH intercalated by the symmetrical Keggin POM, whereas two different gallery heights were found for the cylindrical Dawson (14.5 and 12.8 Å) and Finke (13.3 and 12.6 Å) anions, depending on the preparation temperature [78]. The differences in POM orientations were rationalized in terms of different electrostatic and hydrogen-bonding interactions between the POM pillars and the LDH layers. Because of the strong basicity of LDHs, which may cause decomposition of POM, a strategy to control the pillaring reactions is necessary to successfully intercalate the POM between hydrotalcite layers [80]. Several monographs and review papers addressed the synthesis, properties and applications in catalysis of POM-LDHs materials [49,84-89]. So, we limit here to cite just some recent results which demonstrate the high interest for these materials in liquid phase oxidations.

Hydrotalcites intercalated with $[Mo_7O_{24}]^{6-}$ and $[H_2W_{12}O_{42}]^{10-}$ showed catalytic activities for the shape-selective epoxidation of alkenes with H_2O_2 [80]. Hydrotalcites-supported heptamolybdate $Mo_7O_{24}{}^{6-}$ were prepared by ion exchange from suitable precursors containing terephthalate and tested in liquid phase oxidation of norbornadiene, benzonorbornadiene and cyclohexene using hydrogen peroxide as oxidant and dioxane or dioxane–butyl maleate mixture as solvents [83]. The selectivity to epoxides was found to depend on the hydrotalcite composition and on the reaction solvent. Recycling tests over six successive runs indicated that the catalysts preserved the activity without any loss in their intrinsic activity or selectivity. However, elemental analysis of the liquor indicated some loss of the active component (about 1.2%), and no convinced proof of the true heterogeneous catalysis was provided.

A new type of layered double hydroxides intercalated by peroxo-polyoxometalates was synthesized by treating PdMAl-$[PW_{11}O_{39}]$ (M = Mg^{2+} or Zn^{2+}) with 30% H_2O_2 [90]. XRD,

FT-IR, and ^{31}P NMR MAS revealed that the interlayer anion $[PW_{11}O_{39}]^{7}$ was converted mainly to $\{PO_4[WO(O_2)_2]_4\}^{3}$ and $[W_2O_3(O_2)_4(H_2O)_2]^{2}$. ^{27}Al NMR MAS revealed that the brucite-like layers did not collapse during synthesis. PdMgAl-[PW_4 + W_2] showed 91.5% selectivity for propene oxide at 47.5% conversion during propene epoxidation by O_2 in methanol at 353 K for 10 h. The catalytic ability can be attributed to the presence of highly active Pd particles in the brucite-like layers that generate H_2O_2 in situ, the strong epoxidation ability of interlayer peroxo-polyoxometalates with dilute H_2O_2, and the promotion effect of the basic MgAl-type brucite-like layers toward alkene epoxidation. The catalyst can be separated and reused by filtration without leaching active components into the solvent.

Co-containing POMs have been found to be among the most efficient catalysts for homogeneous aerobic oxidation and co-oxidation processes [91-93]. This prompted many researchers to design solid Co-POM-containing materials [78,94-100]. Thus, various Co-POMs have been deposited on cotton cloth [94] and silica [100], datively [95] or electrostatically [96,97] bonded to NH_2-modified silica surfaces (*vide infra*) as well as intercalated in LDHs [78,98,99]. The resulting materials were successfully used for aerobic oxidation of aldehydes, alkenes, alkanes, alcohols and some other organic substrates.

Recently, Corma *et al.* have patented a process of oxidizing cycloalkane with molecular oxygen to produce cycloalkanol and/or cycloalkanone in the presence of hydrotalcite-intercalated heteropolyanion $[Co^{II}M_{11}Co^{II}(H_2O)O_{39}]^{8-}$ (M = W or Mo), which comprised one cobalt as a central atom and another as a substitute of a W=O fragment in the Keggin structure [98]. At 130 °C and 0.5 MPa, 64 and 24% selectivity to cyclohexanone and cyclohexanol, respectively, was achieved at cyclohexane conversion about 5%. This catalytic system could be of practical importance provided a true heterogeneous nature of catalysis and good catalyst recyclability had been proved. Unfortunately, this information was lacking in [98].

Co-trisubstituted POM, α-$\{SiW_9O_{37}\{Co(H_2O)\}_3\}^{10-}$, was heterogenized by inserting it between the layers of basic Mg_3Al-hydrotalcite [99]. The intercalated Co-POM showed high catalytic activity and reusability in liquid-phase oxidation of cyclohexanol to cyclohexanone using 1 atm of molecular oxygen under mild reaction conditions. The pillared catalyst exhibited similar XRD and FT-IR patterns before and after use in the oxidation reaction; moreover, the catalyst surface area did not change significantly. All these suggested stability of the POM-based catalyst under turnover conditions. Elemental analysis confirmed the presence of trace amounts (at the ppb level) of cobalt species in the solution, but "hot catalyst filtration experiment" [26] pointed out a true heterogeneous nature of catalysis [99].

Electrostatic attachment to a support modified with appropriate spacer ligands containing cationic groups (sometimes these groups are generated *in situ* by protonation of a basic ligand) has been widely used for POM attachment *via* anion exchange or ion pairing [74,101-113]. One of the first attempts dealt with immobilization of $[Mo_7O_{24}]^{6-}$ on the colloidal polymer containing alkylammonium cations [101]. However, the resulting catalyst showed a rather low activity in alkene oxidation with H_2O_2. Keggin type heteropoly acids $H_3PW_{12}O_{40}$ and $H_3PMo_{12}O_{40}$ were incorporated in polyaniline [102] and polyimine [103] *via* protonation of the basic sites of the polymer matrices, which has been confirmed by UV-vis, FT-IR and X-ray studies. The polymer-supported POMs revealed a decreasing acidity function but enhancing thermal stability and redox activity in isopropanol conversion to acetone. Note that neither recycling nor leaching tests were reported for the majority of the polymer-immobilized POM materials, but one might expect a rather low stability of such materials because of the organic nature of the polymer matrix.

In order to overcome the disadvantages of organic polymer-supported POMs, such as small surface area (and as a result, low activity), low thermal stability and nonresistance to oxidation, several research groups used for immobilization inorganic (presumably, silica) supports modified with appropriate spacer ligands containing quaternary ammonium cations [74,104,109,113] or amine groups [95-97,105,106,108,110,111]. Such ligands can be easily grafted onto a silica surface, and their covalent attachment ensures stability of the resulting modified silica materials [114].

In 1995, Neumann and Miller immobilized polyoxometalates $\{PO_4[WO(O_2)_2]_4\}^{3-}$ and $[ZnWMn_2(H_2O)_2(ZnW_9O_{34})_2]^{12-}$ on silicate xerogel covalently modified with phenyl groups and quaternary ammonium cations, prepared by copolymerization of $Si(OEt)_4$, $PhSi(OEt)_3$ and trialkoxyorganosilanes using a sol–gel technique [104]. The resulting materials were employed for heterogeneous alkene oxidation using H_2O_2 in a solvent-free system. The catalytic activities were found to be greatly affected by the nature of the quaternary ammonium cations. A similar approach was applied to support the Venturello complex on $N^+(Et)_3$-modified MCM-41 [74]. The catalytic activity of this material in alkene epoxidation was, however, lower compared to the corresponding Amberlite-supported catalyst. Recently, decatungstate $[W_{10}O_{32}]^{4-}$ was electrostatically bound to silica functionalized *via* anchoring alkylammonium salts [113]. The catalyst demonstrated fairly good catalytic activity, selectivity and reusability in H_2O_2-based sulfoxidation of various thioethers, including allyl phenyl sulfide and dibenzyl sulfide. Turnover numbers as high as 960 were achieved. The heterogeneous nature of catalysis has been demonstrated clearly by carrying out the filtration test as had been suggested by Sheldon [26].

Mizuno and co-workers synthesized an organic-inorganic hybrid support by covalently anchoring *N*-octyldihydroimidazolium cation fragment onto SiO_2 [112]. This support acted as a good anion exchanger, and the catalytically active polyoxometalate anion, $[\gamma\text{-}1,2\text{-}H_2SiV_2W_{10}O_{40}]^{4-}$ was immobilized with the stoichiometric anion exchange. The POM structure was preserved after the anion exchange, which was confirmed by IR and ^{51}V NMR spectroscopy. The supported catalyst was capable of heterogeneously oxidizing a broad range of olefins and sulfides with high selectivity to the desired products. The unique stereospecificity, diastereoselectivity, and regioselectivity of the catalyst were very close to those of the homogeneous analogue under the same reaction conditions, showing that the homogeneous catalysis was heterogenized without loss of the intrinsic catalytic nature. For the competitive epoxidation of *cis*- and *trans*-2-octenes, the formation rate of *cis*-2,3-epoxyoctane was much higher than that of the *trans* isomer. More accessible, but less nucleophilic double bonds in non-conjugated dienes were highly regioselectively epoxidized. The activity of the supported catalyst was about twice lower compared to the homogeneous analogue. The oxidation was immediately stopped by the removal of the solid catalyst, and vanadium and tungsten species could hardly be found in the filtrate after removal of the catalyst. These results could rule out any contribution to the observed catalysis from the vanadium and tungsten species that leached into the reaction solution and suggested that the observed catalysis was truly heterogeneous in nature. In addition, the catalyst was reusable for both epoxidation and sulfoxidation without the loss of the catalytic performance.

The peroxotungstate $[W_2O_3(O_2)_4(H_2O)_2]^{2-}$, immobilized on dihydroimidazolium-based ionic liquid-modifed SiO_2 has been employed for alkene and allylic alcohol epoxidation

[115]. It has been established that the epoxidation of *cis*- and *trans*-alkenes proceeded stereospecifically. The regioselective epoxidation of geraniol took place at the electron-deficient 2,3-allylic double bond to afford 2,3-epoxy alcohol in high yield. The catalyst could be easily recovered by filtration and reused at least three times for the epoxidation of cyclooctene without loss of catalytic activity and selectivity. The epoxidation process was terminated completely by catalyst removal, and no tungsten species were detected in the filtrate. Therefore, any contribution to the observed catalysis from homogeneous tungsten species that might have leached into the reaction solution can be ruled out. Recently, peroxotungstates immobilized on ionic-liquid modified silicas were employed for heterogeneous oxidation of thioethers to sulfoxides with 30% aqueous H_2O_2 [116]. The catalyst showed quite good recyclability, but SEM-EDAX spectra analysis performed on the recovered catalyst after six runs revealed some decrease of the tungstate. Since the question on the activity of the filtrate has not been addressed in [116], the nature of catalysis remains unclear in the system described.

Heteropolyacid $H_6PMo_9V_3O_{40}$ has been attached to NH_2-MCM-41 and used for cyclohexene oxidation with TBHP in MeCN [105]. The catalyst with POM loading as high as 10 wt.% revealed almost the same activity as homogeneous $H_6PMo_9V_3O_{40}$ and was significantly more stable to leaching in polar solvent medium compared to the catalyst supported on NH_2-free MCM-41. Monovanadium molybdovanadophosphoric acid has been supported on amine-functionalized SBA-15 [108]. Small angle X-ray scattering (SAXS) analysis and SEM technique provided evidence for the structural integrity of the immobilized sample. Diffuse reflectance UV–vis, ^{31}P NMR and FT-IR data confirmed the incorporation of the heteropolyacid onto the NH_2-SBA-15 surface. In oxidation of selected alkenes, such as norbornene, cyclooctene, cyclohexene and styrene with aqueous hydrogen peroxide, the catalytic activity of the supported catalyst was, in general, lower than that of homogeneous $PMo_{11}V$ but the selectivity was higher in the former case. The immobilized catalyst demonstrated fairly good reusability and stability to leaching (just 1% of POM was found in the filtrate). The filtration experiment confirmed that catalysis is mostly heterogeneous in nature [108]. Recently, $H_5PMo_{10}V_2O_{40}$ immobilized onto amine-functionalized SBA-15 was examined in the liquid-phase oxidation of anthracene (AN) with 70% aqueous *tert*-butylhydroperoxide (TBHP) in benzene [117]. The catalyst operated with the turnover frequency (TOF) of 21 mole AN converted per mole of catalyst per hour and produced 100% selectivity to 9,10-anthraquinone. Catalyst leaching studies indicated the absence of leaching into the reaction medium, and the hot catalyst filtration experiment at 25% conversion of AN gave a convinced proof of the heterogeneous nature of catalysis.

Immobilization of the cobalt- and cerium-monosubstituted Keggin heteropolyanions, $[PW_{11}CoO_{39}]^{5-}$ [96] and $[SiW_{11}CeO_{39}]^{4-}$ [118] on both amine-modified mesoporous amorphous silica (NH_2-xerogel) [95,118] and well-ordered mesostructured silicate materials NH_2-SBA-15 and NH_2-MCF [118] has been comprehensively studied by our group in collaboration with C.L. Hill and A. Jarzębski. Both Co- and Ce-POMs have been identified as active and selective in homogeneous aerobic oxidation of formaldehyde to formic acid under mild reaction conditions (25-40 °C, 1 atm of air) [96,119]. The NH_2-modified mesoporous silica supports (NH_2-X) were prepared using 10 mol. % of N-[3-(trimethoxysilyl)propyl]ethylenediamine (or 3-aminopropyltrietoxysilane) and ethyl silicate 40 as silica precursors by a sol-gel method [120]. Two different tetra-butylammonium (TBA) salts of $[PW_{11}CoO_{39}]^{5-}$, (TBA_4H) and (TBA_5), were obtained [96] and used for immobilization and catalytic studies. The supported Co-POM catalysts were characterized by N_2-adsorption measurements, elemental analysis, DRS-UV and FT-IR spectroscopy. The effect of protonation of the amine-modified silica surface on the Co-POM loading, activity and stability of the solid catalyst was investigated.

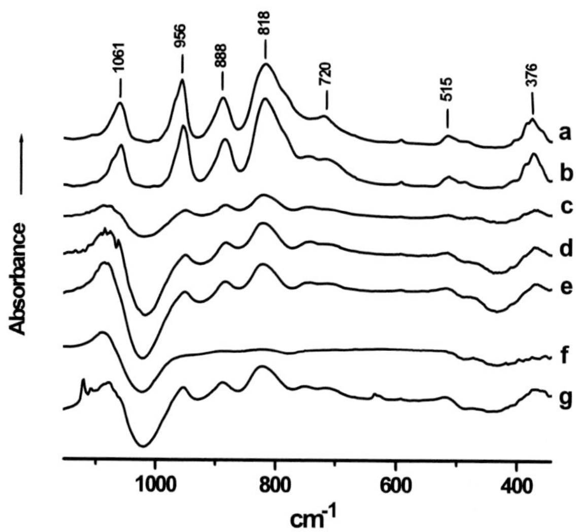

Figure 2. FT-IR spectra of (a) $TBA_4HPW_{11}CoO_{39}$, (b) $TBA_5PW_{11}CoO_{39}$, (c-e) samples of NH_2-silica-supported Co-POM with different loading, (f and g) samples with 13 and 32 wt.% of Co-POM, respectively, after three catalytic cycles of IBA oxidation. In (c-g) the spectrum of the NH_2-silica was subtracted. Reprinted from *J. Catal.* 226, Kholdeeva, O. A., *et al.*, 363–371, Copyright (2004), with permission from Elsevier.

To accomplish the attachment of Co-POM to the surface of the support, two different approaches were explored, namely, anion exchange and ion paring. The first one involved treatment of the support with an acid solution followed by the anion exchange process (Eqs. 1-2)

$$X\text{-}NH_2 + HClO_4 \rightleftharpoons X\text{-}NH_3^+ClO_4^- \quad (1)$$

$$nX\text{-}NH_3^+ClO_4^- + TBA_5PW_{11}CoO_{39} \rightleftharpoons (X\text{-}NH_3^+)_nTBA_{5-n}[PW_{11}CoO_{39}] + nTBAClO_4 \quad (2)$$

The Co-POM loading increased with increasing amount of H^+ added initially to the support. Thus, samples with 13 and 32 wt.% of Co-POM were obtained for n = 1 and 6, respectively [96]. This can be readily rationalized if the high negative charge of Co-POM heteropolyanion, 5-, is taken into account. The more positively charged NH_3^+ groups in the support, the stronger is the expected electrostatic binding of POM. It is well precedented that the monosubstituted Keggin anions $[XW_{11}MO_{39}]^{n-}$, where M = M(II) or M(III), in contrast to many corresponding anions with M(IV) and M(V), are not stable in acid solutions and can not be isolated as free acids [121]. Given this point, the acid should be added to the support before addition of the polyoxometalate. Importantly, the textural properties of the supported catalyst did not suffer significantly even if the maximal amount of Co-POM was introduced.

Indeed, both the surface area and mesopore volume of the sample containing 32 wt.% of Co-POM were about 70-76 % of the corresponding characteristics of the initial support.

In the second approach, M-POMs containing protons as counter ions were used to fulfill electrostatic binding of M-POM to the support in one stage *via* the formation of anion-cation pairs (Eqs. 3 and 4):

$$X\text{-}NH_2 + TBA_4HPW_{11}CoO_{39} \rightleftharpoons (X\text{-}NH_3^+)TBA_4[PW_{11}CoO_{39}] \quad (3)$$

$$3X\text{-}NH_2 + NaH_3SiW_{11}Ce(H_2O)_nO_{39} \rightleftharpoons (X\text{-}NH_3^+)_3Na[PW_{11}Ce(H_2O)_nO_{39}] \quad (4)$$

The presence of protons accelerates anchoring Co-POM to the support. When no H^+ was introduced with the acid salt of POM itself or by adding an extra acid, the Co-POM binding proceeded *via* substitution of H_2O (or MeCN) molecule in the coordination sphere of Co(II) for NH_2 [95] (*vide infra*), the process being significantly slower than the anion exchange or ion paring.

FT-IR study revealed that no damage of the Co-POM structure occurred when its TBA-salts dissolved in MeCN were used for the immobilization. As one can judge from Figure 2, the IR spectra after subtraction of the peaks due to NH_2-X exhibited the principal stretching modes of the Keggin Co-POM unit (956, 888, 818, 752, 720 cm^{-1}).

Utilization of an aqueous solution of a sodium salt resulted in a partial Co-POM destruction. A similar observation has been made in [95]. The Ce-POM appeared to be more stable than the Co-POM and both aqueous solutions of the acid sodium salt $NaH_3SiW_{11}CeO_{39}$ and MeCN solutions of $TBA_3HSiW_{11}CeO_{39}$ could be successfully used for the immobilization.

The ionic character of binding between Co-POM and the support has been confirmed by DRS-UV spectroscopy. The spectra of the NH_2-silica-supported Co-POM catalysts (Figure 3, spectra a and b) are very similar and resemble the spectrum of $Na_5PW_{11}Co(H_2O)O_{39}$ in water [122] or in SiO_2 matrix (Figure 3, spectrum c) as well as that of $TBA_4HPW_{11}Co(H_2O)O_{39}$ in MeCN [97]. Importantly, the spectra of the Co-POM supported on the protonated NH_2-silica differ from the spectra published for Co-POMs datively linked to NH_2-silica [95]. Indeed, no peaks or shoulders were observed in the range of 628-640 nm, indicating no formation of dative bonds between NH_2 groups and cobalt.

Another argument in favor of the electrostatic nature of the interaction between $TBA_4HPW_{11}Co(H_2O)O_{39}$ and NH_2 groups of the support was obtained from anion exchange experiments [97]. The color of the supported catalyst (pale pink) became completely colorless after treatment with a 1 M solution of $TBAClO_4$ in MeCN. Additionally, the elemental analysis data confirmed elimination of Co-POM from the solid matrix after such a treatment.

The catalytic performance of both Co- and Ce-POM supported on NH_2-functionalized mesoporous silica was assessed in the aerobic oxidation of formaldehyde in H_2O under mild conditions (20-40 °C, 1 atm of air) [97,118]. While the Co-POM-based catalyst underwent rapid deactivation, the Ce-POM catalyst could be used repeatedly without significant loss of the catalytic activity as one can judge from Figure 4 [118].

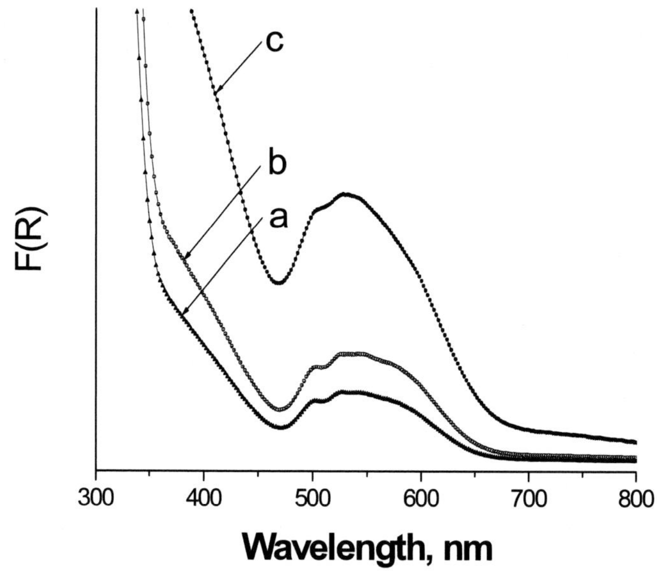

Figure 3. DR-UV-vis spectra of (a) NH2-silica-supported TBA4HPW11Co(H2O)O39, (b) NH3+-silica-supported TBA5PW11Co(H2O)O39 and (c) Na5PW11Co(H2O)O39/SiO2 composite material.

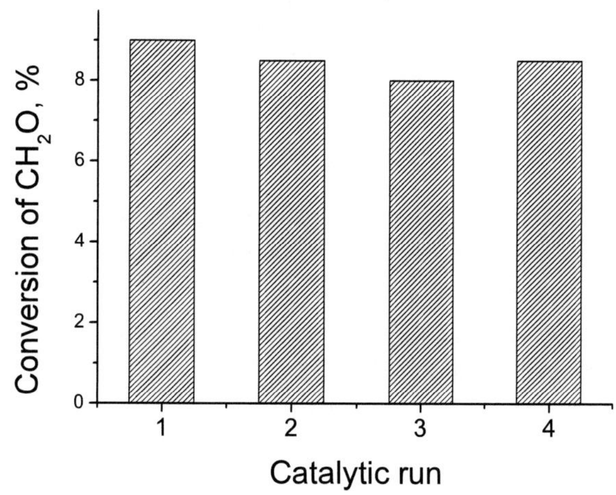

Figure 4. Oxidation of CH2O (0.468 mmol) by air (1 atm) in the presence of [SiW11CeO39]4- supported on NH2-xerogel (200 mg) in H2O (2 ml) at 40°C for 5 h [118].

The catalytic performance of the NH_2-silica-supported Co-POMs was also examined in aerobic oxidation of isobutyraldehyde (IBA) [96] and α-pinene [97] as well as their co-oxidation [97]. In IBA oxidation, the activity of the catalyst containing 13 wt.% of Co-POM was as high as the activity of homogeneous $TBA_4HPW_{11}Co(H_2O)O_{39}$ (TOF = 31 h^{-1}). However, some loss of the catalytic activity was observed already after 3 catalytic cycles because of Co-POM leaching during the oxidation process, which was indicated by both elemental analysis and FT-IR spectroscopy [96]. The increase of the Co-POM loading up to 32 wt.% led to increased stability of the catalyst (compare spectra a and g in Figure 2), but was accompanied by a reduction of the catalytic activity (TOF = 13 h^{-1}).

We attempted to improve the catalytic performance, including stability, of the silica-immobilized Co-POM catalysts by using hydrothermally stable supports, specifically, the mesostructured silicates SBA-15 and MCF, both modified with amino groups by grafting 3-aminopropyltrietoxysilane [97]. The physico-chemical properties of three representative NH_2-X (X = xerogel, SBA-15 and MSF) supported Co-POM catalysts are given in Table 1. The textural properties of the initial, POM-free supports are shown for comparison.

Table 1. Physico-chemical properties of NH_2-silica supported $TBA_4HPW_{11}Co(H_2O)O_{39}$ [97]

NH_2-Support/Catalyst	S [a] (m^2/g)	V [b] (cm^3/g)	d [c] (nm)	NH_2 (mmol/g)	Co-POM (wt. %)	NH_2/POM (mol/mol)
xerogel	540	1.24	12-16 [d]	1.15		
Co-POM/xerogel	510 (520)[e]	0.09 (1.02)[e]	12-16 [d]		20 (5)[e]	21
SBA-15	623	1.34	13	0.64		
Co-POM/SBA-15	418 (420)[e]	1.15 (1.14)[e]	13 (13)[e]		15 (7)[e]	16
MCF	625	2.82	15	2.59		
Co-POM/MCF	440 (400)[e]	1.40 (1.35)[e]	15 (14)[e]		32 (18)[e]	30

[a] Mesopore surface area; [b] mesopore volume; [c] avarage mesopore diameter; [d] broad mesopore distribution; [e] inside parentheses, the values after five catalytic cycles of α-pinene and IBA co-oxidation [97].

The Co-POM loading depended on the surface concentration of NH_2-groups on the support. Both the surface area and pore volume decreased upon Co-POM deposition, unlike the average mesopore diameter did not change.

The catalytic properties of the obtained materials were assessed in α-pinene autoxidation and its co-oxidation with isobutyraldehyde [97].

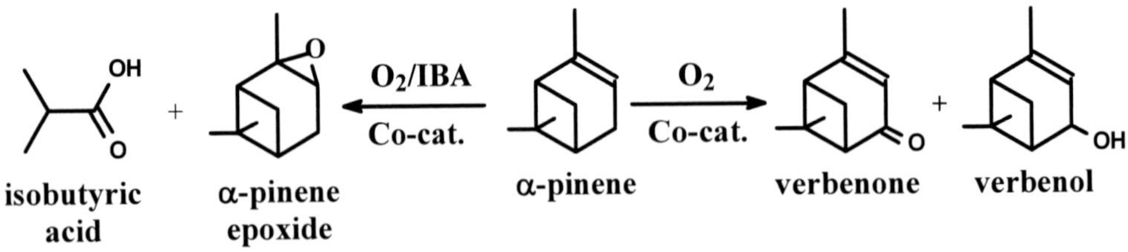

At low conversions, the autoxidation process produced mainly the allylic oxidation products, verbenol and verbenone, along with some amount of camphene and campholenic aldehyde. With increasing conversion the selectivity to verbenol/verbenone went down due to overoxidation processes leading to oligomeriztion/polymerization products. Thus, the verbenol/verbenone selectivity attained 70 and 40% at 20 and 46 % conversion of α-pinene, respectively. The activity of the supported Co-POM catalysts was as high as that of the homogeneous $TBA_4HPW_{11}CoO_{39}$. Significantly, the supported catalysts could be used repeatedly without loss of the activity and selectivity during several catalytic cycles (Figure 5).

Pinene co-oxidation with IBA gave selectively α-pinene epoxide and isobutyric acid. It is noteworthy that the activity of the supported Co-POM catalysts was comparable to the activity of the homogeneous Co-POM, while the epoxide selectivity could increase after the Co-POM immobilization.

In contrast to α-pinene autoxidation, for which no effect of the support on the reaction selectivity was observed, the selectivity of the co-oxidation process could be altered by varying the amount of NH_2-groups on the support. Thus for Co-POM supported on NH_2-MCF and NH_2-SBA-15 (NH_2/Co-POM = 30 and 16 mol/mol, respectively), the α-pinene epoxide selectivity attained 94 and 76%, respectively, at 96% of substrate conversion.

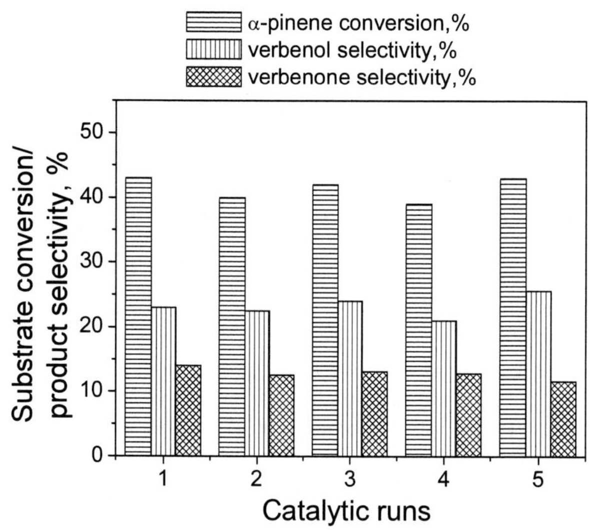

Figure 5. Recycling of NH2-MCF-supported TBA4HPW11CoO39 in α-pinene autoxidation. Reaction conditions: α-pinene, 0.1 mmol; O2, 1 atm; catalyst, 6×10-4 mmol Co-POM; MeCN, 1 mL; 50 °C, 1 h. Reprinted from *J. Catal.* 246, Maksimchuk, N. V., et al., 241–248, Copyright (2007), with permission from Elsevier.

This may be rationalized if we take into account that the excessive surface NH_2-groups may partially neutralize carboxylic acid formed during the co-oxidation process and thus improve the epoxide selectivity.

In contrast to α-pinene autoxidation, the activity of the supported catalysts remained constant during few catalytic cycles and then tended to decrease regardless of the nature of the NH_2-silica support. Recycling of NH_2-MCF-supported Co-POM is shown in Figure 6a. FT-IR study revealed that the Co-POM structure retained after the catalyst use in several catalytic runs, while the nitrogen adsorption data confirmed stability of the textural properties (Table 1). In turn, DR-UV-vis study pointed out some changes in the position and intensity of *d-d* transition bands in the spectra of the supported Co-POM after α-pinene/IBA co-oxidation or treatment with isobutyric acid, indicating that the catalyst deactivation may arise from the chemosorption of the polar reaction product on the cobalt catalytic center [96]. Indeed, both the catalytic activity and selectivity remained constant during several catalytic cycles when catalyst regeneration by evacuation was performed to remove the adsorbed carboxylic acid (Figure 6b).

Catalyst filtration experiments performed at the reaction temperature showed that α-pinene conversion in the filtrate stopped after separation of the catalyst in the case of α-pinene autoxidation (Figure 7a), which corroborates substantially a heterogeneous nature of the oxidation catalysis. Yet, no leaching was found by the elemental analysis after 5 cycles of α-pinene autoxidation over the supported Co-POM catalysts. In contrast, further α-pinene transformation occurred in the filtrate in the case of α-pinene and IBA co-oxidation (Figure 7b), the reaction rate being significantly higher than that in the blank experiment.

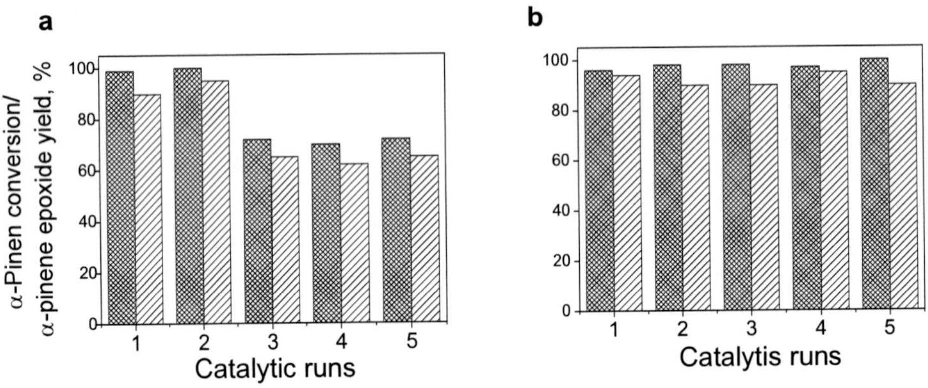

Figure 6. Recycling of NH_2-MCF-supported $TBA_4HPW_{11}CoO_{39}$ in α-pinene co-oxidation with IBA: (a) without regeneration and (b) catalyst regeneration by evacuation at 130 °C for 2 h before re-use. Reaction conditions: *α*-pinene, 0.1 mmol; IBA, 0.4 mmol; O_2, 1 atm; catalyst, 6×10^{-4} mmol Co-POM; MeCN, 1 mL; 25 °C, 2 h.

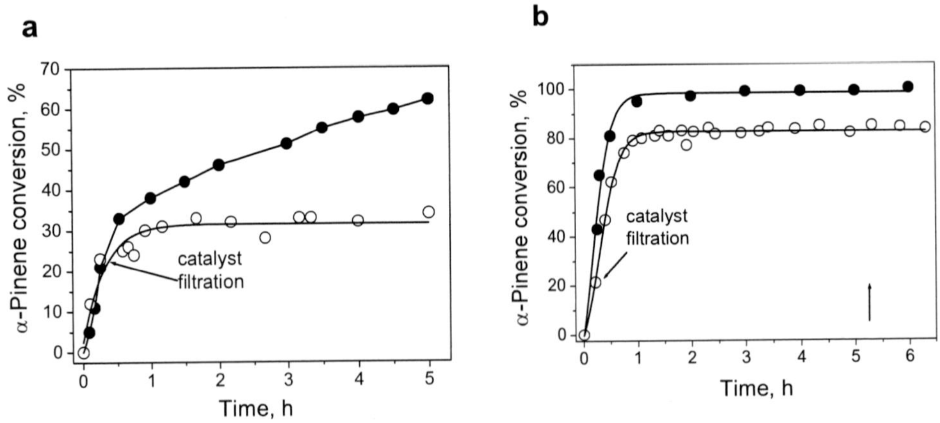

Figure 7. (a) α-Pinene autoxidation and (b) α-pinene co-oxidation with IBA over TBA4HPW11CoO39 supported on NH2-SBA-15. The reaction conditions as in Figs. 5 and 6.

This indicated unambiguously that Co-POM leaching from the solid catalyst into the solution occurred in the co-oxidation process and catalysis, at least, partially is homogeneous in nature. Indeed, the elemental analysis data confirmed substantial leaching of Co-POM during α-pinene/IBA co-oxidation (see Table 1). The difference observed for the autoxidation and co-oxidation processes is not surprising if we keep in mind that less polar products, alcohol and ketone, are produced in the former reaction. Meanwhile, as was demonstrated above (Figure 6b) both the substrate conversion and the product yields may remain unchanged during several runs of the α-pinene/IBA co-oxidation, provided the catalyst is exposed to regeneration before reuse. This seeming contradiction with the leaching and filtration experiments may be explained by a bell-shaped dependence of the oxidation rate *vs.* catalyst concentration, which is typical of chain radical reactions in the presence of Co-containing catalysts [92,93]. As a result, the observed catalytic activity maintains in a wide range of catalyst concentration. Hence, the pinene-IBA co-oxidation reaction in the presence of the supported Co-POM catalysts can serve as an excellent example which demonstrates that good catalyst recyclability itself can be a proof of neither absence of leaching nor heterogeneous nature of catalysis.

Recently, the tetrairon(III)-substituted polytungstates $[Fe_4(H_2O)_{10}(\beta\text{-}XW_9O_{33})_2]^{n-}$ (n = 6, X = As^{III}, Sb^{III}; n = 4, X = Se^{IV}, Te^{IV}) were immobilized on (3-aminopropyl)triethoxysilane-modified SBA-15 and showed fairly good catalytic performance for solvent-free aerobic oxidation of long-chain *n*-alkanes using air as the oxidant under ambient conditions [111]. The catalysts maintained the high catalytic activity even after four cycles of catalyst reuse. Moreover, hot-filtered-solution experiments indicated that there was no contribution from leached, homogeneous, catalytically active POM species.

The novel polyoxometalate $[(Eu_2PW_{10}O_{38})_4(W_3O_8(H_2O)_2(OH)_4)]^{22-}$ has been immobilized inside the channels of NH_2-MCM-41 and characterized by XRD, UV-vis absorption, emission, Raman excitation, Raman, IR, and ^{31}P solid-state NMR measurements [123]. Infrared and Raman spectra of the polyoxometalate/MCM-41 composite systems were interpreted as showing spectral shifts due to site induced electrostatic interactions. Rare earth metal sandwiched Keggin-type heteropolyoxometalates, $K_{11}[RE(PW_{11}O_{39})_2]$ (RE = La, Ce, Pr, Nd, Sm, Eu, Dy and Y), were also anchored onto aminosilylated mesoporous silicate SBA-15 and the resulting materials were characterized by ICP, FT-IR, XRD, N_2 adsorption, ^{31}P NMR MAS and TEM [124]. The POM clusters preserve their structure in the surface-modified mesopores. Their catalytic activity was tested in heterogeneous oxidation of cyclohexene by H_2O_2. The interaction between the rare earth metal centre and amino groups grafted on the channel surface of SBA-15 led to strong immobilization which prevented leaching during the reaction, which is in sharp contrast to the catalysts acquired using an amine-free SBA-15 support. After reaction for 2 h in which the cyclohexene conversion reached *ca.* 41%, the reaction mixture was filtered and then the mother liquor was allowed to react for another 8 h under the same reaction condition. No significant activity was observed, demonstrating that the active species are not the dissolved POM leached from the support. Therefore, it was reasonable to suggest that the present catalysis is heterogeneous in nature.

Hill and co-workers attached $[V_{10}O_{28}]^{6-}$, $[PV_2Mo_{10}O_{40}]^{5-}$ and $[(Fe^{III}(OH_2)_2)_3(A\text{-}\alpha\text{-}PW_9O_{34})_2]^{9-}$ on cationic silica nanoparticles (Figure 8a) covered with alumina $(Si/AlO_2)^{n+}$ [107]. The POMs form an approximately single layer of monoanions on the surface of each nanoparticle. Their application as heterogeneous catalysts for efficient aerobic oxygenation of organic sulfides and autoxidation of acetaldehyde to acetic acid under ambient conditions has been reported [107]. The recovered solid catalysts were almost as active as in the initial reaction, and the filtrate was completely inactive. This strongly supports the facts that the solid is the actual active catalyst and these new catalytic materials are quite stable. The number of turnovers for all three POMs was enhanced by electrostatic immobilization on the silica nanoparticles.

Recently, Férey and co-workers have demonstrated a successful incorporation of a Keggin heteropolytungstate within the cages of the nanoporous chromium terephthalate polymer MIL-101, which possesses a rigid zeotype crystal structure, large pore sizes (~3.4 nm), outstanding surface area (~6000 m^2g^{-1}) and, furthermore, is resistant to air, water, common solvents and thermal treatment (up to 320 °C) [125]. To probe the presence of the polynanion within the pores, the resulting solid was analyzed by thermal gravimetric analysis (TGA), N_2 sorption measurement, XRD, ^{31}P solid state NMR, and infrared spectroscopy.

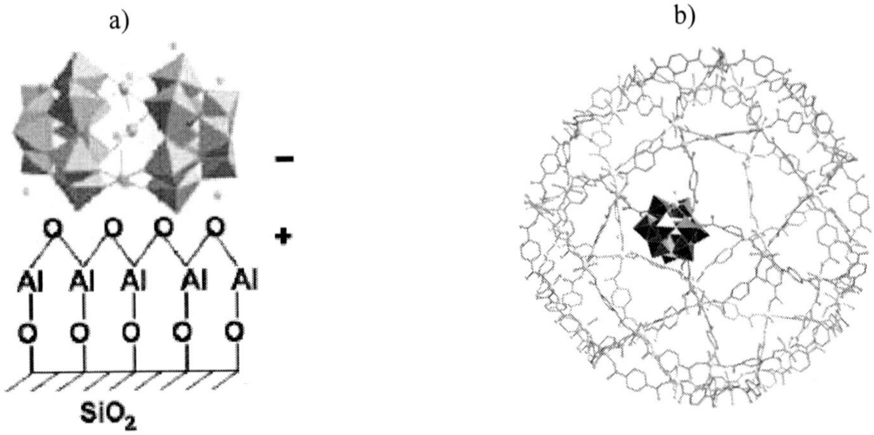

Figure 8. Examples of electrostatic immobilization of POMs: (a) on the surface of $(Si/AlO_2)^{n+}$ particles [107] and (b) within nanocage of on MIL-101 coordination polymer [126].

All of these techniques confirmed the presence of a large amount of POM ions within the pores. Theoretical estimation gave the value of ~5 Keggin anions per nanocage.

Catalytically active titanium- and cobalt-monosubstituted Keggin heteropolyanions, $[PW_{11}CoO_{39}]^{5-}$ and $[PW_{11}TiO_{40}]^{5-}$, have been introduced into the MIL-101 nanocages by adsorption of the corresponding Na- or TBA-salts from acetonitrile solutions at room temperature [126]. An irreversible character of the sorption of about 10 wt. % of POM (~1 POM anion per MIL-101 nanocage, Figure 8b) indicated a strong interaction between POM and the support. Importantly, the immobilized POM could be re-extracted using a 1 M solution of Bu_4NClO_4 in MeCN thus pointing to the electrostatic character of binding between POM anions and the positively charged surface of MIL-101 (Eq. 5):

$$5[Cr_3O(bdc)_3(H_2O)_2F] + POM^{5-} \rightarrow 5[Cr_3O(bdc)_3(H_2O)_3]\cdot POM + 5F^- \quad (5)$$

The MIL-supported POM catalysts were characterized by elemental analysis, XRD, N_2 adsorption, and FT-IR-spectroscopy, which indicated the preservation of both MIL and POM structures after immobilization. The composite M-POM/MIL-101 materials demonstrated fairly good catalytic performance in α-pinene allylic oxidation (81-84 % verbenol/verbenone selectivity at 15-25 % substrate conversion) and caryophyllene epoxidation (100 % selectivity at 88 % conversion) with green oxidants – H_2O_2 (Ti-POM) and O_2 (Co-POM) [126].

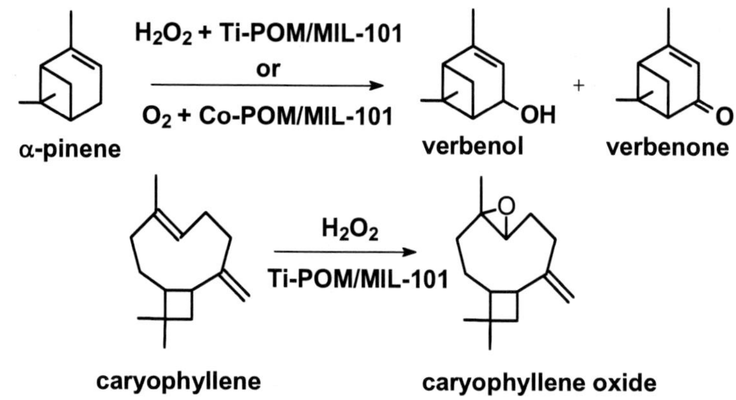

It is worth of noting that the activity of the M-POM/MIL catalysts expressed in TOF values determined from the initial rates of substrate consumption was close to the activity of the corresponding homogeneous M-POMs; therefore, the activity did not decline after immobilization. Interestingly, the MIL-supported M-POMs appeared to be more selective in α-pinene allylic oxidation than the NH_2-silica-supported Co-POM [97], probably, due to the MIL-101 zeotype porous structure and smaller pore diameter, which may prevent allylic oxidation products from further oxidative polymerization. Importantly, the selectivity of caryophyllene epoxide formation was significantly higher for MIL-supported Ti-POM than for homogeneous Ti-POM (80 and 52%, respectively, under the same reaction conditions). This may result from deprotonation of the Ti-POM anion upon immobilization within the MIL-101 cages *via* anion exchange (Eq. 5), which protected the acid-sensitive epoxide from ring opening.

The Co-POM/MIL-101 catalyst could be used repeatedly without suffering a loss of the activity and selectivity in the oxidations with molecular oxygen. However, when H_2O_2 was used as the oxidant, the reaction conditions had a strong impact on the Ti-POM/MIL-101 catalyst stability. Under rather mild conditions (< 50 °C, $[H_2O_2]$ < 0.2 M), the Ti-POM/MIL-101 composite was rather stable according to FT-IR, N_2 adsorption and XRD measurements, behaved as a true heterogeneous catalyst (Figure 9a), did not undergo POM leaching according to elemental analysis, and could be reused without loss of the catalytic properties (Figure 9b). At higher H_2O_2 concentrations and higher temperatures, the MIL-101 matrix was destroyed and the POM/MIL-101 catalyst underwent deactivation.

Before closing this section devoted to the electrostatic attachment of POMs to supports, we would like to come back to the "simple" silica- and carbon-supported POM catalysts. As was mentioned in the Introduction, a wide majority of these catalysts are not stable to leaching and, in effect, behave as homogeneous rather than heterogeneous catalysts. However, in few cases, a careful control of the POM loading, the solvent used for impregnation as well as the textural and surface properties of the support allowed the preparation of solid catalysts which were relatively stable to leaching, provided the reaction mixture polarity was also under control [105,127-129].

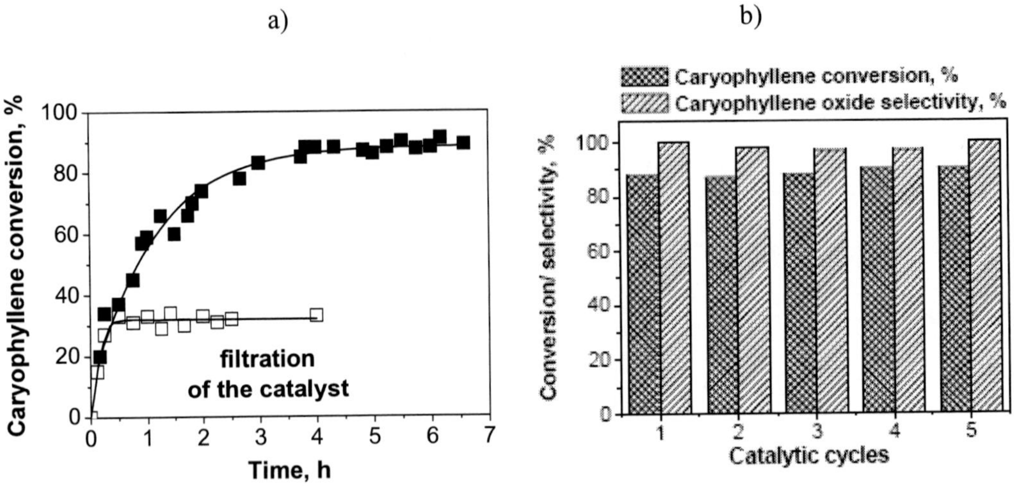

Figure 9. Oxidation of caryophyllene with H2O2 over [PW11TiO40]5- supported on MIL-101 [126]: (a) hot catalyst filtration experiment and (b) catalyst recycling. Reaction conditions: substrate, 0.1 mmol; oxidant, 0.1 mmol; catalyst, 14 mg; MeCN, 1 mL; 50°C.

Actually, the mechanism of the strong adsorption of POMs on some silica and carbons at low loadings is not fully understood yet. ^{1}H and ^{31}P NMR-MAS studies indicated a chemical interaction between heteropoly anion and the support. Thus a significant broadening of the ^{31}P NMR MAS signal was observed for Aerosil- [130], MCM-41- [105] and carbon- [131] supported POMs. The formation of $(SiOH_2^+)(H_2PW_{12}O_{40}^-)$ species on the silica surface was supposed [130]. The adsorption on carbon was also believed to involve proton transfer from POM to the support [132]. Hence, it seems very likely that electrostatic interaction may contribute significantly in binding POMs to the surface of some carbon and silicate materials. Note that strong interaction with a support surface may result in polyanion destruction [133], so careful control of the POM state after immobilization should be done using spectroscopic techniques, first of all, FT-IR and, if possible, solid state ^{31}P NMR. The question on the nature of catalysis which occurs over such materials also needs clarification and must be carefully re-investigated using the methodology suggested by Sheldon [26,26].

2.2. Dative Binding

The examples of dative binding of catalytically active POMs to acquire heterogeneous POM-based catalysts can be found in the literature quite rarely compared to the examples of electrostatic attachment which were discussed in the previous section.

Transition-metal-substituted polyoxometalates $[PW_{11}M^{II}(H_2O)O_{39}]^{5-}$ (M = Co, Zn) and $[SiW_9O_{37}\{Co^{II}(H_2O)\}_3]^{10-}$ (Co$_3$-POM) have been anchored to NH$_2$-modified macroporous (400 nm pores), mesoporous (2.8 nm pores), and amorphous fumed silica surfaces [95]. The materials were characterized by solid-state ^{31}P NMR MAS, UV-vis, FT-IR spectroscopy, and N_2 adsorption experiments to verify cluster attachment and the structure of POM on the support. The integrity of the $[PW_{11}CoO_{39}]^{5-}$ anion was maintained for nonaqueous impregnation with $TBA_5PW_{11}CoO_{39}$, while partial degradation of the cluster occurred when it was impregnated from aqueous solution using the corresponding K-salt. DRS-UV-vis study revealed splitting of the $^4T_{1g}$ band of the Co^{II} absorbance and appearance of a shoulder at 628-640 nm which indicated the formation of a donor-acceptor bond between cobalt and surface-anchored NH_2-ligand *via* substitution of H_2O in the coordination sphere of $[PW_{11}Co^{II}(H_2O)O_{39}]^{5-}$ 95] (Figure 10).

When the Co-POM was impregnated by incipient wetness onto unmodified silica [95] or embedded in a silica matrix or attached electrostatically to NH_3^+-silica [96,97], no splitting of the absorption occurred (Figure 3). Recently, the formation of coordination bond was found for the sandwich-type polyanion, $[(PW_9O_{34})_2Co_4^{II}(H_2O)_2]^{10-}$, supported on NH_2-silica [134].

Using DR-UV-vis spectroscopy the authors confirmed also the coordination of cobalt centers in the monosubstituted $[SiW_{11}Co^{II}(H_2O)O_{39}]^{6-}$ with the nitrogen atom of amine groups of NH_2-modified silica at pH $\geq$ 5.5 or the electrostatic bonding between the polyoxoanion and protonated $C_3H_6NH_3^+$-group at pH = 3.5 [134]. So, these results are consistent with the conclusions on the type of Co-POM binding on NH_2-silica made previously by Stein's and our groups.

The catalytic activities of the supported Co-POM catalysts were tested by the epoxidation of cyclohexene to cyclohexene oxide in the presence of isobutyraldehyde [95].

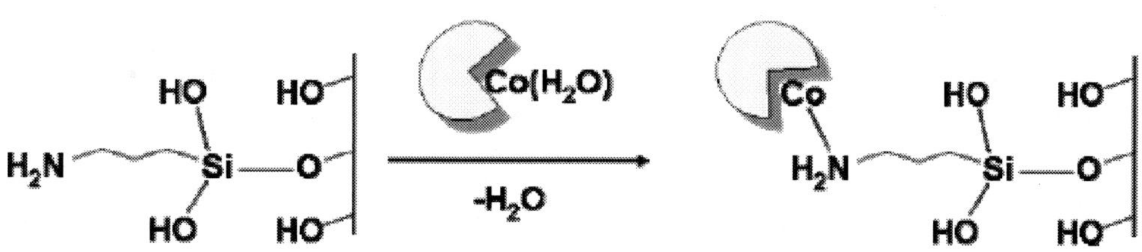

Figure 10. Dative binding of $[PW_{11}CoO_{39}]^{5-}$ to NH_2-modified silica [95].

The conversion of the substrate and the amount of product formed per unit time were similar for Co-POM supported on each of the three silica supports, but slightly lower than for purely homogeneous reaction. This contrasts to the electrostatically bound Co-POM (*vide supra*) and may arise from the occupation of the sixth position in the cobalt coordination sphere by NH_2-ligand, which reduces the accessibility of the active center for activation of the reagents. Significantly, the trisubstituted polyanion $[SiW_9O_{37}\{Co^{II}(H_2O)\}_3]^{10-}$ with three available cobalt centers exhibited higher catalytic activity with nearly identical performance on a silica support or in homogeneous solution. In the case of monosubstituted Co-POM, a small fraction of the POM leached into solution during the reaction, but the immobilized Co_3-POM was stable thus showing a clear advantage of binding to the support *via* three (or possibly, two) cobalt centers.

A new concept of "solvent-anchored supported liquid phase catalysis" has been introduced by Neumann and co-workers [135,136]. Using the sol–gel method they prepared silicates containing on the surface various combinations of hydrophilic polyethylene oxide, PEO, and hydrophobic polypropylene oxide, PPO. These functionalized silicates were coupled with polyoxometalates to produce catalytically active assembles, in which PEO and PPO acted as a solvent and/or complexing agent for POMs. The PEO/PPO–SiO_2-anchored Venturello complex catalyzed selectively (99%) cyclooctene epoxidation with 30% aqueous H_2O_2 without organic solvent at room temperature [135]. The correct balance of polyethers was found to be critical for the catalytic activity, a mixture of 10% PEO and 10% of PPO being produced the optimal catalyst particle. The solvent-anchored catalyst was recycled five times without loss of the catalytic activity. Use of a reaction filtrate showed no catalytic species and no catalytic activity in the filtrate. Heteropolyacid $H_5PV_2Mo_{10}O_{40}$ was complexed to 25% PEO–SiO_2 (schematic representation is shown in Figure 11) and used as a recyclable catalyst in oxydehydrogenation of dihydroanthracene using *tert*-butylhydroperoxide as oxidant with improved activity and selectivity compared to the non-supported catalyst [136]. Anthracene was produced with a 96% yield.

Introducing cationic quaternary ammonium groups Q^+ additionally to PEO-PPO groups allowed a contribution of electrostatic binding to the POM immobilization process. Thus $[ZnWMn_2(ZnW_9O_{34})_2]^{12-}$, bound by a Q^+ moiety to the silicate surface modified also with polyethers formed a catalytically active assembly, which demonstrated excellent recycling for the epoxidation of alkenes with aqueous 30% H_2O_2 [136]. A silicate with a hydrophobic (PPO)/hydrophilic (PEO) balance showed the highest activity due to the optimal contact of the apolar hydrocarbon and the aqueous oxidant

2.3. Covalent Binding

The preparation of solids with covalently attached POM complexes is a serious and worthwhile research target because these materials might be expected to be rather stable to POM leaching in solutions. Many new materials of this type have been reported in the literature [16,49,74,137-143]; however, catalytic studies on covalently bound POMs still remain a rare event. In 1992, Judeinstein reported the first POM-polymer hybrid where a lacunary Keggin POM cluster was covalently linked to polystyrene or polymethacrylate backbone through Si-O bonds [137]. This approach has been further developed by several research groups.

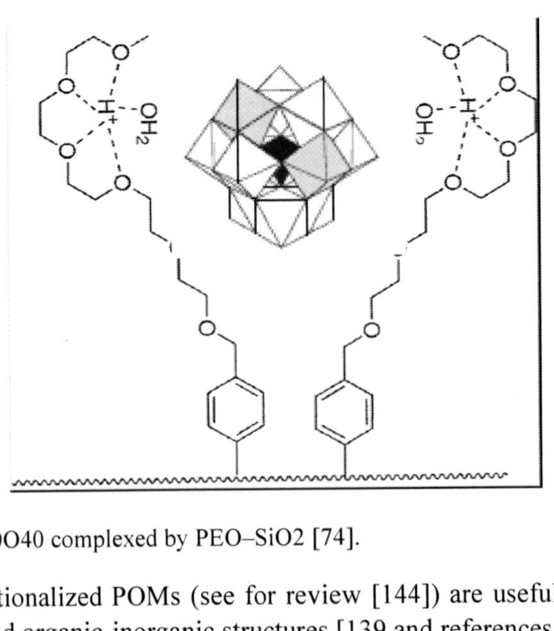

Figure 11. H5PV2Mo10O40 complexed by PEO–SiO2 [74].

Organically functionalized POMs (see for review [144]) are useful hybrid nanobuilding blocks to obtain hybrid organic-inorganic structures [139 and references therein].

The lacunary polyoxometalate $SiW_{11}O_{39}^{8-}$ reacted with trichloro or trialkoxysilanes to produce $[SiW_{11}O_{40}(SiR)_2]^{4-}$ (R = vinyl, allyl, methacryl, styryl), which were further polymerized in the presence of a radical initiator to yield hybrid polymers. Lacunary polyanions $XW_{11}O_{39}^{n-}$ (X = P, Si, Ge, B) were incorporated into the wall structure of macroporous silica resulting in hybrid XW_{11}-SiO_2 composites through the chemical grafting of organosilanol groups from the silica network onto the surface oxygen atoms at vacant sites of XW_{11} clusters [141]. Lin Xu *et al.* elaborated an efficient method for the selective synthesis of bifunctionalized organoimido hexamolybdates, which allows facile introduction of two functional groups to a hexamolybdate cluster, and thus paves the way for the synthesis of main chain-POM-containing hybrid polymers [140]. An example of well defined covalent attachment of POM to a surface is that of a thiol-functionalized POM on gold nanoparticles [142]. Functionalization of γ-$[SiW_{10}O_{36}]^{8-}$ with the sulfanylalkylsilane $HSC_3H_6Si(OMe)_3$ afforded difunctionalized γ-$[SiW_{10}O_{36}(HSC_6H_3Si)_2O]^{4-}$, which on addition to colloidal solutions of gold nanoparticles (4-12 nm) formed a novel hybrid nanosystem. Recently, Errington and co-workers reported immobilization of the Linquist type polyoxometalate $[(OMe)TiW_5O_{18}]^{3-}$ on alkanol-derivatized silicon surfaces through the formation of covalent M-O-C alkoxide bonds by alcoholysis of the Ti-OR bond in the Ti-POM [142]. Samples of single-crystal and porous silicon were functionalized with ω-hydroxyundecyl monolayers and then treated with acetonitrile solutions of $(TBA)_3[(MeO)TiW_5O_{18}]$ at 85°C. To the best of our knowledge, catalytic properties of the majority of these interesting materials have not been studied yet.

Various supports containing phosphorylated ligands were synthesized and used to form covalent bonds with peroxotungstate species [74,145,146]. Jacobs and co-workers developed an original alternative route to a supported Venturello-like complex, in which the formation of P-O-W bonds is performed directly on the inner wall of MCM-41 (Figure 12) [74]. For that, a phosphoramide-grafted MCM material was prepared by functionalization of the surface silanols of MCM-41 *via* reaction with alkoxysilyl amino compounds followed by the reaction with $POCl_3$ in order to generate P–N covalent bonds. Finally, this material was treated with H_2O_2 and H_2WO_4. ^{31}P NMR study confirmed that the immobilized $-PO_3H_2$ groups react with the peroxo–W complexes with formation of P–O–W bonds. IR and Raman spectroscopy gave clear evidence for the existence of W–peroxo species in the synthesized material, which is capable of epoxidizing a broad variety of olefins with H_2O_2 in a heterogeneous way. Tungsten leaching could be limited to less than 2% of the total amount of tungsten present in the catalyst [16,74].

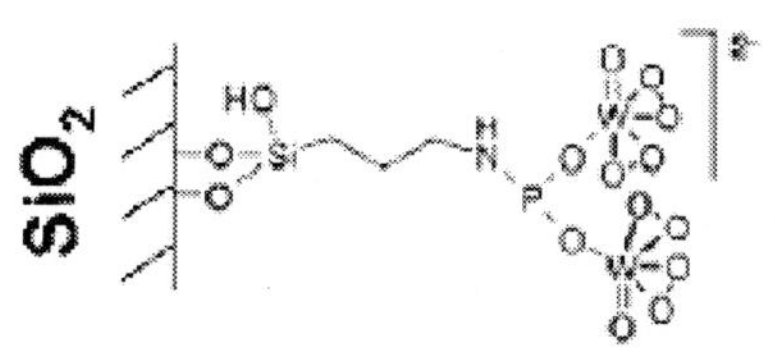

Figure 12. Schematic representation of the active peroxotungstate catalyst covalently anchored to silica via phosphoramide group [136].

Using a similar approach, the peroxo compound $[HPO_4\{W(O)(O_2)_2\}_2]^{2-}$ (PW_2) was synthesized on the surface of another mesoporous material, HMS, by reacting HMS–

$(CH_2)_3NH(PO_3H_2)$ with peroxotungstate $[W_2O_3(O_2)_4(H_2O)_2]^{2-}$ [147]. Then palladium ions were exchanged into the channels of HMS to form a hybrid catalyst. The catalyst was active in the oxidation of propylene in methanol using O_2 as an oxidant to produce propylene oxide in 83% selectivity at 34% conversion. The actual epoxidation catalyst is suggested to be the HMS-bound peroxo species PW_2, regenerating by H_2O_2, which is likely produced by the reaction of methanol and O_2 over the palladium catalyst. The active components, palladium and tungsten, did not leach into the reaction medium, and the catalyst could be reused.

3. Composite POM Materials

A promising approach to designing heterogeneous POM-based catalysts is occluding a catalytically active POM in silica or some other inert matrices during the matrix synthesis. Izumi and co-workers were the first who suggested embedding a catalytically active POM, $Cs_{2.5}H_{0.5}PW_{12}O_{40}$, into a silica by means of a sol-gel technology using tetraethyl orthosilicate (TEOS) as a silica source [58 and references therein]. A composite mesoporous material with a surface area 364 m^2/g has been prepared and successfully used as acid solid catalyst in liquid phase. Later on, a preparative method for silica-included $H_3PW_{12}O_{40}$ heteropolyacid has been developed by alteration of the sol-gel procedure, first of all, by increasing silica content to make the matrix denser and thus to prevent POM leaching [58]. Jarzebski and co-workers found that replacing, at least partially, TEOS by ethylsilicate 40 (ES-40, a commercial form of ethoxypolysiloxane), results in POM/SiO_2 composite materials with higher surface areas and improved adsorption properties [148].

Two composite materials, Co-POM/SiO_2 [96,97] and Ti-POM/SiO_2 [149], have been prepared by the sol-gel method using $Na_5PW_{11}CoO_{39}$ and $H_5PW_{11}TiO_{40}$, respectively. Tetramethoxysilane (TMOS) was used as the silica precursor. The molar ratio of reagents was as follows: Si : H_2O : MeOH : HCl = 1 : 9-12 : 6-9 : 0.0016. First, TMOS was prehydrolized using ½ of the total amount of methanol, 0.2 M solution of HCl and 2 mole of water per 1 mole of Si at 50 °C for 1 h. Then the solution containing the remaining half of methanol, the rest of water and M-POM (10 wt.%) was added to the first solution under vigorous stirring. After gelation and aging during 7 days at room temperature (Co-POM) or 50 °C (Ti-POM), the resulting material was dried in vacuum, ground, washed several times with water, and dried again. The Co- and Ti-POM/SiO_2 composites were characterized by elemental analysis, FT-IR, DRS-UV-vis and N_2 adsorption. Their physico-chemical properties are presented in Table 2.

The Co-POM/SiO_2 material possessed microporosity with textural properties being similar to those found earlier for the silica-included $H_3PW_{12}O_{40}$ and $H_3PMo_{12}O_{40}$ [58,148], while Ti-POM/SiO_2 appeared to be a non-porous material. The DR-UV-vis spectrum of Co-POM/SiO_2 (Figure 3c) was very similar to that of $Na_5PW_{11}CoO_{39}$ in aqueous solution [122] and confirmed preservation of the Co-POM structure and ligand surroundings in the sol-gel composite.

The catalytic properties Co-POM/SiO_2 were assessed in oxidation of IBA to isobutyric acid [96] and α-pinene to verbenol/verbenone [97] with molecular oxygen as well as in α-pinene/IBA co-oxidation to produce the corresponding epoxide and carboxylic acid [97]. In turn, the catalytic performance of Ti-POM/SiO_2 was examined in α-pinene oxidation to verbenol/verbenone [149] using aqueous H_2O_2 as oxidant. Some important characteristics of their catalytic behavior are given in Table 2. Co-POM/SiO_2 showed high activity and selectivity in the aerobic oxidation of IBA (conversion 93%, selectivity to isobutyric acid 98% after 6 h at room temperature). Significantly, the activity of the composite catalyst was as high as that of the corresponding homogeneous catalyst ($TBA_4HPW_{11}CoO_{39}$), while the selectivity of the latter was lower (just 57%).

Table 2. Physico-chemical and catalytic properties of M-POM/SiO_2 sol-gel composites

M-POM	S [a] (m^2/g)	V [b] (cm^3/g)	d [c] (nm)	M-POM (wt. %)	Substrate/ Oxidant	TOF_{av} [d] (h^{-1})	Selectivity [e] (%)
$H_3PW_{12}O_{40}$ [58]	554	0.2	1.4	11	- [f]	-	-
$Na_5PW_{11}CoO_{39}$ [96]	442	0.17	1.5	4	IBA/O_2	30 (31)	98 (57)
					α-Pinene/O_2	135 (50)	62 (60)
					α-Pinene/ IBA/O_2	220 (130)	73 (87)
$H_5PW_{11}TiO_{40}$ [149]	<0.5 [g]	-	-	9	α-Pinene/ H_2O_2	24 (24)	81 (40)

[a] Langmuir surface area; [b] micropore volume; [c] average pore size $d = 4V_p/S$ by Langmuir; [d] TOF_{av} = (moles of substrate consumed)/[(moles of M) × time]; [e] inside parentheses, the values for the corresponding homogeneous catalyst; [f] no liquid phase oxidation has been reported; [g] nonporous material.

On the contrary, the same level of selectivity to the allylic oxidation products, verbenol and verbenol, was achieved in α-pinene autoxidation with both types of the catalysts, but the activity of Co-POM/SiO_2 expressed in TOF values was remarkably higher (Table 2). A higher activity but a bit lower selectivity to epoxide was found for α-pinene/IBA co-oxidation over Co-POM/SiO_2. Interestingly, Ti-POM/SiO_2 despite its nonporous structure revealed the same activity as homogeneous $H_5PW_{11}TiO_{40}$ in H_2O_2-based α-pinene oxidation. Furthermore, a remarkable increase of the selectivity to verbenol/verbenone was observed for the sol-gel catalyst. These examples clearly demonstrate that the catalytic performance of sol-gel/POM composites is still hardly predictable and only experiment can show whether the composite have advantages in activity and selectivity compared to a homogeneous analogue or not. The considerable differences in the catalysts activities and selectivities, which are observed for POM/SiO_2 composite materials and homogeneous POMs, can be explained by several reasons. First, it was suggested that POM is entrapped in the silica network as a highly concentrated aqueous solution [58], which may account for an increased catalytic activity. Second, the microporous structure of the matrix may enhance the reaction selectivity by suppressing overoxidation and polymerization processes. Third, it is a matter of common observation that the acidity of heteropolyacids is reduced by silica supports [59,63,59]. This may also lead to enhancing selectivity in some reactions, especially those which are accompanying by acid-catalyzed rearrangements of substrate and/or target product.

In contrast to Co-POM supported on NH_2-silica, the silica-entrapped Co-POM catalyst did not require regeneration and could be used repeatedly without suffering a loss in the catalytic activity and selectivity [96]. No Co-POM leaching was established by elemental analysis. The superior stability of Co-POM/SiO_2 is likely due to its microporous structure, which may prevent catalyst deactivation and leaching caused by the carboxylic acid product, producing diffusion limitations for penetrating acid molecules to the active Co centers. No wonder that nonporous Ti-POM/SiO_2 catalyst also revealed a very good stability and recyclability in H_2O_2-based oxidations (Figure 13a) and catalysis over it has a true heterogeneous nature as is indicated by the catalyst filtration experiment (Figure 13b). The protonation of surface silanols under the acidic sol–gel conditions to form cationic Si-OH_2^+ groups, which may act as counter cations for POM anions, has been suggested in [66]. Hence, additionally to the occluding effect, electrostatic binding between protonated Si-OH_2^+ groups and polyanions might be expected to contribute to the high stability of the composite POM/SiO_2 materials.

During the last two decades the sol-gel approach has been further developed by material chemists to insert POMs within various inorganic matrices. Many sol-gel/POM composites have been reported [66,67,150-152]; however, the number of their applications as liquid-phase oxidation catalysts is still not as large as one might have anticipated based on the advantages that POMs offer in the ease of incorporation into sol-gel matrixes and on their unique blend of properties. One of possible reasons is the lack of information on the stability of the composite materials with respect to POM leaching in polar media.

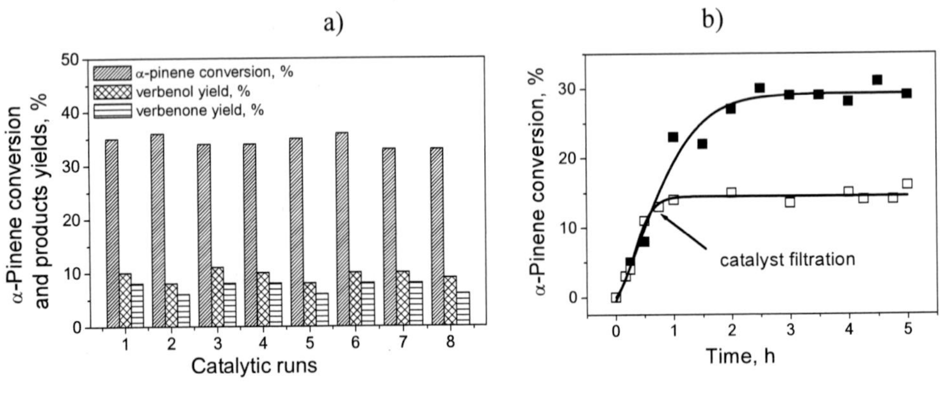

Figure 13. Oxidation of α-pinene with 30% H_2O_2 over sol-gel $H_5PW_{11}TiO_{40}/SiO_2$ composite [149]: (a) catalyst recycling and (b) hot catalyst filtration experiment. Reaction conditions: α-pinene, 0.1 mmol, H_2O_2, 0.12 mmol, catalyst, 14 mg, MeCN, 1 mL, 30 °C.

4. CONCLUSION

Polyoxometalates undoubtedly have enormous catalytic potential in liquid phase selective oxidation of organic compounds. Various strategies for immobilization of POMs on solid matrices have been developed during the past two decades and opened new opportunities for practical applications. The most developed and widely used technique is electrostatic attachment of polyanions to positively charged supports, such as anion resins, hydrotalcite-like materials, cation-modified silica and other matrices, including the sophisticated coordination polymers of the MIL family. The number of materials, where POM is bound *via* dative or covalent bonding, is still less but these families are also growing rapidly because many material and synthetic chemists are working in this field. There are many POM-based materials, for which different types of interactions may contribute to the immobilization process and one can hardly classify these materials in terms of chemical binding. Synthetic chemists certainly outstrip catalytic researchers, and the catalytic potential of many new POM-based materials still remains unstudied. Joint efforts of specialists from different fields of chemical science would lead to a rapid advancement in this area.

The problem of decreasing catalytic activity upon immobilization of POM seems to be solved to a considerable degree by using high surface area supports and employing methods which allow creation on the surface of spatially well separated active centers, keeping the POM structure intact. The catalytic activity of electrostatically immobilized POMs is often as high as that of the corresponding homogeneous POM, because the ligand composition of the active metal remains unchanged after immobilization. However, strong interaction with surface (especially, when it is not mediate by a spacer ligand) may result in destruction of POM. Hence, the maintenance of the POM structure after immobilization has to be carefully controlled by spectroscopic techniques, first of all, FT-IR, DRS-UV-vis and ^{31}P NMR MAS.

As illustrated by the examples, the catalytic properties of POMs may benefit from the association with the matrix, and in some cases both enhanced activity and selectivity have resulted from immobilization. That is often the case of POMs embedding in silica matrix and may be attributed to a "constrained environment effect". However, it is still very difficult to predict the catalytic performance after immobilization, and only experiment can clarify this matter. Meanwhile, a rational modification of acid-base and/or hydrophobic-hydrophilic properties of a support is possible in some cases and may tune the catalytic properties.

Many supported POMs demonstrate fairly good recyclability in liquid phase oxidations. The use of hydrothermally stable supports, such as SBA-15, MCF, etc., ensures stability of textural properties of the supported catalysts. However, good recyclability does not mean that no leaching of the active POM occurs and that catalysis has a true heterogeneous nature rather than is due to leached, homogeneous species. Indeed, some electrostatically attached and especially silica-embedded POMs are quite stable in reaction media of a medium polarity, but leaching remains a serious problem when highly polar products, such as carboxylic acids, are formed during the oxidation process. Such products may induce POM leaching *via* anion exchange or ligand substitution mechanisms, the latter being expected for datively bound POMs. In turn, covalently attached POMs are anticipated to be stable with respect to anion exchange or ligand substitution, but one has to remember that hydrolysis of covalent bonds, e.g. M-O-Si or Si-O-Si, is another process which might be responsible for leaching, particularly, when aqueous hydrogen peroxide is used as oxidant. It should be noted, however, that the number of catalytic studies on datively and covalently attached POMs is still very restricted to make a general conclusion on their stability. To the moment, microporous silica-entrapped POMs seem to be more resistant to POM leaching in polar media compared to other POM-based materials, provided the structure of the POM is maintained under turnover conditions. Indeed, POM destruction during immobilization or catalysis is another process which may account for leaching of the active species from supported POM catalysts. Sometimes, immobilization allows increasing stability of the POM structure with respect to solvolytic destruction which may occur in the presence of hydrogen peroxide. That is the case of $PW_{12}O_{40}^{3-}$ polyanion, which is easily transformed in solutions to lower nuclearity species, including the Venturello complex [40,48,71], but may keep its structure in supported catalysts. In general, the stability of POMs with respect to H_2O_2 increases with increasing the negative polyanion charge but, any way, the retention of POM structure after several cycles of use of the supported catalyst must be verified by means of spectroscopic techniques.

While at the end of 1990th – in the beginning of 2000th very few researches addressed the crucial issues of POM leaching under the reaction conditions and nature of catalysis following the methodology suggested by R. A. Sheldon [25,26], one should establish a fact that a substantial part of papers published in 2004-2007 in the area of liquid phase oxidations over supported POMs provided the required experiments and discussions on the matter. Although many supported catalysts still remain "Trojan Horses" in liquid phase oxidation processes, a considerable number of novel catalytic materials do behave as efficient and true heterogeneous catalysts. Even if they are not "Philosophers' Stones" yet the perspective of heterogeneous liquid phase selective oxidations today seems to be much more optimistic (at least, for medium polarity solutions) than ten years ago. The progress in this field is to a great extent related to the creation of new families of immobilized POM catalysts. It is, however, apparent that a great number of opportunities, challenges, and applications are yet to come in the near future.

Acknowledgments

The author is grateful to all the co-workers, especially, M.N. Timofeeva and N.V. Maksimchuk, who contributed significantly to our work in the field of oxidation catalysis by supported POMs. Collaboration of C.L. Hill, A.B. Jarzebski and J. Mrowiec-Bialon is highly appreciated. The author thanks the U.S. Civilian Research and Development Foundation for the financial support of the research on supported POMs (CRDF grant RC1-2371-NO-02).

References

[1] Sheldon, R. A.; Kochi, J. K. *Metal-Catalyzed Oxidations of Organic Compounds.* Academic Press: New York, 1981.

[2] *Fine Chemicals through Heterogeneous Catalysis.* Sheldon, R. A.; van Bekkum, H.; Eds.; Wiley-VCH: Weinheim, 2001.

[3] Centi, G.; Cavani, F.; Trifirro, F. *Selective oxidation by heterogeneous catalysis.* Kluwer Academic/Plenum Publishers: New York, 2001.

[4] Centi, G.; Perathoner, S. In *Encyclopedia of Catalysis*; Horvath, I. T.; Ed.; Wiley-Interscience: New Jersey, 2003, Vol. 6, p. 239.

[5] Brink, G.-J.; Arends, I. W. C. E.; Sheldon, R. A. In *Encyclopedia of Catalysis*; Horvath, I. T.; Ed.; Wiley-Interscience: New Jersey, 2003, Vol. 6, p.189.

[6] *Ulmann's Encyclopedia of Industrial Chemistry*; *5th Edition*, Elvers, B.; Hawkins, S.; Shulz, G.; Eds.; Vol. A(18), VCH: Weinheim, 1991.

[7] Parshall, G. W.; Ittel, S. D. *Homogeneous Catalysis: The Applications and Chemistry of Catalysis by Soluble Transition Metal Complexes*; *2nd Edition.* Wiley-Interscience: New York, 1992.

[8] Centi, G.; Misono, M. *Catal. Today,* 1998, *41*, 287-296.

[9] Clark, J. H.; Rhodes, C. N. *Clean syntheses using porous inorganic solid catalysts and supported reagents,* Royal Society of Chemistry: Cambridge, UK, 2000.

[10] Buijs, W. *Topics Catal.* 2003, *24*, 73-78.

[11] Thomas, J. M.; Raja, R. Topics Catal. 2006, *40*, 3-17.

[12] Notari, B. *Adv. Catal.* 1996, *41*, 253-334.

[13] Thomas, J. M. *Angew. Chem. Int. Ed.* 1999, *38*, 3589-3628.

[14] Lefebvre, F.; Basset, J.-M. *Current Topics Catal.* 2002, *3*, 215-236.

[15] Don Tilley, T. *J. Mol. Catal. A: Chem.* 2002, *182-183*, 17-24.

[16] De Vos, D. E.; Dams, M.; Sels, B. F.; Jacobs, P.A. *Chem. Rev.* 2002, *102*, 3615-3640.

[17] Taguchi, A.; Schuth, F. *Micropor. Mesopor. Mater.* 2005, *77*, 1-45.

[18] Barteau, M. A.; Lyons, J. E.; Song, In K. *J. Catal.* 2003, *216*, 236-245.

[19] Tanev, P. T.; Chibwe, M.; Pinnavaia, T. *Nature*. 1994, *368*, 321-323.

[20] Sayari, A. *Chem. Mater.* 1996, *8*, 1840-1852.

[21] Corma, A. *Chem. Rev.* 1997, *97*, 2373-2419.

[22] Biz, S.; Occelli, M. L. *Catal. Rev.-Sci. Eng.* 1998, *40*, 329-407.

[23] Selvam, P.; Bhatia, S. K.; Sonwane, C. G. *Ind. Eng. Chem. Res.* 2001, *40*, 3237-3261.

[24] Viswanathan, B.; Jacob, B. *Catal. Rev.* 2005, *47*, 1-82.

[25] Sheldon, R. A. *Stud. Surf. Sci. Catal.* 1997, 110, 151-175.

[26] Sheldon, R. A.; Wallau, M.; Arends, I. W. C. E.; Schuchardt, U. *Acc. Chem. Res.* 1998, *31*, 485-493.

[27] Lempers, H. E. B.; Sheldon, R. A. *J. Catal.* 1998, *175*, 62-69.

[28] Arends, I. W. C. E.; Sheldon, R. A. *Appl. Catal. A: General* 2001, *212*, 175-187.

[29] Deng, Y.; Lettmann, C.; Maier, W. F. *Appl. Catal. A: General* 2001, *214*, 31-45.

[30] Ziolek, M. *Catal. Today*. 2004, *90*, 145-150.

[31] Kholdeeva, O. A.; Trukhan, N. N. *Russ. Chem. Rev.* 2006, *75*, 411-432.

[32] Timofeeva, M. N.; Kholdeeva, O. A.; Jhung, S. H.; Chang, J. S. *Appl. Catal. A: General, in press.*

[33] Taramasso, M.; Perego, G.; Notari, B. *US Patent* 4 410 501 (1983).

[34] Clerici, M. G. In *Proceedings of the DGMK/SCI Conference "Oxidation and Functionalisation: Classical and Alternative Routes and Sources" October 12-14, 2005, Milan, Italy,* 165-176.

[35] Sanderson, W. R. *Pure and Appl. Chem.* 2000, *72*, 1289-1304.

[36] Pope, M. T. *Heteropoly and Isopoly Oxometalates*, Springler: Berlin, 1983.

[37] Moffat, J. B. *Metal-Oxygen Clusters: The Surface and Catalytic Properties of Heteropoly Oxometalates*; Kluwer/Plenum: New York, 2001, pp. 1-320.

[38] Hill, C. L.; Prosser-McCartha, C. M. *Coord. Chem. Rev.* 1995, *143*, 407-455.

[39] *Chem. Rev.* 1998, *98, special issue on POMs*; Hill, C. L.; Ed.

[40] Neumann, R. *Prog. Inorg. Chem.* 1998, *47*, 317-370.

[41] Eds.; Kluwer: Dordrecht, The Netherlands, 1993.

[42] *Polyoxometalate Chemistry for Nanocomposite design*; Yamase, T.; Pope, M. T., Eds.; Kluwer Academic/Plenum Publishes, 2002, pp. 1-235.

[43] *Polyoxometalate Chemistry: From Topology via Self-Assembly to Applications*; Pope, M. T.; Müller, A.; Eds.; Kluwer: Dordrecht, The Netherlands, 2001.

[44] Kozhevnikov, I. V. *Catalysis by Polyoxometalates*, In Catalysts for Fine Chemical Synthesis; Roberts, S. N.; Kozhevnikov, I. V.; Derouane, E.; Eds.; Wiley: Chichester, 2002, Vol. 2.

[45] Neumann, R. In *Transition Metals for Organic Synthesis; 2nd Edditon.* Beller, M.; Bolm, C.; Eds.; Wiley-VCH: Weinheim, 2004, Vol. 2, pp. 415-426.

[46] Neumann, R. In *Modern Oxidation Methods*; J.-E. Baeckvall; Ed.; Wiley-VCH: Weinheim, 2004, pp. 223-251.

[47] Pope, M. T. *In Comprehensive Coordination Chemistry II*; Wedd, A. G.; Ed.; Elsevier Science: New York, 2004, Vol. 4, p. 635.

[48] Hill, C. L. *In Comprehensive Coordination Chemistry II*; Wedd, A. G.; Ed.; Elsevier Science: New York, 2004, Vol. 4, p. 679.

[49] Mizuno, N.; Kamata K.; Yamaguchi, K. In *Surface and Nanomolecular Catalysis*; Richards, R.; Ed.; CRC Press LLC: Boca Raton, Fla, 2006; pp 463-492.

[50] *J. Mol. Catal. A: Chem.* 2007, *262*, *special issue Polyoxometalates in Catalysis*; Hill, C. L.; Ed.

[51] Baiker, L. C. W. In *Advances in the Chemistry of Coordination Compounds*; Kirschner, S.; Ed.; Macmillan: New York, 1961.

[52] Day, V. W.; Klemperer, W. G. *Science* 1985, *228*, 533-541.

[53] Finke, R. G.; Droege, M. W. *J. Am. Chem. Soc.* 1984, *106*, 7274-7277.

[54] Fournier, M.; Louis, C.; Che, M.; Chaquin, P.; Masure, D. *J. Catal.* 1989, *119*, 400-414; 415-425.

[55] Chen, Q.; Zubieta, J. *Coord. Chem. Rev.* 1992, *114*, 107-167.

[56] Kholdeeva, O. A. *Topics Catal.* 2006, *40*, 229-243.

[57] Kholdeeva, O. A.; Maksimovskaya, R. I. *J. Mol. Catal. A: Chemical*, 2007, *262*, 7-24.

[58] Izumi, Y. *Res. Chem. Intermed.* 1998, *24*, 461-471.

[59] Mizuno, N.; Misono, M. *Chem. Rev.* 1998, *98*, 199-217.

[60] Kozhevnikov, I. V. *Chem. Rev.* 1998, *98*, 171-198.

[61] Okuhara, T. *Chem. Rev.* 2002, *102*, 3641-3666.

[62] Rhule, J. T.; Neiwert, W. A.; Hardcastle, K. I.; Do, B. T.; Hill, C. L. *J. Am. Chem. Soc.*, 2001, *123*, 12101-12102

[63] Yamaguchi, K.; Mizuno, N. *New J. Chem.* 2002, *26*, 972-974.

[64] Vasylyev, M. V.; Neumann, R. *J. Amer. Chem. Soc.* 2004, *126*, 884-890.

[65] Vazylyev, M.; Sloboda-Rozner, D.; Haimov, A.; Maayan, G.; Neumann, R. *Topics Catal.* 2005, *34*, 93-99.

[66] Guo, Y.-H.; Hu, C.-W. *J. Cluster Sci.* 2003, *14*, 505-526.

[67] Guo, Y.-H.; Hu, C.-W. *J. Mol. Catal. A: Chemical* 2007, *262*, 136-148.

[68] Baba, T.; Ono, Y.; Ishimoto, T.; Morikawa, S.; Tanooka, S. *Bull. Chem. Soc. Jpn.* 1985, *58*, 2155.

[69] Venturello, C.; D'Aloisio, R.; Bart, J. C. J.; Ricci, M. *J. Mol. Catal.* 1985, *32*, 107-110.

[70] Venturello, C.; D'Aloisio, R. *J. Org. Chem.* 1988, *53*, 1553-1557.

[71] Bregeault, J.-M. *Dalton Trans.* 2003, 3289-3302.

[72] Lane, B. S.; Burgess, K. *Chem. Rev.* 2003, *103*, 2457-2473.

[73] De Villa, P. A. L.; Sels, B. F.; De Vos, D. E.; Jacobs, P. A. *J. Org. Chem.* 1999, *64*, 7267-7270.

[74] Hoegaerts, D. H.; Sels, B. F.; De Vos, D. E.; Verpoort, F.; Jacobs, P. A. *Catal. Today* 2000, *60*, 209-218.
[75] Sels, B. F.; de Villa, P. A. L.; Hoegaerts, D. H.; De Vos, D. E.; Jacobs, P.A. *Topics Catal.* 2000, *13*, 223-229.
[76] Kwon, T.; Tsigdinos, G. A.; Pinnavaia, T. J. *J. Am. Chem. Soc.* 1988, *110*, 3653-3654.
[77] Drezdzon, M. *Inorg. Chem.* 1988, *27*, 4628-4632.
[78] Yun, S. K.; Pinnavaia, T. J. *Inorg. Chem.* 1996, *35,* 6853-6860.
[79] Kwon, T.; Pinnavaia, T. J. *J. Mol. Catal.* 1992, *74*, 23-33.
[80] Tatsumi, T.; Yamamoto, K.; Tajima, H.; Tominaga, H. *Chem. Lett.* 1992, 815-816.
[81] Evans, J.; Pillinger, M.; Zhang, J. *J. Chem. Soc. Dalton Trans.* 1996, 2963-2974.
[82] Hu, C.; Zhang, X; Xu, L.; Mu, B.; Zu, W.; Wang, E. *Appl. Clay Sci.* 1998, *13*, 495-511.
[83] Carriazo, D.; Martin, C.; Rives, V.; Popescu, A.; Cojocaru, B.; Mandache, I.; Parvulesku, V. I. *Micropor. Mesopor. Mater.* 2006, *95*, 39-47.
[84] Izumi, Y.; Urabe, K.; Onaka, A. *Zeolite, Clay, and Heteropolyacids in Organic Reactions*, Kodansha, Tokio-VCH: Weinheim, 1992, pp. 1-106.
[85] Cavani, F.; Trifiro, F.; Vaccari, A. *Catal. Today* 1991, *11*, 173-301.
[86] Vaccari, A. *Catal. Today* 1998, *41*, 53-71.
[87] Rives, V.; Ulibarri, M. A. *Coord. Chem. Rev.* 1999, *181*, 61-120.
[88] Sels, B. F.; De Vos, D. E.; Jacobs, P.A. *Catal. Rev. Sci. Eng.* 2001, *43*, 443-488.
[89] Centi, G.; Perathoner, S. *Micropor. Mesopor. Mater.* 2008, *107*, 3-15.
[90] Liu, Y.; Murata, K.; Hanaoka, T.; Inaba, M.; Sakanishi, K. *J. Catal.* 2007, *248*, 277-287.
[91] Mizuno, N.; Hirose, T.; Tateishi, M.; Iwamoto, M. *Chem. Lett.* 1993, 1839.
[92] Kholdeeva, O. A.; Grigoriev, V. A.; Maksimov, G. M.; Fedotov, M. A.; Golovin, A. V.; Zamaraev, K. I. *J. Mol. Catal. A, Chemical*, 1996, *114*, 123-130.
[93] Kholdeeva, O. A.; Khavrutskii, I. V.; Romannikov, V. N.; Tkachev, A. V.; Zamaraev, K. I. *Stud. Surf. Sci. Catal.* 1997, *110*, 947-955.
[94] Xu, L.; Boring, E.; Hill, C. L. *J. Catal.* 2000, *195*, 394-405.
[95] Johnson, B. J. S.; Stein, A. *Inorg. Chem.* 2001, *40*, 801-808.
[96] Kholdeeva, O. A.; Vanina, M. P.; Timofeeva, M. N.; Maksimovskaya, R. I.; Trubitsina, T. A.; Melgunov, M. S.; Burgina, E. B.; Mrowiec-Bialon, J.; Jarzebski, A. B.; Hill, C. L. *J. Catal.* 2004, *226*, 363-371.
[97] Maksimchuk, N. V.; Melgunov, M. S.; Mrowiec-Białoń, J.; Jarzębski, A. B.; Chesalov, Yu.; Kholdeeva, O. A. *J. Catal.* 2007, *246*, 241-248.
[98] Corma, A.; Lopez Nieto J. M.; Domine, M. E. *EP Appl. Patent* 1707553, 2006.
[99] Jana, S. K.; Kubota, Y.; Tatsumi, T. *J. Catal.* 2008, *255*, 40-47.
[100] Kharat, A. N.; Pendleton, P.; Badalyan, A.; Abedini, M.; Amini, M. M. *J. Mol. Catal. A: Chem.* 2001, *175*, 277-283.
[101] Strinvasan, S.; Ford, W. T. *New J. Chem.* 1991, *15*, 693.
[102] Hasik, M.; Turek, W.; Stochmal, E.; Lapkowski, M.; Pron, A. *J. Catal.* 1994, *147*, 544-551.
[103] Stochmal-Pomarzanska, E.; Hasik, M.; Turek, W.; Pron, A. *J. Mol. Catal. A: Chem.* 1996, *114*, 267-275.
[104] Neumann, R.; Miller, H. *J. Chem. Soc. Chem. Commun.* 1995, 2277-2278.
[105] Khenkin, A. M.; Neumann, R.; Sorokin, A. B.; Tuel, A. *Catal. Lett.* 1999, *63*, 189-192.
[106] Kaleta, W.; Nowinska, K. *Chem. Commun.* 2001, 535-536.

[107] Okun, N. M.; Anderson, T. M.; Hill, C. L. *J. Amer. Chem. Soc.* 2003, 125, 3194-3195.
[108] Kala Raj, N. K.; Deshpande, S. S.; Ingle, R. H.; Raja, T.; Manikandan, P. *Catal. Lett.* 2004, *98*, 217-223.
[109] Kumar, D.; Landry, C. C. *Micropor. Mesopor. Mater.* 2007, *98*, 309-316.
[110] Zhou, Y.; Bao, R.; Yue, B.; Gu, M.; Pei, S.; He, H. *J. Mol. Catal. A: Chem.* 2007, *270*, 50-55.
[111] Chen, L.; Zhu, K.; Bi, L.-H.; Suchopar, A.; Reicke, M.; Mathys, G.; Jaensch, H.; Kortz, U.; Richards, R. M. *Inorg. Chem.* 2007, *46*, 8457-8459.
[112] Kasai, J.; Nakagawa, Y.; Uchida, S.; Yamaguchi, K.; Mizuno, N. *Chem. Eur. J.* 2006, *12* 4176-4184.
[113] Bigi, F.; Corradini, A.; Quarantelli, C.; Sartori, G. *J. Catal.* 2007, *250*, 222-230
[114] Wight, A. P., Davis, M. E., *Chem. Rev.*, *102*, 3589 (2002).
[115] Yamaguchi, K.; Yoshida, C.; Uchida, S.; Mizuno, N. *J. Am. Chem. Soc.* 2005, *127*, 530-531.
[116] Xian-Ying, S.; Jan-Fa W. *J. Mol. Catal. A: Chem.* 2008, *280*, 142-147.
[117] Bordoloi, A.; Lefebvre, F.; Halligudi, S. B. *J. Catal.* 2007, *247*, 166-175.
[118] Kholdeeva, O. A.; Timofeeva, M. N.; Maksimov, G. M.; Maksimovskaya, R. I.; Rodionova, A. A.; Hill, C. L. In *Catalysis of Organic Reactions*; Sowa, J. R., Jr.; Ed.; Taylor & Francis Group, Boca Raton, 2005, pp. 429-433.
[119] Kholdeeva, O. A.; Timofeeva, M. N.; Maksimov, G. M.; Maksimovskaya, R. I.; Neiwert, W. A.; Hill, C. L. *Inorg. Chem.* 2005, *44*, 666-672.
[120] Alie, C.; Pirard, R.; Lecloux, A. J. Pirard, J.-P. *J. Non-Cryst. Solid.* 1999, 246, 216-228.
[121] Maksimov, G. M. *Russ. Chem. Rev.* 1995, *64*, 445-461.
[122] Weakley, T. J. R. *J. Chem. Soc. Dalton Trans.* 1973, 341-346.
[123] Zhang, X.; Zhang, C.; Guo, H.; Huang, W.; Polenova, T.; Francesconi, L. C.; Akins, D. L. *J. Phys. Chem. B* 2005, *109,* 19156-19160.
[124] Zhou, Y.; Bao, R.; Yue, B.; Gu, M.; Pei, S.; He, H. *J. Mol. Catal. A: Chem.* 2007, *270*, 50-55.
[125] Férey, G.; Mellot-Draznieks, C.; Serre, C.; Millange, F.; Dutour, J.; Surble, S.; Margiolaki, I. *Science* 2005, *309*, 2040-2042.
[126] Maksimchuk, N. V.; Timofeeva, M. N.; Melgunov, M. S.; Shmakov, A.N.; Chesalov, Yu. A.; Dybtsev, D. N.; Fedin, V. P.; Kholdeeva, O. A. *J. Catal. in press.*
[127] Izumi, Y.; Urabe, K. *Chem. Lett.* 1981, 663-664.
[128] Neumann, R.; Levin, M. *J. Org. Chem.* 1991, 56, 5707-5710.
[129] Gall, R. D.; Hill, C. L.; Walker, J. E. *J. Catal.* 1996, *159*, 473-478.
[130] Lefebvre, F. *J. Chem. Soc., Chem. Commun.* 1992, 756-757.
[131] Kozhevnikov, I. V.; Sinnema, A.; Jansen, R. J. J.; van Bekkum, H. *Catal. Lett.* 1994, *27*, 187-197.
[132] Schwegler, M. A.; Vinke, P.; van der Eijk, M.; van Bekkum, H.; *Appl. Catal.*1992, *80*, 41-57.
[133] Bruckman, K.; Che, M.; Haber, J.; Tatibouet, J. M. *Catal. Lett.* 1994, *25*, 225-240.
[134] Gamelas, J. A. F.; Evtuguin, D. V.; Esculcas, A. P. *Transit. Metal. Chem.* 2007, 32, 1061-1067.
[135] Neumann, R.; Cohen, M. *Angew. Chem. Int. Ed. Engl.* 1997, *36*, 1738-1740.
[136] .Neumann, R.; Cohen, M. *J. Mol. Catal. A: Chem.* 1999, *146*, 291-298
[137] .Judeinstein, P. *Chem. Mater.* 1992, *4*, 4-7.

[138] .Mayer, R. C.; Thouvenot, R. *Chem. Mater.* 2000, *12*, 257-260.
[139] Sanchez, C.; Lebeau, B.; Ribot, F.; In, M. *J. Sol-Gel Sci. Technol.* 2000, *19*, 31-38.
[140] Xu, L.; Lu, M.; Xu, B.; Wei, Y.; Peng, Z.; Powell, D. R. *Angew. Chem. Int. Ed.* 2002, *41*, 4129-4132.
[141] Guo, Y.; Yang., Y.; Hu, C.; Guo, C.; Wang, E.; Zou, Y.; Feng, S. *J. Mater. Chem.* 2002, *12*, 3046-3052.
[142] Mayer, C. R.; Neveu, S.; Cabuil, V. *Angew. Chem. Int. Ed.* 2002, *41*, 501-503.
[143] Errington, R. J.; Petkar, S. S.; Horrocks, B. R.; Houlton, A.; Lie, L. H.; Patole, S. N. *Angew. Chem. Int. Ed.* 2005, *44*, 1254-1257.
[144] Gouzerh, P.; Proust, A. *Chem. Rev.* 1998, *98*, 77-111.
[145] Gelbard, G.; Breton, F.; Quenard, M.; Sherrington, D. C. *J. Mol. Catal. A: Chem.* 2000, *153*, 7-18.
[146] Gelbard, G.; Gauducheau, T.; Vidal, E.; Parvulescu, V. I.; Crosmanb, A.; Pop, V. M. *J. Mol. Catal. A: Chem.* 2002, *182-183*, 257-266.
[147] Liu, Y., Murata, K., Inaba, M., *Green Chem.* 2004, *6*, 510-515.
[148] Mrowiec-Bialon, J.; Turek, W.; Jarzebski, A. B. *React. Kinet. Catal. Lett.* 2002, *76*, 213-219.
[149] Maksimchuk, N. V.; Melgunov, M. S.; Mrowiec-Bialon, J.; Jarzebski, A. B.; Kholdeeva, O.A. *J. Catal.* 2005, *235*, 175-183.
[150] Katsoulis, D. E. *Chem. Rev.* 1998, *98,* 359-387.
[151] Guo, Y.; Yang, Y.; Hu, C.; Guo, C.; Wang, E.; Zou, Y.; Feng, S. *J. Mater. Chem.* 2002, *12*, 3046-3052.
[152] Wu, Q. *Mater. Chem. Phys.* 2002, *77*, 204-208.

In: Heterogeneous Catalysis Research Progress
Editor: Mathias B. Gunther, pp. 299-344

ISBN 978-1-60456-979-7

Chapter 8

NEW APPROACHES FOR REUSABLE CHIRAL HETEROGENEOUS CATALYSTS FOR EPOXIDE RING OPENING REACTION

R. I. Kureshy*[1]*, N. H. Khan, S. H. R. Abdi, Santosh Agrawal, K. Jeya Prathap and R. V. Jasra

Discipline of Inorganic Materials and Catalysis, Central Salt and Marine Chemicals Research Institute (CSMCRI), Bhavnagar- 364 002, Gujarat, India

ABSTRACT

The heterogenization of chiral homogeneous catalysts, which endows homogeneous systems with attractive features such as easy product separation and catalyst recovery by simple filtration, constitutes a rapidly expanding research area in asymmetric catalysis. Among various organic transformations, chiral transition-metal complexes catalyzed asymmetric ring opening reaction is one of the most fascinating areas because the enantiopure end-products have wider application in pharmaceuticals, fine chemicals and as chiral auxillaries. In this direction attempts were made to develop highly active and enantioselective catalysts based on chiral ligands viz. BINOLs, SALENs etc., with various transition metals. As chiral catalysts are expensive, focuses on low catalyst loading and/ or their recovery and re-use are important aspects. Since various homogeneous catalysts have been found to be efficient in asymmetric catalysis their immobilization on solid supports is of great interest. This led researches to make these systems recyclable by way of supporting the catalyst on organic and inorganic polymeric materials or making use of ionic liquids or manipulating solubility of the catalyst by increasing the molecular weight of catalyst with simultaneous increase in active sites so that the catalyst is easily recovered by simple precipitation method in a post catalytic workup process. Therefore, this chapter will give an in-sight of epoxide asymmetric ring opening reaction based on chiral recyclable catalysts and would bring about the latest trends in this area of research.

[1] Fax: +91-0278-2566970.E-mail: rukhsana93@yahoo.co.in.

INTRODUCTION

The growing demand for enantiopure compounds in pharmaceuticals, food and agrochemicals, fragrance and flavors, biochemicals and fine chemicals industries has stimulated interest in asymmetric catalysis [1]. Asymmetric catalysis strongly relies on ready access to enantiomerically pure ligands with precious metals (metalloenzyme mimics), chirally pure organic molecules with extensive hydrogen bonding (enzyme mimics) and enzymes as catalyst. High chemo and enantioselectivity together with high turnover number are requisite properties for an efficient chiral catalyst. Last two decades have shown tremendous growth in the area of asymmetric catalysis under homogeneous condition to achieve target chiral molecules in their high optical purity. In majority of the homogeneous catalysis, a highly specialized chiral ligand is required as pre-catalyst, which is often more expensive than most precious metals. Thus recovery and recyclability of the catalyst is an unavoidable issue for such processes. On the other hand, the heterogeneous catalysis has been very successful in terms of industrial applications for the production of racemic and achiral compounds. Therefore, the heterogenization of chiral homogeneous catalyst was considered as logical option for the practical development of chiral technology.

In asymmetric catalysis, the epoxide ring opening reaction provides an elegant route to develop synthetically useful and biologically interesting molecules in optically pure form [2-4]. Some of the useful racemic / *meso* epoxide ring opening reactions and their resultant end products are depicted in Scheme 1.

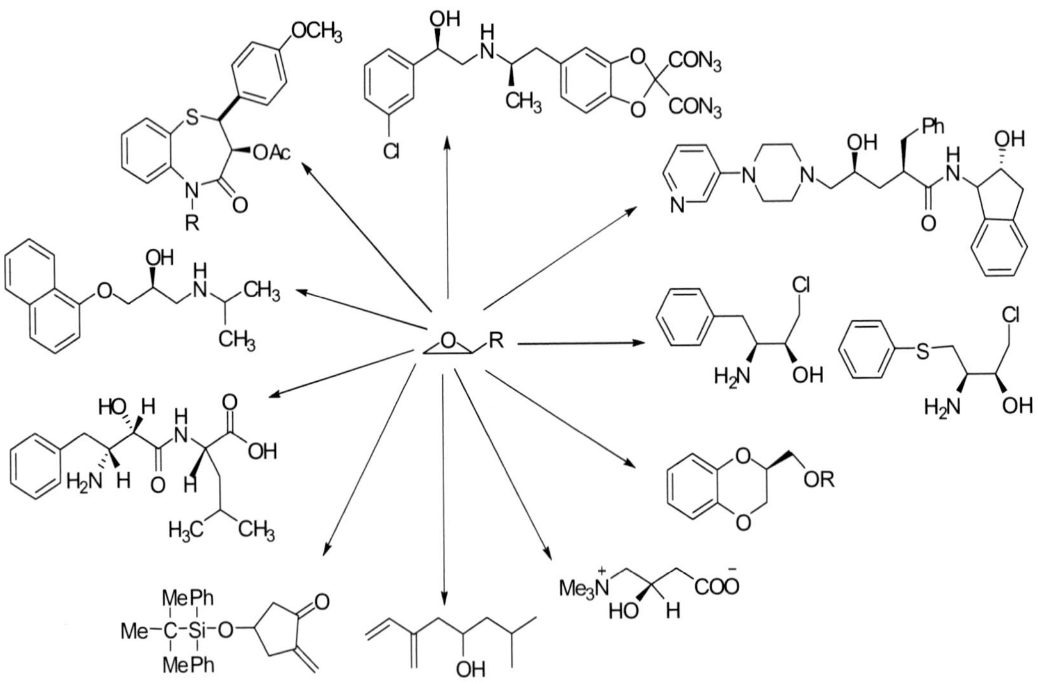

Scheme 1. Stereospecific ring opening of epoxide.

Typically chiral metal complexes catalyzed asymmetric ring opening of achiral / racemic and chiral epoxides with various nucleophiles conveniently produce enantioenriched 1,2-

bifunctional compounds viz., β-azido alcohols [5-11], β-halohydrins [12-19], β-cyanohydrins [10,20-22], β-hydroxysulfides [23-26], β-benzoyloxy alcohols [27], β-aryloxy alcohols [28, 29], and β-aminoalcohols [30-35]. The entire literature available on these transformations can be classified as under (Scheme 2).

- Kinetic resolution (KR) of racemic epoxides
- Desymmetrization of *meso*-substrates
- Dynamic kinetic resolution (DKR) of racemic epoxides
- Asymmetric ring opening (ARO) reaction of enantiopure epoxide

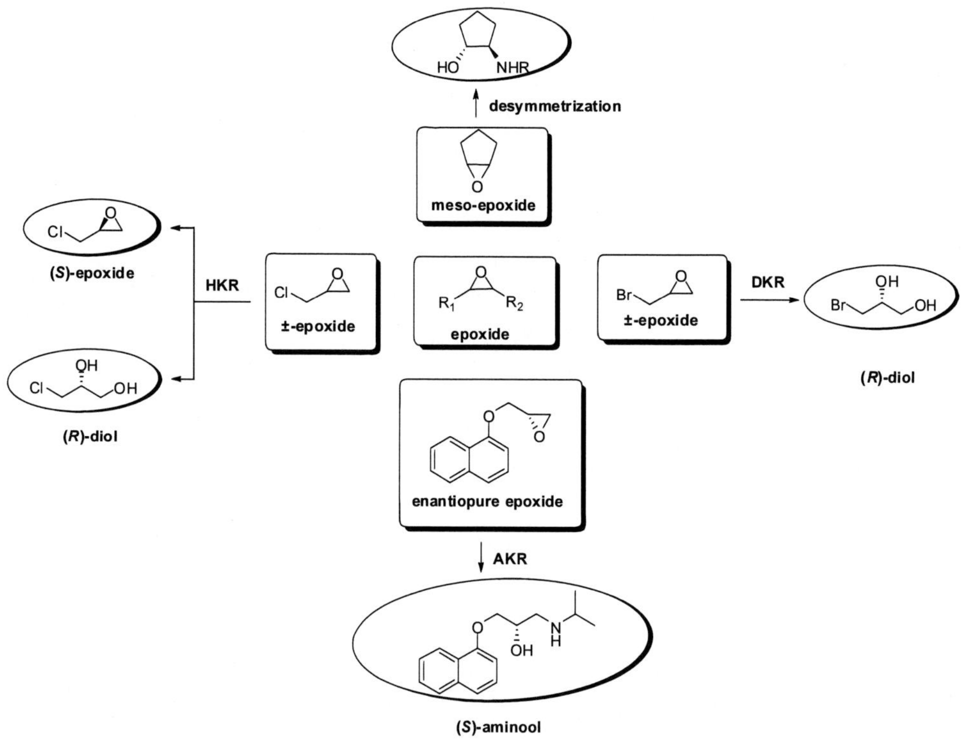

Scheme 2 Various strategies for enantioselective epoxide ring opening reactions.

The above transformations have been extensively studied under homogeneous condition [36-41], however, their heterogeneous counter part is still in developmental stage [42-47].

Considering the importance of recoverable catalyst, this chapter, would provide an overview on asymmetric ring opening of epoxides using heterogeneous catalysis, however, we would also include some of the homogeneous chiral catalysts which are recoverable and have potential for heterogenization. Several strategies used for heterogenization/recycling of chiral catalysts discussed in this chapter include; immobilization of the soluble catalysts; on organic and inorganic polymers, on ionic liquids, use of differential solubility of the catalyst in various solvents and two-phase systems. In principle, all the methods presented here have capability of repeated use of a chiral catalyst without loss of activity and/or enantioselectivity. While the chapter will present up to the date information on epoxide ring opening reaction available in the literature, an attempt would be made to discuss highlights and limitations of the existing systems and bring about recent trends in heterogeneous catalysis.

Kinetic Resolution of Racemic Epoxides

Jacobsen et al. [48], in 1997 for the first time demonstrated KR of racemic terminal epoxides with water as nucleophile for the production of optically pure epoxides and corresponding 1,2-diols. Since then, various other nucleophiles viz., carboxylic acids, phenols, thiols, amines, carbamates and indols were used in KR to produce optically pure epoxides with concomitant production of corresponding enantioenriched 1,2-bifunctional moieties [49-52].

Hydro Kinetic Resolution (HKR) of Racemic Epoxides

Kinetic resolution of racemic terminal epoxide with water (HKR) is an attractive strategy for the synthesis of valuable enantiopure terminal epoxide and corresponding diol. Easy availability of terminal epoxides at cheaper price and water as sole reagent with a recoverable chiral catalyst makes this solvent free protocol very attractive for its commercial exploitation [53, 54]. Both terminal epoxides and respective diols in their chirally pure form have wider applications in academics and industry [48, 50]. For the efficient resolution the reaction rates of the two enantiomers must be unequal and the reaction must be stopped when only one enantiomer reacts to give a maximum of 50% product leaving behind the other enantiomer unreacted.

In the case of Jacobsen's Co(III) salen complexes 1, 2 (Figure 1) [48], (*R, R*) form of the catalyst selectively allows water to react with (*S*)-form of epoxide to generate (*S*)-1,2-diol leaving behind (*R*)-epoxide in high enantioselectivity (ee) (Scheme 3, PATH A). By simply reversing the chirality of the catalyst from (*R,R*) to (*S,S*), epoxides and 1,2-diols in their respective (*S*) and (*R*) forms (Scheme 3, PATH B) were achieved in high optical purity. This method was found to be highly effective and economical for the low boiling racemic epoxides where catalyst separation was affected by simple distillation process. However, this process was less effective for high boiling racemic epoxides due to the decomposition of the desired products at higher temperatures.

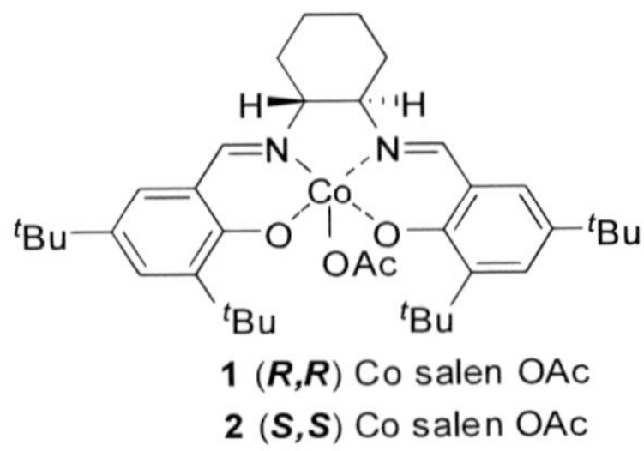

1 (***R,R***) Co salen OAc
2 (***S,S***) Co salen OAc

Figure 1. Structure of the catalysts 1, 2.

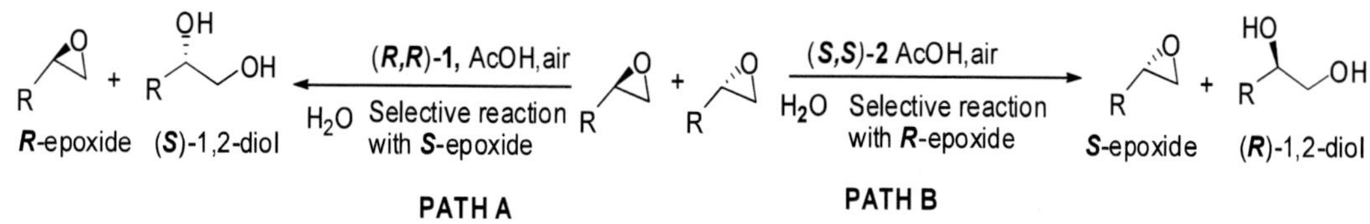

Scheme 3. HKR of terminal epoxide using Jacobsen Co(salen) complexes 1, 2.

To overcome this issue Kureshy et al. [55, 56] reported dimeric form of Jacobsen's catalysts 3, 4. They used the concept of solubility modification by altering the molecular weight of the catalyst so that in a post catalytic work-up procedure the catalyst is precipitated, filtered and used for subsequent catalytic runs. The complexes 3, 4 (0.2 mol % of Co(III)–salen unit) (Figure 2) were effectively used for HKR of racemic epoxides, e.g., styrene oxide, epichlorohydrin, 1,2-epoxypropane, 1,2-epoxyhexane, 1,2-epoxyoctane, and 1,2-epoxydodecane to achieve corresponding epoxides and 1,2-diols in high optical purity and isolated yields. In this process, once the catalytic reaction is complete the product epoxides were collected by reduced pressure distillation. Addition of diethylether to the residue precipitated the catalyst which was removed by filtration. However, the recovered catalyst was required to be reactivated by its treatment with acetic acid in air. The catalysts were reused 4 times with complete retention of its performance.

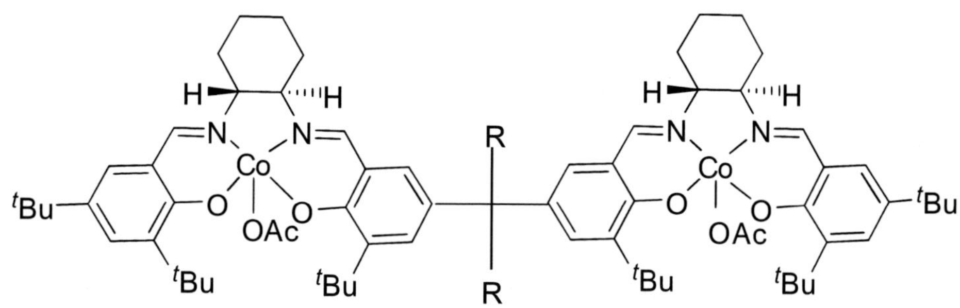

3 R= H
4 R=CH_3

Figure 2. Structure of dimeric Co(salen)complexes 3,4

Ironically, oligomeric Co(III) salen complexes **5** (Figure 3) developed by Jacobsen et al. [57, 58] though highly active was non-recyclable. These complexes degraded during HKR of terminal/meso epoxides under the reaction conditions used.

Keeping in mind the recovery of the catalyst issue, Pozzi et al. [59] specifically tailored the salen ligand to suit its application in fluorous biphasic (FB) system (Figure 4). Accordingly, authors made modification at 5 and 5' position of Jacobsen catalyst by replacing *tert*-butyl group with perfluoroalkyl chain **6** or 3,5-bis heptadecafluorooctylphenyl 7, 8 [60].

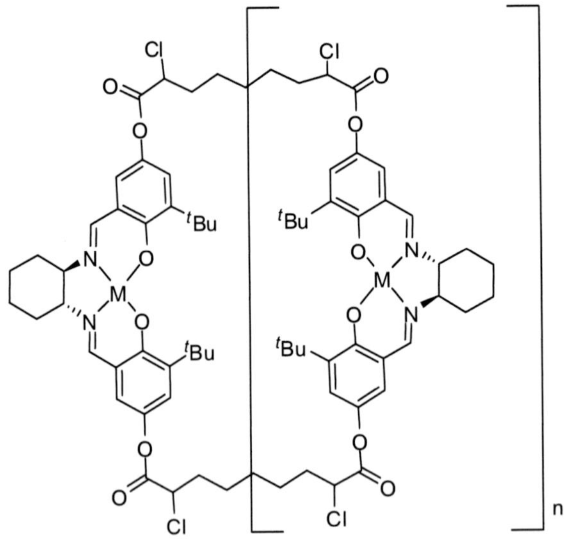

5

M = Co(OTs)
n = 1-5

Figure 3. Oligomeric Co(salen) complex 5.

The catalyst 6 afforded 46% diol with 99% ee and epoxide (ee, 91.2%) in 6h for the HKR of racemic 1-hexene oxide. The catalyst 6 was recovered by liquid-liquid extraction at room temperature in the presence of a fluorinated solvent; however there was a loss of catalyst by 16% in the first reuse experiment. Authors also tried solid phase extraction of 6 by fluorous reverse phase silica with toluene. However, only 50% of the catalyst was recovered while remaining material was adsorbed on silica. The recovered catalyst still showed catalytic activity (47% conversion with diol ee >99%) but reaction rate was reduced considerably (50h). Ironically silica-bound material which contained 0.4 mol% of 6 was inactive for HKR of 1-hexene oxide. Site isolation resulting from low catalyst loading on the fluorous silica was explained to be the main reason for the absence of catalytic activity of this material. The heavily fluorinated catalyst 7, 8 when activated with $C_8F_{17}COOH$ was found to be highly active (50% conversion in 2h) and enantioselectivity for both epoxide and diol (ee, >99%) under fluorous biphasic condition. The recycling of the catalysts 7, 8 caused sever loss of activity and enantioselectivity of epoxide however, ee of diol was retained. It is to be noted that recycling of the heavily fluorinated chiral complexes 7, 8 was less effective than that of the light-fluorous complex 6.

Kim et al. [61] demonstrated that with the change in counter ion in Co(III)-X where (X= 9-17), the catalysts could be reused ten times after simple distillation of products without observable loss in activity and enantioselectivity for HKR of epichlorohydrin. Interestingly the catalyst-regeneration step was not required with the use of PF_6 11 and BF_4 12 as counter ion in this system (Scheme 4).

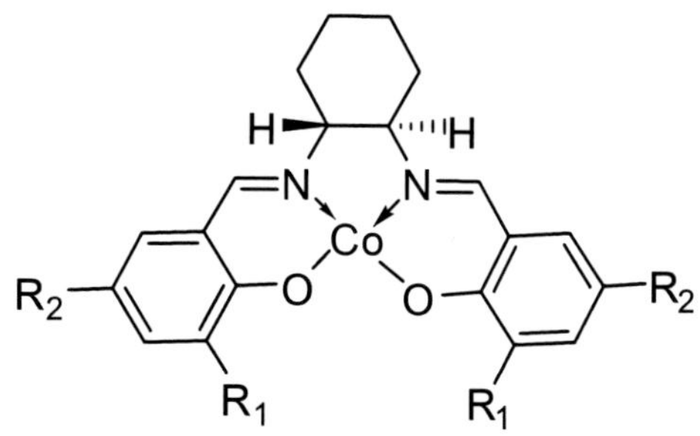

	R_1	R_2
(*R,R*)-**6**	Bu^t	C_8F_{17}
(*R,R*)-**7**	C_8F_{17}	C_8F_{17}
(*R,R*)-**8**	Bu^t	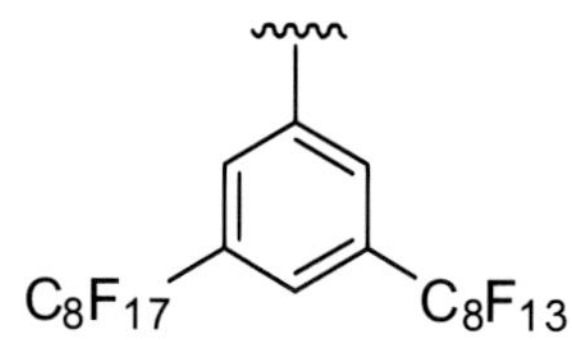

Figure 4. Structure of fluorous chiral Co(salen) catalysts 6-8.

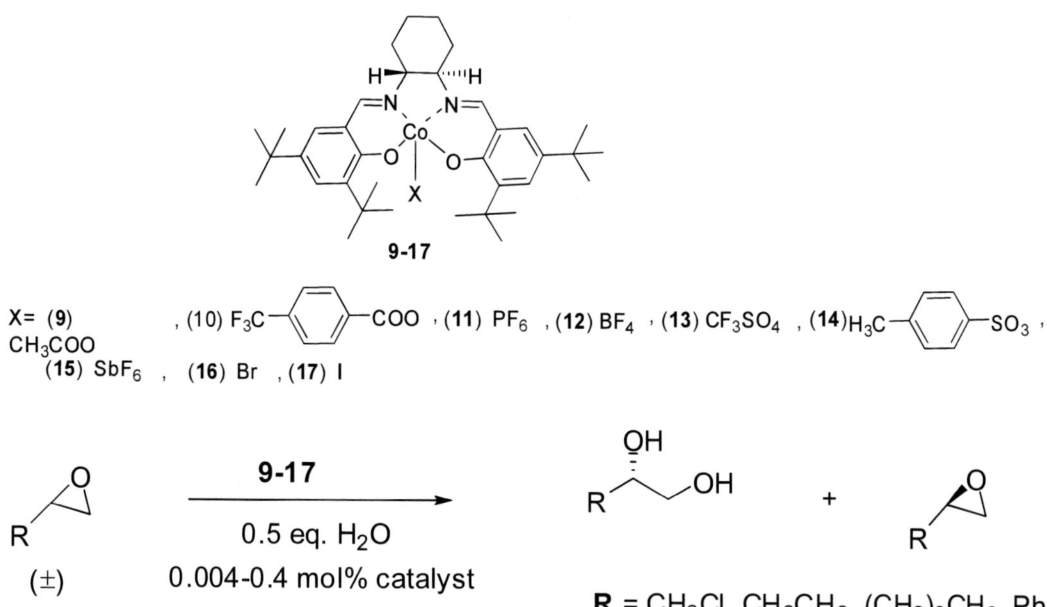

Scheme 4. HKR of terminal epoxide using Co(salen) complexes 9-17.

Song et al. [62] reported poly-salen Co(III) complexes 18, 19 as catalyst for HKR (Figure 5) of terminal alkene epoxides. The polymeric catalysts provided product epoxides with excellent conversion (>49%) and high chiral purity (ee's, 98%) and the catalytic system could be recycled once with retention of activity and enantioselectivity.

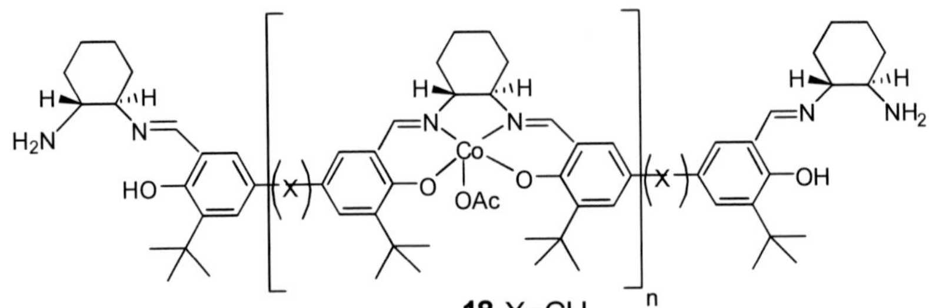

18 X=CH_2
19 X= $C(CH_3)$

Figure 5. Structure of the catalysts 18, 19.

Later on same authors [63] developed several crosslinked polymeric salen-Co(III) complexes 20, 21 with different pore sizes by using the trialdehyde and the dialdehyde in different proportions with (*R*,*R*)-1,2-diaminocyclohexane (Figure 6). The molecular weight of these complexes was found to be in the range of 4000-10000 by (GPC). The crosslinked polymeric salen–Co(III) complexes 20, 21 (0.02 mol %) were employed in the HKR of terminal epoxides viz. epichlorohydrin, styrene oxide and phenyl glycidyl ether. Excellent activities and enantioselectivities were achieved with all catalysts in different ratios of trialdehyde and dialdehyde. Most of the crosslinked polymer catalysts showed better activities than oligomeric catalyst with trialdehyde:dialdehyde::0:100, which means the crosslinker had positive effect due to better cooperation between the metal centers of the salen units. Unfortunately none of these catalysts were recyclable, possibly due to the hydrolysis of ester linkage under the HKR condition.

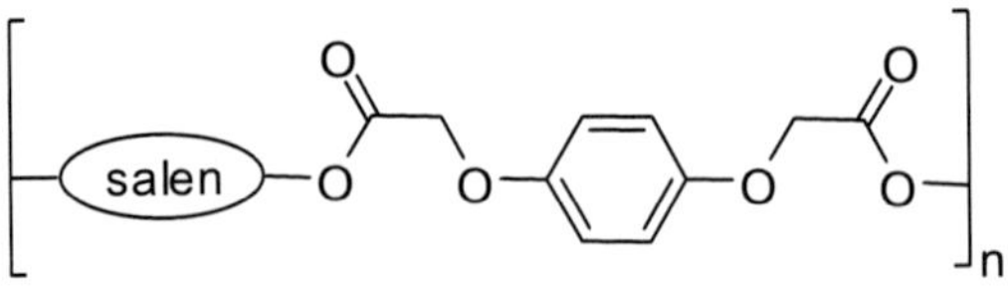

n = 6

20

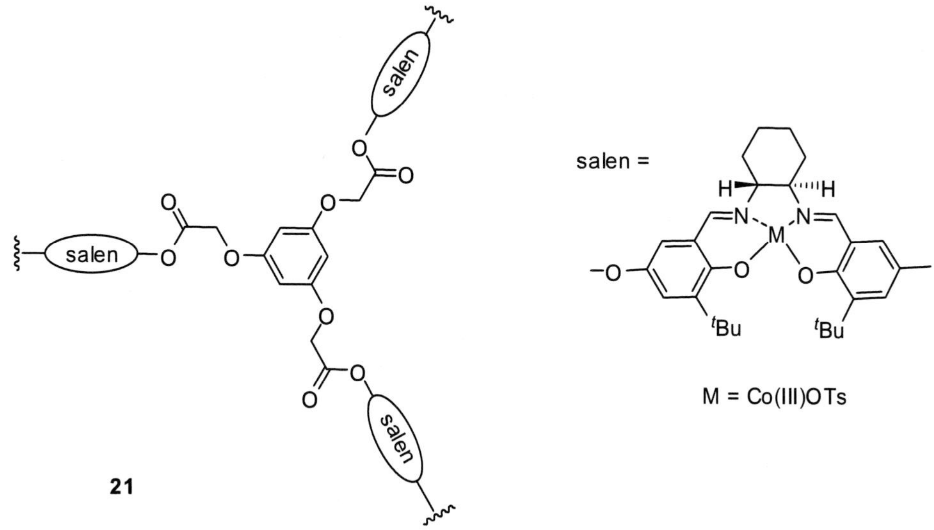

21

Figure 6. Structure of the crosslinked polymeric complexes 20, 21.

Weck et al. [64] reported synthesis of poly-(styrene)-supported Co-salen complexes by free-radical co-polymerization of vinyl-salen derivatives with styrene **22** (Figure 7). These catalysts (0.5 % of catalyst loading) were tested for HKR of racemic epychlorohydrin with water. The reaction was run neat since the polymers were soluble in the epoxide. Further, an improvement in the activity and selectivity was reported by diluting the Co–salen units along the polymer backbone by increasing the ratio between styrene and vinyl-salen monomer. The copolymerized catalysts showed better catalytic performance (>99% ee, 54% conversion, 1h) in comparison to the homopolymeric analogues and small molecule Co-salen complex. This difference in the performance might be due to the dilution effect or better accessibility of the reactants to the catalytic sites in more flexible co-polymeric catalysts. The polymeric catalyst could be recovered by precipitation with diethylether and was reused three times with a slight decrease on reactivity but with consistent conversion and enantioselectivity.

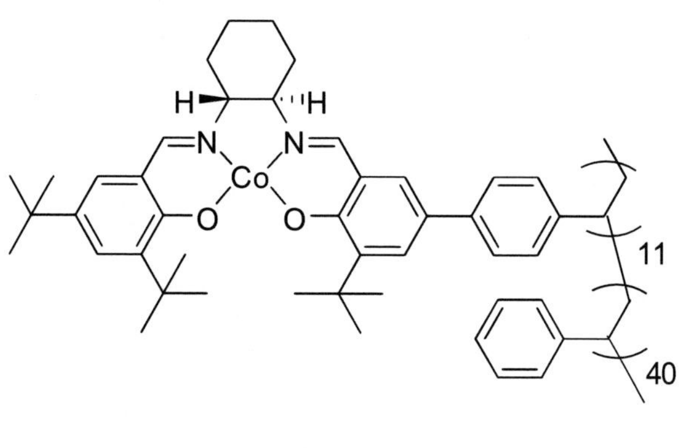

22

Figure 7. Structure of the complex 22.

Weck et al. [65] further reported the Co salen complex supported on norbornene polymers (23, 24) with stable phenylene-acetylene linker (Figure 8). The polymer-supported salen catalysts were investigated for HKR of the racemic terminal epoxides that showed outstanding catalytic activities and comparable selectivities to the original catalysts reported by Jacobsen. However, the polymeric catalyst was recycled only once after its precipitation with diethylether as the catalyst became less soluble and less reactive in subsequent catalytic runs.

Davis et al. [66] developed an oligo(cyclooctene)-supported Co-OAc salen complex **25** (Figure 9) which efficiently catalyzed HKR of epichlorohydrin. The catalyst was recycled many times with negligible deactivation. The oligomeric complex was found to be 25 times more reactive than monomeric Co-salen. This study also gives possible modes of deactivation of Jacobsen's Co-salen catalyst during HKR of epichlorohydrin by spectroscopy, combined with recycling studies.

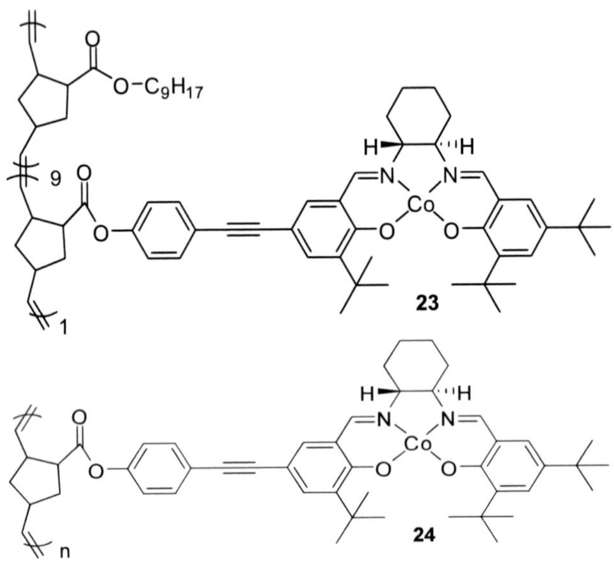

Figure 8. Structure of norbornene-functionalized Co(salen) complexes 23, 24.

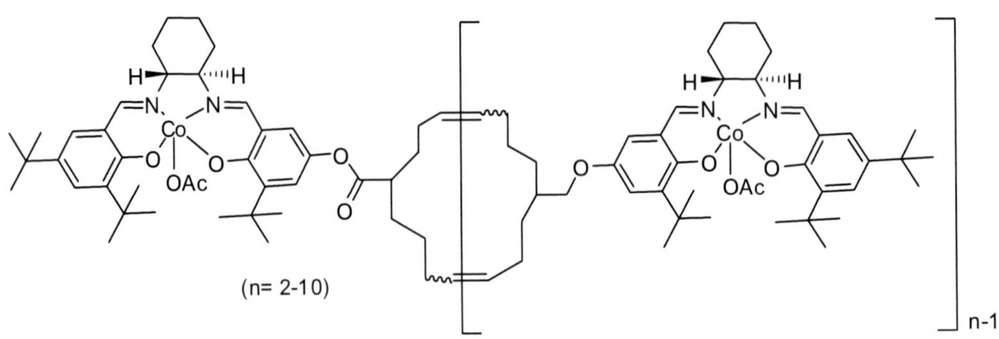

25

Figure 9. *R,R*-Oligo(cyclooctene) supported Co-OAc salen catalyst 25.

Kim et al. [67] recently reported the synthesis of heterometallic chiral polymer (salen) Co-(Al, Ga, In)Cl_3 complexes **26-32** (Figure 10) and their use in the HKR of racemic epoxides. Polymeric salen catalysts showed very high reactivity and enantioselectivity at substantially lower catalyst loadings for the asymmetric ring opening of terminal epoxide to obtain the enantio-enriched products. The performance of catalysts is retained on multiple-use and do not suffer the problems of solubility and deactivation (Scheme 5).

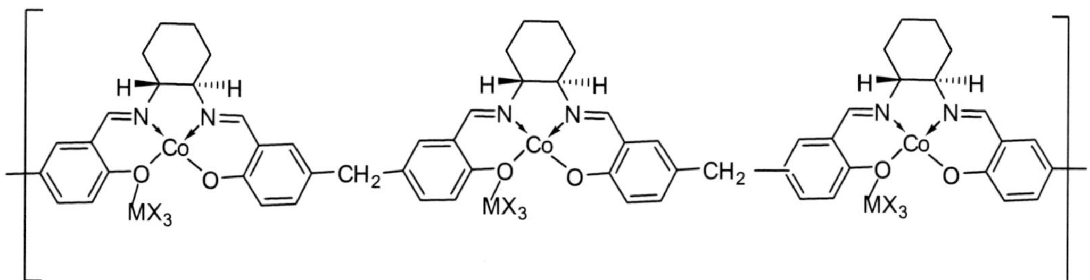

Polymer (Salen) Cata. **26** - **28**

MX_3 ; **26** = $AlCl_3$
27 = $GaCl_3$
28 = $InCl_3$

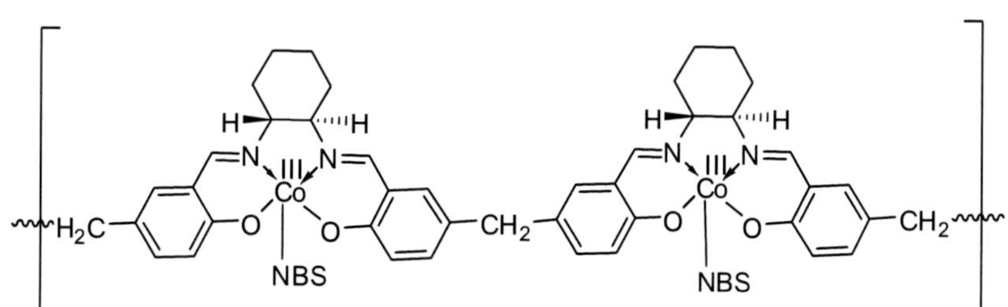

Polymer (Salen) Cata. **29**

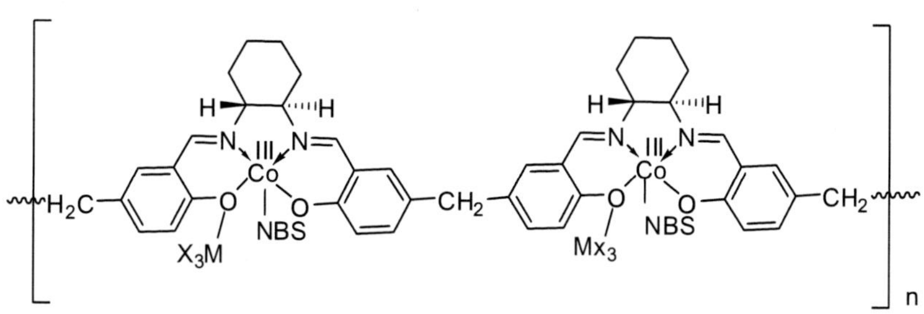

Polymer (Salen) Cata. **30** - **32**

MX_3 ; **30** = $AlCl_3$
31 = $GaCl_3$
32 = $InCl_3$

Figure 10. Structure of chiral heterometallic salen complexes 26-32.

Further, the effects of metal halide source on the catalytic activity in HKR of epichlorohydrin was also studied with the following activity trend; Co-$InCl_3$ 32> Co-$GaCl_3$ 31> Co-$AlCl_3$ 30> Co-$GaCl_3$ 27> catalyst 29> Co-$AlCl_3$ 26. It was claimed that neither the pre-catalyst (salen)Co 29 nor 13-group MX_3 salts alone exhibits any epoxide ring opening reactivity in the presence of nucleophiles.

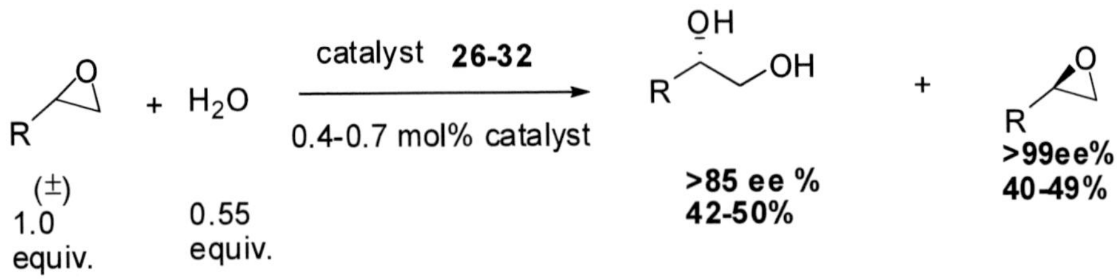

Scheme 5. HKR of terminal epoxides catalyzed by polymeric salen catalysts 26-32.

Breinbauer and Jacobsen [68] reported the synthesis of dendrimer-bound Co(salen) complexes 33-35 (Figure 11) derived from commercially available NH_2-terminated PAMAM (polyamidoamine) dendrimers. The dendritic catalysts were used for HKR of (*rac*)-1,2-epoxyhexane (Scheme 6) and (*rac*)-vinylcyclohexane epoxide as substrates where highly enantioenriched (>98% ee) epoxide at 50% conversion was achieved. The best results were obtained with the first generation (4-branch) 33 metallodendrimer and the efficiency of catalyst on a per-metal basis was in the following order: 4-Co(salen)-PAMAM 33 > 8-Co(salen)-PAMAM 34 > 16-Co-(salen)-PAMAM 35. This "dendrimer effect" was thought to arise from restricted conformation imposed by the dendrimer structure, which enhanced the cooperative interactions between Co-salen units (Figure 12). However, no catalyst recycle data was provided.

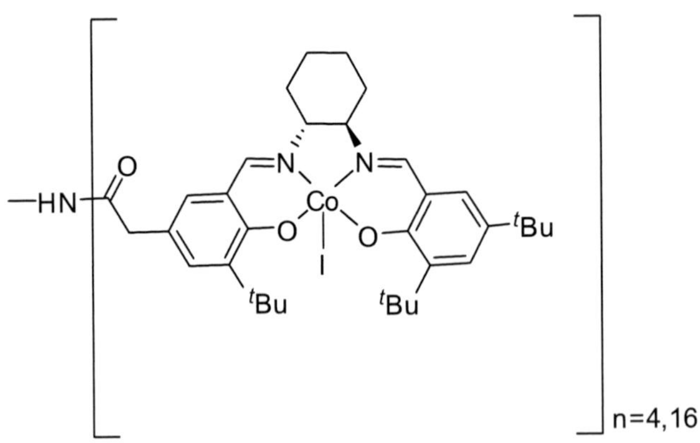

PAMAM-4-Co(Salen) 33
PAMAM-16-Co(Salen) 35

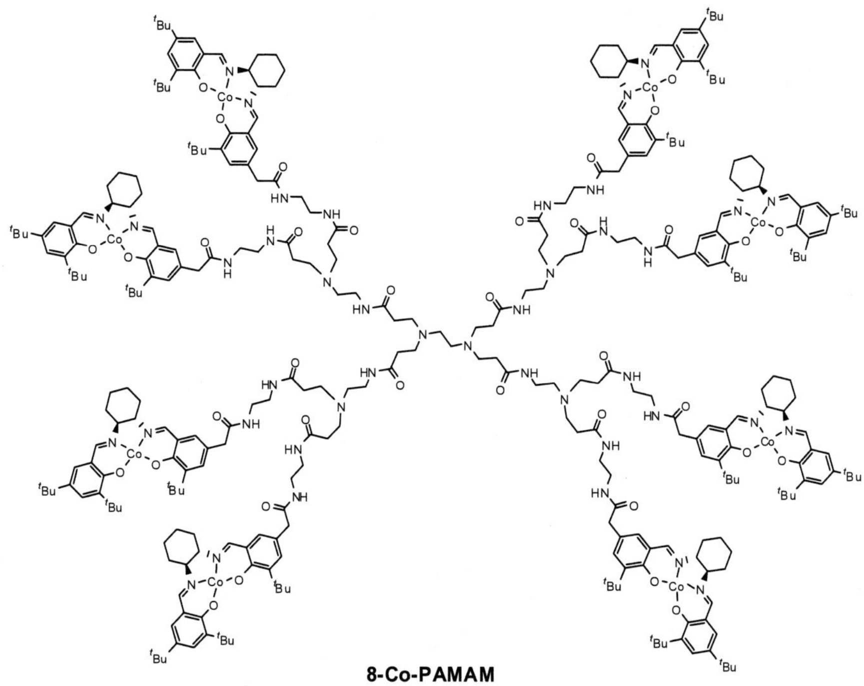

8-Co-PAMAM

34

Figure 11. Structure of dendrimeric [Co(salen)] complexes 33,34.

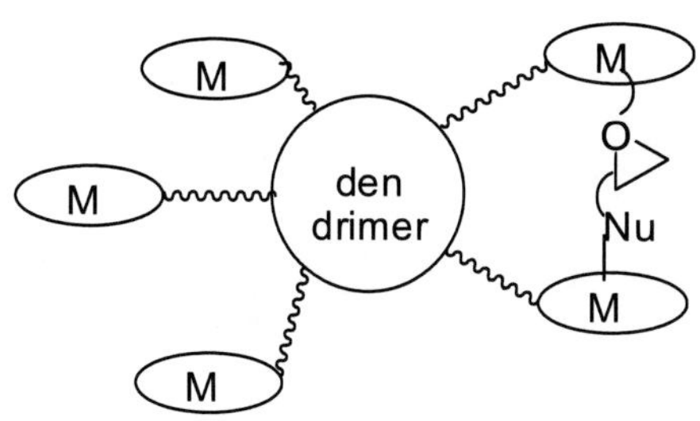

Figure 12. Proposed mechanism for cooperative catalysis in asymmetric ring opening of epoxide by dendimeric frame work.

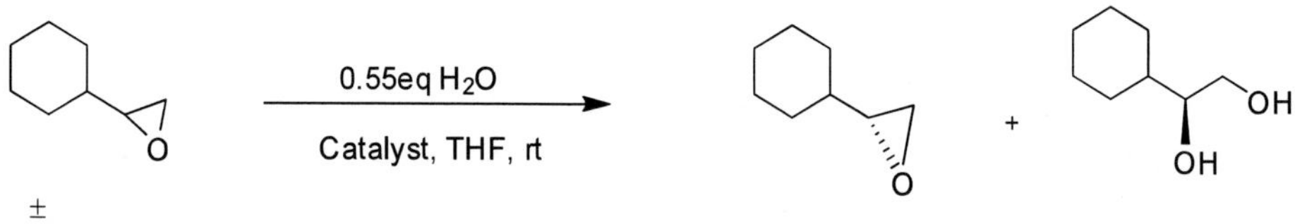

Scheme 6. Reaction for HKR of racemic cyclohexeneepoxide using dendimeric Co(salen) complex 34.

Annis et al. [69] reported the synthesis of polystyrene- 36 and silica bound Co(salen) 37 and their use in HKR of racemic epoxides (Scheme 7-9). Polystyrene bound systems 36 demonstrated highly practical solutions to certain technical difficulties associated with the isolation of reaction products from HKR especially problematic substrates like epichlorohydrin and other high boiling epoxides (Table 1).

Table 1. Recycling data of HKR of epichlorohydrin using the catalyst 36

Cycle	Conversion[a] (%)	Ee epoxide(%)	Ee Diol (%)	k_{rel}
1	52	>99	92.4	133
2	51	>99	95.0	206
3	51	>99	93.6	159
4	51	>99	93.4	154
5	52	>99	93.0	145

[a]Estimated based on the ee of epoxide and diol product

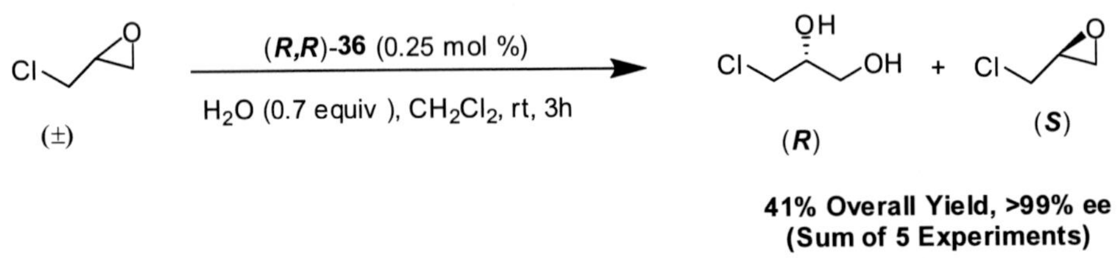

Scheme 8 HKR of epichlorohydrin using supported catalyst 36.

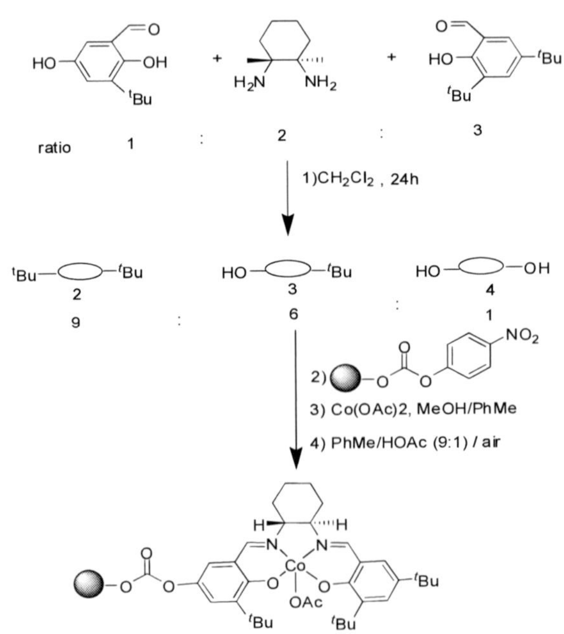

Scheme 7. Resin captured synthesis of polystyrene-bound chiral Co(salen) complex **36.**

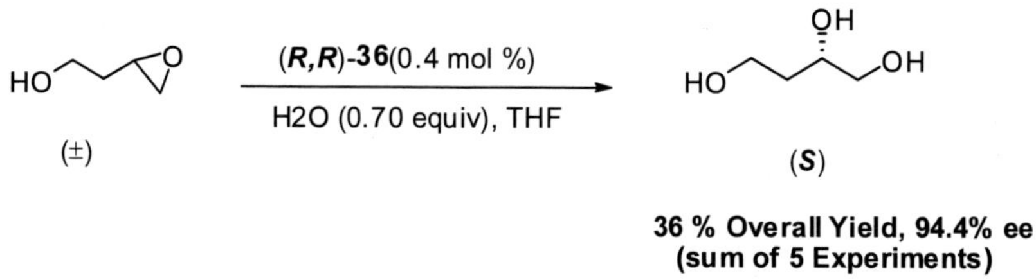

Scheme 9. HKR of 4-hydroxy-1-butene oxide with solid phase catalyst.

Similarly Silica-Bound Co(salen) 37 (Scheme 10) [69] was also effectively used in the HKR of styrene oxide (Scheme 11) and 4-hydroxy-1-butene oxide (Scheme 12). The immobilized catalysts were adapted to a continuous flow process for the generation of reaction products in high yield and ee, requiring only very simple techniques for product purification (Scheme 13).

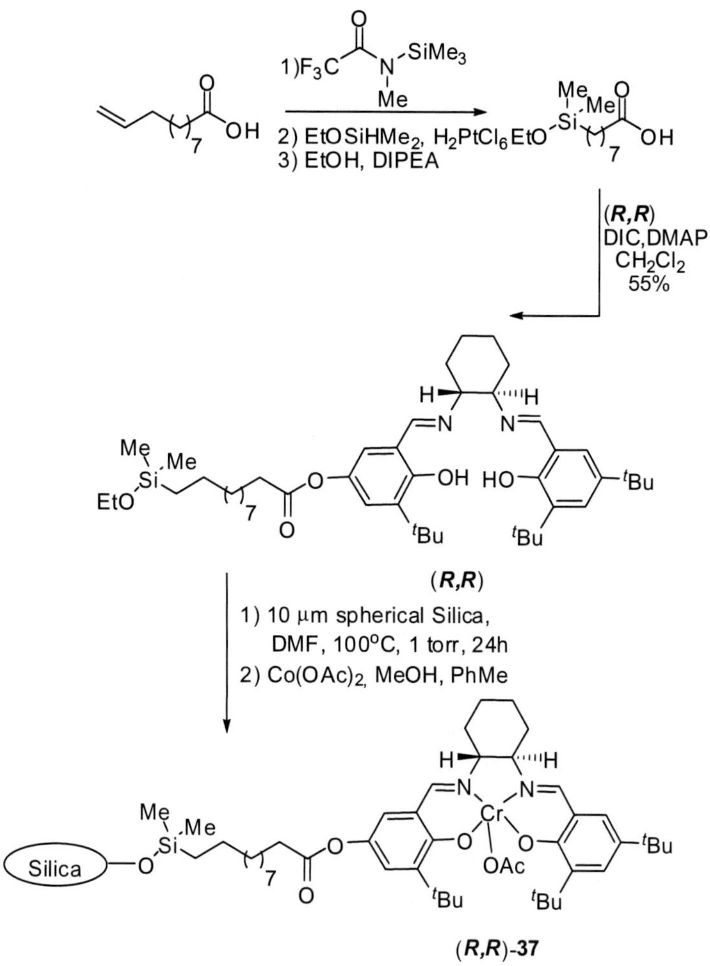

Scheme 10. Silica bound Co(salen) complex 37.

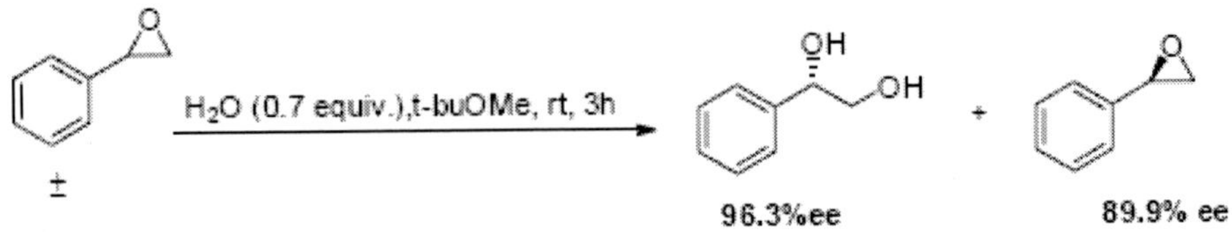

Scheme 11. HKR of styrene oxide with silica bound Co(salen) complex 37.

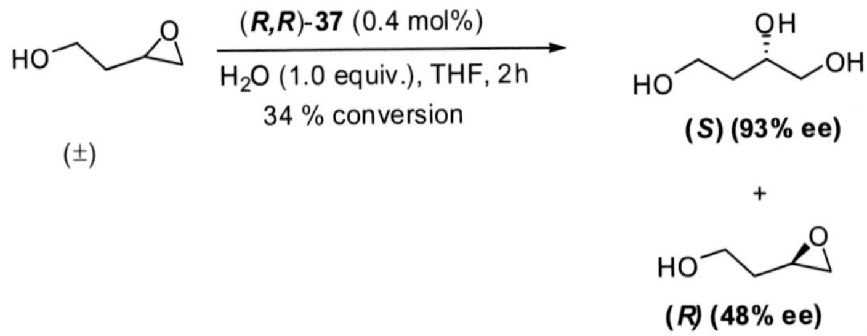

Scheme 12. HKR of 4-hydroxy butene oxide with silica bound Co(salen) complex 37.

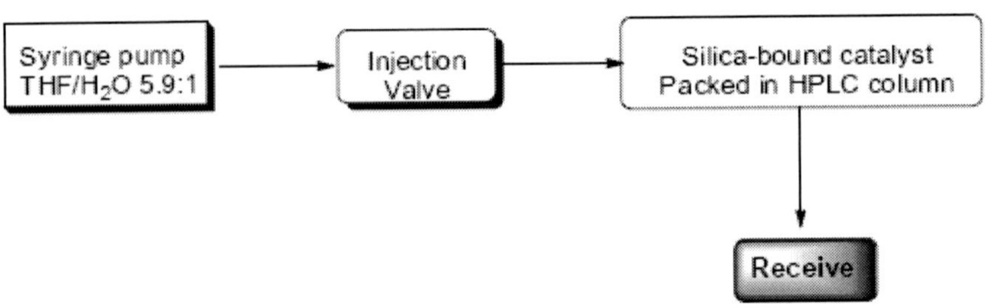

Scheme 13. Continuous flow apparatus for the HKR of 4-hydroxy butene oxide over silica bound Co(salen) complex 37.

The same immobilized catalyst **36** [69] was also effectively used for dynamic hydrolytic kinetic resolution of epibromohydrin. Five reaction cycles performed with single catalyst batch provided combined yield of 94% with 96% ee and in 90% chemical purity for the product 1-bromo-2,3-propandiol (Scheme 14).

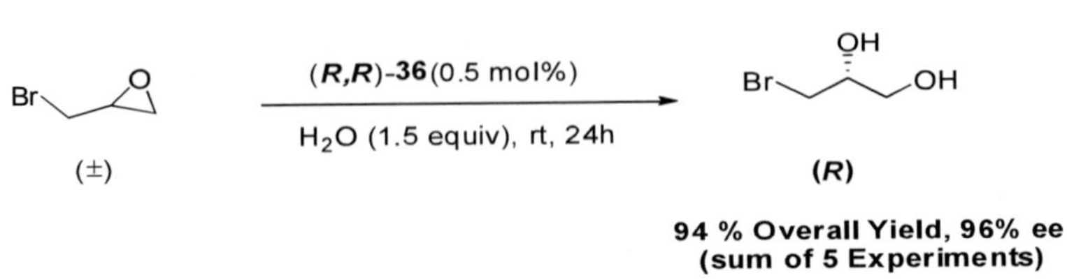

Scheme 14. Dynamic kinetic resolution of epibromohydrin with polymer bound catalyst 36.

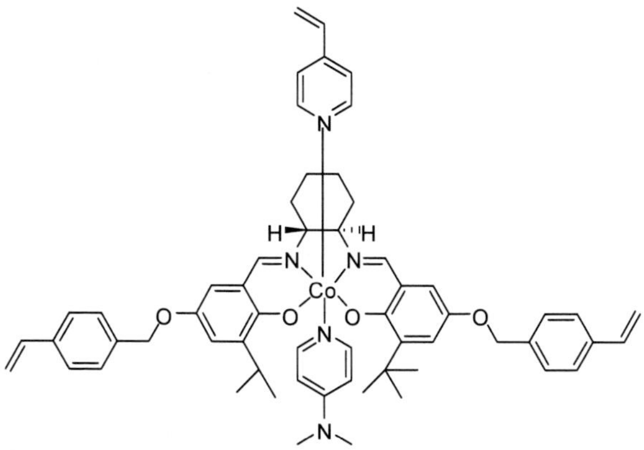

38

Figure 13. Structure of Co(salen) 38 supported on porous materials.

Borovik et al. [70] prepared a highly crosslinked polymeric porous material containing Co-salen units **38** (Figure 13) by template copolymerization method. The authors reported that as the cross-linking degree increases from 5 % to 50 %, the catalyst become more efficient in terms of reactivity, possibly due to the improved proximity of metal centers that work in cooperation. Unfortunately low enantioselectivity for the product epoxide was observed (<42 % ee) while the ee for concomitantly produced diol did not go above 86%. Reusability of the catalyst containing 50 mol% template showed consistent activity and enantioselectivity for three consecutive recycle experiments.

Kim and Park [71] reported multi-step synthesis of various unsymmetrical chiral salen Co(III)(OAc) complexes co-valently bonded onto MCM-41 type mesoporous Al-Si material (39-44) (Figure 14). Authors developed a new approach of anchoring method where the reaction of a functionalized ligand, diformylphenol, was carried out with 3-aminopropyltrimethoxysilane modified Al-MCM-41. These supported catalysts were used in the HKR of racemic epichlorohydine, 1,2-epoxyhexane, epoxystyrene and epoxycyclohexane under mild conditions to produce respective epoxides and diols in high yield and ee (Table 2).

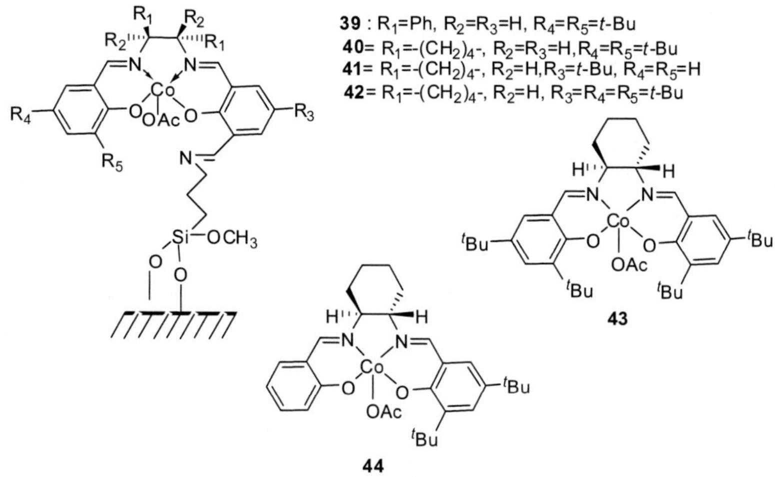

Figure 14. Structure of Co(salen) homogeneous and heterogeneous complexes 39-44.

The substituents on salen ligand have strong influence on the activity and selectivity of catalyst. It was observed that salen ligands bearing bulkier substituents are less effective, this observation is contrary to the one observed in the case of monomeric salen ligand used for the same reaction. The reaction using the Co(III)(OAc) salen catalyst immobilized on MCM-41 gave the almost same enantioselectivity as against homogeneous salen catalysts, however, the reaction rate was low (prolonged reaction time was required).

Table 2. HKR of terminal and meso epoxide catalyzed by the immobilized chiral Co(III) salens[a]

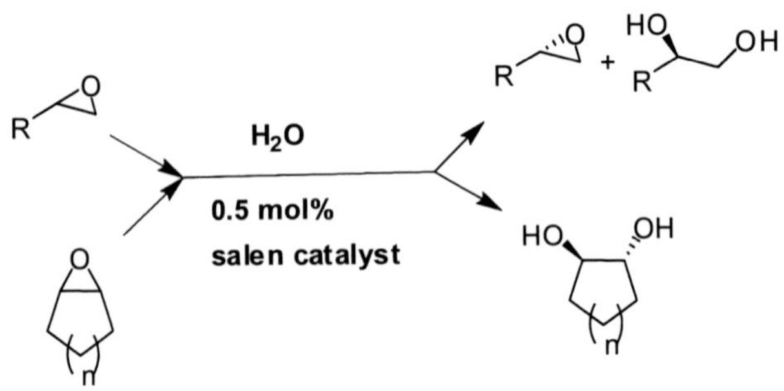

Catalyst	Substrate	Time (h)	Solvent	Diol yield (%)	Diol Ee (%)
39	Epichlorohydrin	24	None	Trace	-
40	Epichlorohydrin	24	None	34	86
41	Epichlorohydrin	24	None	35	92
42	Epichlorohydrin	24	None	31	87
43	Epichlorohydrin	12	None	35	86
44	Epichlorohydrin	12	THF	34	86
43	Epichlorohydrin	12	Acetonitrile	18	85
44	Epichlorohydrin	12	None	34	96
39	Styrene oxide	48	None	Trace	-
40	Styrene oxide	48	None	35	98
41	Styrene oxide	48	None	35	97
43	Styrene oxide	24	None	38	98
44	Styrene oxide	24	None	35	97
40	1,2 Epoxyhexane	24	None	42	97
42	1,2 Epoxyhexane	24	None	48	96
43	1,2 Epoxyhexane	12	None	46	98
40	Epoxycyclohexane	24	None	34	83
42	Epoxycyclohexane	24	None	29	83
42	Epoxycyclohexane	24	MeOH	11	82
42	Epoxycyclohexane	24	THF	37	83
44	Epoxycyclohexane	24	None	39	84
40	Epoxycyclopentane	48	None	29	61
42	Epoxycyclopentane	48	None	18	61
42	Epoxycyclopentane	48	MeOH	7	60
42	Epoxycyclopentane	48	THF	29	62
44	Epoxycyclopentane	48	None	39	64

[a]Olefins, 10mmol; water,5.5mmol; chiral catalyst, 0.5 mol%; reaction temperature 20 ^{0}C

Can Li et al. [72] synthesized a chiral Co(salen) complex in the mesoporous cage of SBA-16 through the flexible ligand method which is akin to "ship in a bottle" method. It was based on the concept that the pore entrance size should be larger than the molecular size of the fragments used for constructing the metal complex and smaller than the metal complex catalyst formed inside the cage. To achieve this authors tailored the pore entrance size (4.9, 5.4, and 5.9 nm) of SBA-16 precisely by varying the autoclaving time and silylation with phenyltrimethoxysilane. Flexible chiral salen ligand was then diffused inside the pores of thus modified SBA-16 which was trapped inside the cage of SBA-16 on complexation with cobalt metal ion (Scheme 15). Chiral Co(salen) trapped in SBA-16 45 showed high enantioselectivity (87–96% ee, at par with the homogeneous catalyst) for the asymmetric ring opening of terminal epoxides and can be recycled at least 10 times with no apparent loss of activity.

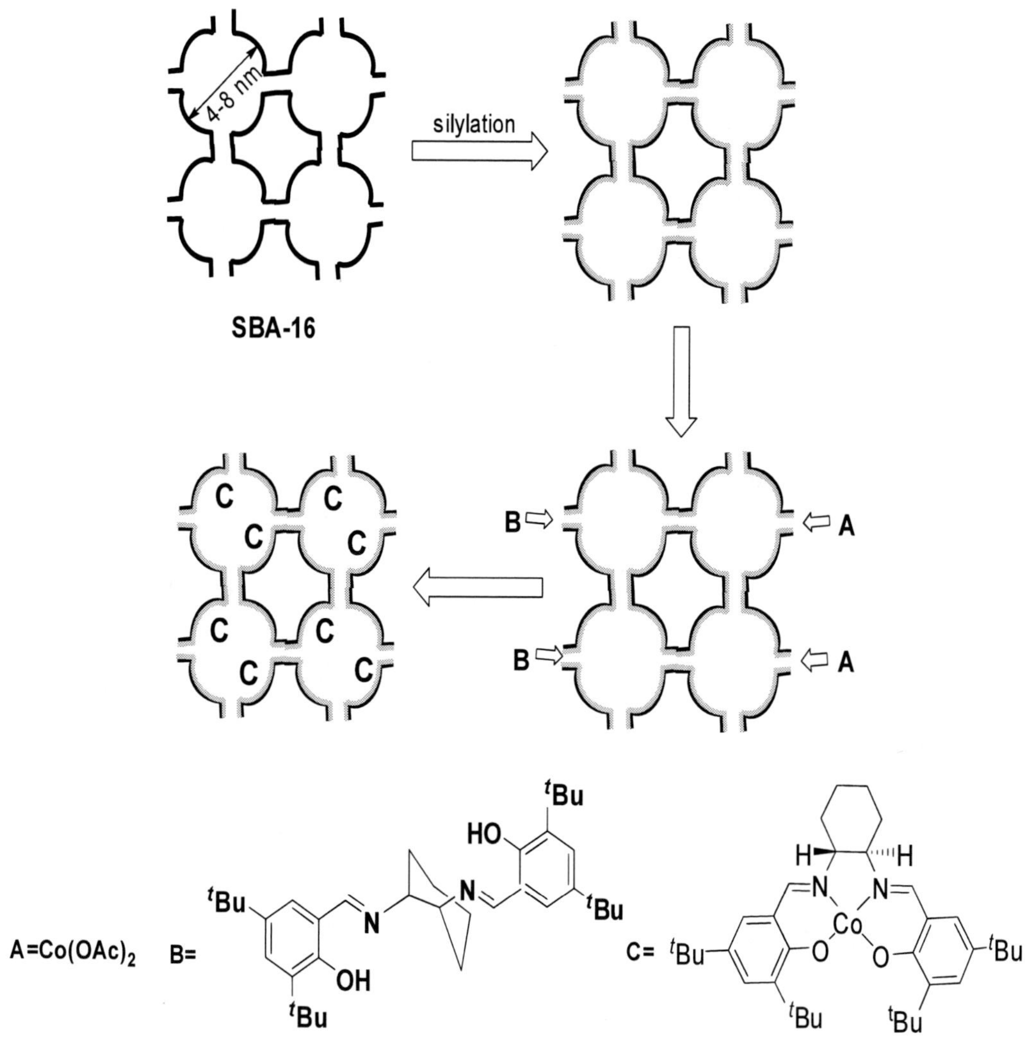

Scheme 15. Schematic description of chiral Co(salen) **45** synthesized in the cages of phenyl-modified SBA-16 through the "ship in a bottle"method.

Choi and Kim [73] immobilized Co(III)-salen complexes in a ZSM-5 46 and Anodisc membrane 47 for their use as catalyst in HKR of terminal epoxides with water. The supported ZSM-5 film for this purpose was synthesized hydrothermally on the porous supports such as Anodisc and alumina tubes. The method for impregnation of the homogeneous complex at the interface between macroporous matrix (Anodisc 47) and ZSM-5 film layer is shown in Figure 15. The Co(III)-salen complex 46 was loaded into the macropore of Anodisc 47 by impregnation under vacuum. The authors suggested that the salen catalyst must exist near the interface of ZSM-5 film so as the contact of the catalyst with reactants is efficient. By using the immobilized chiral salen membrane catalysts, the product separation became easier and the catalyst could be recycled without observable loss in activity. When the organic-phase feed stream containing a reactant epoxide was contacted with the catalyst at the interface of membrane, the epoxide remained in an organic phase and the converted hydrophilic diol diffused into the aqueous phase on the other side. Enantioselectivities up to 98 % were achieved in this way for several epoxides such as epichlorohydrin, styrene oxide, 1,2-epoxihexane and 1,2-epoxybutane in 27-100 h.

Jacobsen group [74] recently reported the immobilization of chiral salen ligands into self-assembled thiolate monolayers (SAMs) on 3.4 nm gold colloids. Treatment of the immobilized ligand with $Co(OAc)_2{\cdot}4H_2O$ yielded the corresponding [(salen)Co(II)] complex **48**, which on aerobic oxidation in the presence of triflic acid afforded the catalytically active [(salen)Co (III)] complex. The synthesis of the gold colloids was straightforward, and catalyst immobilization took place under mild conditions. A single functionalized gold colloid can carry several hundred catalyst moieties on its surface, and the thiolate chains self-assemble to minimize steric interactions, leading to a uniform catalyst distribution on the colloid surface (Figure 16). The colloid-bound catalysts were shown to catalyze the HKR of 1,2-epoxyhexane with significant rate enhancements (>99.9% ee of recovered epoxide in 5h) relative to monomeric [(salen)Co(III)] complexes which took 24h that too with higher catalyst loading for getting similar results.

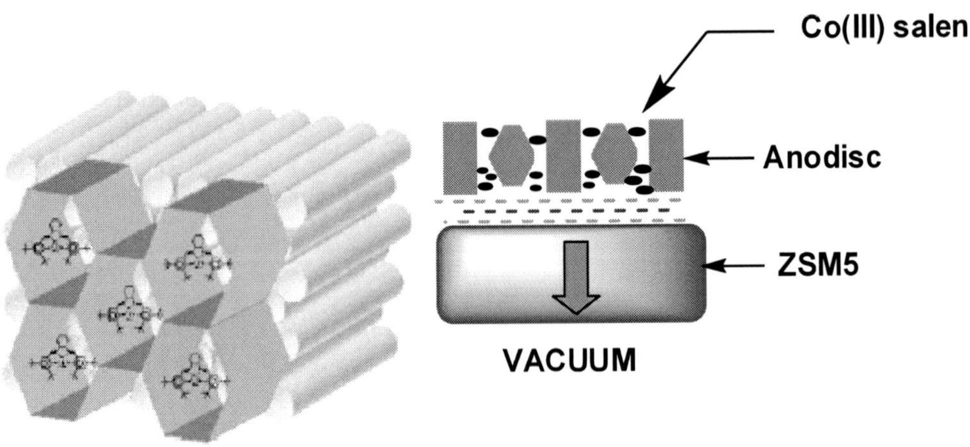

46 47

Figure 15. Representative construction of composite membrane 47 for impregnation of chiral Co(salan) 46.

Recovery of the immobilized catalyst **48** was accomplished by simple filtration, and catalyst reoxidation and repeated recycling (seven times) was possible with no loss of reactivity or enantioselectivity.

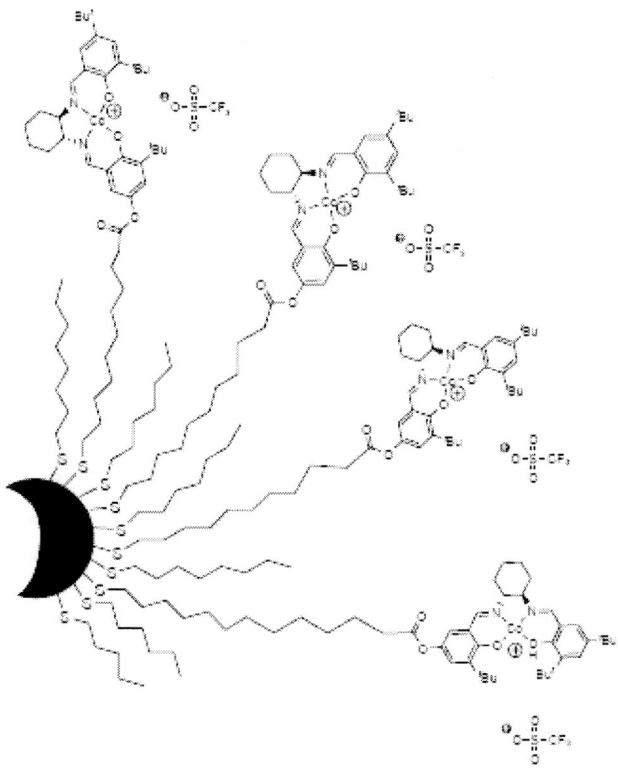

Figure 16. Structure of chiral Co(salen) 48 immobilized on gold colloids.

In an another strategy for supporting chiral Co(III) salen complex 49, ionic liquid 50 was used by Song et al. [75] to carry out HKR of racemic epoxides. They used chiral Co(III) salen complex (Figure 17) in a mixture of ionic liquid [bmim][PF_6] in THF (1:4). Excellent enantioselectivities (>99% ee) for epichlohydrin and 92% ee for its corresponding diol with 0.1 mol% catalyst was achieved with this system. Interestingly, in this system no reduction of Co(III) species was observed during HKR, which is normally the case in the HKR reactions using Jacobsen's catalyst under homogeneous condition. The supported catalyst was retained in the ionic phase, which re-used for ten more cycles without any loss in its performance. Authors also reported that Co(II)salen catalyst 49 can be directly used for the HKR and does not require oxidation in the presence of acetic acid. Very interestingly, with this system the catalytic activity of the recovered ionic liquid phase in the HKR of racemic epichlorohydrin increased upon reuse (reaction time; 22 h for the first run, 22 h for the second run, 15 h for the third run, 4 h for the fourth run, 3 h for the fifth run, 2 h for the sixth run, 2 h for the seventh run, 2 h for the eighth run, 2 h for the ninth run, 2 h for the tenth run). The reason for the increase in activity upon reuse was ascribed to increasing concentration of catalytically active Co(III) complex in the reaction mixture.

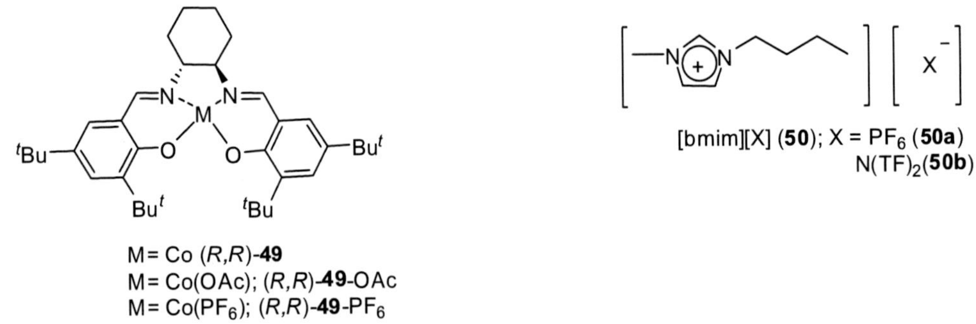

Figure 17. Structure of Co(salen) complex **49** in presence of ionic liquid **50** for HKR reaction.

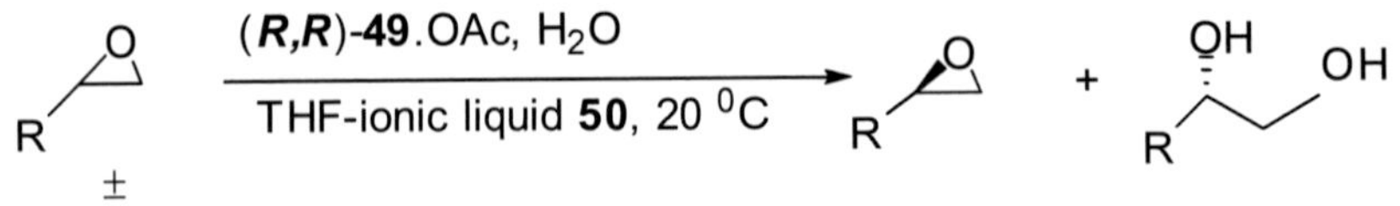

Scheme 16. Reaction for HKR for terminal epoxides using chiral Co(salen) **49** in presence of ionic liquid 50.

ARO of Epoxides with Phenols and Alcohols as Nucleophiles

ARO reaction with phenols and alcohols as nucleophiles is a logical extension of HKR of epoxides to synthesize libraries of stereochemically defined ring-opened products in high optical purity. To this effect Annis and Jacobsen [69] used their polymer-supported Co(salen) complex **36** as catalyst for kinetic resolution of epoxides with phenols to give 1-aryloxy-2-alcohols in high yield, purity and ee (Scheme 17). Conducting the same reaction in the presence of tris(trifluoromethyl)methanol, a volatile, nonnucleophilic protic acid additive accelerates KR reaction with no compromise with enantioselectivity and yield. Presumably the additive helped in maintaining the Co(III) oxidation state of the catalyst.

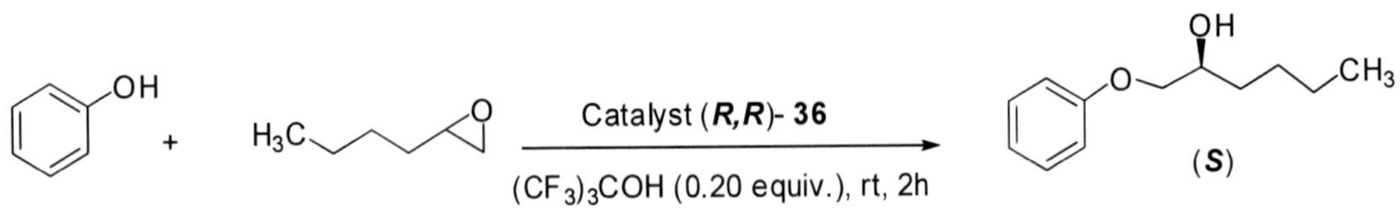

Scheme 17. KR of racemic terminal epoxide with phenol using chiral catalyst 36.

ARO of epibromohydrin with phenols using polymer-bound catalysts, 36 [69] proceeded with high efficiency and enantioselectivity, but the bromohydrin product (A) underwent ring-closure reaction to produce phenyl glycidyl ether (B) under the ring-opening reaction conditions. Incidentally, the aryl glycidyl ether thus produced is useful in the synthesis of combinatorial libraries of the general structure (C). For instance, Propranolol, (D) is commercial antihypertensive agent (Scheme 18). The catalyst 36 is recyclable and its performance was found to be persistent over 5 cycles.

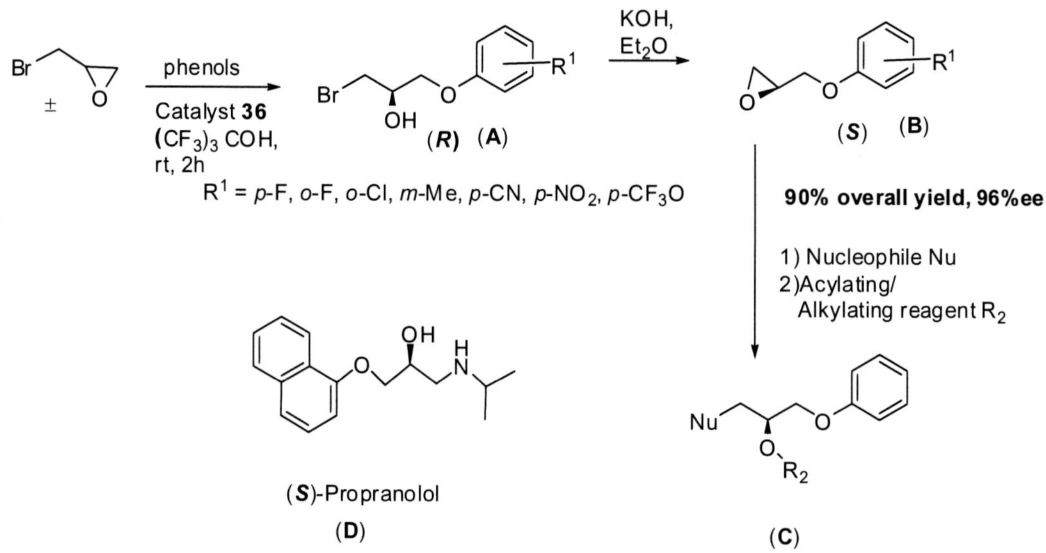

Scheme 18. A possible application of the enantioselective ring opening of epibromohydrin with phenols in combinatorial synthesis using the catalyst 36

Cycle	Ee of **A** %
1	97.1
2	97.1
3	95.5
4	95.8
5	94.6

Kim et al. [67], used the self-polymerized heterometallic polymeric salen complexes **26-32** as efficient catalysts for kinetic resolution of terminal epoxides with phenols to give α-aryloxy alcohols in high yields (38-43%) and ee (92-99%) (Scheme 17). These catalysts were recycled up to three times without any loss in their performance.

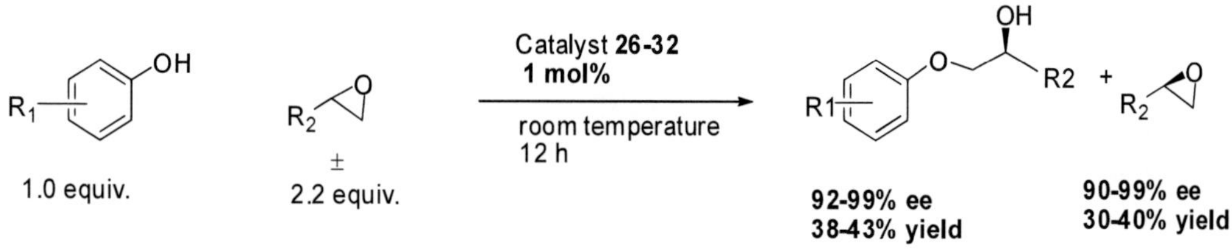

Scheme 19. ARO of terminal epoxides with phenol using heterobimetallic chiral catalysts 26-32.

Unfortunately there is no report on ARO reaction of epoxides with alcohols using heterogeneous or recyclable homogeneous catalysts. In this direction, it is worth mentioning the potentiality of Jacobsen's oligomeric complex **5** [57, 58] which though not recyclable is highly active and enantioselective for ARO of *meso* and terminal epoxides with alcohols as well as phenols (Scheme 20). Stability of this oligomeric complex was the main issue, which if addressed can provide highly practical protocol for the synthesis of chirally pure alkoxy and aryloxy alcohols.

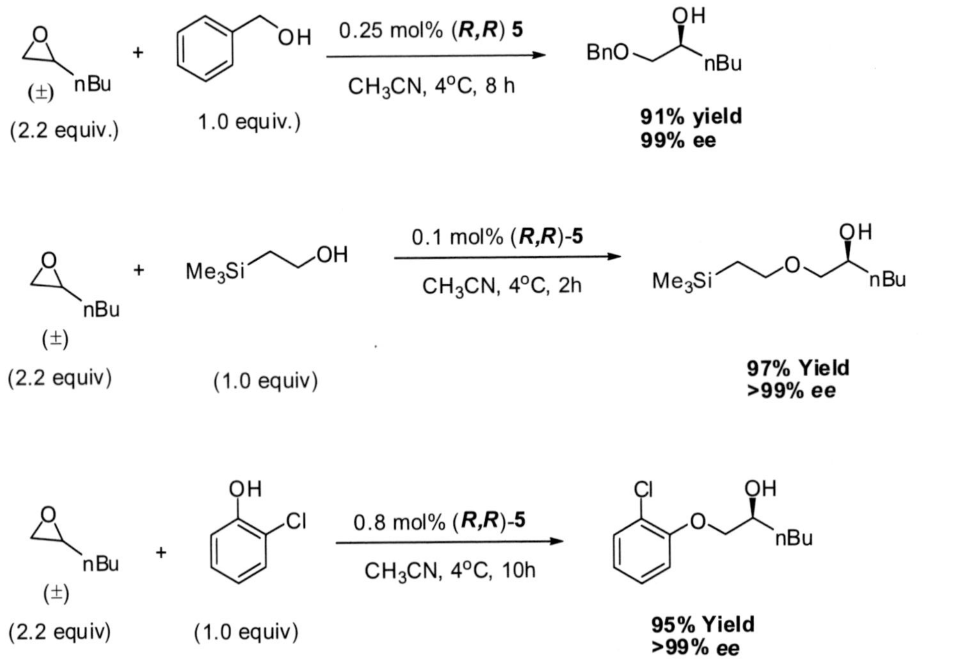

Scheme 20. ARO of terminal epoxides with phenol and alcohol using oligomeric Co(salen) complex **5.**

C-C and C-S bond formation reactions are important organic reaction and are usually accomplished by Lewis acid-based catalysts. However, this reaction is rarely achieved by epoxide ring opening reaction. One such example of C-C and C-S bond formation was reported by Bandini et al. [76]. They reported polymer-supported indium Lewis acid **51** (Amberlyst 15/indium complex) (Figure 18) as catalyst for the formation of β-indolyl alcohols and β-hydroxy sulphides through the highly regio- and stereoselective ring-opening reaction of enantiomerically pure internal as well as terminal epoxides under ambient and atmospheric conditions (Scheme 21-23). This catalyst was recycled five times without a remarkable loss of activity.

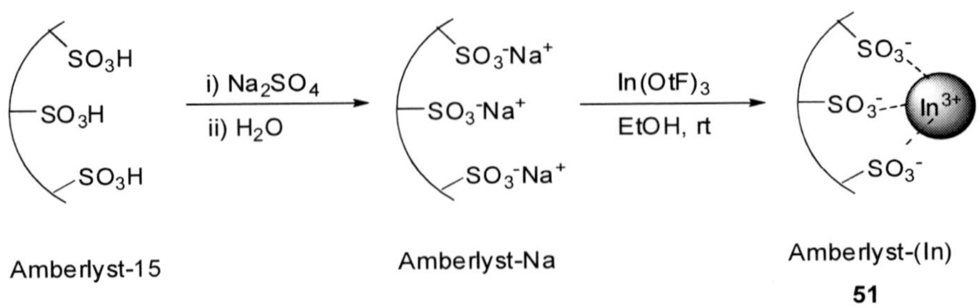

Figure 18. Synthesis of the resin-supported indium Lewis acid catalyst 51.

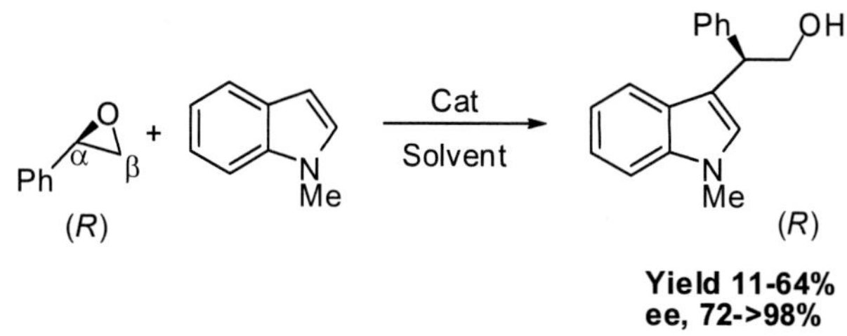

Scheme 21. ARO reaction for optimization of conditions using the catalyst 51.

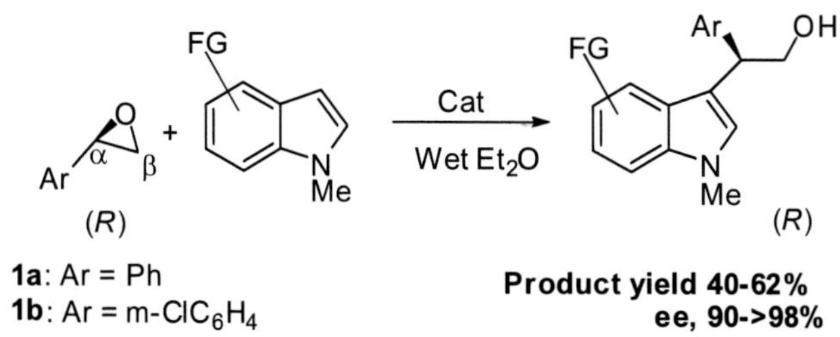

Scheme 22. ARO of enantiomerically pure epoxide with various indoles using catalyst **51.**

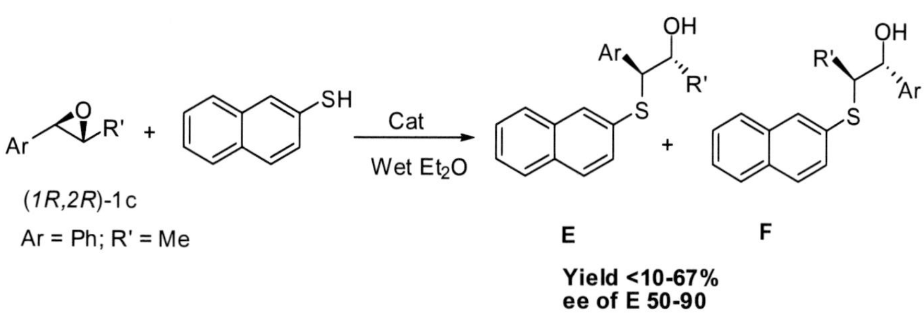

Scheme 23. Reaction for Thiolysis of enantiomerically pure epoxide mediated by **51,**

Asymmetric Ring Opening Reaction of Epoxides with Trimethylsilyl Azide ($TMSN_3$)

Optically pure β-amino alcohols are important structural entities in many biologically active compounds and intermediates in the preparation of many chiral auxiliaries and ligands [3, 77-79]. Among many other methods, asymmetric ring opening reaction (ARO) of epoxides with $TMSN_3$ catalyzed by Cr(salen) complex is an attractive approach for the synthesis of optically pure β-aminoalcohols. Cr(salen) complex is very stable under the homogeneous reaction condition used for ARO of epoxides and can be recycled a number of times without loss of activity and enantioselectivity. However, this catalyst recycling procedure involves the potentially hazardous distillation of neat liquid azides, which may prove a limitation for large scale applications.

Angelino and Laibinis [80] reported polystyrene-supported salen complex 52 (Scheme 24) in the asymmetric ring opening of epoxides. At 0 °C with 1 mol % catalyst loading, the asymmetric ring opening of epoxyhexane, propylene oxide, and cyclohexene oxide gave the corresponding products with 34%, 36%, and 6% ee, respectively, and in 40-47% yield (Scheme 25). The recycled catalysts were stable and were reused three times without loss of activity or enantioselectivity. However, during the reaction the leaching of Cr to the extent of <0.1% was reported, suggesting the possible use of the catalyst over hundreds of cycles before reloading the polymer-supported salen ligand with metal would be necessary.

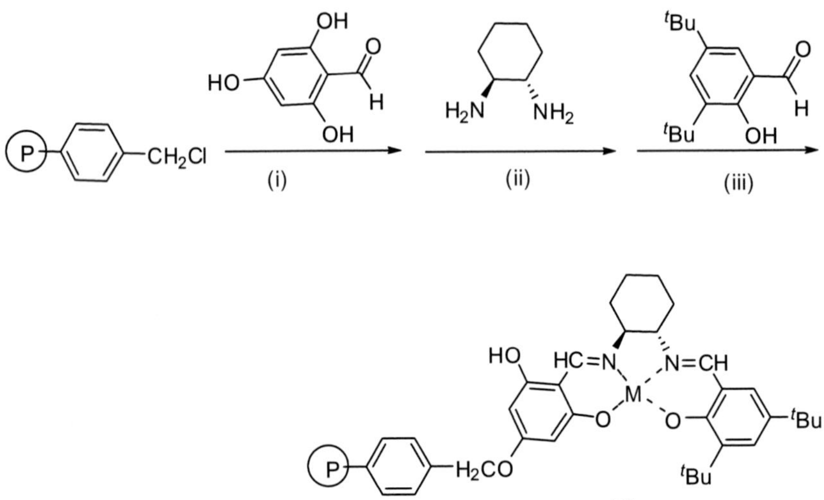

52

M = Cr-Cl

Scheme 24. Synthesis of chiral polymeric Cr(salen complex **52.**

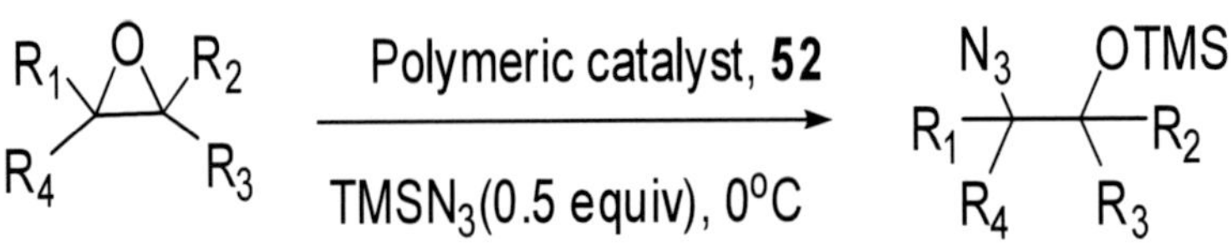

Scheme 25. ARO of epoxides with $TMSN_3$ using the complex **52.**

Corma group [81] reported incorporation of different chiral Cr(III)-salen complexes within the cavities of zeolites Y **53**, EMT **54** and into the interlamellar region of K-10 montmorillonite **55** (Scheme 26). These heterogeneous catalysts promoted the asymmetric ring opening of epoxides viz., cyclohexane oxide and cyclopentene oxide with trimethylsilylazide to afford chiral azido trimethylsilyl ethers and azido alcohols with modest ees that vary depending on the inorganic support used (Scheme 27). The enantioselectivity of the ring-opened azide however, was inferior than the ees achieved with the unsupported complexes under homogeneous condition.

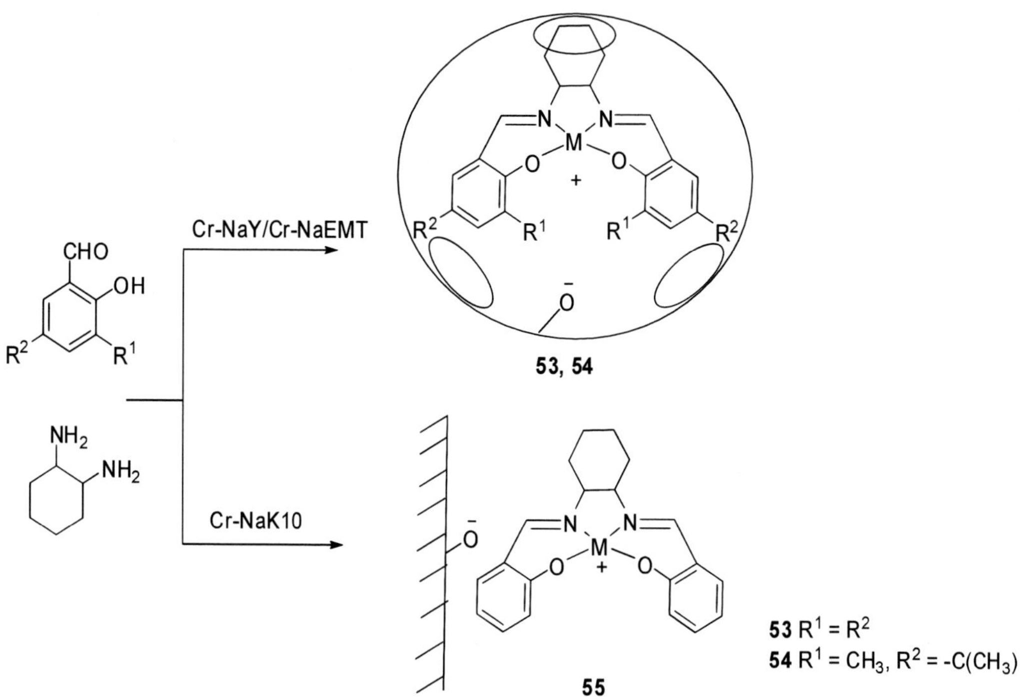

Scheme 26. Preparation of Cr(III) Schiff base complexes into the cavities of zeolites Y **53**, EMT **54** and in the montmorillonite clay K-10 **55.**

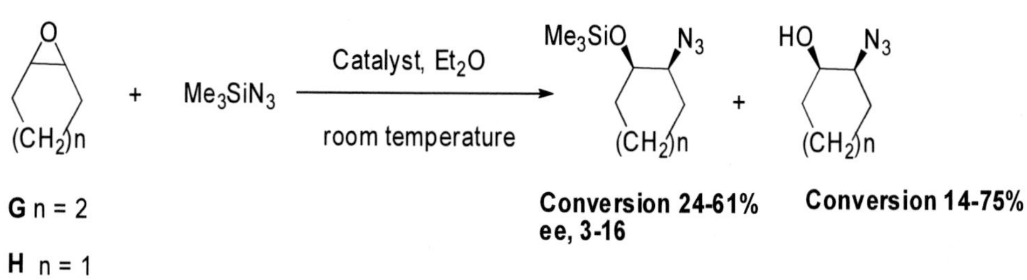

Scheme 27. Enantioselective ring opening of cyclic oxides **G** and **H** by $TMSN_3$ catalyzed by chiral Cr(III) Schiff base complexes within the cavities of zeolites Y **53**, EMT **54** and into the interlamellar region of K-10 montmorillonite **55**.

Later on, Corma et al. [82] reported covalent anchoring of chiral chromium salen complexes on SiO2, ITQ-2 or MCM-41 via aminopropyl functionalization. The catalysts in which anchoring of the complex was accomplished through coordination with the metal 56, 57 (Figure 19) showed high enantiomeric excesses (ee up to 70%) for ARO of cycloheneoxide with TMSN3 (Scheme 28), but there was considerable leaching of the complex into the reaction medium. In contrast, the complexes 58-62 which were attached to the surface of silica through covalent linkage do not leach out, but at the same time they induct modest ee (below 20%) in the products as compared to the homogeneous catalysis under comparable conditions (>50% ee) (Table 3).

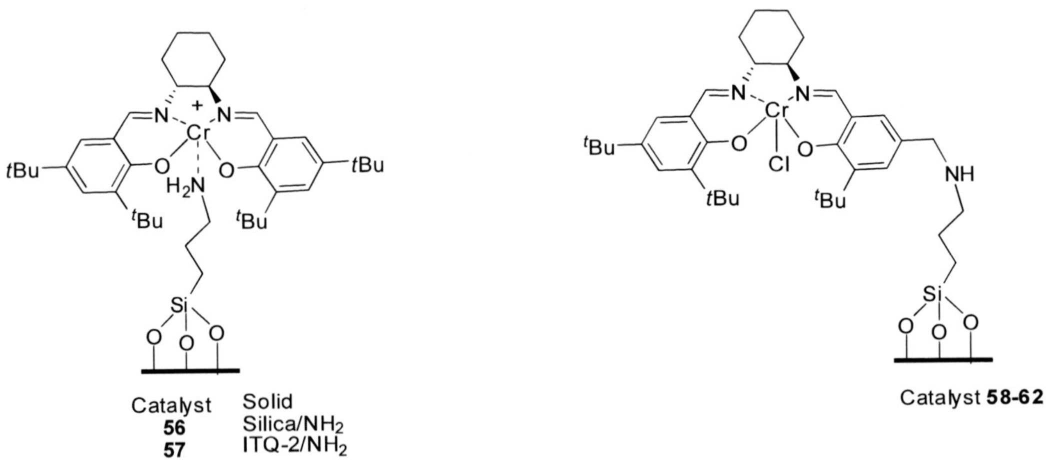

Figure 19. Structure of the catalysts 56-62.

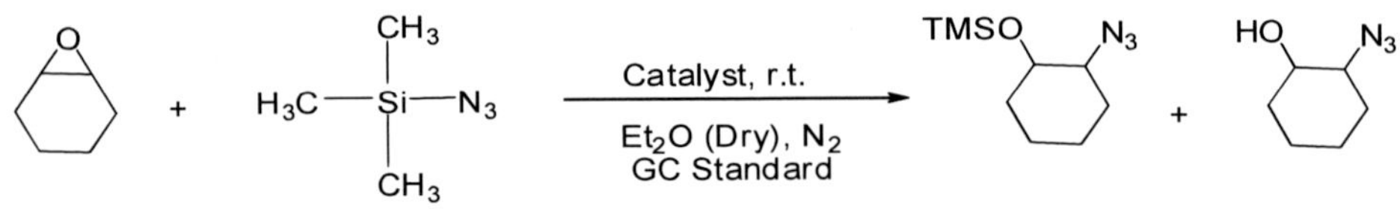

Scheme 28. ARO of cyclohexene oxide with $TMSN_3$ using immobilized catalysts **56-62.**

Table 3 Enantioselective ring opening of cyclohexeneoxide with $TMSN_3$ catalyzed by chromium salen complexes[a]

Catalyst	Time (h)	Conversion (%)[b,c]	E.e (%)[b,d]
Homo (2%)	18	83	84
Homo (1%)	42	56	59
Homo (0.3)	88	51	50
56	48	100(45)	70
57	48	93(20)	52
58	48	66	8
59	48	43	9
60	48	57	13
61	48	47	14
62	48	61	18

[a]All reactions were performed at room temperature with 3 mmol of cyclohexeneoxide, 3 mmol of *cis* decahydronaphthalene (internal standard), unsupported catalyst (homo) or 100 mg of heterogeneous complex and 3.15 mmol $TMSN_3$ in 1 ml of dry Et_2O.

[b]Determined by capillary GC column (Cyclodex B, 30 x 25mm).

[c]The values between brackets correspond to the percentage of leaching; no leaching was observed unless specified.

[d]Ee of azido silyl ether

Dioos and Jacobs [83] impregnated silica with a Cr(salen) complex **63** (1.25 wt%) (Figure 20) for ARO of 1,2-epoxyhexane and 1,2-epoxyoctane with $TMSN_3$ to give the ring opening products in high ees (>97%) and conversions (~50%). For other substrates like styrene oxide and the *meso* epoxides cyclohexeneoxide and cyclopenteneoxide, the ARO displayed lower diastereoselectivities. The recycling experiments with Cr(salen) impregnated with silica catalyst displayed reproducible activities for 10 consecutive ARO reactions (Table 4) of epoxyhexane however, in every run 1.83-0.01% leaching of the complex was observed.

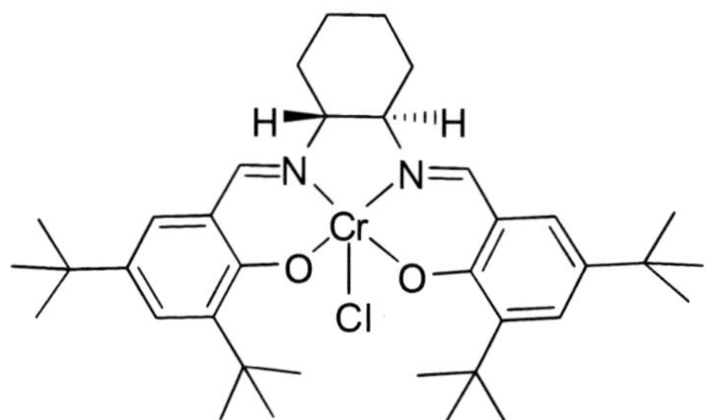

63

Figure 20. Structure of the monomeric Cr(III) salen complex **63** used for impregnation on silica.

Table 4 Recycling experiments with Cr(salen) 63-silica in the ARO reaction of epoxyhexane

Run	Time (h)	E.e[a] (%)	Conversion[b] (%)	Leaching (%)
1	21	84.9	53.3	0.59
2	21	87.2	46.3	1.83
3	23	91.5	48.4	1.07
4	23	91.6	53.4	0.17
5	23	92.8	55.6	<0.01
6	23	92.9	52.8	0.01
7	24	90.9	50.2	0.08
8	26	82.3	51.4	<0.01
9	33	84.9	50.9	0.30
10	41	86.2	50.8	1.22

[a]Ee of (R)-1,2-epoxyhexane

[b]The optimum conversion should be 50%

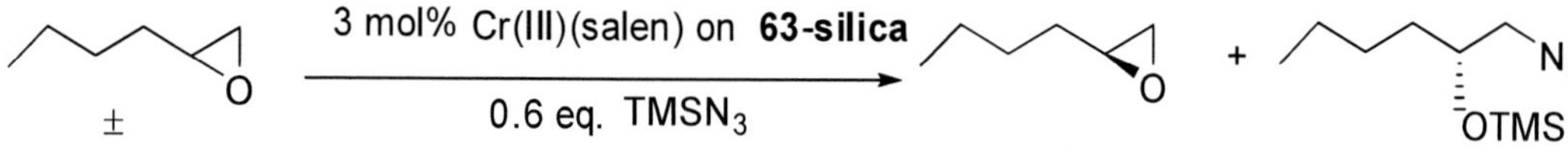

Scheme 29. ARO of 1,2 epoxyhexane using catalyst 63-silica.

Cr(salen) supported on amino-modifed silica **56** was reported by Corma at al. [82] that showed high activity and selectivity for ARO of *meso* epoxides with $TMSN_3$ but suffered acute catalyst leaching problem. To overcome this problem, Jacobs et al. [84] made an improvement over the above catalytic system by the careful choice of the solvent for epoxide ring opening reaction. Authors found that hexane is the right choice of solvent and demonstrated that the supported catalyst worked efficiently for the ARO of cyclohexeneoxide with $TMSN_3$. Additionally, in this protocol the problem of catalyst leaching was minimal (<1%) (Table 5).

Table 5 Catalytic results of the recycling experiments using the catalyst 56

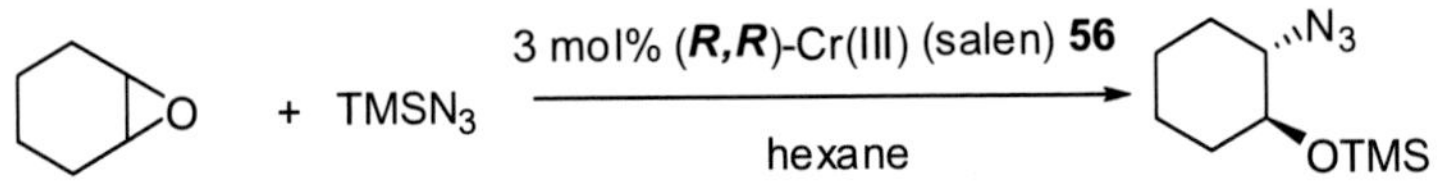

Run	Reaction time (h)	Conversion (%)	Ee (%)	*Leaching (%)*
1	12	99.5	65.4	*0.9*
2	18	100.0	67.9	*0.8*
3	18	100.0	68.7	*0.5*
4	18	100.0	73.3	*1.0*
5	19	99.3	73.1	*1.0*
6	21	99.5	75.0	*0.7*
7	21	99.5	75.5	*0.6*
8	23	99.0	76.6	*0.6*
9	24	99.1	77.2	*0.8*
10	*24*	*99.2*	*76.9*	*0.9*

50 mg catalyst; 3 mL hexane; toluene (75 μL) = internal standard; catalyst/substrate ratio= 3 mol%; 1.1 eq. $TMSN_3$; stirred batch reactor ar RT

To further minimize leaching of the Cr(salen) catalyst Jacobs et al. [85] impregnated relatively less soluble dimeric Cr(salen) catalyst on a silica material 64 (Figure 21). The silica impregnated dimeric Cr(salen) catalyst showed high enantioselectivities for epoxide (ee, >97.5%) and ring opened product (ee, 80%) with good conversions for ARO of 1,2-epoxyhexane with trimethylsilylazide. This catalyst system work well both in batch (12 recycling experiments) and continuous flow experiments and suffers even less catalyst leaching problem than monomeric Cr(salen) supported on silica (63-silica).

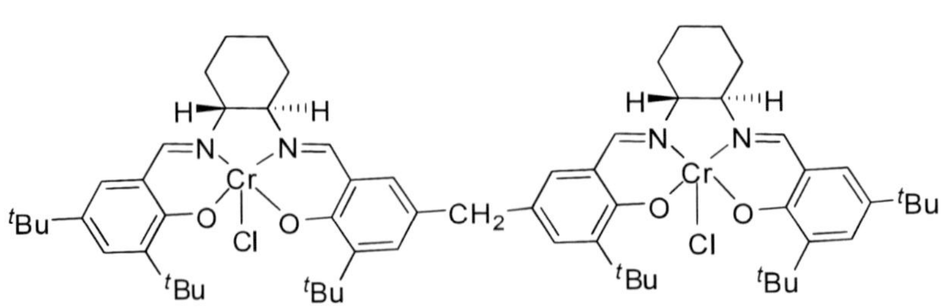

64

Figure 21. Structure of dimeric Cr(III) salen complex **64** used for impregnation on silica.

Interestingly, the dimeric Cr(salen) catalyst **64** supported on silica showed enhanced activity for ARO of 1,2-epoxyhexane and cyclohexene oxide in the presence of ionic liquids particularly with [BMIM][PF6] (**64-IL**) [86] (Table 6). A significant increase in the product selectivity was also observed with silica supported ionic liquid (**64-SILP**) for ARO of 1,2-epoxyhexane and cyclohexene oxide (ee, of 87% and 75% respectively) as compared to silica supported catalyst minus the ionic liquid (Table 6, entries 5,6). However, after repeated recycling, the silica support material deteriorates due to the abrasive forces in the stirred reactor. As a result, silica material was non-recoverable, but the expensive dimeric Cr(salen) catalyst **64** and the ionic liquid was recovered quantitatively by Soxhlet extraction with acetone. SILP-catalyst system was also used in a continuous-flow reactor.

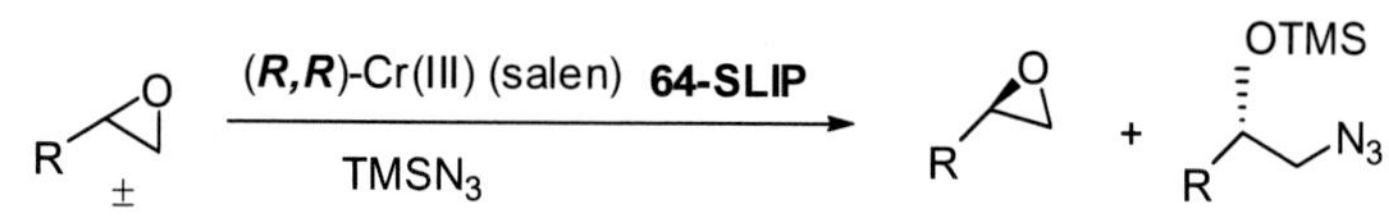

Scheme 30. ARO of epoxides catalysed by (*R,R*)-Cr(III)(salen) complex 64-SLIP.

Table 6 Stirred batch reactor: ARO of epoxides catalyzed by a dimeric (*R,R*)-Cr(III)(salen) complex 65 immobilized in a supported ionic liquid phase compared to the dimeric complex impregnated on silica 64-silica and the reported homogeneous reactions with the monomeric Cr(salen) complex 63

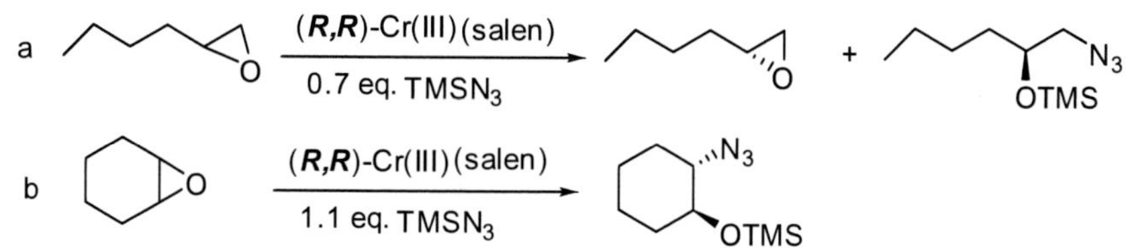

Run[a]	Reaction	Time (h)	Conversion (%)	Ee epoxide (%)	Ee product (%)	Leaching (%)
1	a	3	52	95	84	0.4
2	a	3	48	96	87	0.5
3	b	20	93	-	75	1.1
4	b	20	92	-	73	1.1
Impregnation[b]	a	10	59	96	66	1.0
Impregnation[b]	b	70	98	-	65	1.1
Homogeneous[c]	a	27	45	-	97	-
Homogeneous[d]	b	18	83	-	84	-

The catalyst/substrate ratio is 1.5 mol% for the supported ionic liquid phase (SILP) catalyst, 3 mol% for the impregnated catalyst and 2 mol% for the homogeneous reaction

aRuns 1-4 are consecutive experiments with the same catalyst in a stirred batch reactor.

bDimeric Cr (salen) catalyst impregnated on silica

cHomogeneous reaction at 0-2 0C optimized for product selectivity

dHomogeneous reaction at room temperature optimized for product selectivity

Prior to the study of Jacobs, Song et al. [87] had already reported monomeric Cr-salen **63** catalyzed ARO of epoxides with $TMSN_3$ in ionic liquid 1-butyl-3-methylimidazolium salts [bmim][X]. The ARO reaction proceeded readily at room temperature with easy catalyst/solvent recycling and does not include hazardous workup stages such as distillation of the azide product.

Table 7 Cr-salen 63, 66 catalyzed ring opening of epoxides carried out in ionic liquids

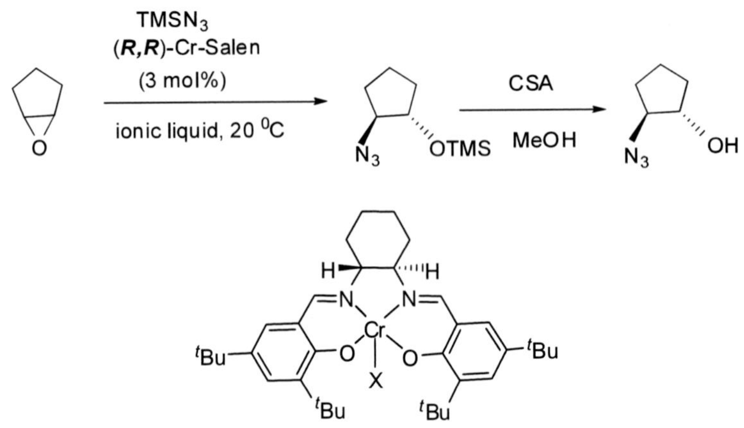

63 X = Cl⁻,
66 X= N3

Ionic liquid	Time (h)	Yield (%)	Ee (%)
[bmim][PF_6]	28	76	94
[bmim][SbF_6]	28	75	87
[bmim][BF_4]	28	5	3
[bmim][OTf]	28	trace	-

Aminolytic Kinetic Resolution of Racemic Epoxides with Amines as Nucleophile

Enantiomerically pure β-amino alcohols which are important intermediates for many bioactive compounds can be directly synthesized by the ARO reaction of readily accessible racemic and meso epoxides with appropriate amines. Indeed, some simple and multifunctional β-amino alcohols have been obtained using this strategy by the promotion of chiral BINOL [30-32,88,89], salen [35,52], bipyridine [33,40,90-94] and proline-*N,N*-dioxide based metal complexes [95]. However, none of these systems demonstrated the recyclability of the precious chiral catalyst.

Kureshy et al. [96], for the first time reported recyclable catalyst based on polymeric Cr(III)(X) salen complexes derived from (1*R*,2*R*)-(-)-cyclohexanediamine with 5,5'-methylene di-3-*tert*-butylsalicylaldehyde and X = Cl, NO_3, and ClO_4 **67-69** (Figure 22). These complexes were used in regio-, diastereo-, and enantioselective aminolytic kinetic resolution (AKR) of *trans*-stilbene oxide, *trans*-β- methyl styrene oxide, and 6-CN-chromene oxide at room temperature, providing the desired anti-β-amino alcohols in high yields (49% out of maximum 50% theoretical yield) and ee (up to 100%) leaving behind the corresponding valuable epoxide in high ee (up to 98%) (Scheme 31). These polymeric catalysts were reportedly more active than their monomeric counterparts due to higher number of repeat salen units that worked in cooperation. The polymeric Cr(III)(X) salen catalysts were recovered by taking advantage of solubility difference and were reused 4 times in the AKR of trans-stilbene oxide without any apparent loss in its performance.(Table 8).

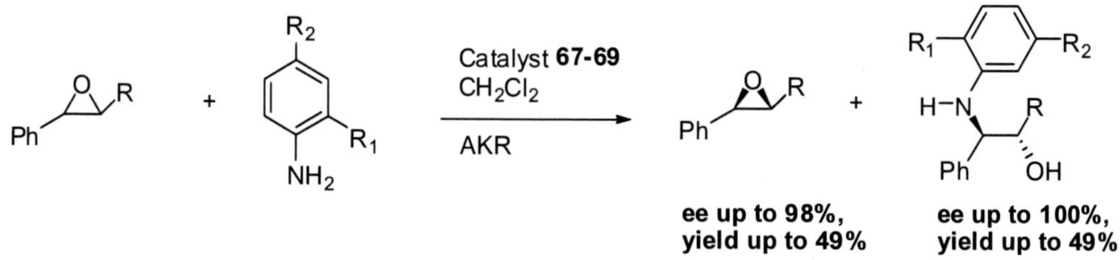

Scheme 31. Reaction for AKR of *trans* epoxides with anilines as nucleophiles using the catalysts **67-69.**

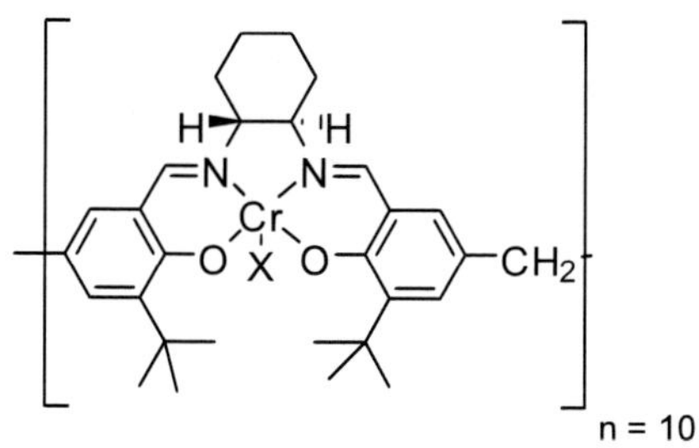

67, X = Cl
68, X = NO_3
69, X = ClO_4

Figure 22. Structure of the catalysts 67-69.

Table 8 Recycling data of catalyst 67 for AKR of *trans* stilbene oxide using aniline as nucleophile

Catalytic run	Time (h)	Ee of epoxide	Ee of amino alcohols	Yield of amino alcohols
1	14	80	87	49
2	14	80	86	48
3	15	79	85	47
4	17	75	87	46
5	20	71	86	45

Kureshy et al. [97] in their subsequent study demonstrated the use of above polymeric **67** and analogous dimeric catalyst **64** in AKR of several epoxides under microwave (MW) irradiations in order to speed up the sluggish chemical reactions under conventional heating. Accordingly, AKR of *trans*-stilbene oxide and *trans-β*-methyl styrene oxide with aniline and substituted anilines were found to proceed smoothly under MW within 2 min using **67, 64** complexes as catalysts giving *anti-β*-amino alcohols. High yields (49%) and chiral purity (ee up to 94%) was achieved in case of 4-methylaniline. This AKR system under MW irradiation was found to be ~5 times faster than traditional oil bath heating at 70 °C and 420 times faster than the reaction conducted at room temperature with concomitant recovery of respective chirally enriched epoxides (ee, 92%) in excellent yields (up to 48%).

Table 9. Aminolytic kinetic resolution (AKR) of *trans*-stilbene epoxides using recyclable catalysts with aniline under different reaction conditions

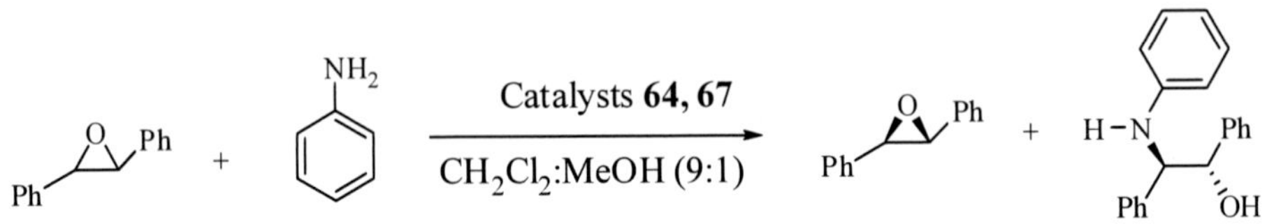

Temp. °C	Time (min).	MW (W)	*Trans-epoxide* Ee(%) Yield(%)		Anti-β-amino Alcohols Yield(%) Ee(%)	
40	2	900	78 (73)	78 (83)	94 (81)	22 (16)
50	2	900	81 (74)	62 (65)	94 (80)	35 (30)
60	2	900	84 (72)	63 (62)	92 (78)	42 (37)
70	2	900	90 (72)	48 (48)	92 (78)	49 (49)
80	2	900	81 (69)	35 (37)	92 (67)	65 (62)
70	2	800	35 (25)	65 (67)	42 (37)	35 (32)
70	2	1000	40 (31)	37 (26)	55(55)	72(74)
70	2	900	75 (68)	65(67)	96(80)	36(32)
70	2	900	92 (79)	45(46)	68 (70)	60 (57)
70	2	900	96 (83)	39 (41)	61 (62)	69(68)
70	2	900	30	72	75	27
70	5	-	42(40)	65(68)	75 (73)	35(32)
70	10	-	79((73)	51(50)	61 (67)	48(49)
27	840	-	85(80)	47(48)	93 (87)	49(49)
70 (2nd cycle)	2.0	900	89 (70)	48 (48)	92 (78)	49(49)
70 (3rd cycle)	2.0	900	89 (71)	47 (49)	91 (78)	48(48)
70 (4thcycle)	2.0	900	90 (72)	46 (48)	92 (77)	48(48)
70 (5th cycle)	2.0	900	90 (72)	46 (48)	91 (76)	47(47)

Very recently Kureshy et al. [98] further reported non-salen chiral Schiff base derived Ti complexes as catalysts 70, 71 (Figure 23) in the KR of *meso*-stilbene oxide, cyclohexene oxide, cyclooctene oxide and *cis*-butene oxide with anilines. The study deliberated upon the role of several chiral and achiral additives with these catalysts to give chiral *β*-amino alcohols with high enantioselectivity (*ee*, >99%) in excellent yield (>99%) at 0 °C in 10h. Unlike the monomeric version 72 the chiral catalyst 70 used in this study was recoverable and recyclable several times with retention of its performance (Table 10)

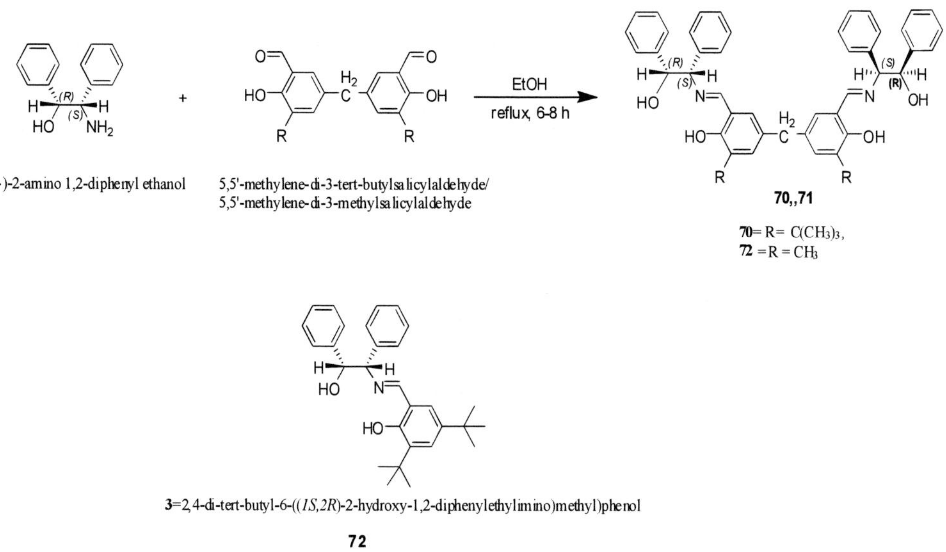

Figure 23. Synthesis of Chiral aminoalcohol ligands 70-72.

Table 10. Products Yield and ee's of asymmetric ring opening reaction *of meso*-stilbene oxide with aniline using complexes 70-72

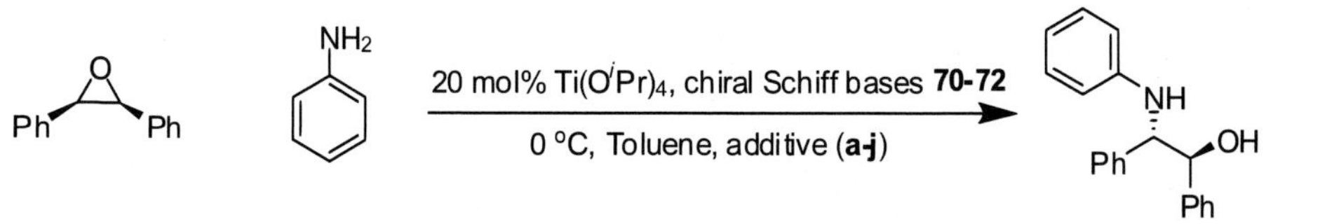

Catalyst	Additive	Time (h)	Yield [%]^g	Ee [%]	Confign. of *β*-aminoalcohol
70	None	10	79	60	1*S*,2*S*
71	None	10	66	50	1*S*,2*S*
70	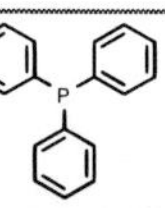 **(a)**	10	89	65	1*S*,2*S*

Table 10. (Continued)

Catalyst	Additive	Time (h)	Yield [%][g]	Ee [%]	Confign. of β-aminoalcohol
71	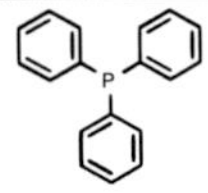 **(a)**	10	82	55	1*S*,2*S*
70	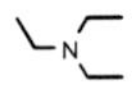 **(b)**	10	90	68	1*S*,2*S*
71	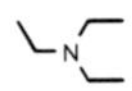 **(b)**	10	80	59	1*S*,2*S*
70	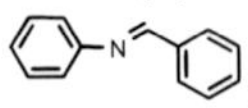 **(c)**	10	94	70	1*S*,2*S*
71	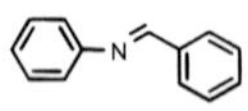 **(c)**	10	88	59	1*S*,2*S*
70	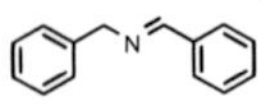 **(d)**	10	92	69	1*S*,2*S*
71	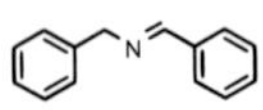 **(d)**	10	87	58	1*S*,2*S*
70	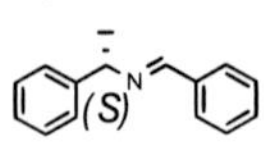 **(e)**	10	99	94	1*S*,2*S*
71	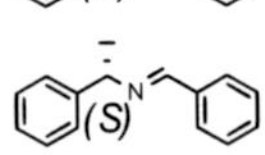 **(e)**	10	92	78	1*S*,2*S*
70	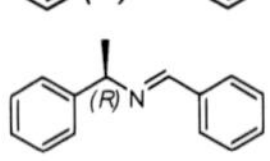 **(e')**	10	80	65	1*S*,2*S*
71	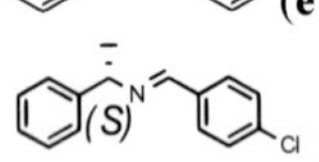 **(f)**	10	84	88	1*S*,2*S*
70	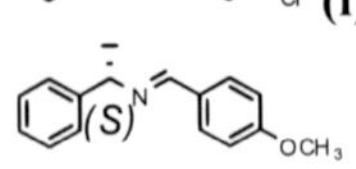 **(g)**	10	94	86	1*S*,2*S*
70	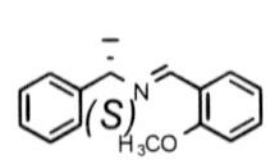 **(h)**	10	99	>99	1*S*,2*S*
70	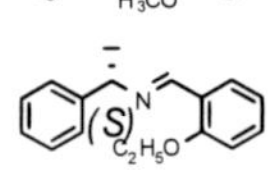 **(i)**	10	70	62	1*S*,2*S*
70	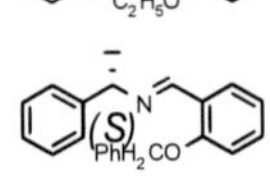 **(j)**	10	74	55	1*S*,2*S*
70	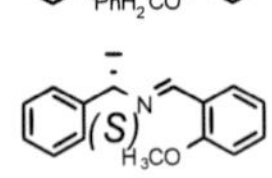 **(h)**	24	75	64	1*S*,2*S*
72	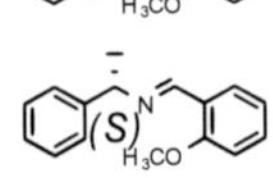 **(h)**	24	86	76	1*S*,2*S*
70 (2nd cycle)	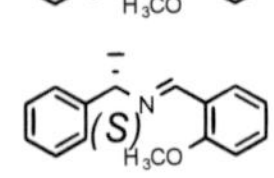 **(h)**	10	97	>99	1*S*,2*S*
70 (3rd cycle)	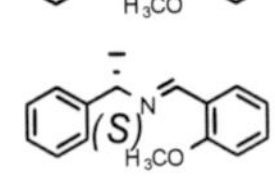 **(h)**	10	95	>99	1*S*,2*S*
70 (4th cycle)	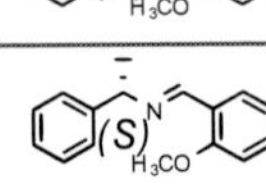 **(h)**	10	95	>99	1*S*,2*S*

The literature on KR of epoxides in the synthesis of chirally pure 1,2-amino alcohols is dominated by salen based chiral catalysts. Another important class of ligand used in various asymmetric organic transformations is BINOL. However, its use in ARO of epoxides is limited and under homogeneous condition with no catalyst recycling data available [30-32,88,89]. In a one off study, Kureshy et al. [99] reported (*S*)-(–)-BINOL-Ti complexes **73-76** as recyclable catalysts in enantioselective ring-opening reaction of *meso*stilbene oxide and cyclohexene oxide with anilines to obtain β-amino alcohols in high yield (95%) and enantioselectivity (*ee*, 78%) at ambient temperature [Table 11]. The catalyst was recovered after first use and was recycled four times with retention in its performance.

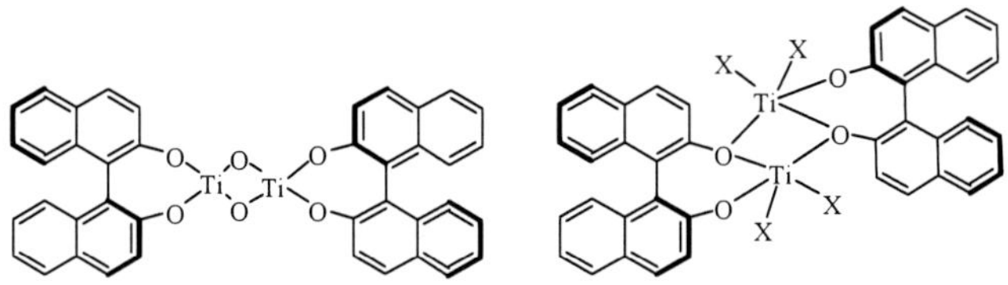

73; OR = Isopropoxide
74; OR = Ethoxide

75; X = Chloride
76; X = Bromide

Figure 24. Structure of the (*S*)- BINOL catalysts 73-76.

Table 11. Asymmetric catalytic ring opening of *meso* stilbene oxide with aniline using complexes 73-76

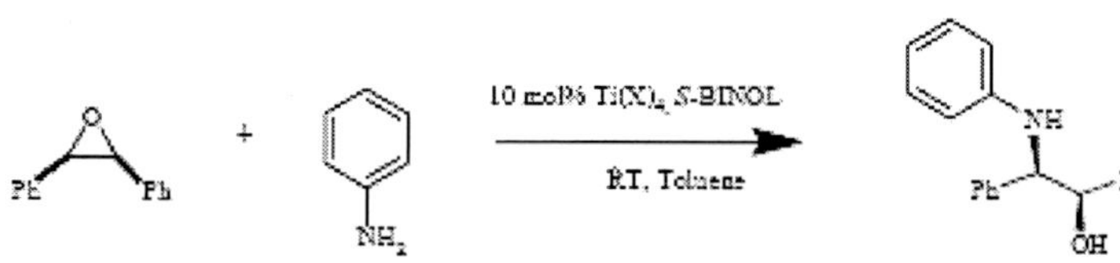

Catalyst	*Time (h)*	*Yield of 3a [%]*	*ee [%]*
73	7	90	67
74	8	85	63
75	5	20[c]	35
76	5	40[c]	51
73 (2nd cycle)	7	75	65
73 (3rd cycle)	7	70	64
67 (4th cycle)	7	60	64

The same catalyst **73** also worked efficiently for catalytic asymmetric ring opening of cyclohexene oxide and meso-stilbene oxide [34] with anilines under microwave irradiation to afford β-amino alcohols in high yield (up to 95%) and good enantioselectivities (ee up to 55%). The reaction under microwave was found to be 10 times faster than traditional oil-bath heating without compromising enantioselectivity. Also the ee for the product β-amino alcohols was comparable with the values obtained at room temperature (Table 12).

Table 12. Asymmetric ring opening of cyclohexene oxide with anilines catalyzed by complex 73 under microwave irradiation

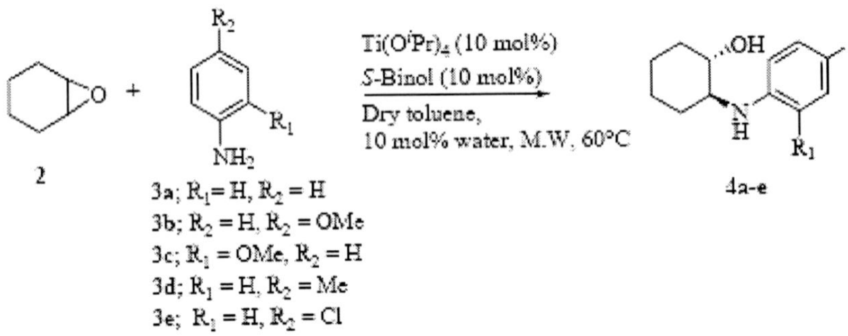

Amine	Additive	Temperature	Reaction time	Isolated Yield **4a-e** [%]	*ee* [%]	TON	TOF
3a	TPP	27	2h	92	39	9.2	4.6
3a	TPP	40	30s	80	50	8.0	960
3a	TPP	60	30s	90	49	9.0	1080
3a	-	60	30s	85	17	8.5	1020
3a	TPP	60	6m	93	54	9.3	93
3a	TPP	70	30s	95	26	9.5	1140
3a	TPP	70	5m	94	55	9.4	112
3b	TPP	60	30s	90	46	9.0	1080
3c	TPP	60	30s	92	50	9.2	1104
3d	TPP	60	30s	95	47	9.5	1140
3e	TPP	60	30s	94	48	9.4	1128

Rao et al. [100] for the first time report the biomimetic approach for the synthesis of a single enantiomer of β-aminoalcohol. In this approach β-cyclodextrin formed by the inclusion of complex **77** with racemic aryloxyepoxide which reacted enantioselectively with amines under solid state condition to give the product in 100% ee and 70-79% isolated yield. The yield which was above 50% was explained in terms of continuous racemization of the epoxide by β−cyclodextrin, hence termed as dynamic kinetic resolution. Ironically, no data on recycling of the catalyst was reported.

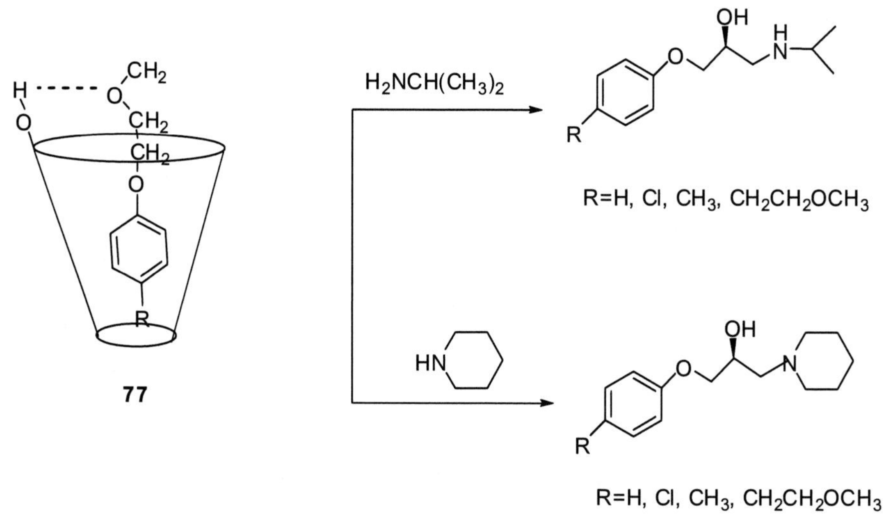

Scheme 32 Dynamic kinetic resolution of racemic aryloxy epoxides with amine using the catalysts 77

Conclusion and Out Look

The primary goal for heterogenization of chiral catalyst is to make it recoverable and reusable for economic and environmental reasons. The present chapter provides an up to date literature on this aspect for epoxide ring opening reaction which has great potential in industry and academia. Among various epoxide ring opening reactions presented in this chapter, HKR of racemic epoxide is one of the most studied reactions and some of the strategies for this transformation have already been practiced in industry at multi-ton level. There may be historical reason for that as HKR of racemic epoxides was one of the earliest epoxide ring opening reaction for achieving desired epoxide and diols in a very high optical purity and yield. What surprises authors that one of the very important epoxide ring opening reaction (ARO) of racemic and meso epoxide with amine has been neglected for heterogenization. Nevertheless Kureshy et al. have reported complexes of recyclable catalysts for this organic transformation under homogeneous system. There remain lot of scope to take up this very exciting organic transformation to the process level where it can be economically viable at process level in industry. Some of the potential candidates for heterogenization are BINOL/SALEN and Bipyridyl complexes for ARO of racemic, terminal and meso epoxides with amines.

Acknowledgements

RIK is thankful to DST and CSIR Network project on Catalysis for financial assistance and

also thankful to Dr. P. K. Ghosh, the Director, of the Institute for providing instrumentation
facilities.

REFERENCES

[1] Stinson, S. C., (2001) Chiral Pharmaceuticals, *Chem. Eng. News.* 79 79-97.

[2] Kassab, D. J.;Ganem, B., (1999) An enantioselective synthesis of (-)-Allosamidin by asymmetric desymmetrization of a highly functionalized meso-epoxide. *J. Org. Chem.*, 64 1782-1783.

[3] Hodgson, D. M.; Gibbs, A. R.; Lee, G. P., (1996) Enantioselective desymmetrization of achiral epoxides. *Tetrahedron*, 52 14361-14384.

[4] Annis, D. A.; Helluin, O.; Jacobsen, E. N. (1998) Stereochemistry as a diversity element: solid-phase synthesis of cyclic RGD peptide derivatives via asymmetric Catalysis., *Angew. Chem., Int. Ed. Engl.*, 37, 1907-1909.

[5] Nugent, W. A. (1992) Chiral Lewis acid catalysis: Enantioselective addition of azide to meso-epoxides., *J. Am. Chem. Soc,* 114 2768–2769.

[6] Adolfsson, H.; Mobrg, C. (1995) Chiral Lewis acid catalyzed asymmetric nucleophilic ring opening of cyclohexene oxide., *Tetrahedron: Asymmetry,* 6 2023–2031.

[7] Martinez, L. E.; Leighton, J. L.; Carsten, D. H.; Jacobsen, E. N. (1995) Highly enantioselective ring opening of epoxides catalyzed by (salen)Cr(III) complexes., *J. Am. Chem. Soc.*, 117 5897–5898.

[8] Eppley, A. W.; Totah, N. I. (1997) A new chiral titanium species for the ring opening reactions of meso-epoxides., *Tetrahedron*, 53 16545–16552.

[9] Schaus, S. E.; Larrow, J. F.; Jacobsen, E. N. (1997) Practical synthesis of enantiopure cyclic 1,2-amino alcohols via catalytic asymmetric ring opening of meso-epoxides., *J. Org. Chem.,* 62 4197–4199.

[10] Pakulskia, Z.; Pietrusiewicz, K. M. (2004) Enantioselective desymmetrization of phospholene meso-epoxide by nucleophilic opening of the epoxide., *Tetrahedron: Asymmetry,* 15 41–45.

[11] Dioos, B. M. L.; Jacobs, P. A. (2005) Microwave-assisted Cr(salen)-catalysed asymmetric ring opening of epoxides., *J. Catal.,* 235 428–430.

[12] Denmark, S. E.; Barsanti, P. A.; Wong, K.-T.; Stavenger, R. A. (1998) Enantioselective ring opening of epoxides with silicon tetrachloride in the presence of a chiral lewis base., *J. Org. Chem.,* 63 2428–2429.

[13] Nugent, W. A. (1998) Desymmetrization of meso-epoxides with halides: A new catalytic reaction based on mechanistic insight, *J. Am. Chem. Soc.,* 120 7139–7140.

[14] Bruns, S.; Haufe, G. (1999) Catalytic asymmetric ring opening of epoxides to chlorohydrins with mild chloride donors and enantiopure titanium complexes., *Tetrahedron: Asymmetry*, 10 1563–1569.

[15] Reymond, S.; Brunel, J. M.; Buono, G. (2000) New development in the enantioselective ring opening of meso-epoxides by various ion silicon sources catalyzed by an o-methoxyaryldiazaphosphonamide Lewis base., *Tetrahedron: Asymmetry,* 11 4441–4445.

[16] Nakajima, M.; Saito, M.; Uemura, M.; Hashimoto, S. (2002) Enantioselective ring opening of meso-epoxides with tetrachlorosilane catalyzed by chiral bipyridine N,N′-dioxide derivatives., *Tetrahedron Lett.,* 43 8827–8829.

[17] Wang, L.-S.; Hollis, T. K. (2003) Demonstration of a phosphazirconocene as a catalyst for the ring opening of epoxides with TMSCl., *Org. Lett.,* 5 2543–2545.

[18] Tokuoka, E.; Kotani, S.; Matsunaga, H.; Ishizuka, T.; Hashimoto, S.; Nakajima, M. (2005) Asymmetric ring opening of meso-epoxides catalyzed by the chiral phosphine oxide BINAPO., *Tetrahedron: Asymmetry* 16 2391–2392.

[19] Roy, C. D.; Brown, H. C. (2006) Asymmetric ring opening of *meso*-epoxides with β-halobis(2-isocaranyl)boranes 2-^{d}Icr$_2$BX, Tetrahedron: Asymmetry 17 1931–1936.

[20] Schaus, S. E.; Jacobsen, E. N. (2000) Asymmetric ring opening of meso- epoxides with TMSCN catalyzed by (pybox)lanthanide complexes., *Org. Lett.*, 2 1001–1004.

[21] Zhu, C.; Yuan, F.; Gu, W.; Pan, Y. (2003) The first example of enantioselective isocyanosilylation of meso-epoxides with TMSCN catalyzed by novel chiral organogallium and indium complexes., *Chem. Commun.*, 692–693.

[22] Belokon, Y. N.; Chusov, D.; Borkin, D. A.; Yashkina, L. V.; Dmitriev, A. V.; Katayeva, D.; North, M. (2006) Chiral Ti(IV) complexes of hexadentate Schiff bases as precatalysts for the asymmetric addition of TMSCN to aldehydes and the ring opening of cyclohexane oxide., *Tetrahedron: Asymmetry*, 17, 2328–2333.

[23] Yamashita, H. (1988) Metal(II) *d*-tartrates catalyzed asymmetric ring opening of oxiranes with various nucleophiles., Bull. *Chem. Soc. Jpn.*, 61 1213–1220.

[24] Iida, T.; Yamamoto, N.; Sasai, H.; Shibasaki, M. (1997) New asymmetric reactions using a gallium complex: A highly enantioselective ring opening of epoxides with thiols catalyzed by a Gallium·Lithium·Bis(binaphthoxide) complex., *J. Am. Chem. Soc.,* 119 4783–4784.

[25] Wu, M. H.; Jacobsen, E. N. (1998) Asymmetric ring opening of meso-epoxides with thiols: enantiomeric enrichment using a bifunctional nucleophile., *J. Org. Chem.,* 63 5252–5254.

[26] Wu, J.; Hou, X.-L.; Dai, L.-X.; Xia, L.-J.; Tang, M.-H. (1998) Enantioselective ring opening of meso-epoxides with thiols catalyzed by a chiral (salen)Ti(IV) complex ., *Tetrahedron: Asymmetry*, 9 3431–3436.

[27] Jacobsen, E. N.; Kakiuchi, F.; Konsler, R. G.; Larrow, J. F.; Tokunaga, M. (1997) Enantioselective catalytic ring opening of epoxides with carboxylic acids., *Tetrahedron Lett.*, 38 773–776.

[28] Iida, T.; Yamamoto, N.; Matsunaga, S.; Woo, H.-G.; Shibasaki, M. (1998) Enantioselective ring opening of epoxides with 4-methoxyphenol catalyzed by gallium heterobimetallic complexes: An efficient method for the synthesis of optically active 1,2-diol monoethers., *Angew. Chem., Int. Ed.*, 37, 2223–2226.

[29] Matsunaga, S.; Das, J.; Roels, J.; Vogl, E. M.; Yamamoto, N.; Iida, T.; Yamaguchi, K.; Shibasaki, M. (2000) Catalytic enantioselective meso-epoxide ring opening reaction with phenolic oxygen nucleophile promoted by gallium heterobimetallic multifunctional complexes., *J. Am. Chem. Soc.,* 122 2252–2260.

[30] Sekine, A.; Ohshima, T.; Shibasaki, M. (2002) An enantioselective formal synthesis of 4-demethoxydaunomycin using the catalytic asymmetric ring opening reaction of meso-epoxide with p-anisidine., *Tetrahedron,* 58 75–82.

[31] Carrée, F.; Gil, R.; Collin, J. (2004) Samarium iodides catalyzed meso-epoxides ring opening by aromatic amines., *Tetrahedron Lett.*, 45 7749–7751.

[32] Carrée, F.; Gil, R.; Collin, J. (2005) Enantioselective ring opening of meso-epoxides by aromatic amines catalyzed by lanthanide iodo binaphtholates., *Org. Lett.,* 7 1023–1026.

[33] Azoulay, S.; Manabe, K.; Kobayashi, S. (2005) Catalytic asymmetric ring opening of meso-Epoxides with Aromatic Amines in Water., *Org. Lett,* 7 4593–4595.

[34] Kureshy, R. I.; Singh, S.; Khan, N. H.; Abdi, S. H. R.; Agrawal, S.; Mayani, V. J.; Jasra, R. V. (2006) Microwave-assisted asymmetric ring opening of meso-epoxides with aromatic amines catalyzed by a Ti-S-(−)-BINOL complex., *Tetrahedron Lett.,* 47 5277– 5279.

[35] Bartoli, G.; Bosco, M.; Carlone, A.; Locatelli, M.; Melchiorre, P.; Sambri, L. (2004) Asymmetric catalytic synthesis of enantiopure N-protected 1,2-amino alcohols., *Org. Lett.***,** *6* 3973–3975.

[36] Jacobsen, E. N. (2000) Asymmetric catalysis of epoxide ring-opening reactions, *Acc. Chem. Res.,* 33 421-431.

[37] Willis, M. C. (1999) Enantioselective desymmetrization., *J. Chem. Soc.*, *Perkin Trans.* 1 1765–1784.

[38] Robinson, D. E. J. E.; Bull, S. D. (2003) Kinetic resolution strategies using non-enzymatic catalysts., *Tetrahedron: Asymmetry* 14 1407–1446

[39] Larrow, J. F.; Jacobsen E. N. (2004) Asymmetric processes catalyzed by chiral (salen) metal complexes., *Topics in Organometallic Chemistry*, 123-152.

[40] Pastor, I. M.; Yus, M. (2005) Asymmetric ring opening of epoxides., *Curr. Org. Chem.*, 9 1-29.

[41] Shibasaki M.; Matsunaga, S. (2006) Design and application of linked-BINOL chiral ligands in bifunctional asymmetric catalysis., *Chem. Soc. Rev.*, 35 269–279.

[42] Bianchini, C.; Barbaro, P. (2002) Recent aspects of asymmetric catalysis by immobilized chiral metal catalysts., *Topics in Catalysis,* 19 17-32.

[43] Heitbaum, M.; Glorius, F.; Escher, I. (2006) Asymmetric heterogeneous catalysis., *Angew. Chem. Int. Ed.,* 45 4732 – 4762.

[44] Fan, Q.-H.; Li, Y.-M.; Chan, A. S. C. (2002) Recoverable catalysts for asymmetric organic synthesis., *Chem. Review,* 102 3385-3466.

[45] Baudequin, C.; Baudoux, J.; Levillain, J.; Cahard, D. Gaumontb, A.-C.; Plaqueventa, J. –C. (2003) Ionic liquids and chirality: opportunities and challenges., *Tetrahedron: Asymmetr,y* 14 3081–3093.

[46] Jorapur,Y. R.; Chi, D. Y. (2006) Ionic liquids: An environmentally friendly media for nucleophilic substitution reactions., *Bull. Chem. Soc.*, 27 345-354.

[47] Osburn, P. L.; Bergbreiter, D. E. (2001) Molecular engineering of organic reagents and catalysts using soluble polymers., *Prog. Poly. Sci.*, 26 2015-2081.

[48] Tokunaga, M.; Larrow, J. F.; Kakiuchi, F.; Jacobsen E. N. (1997) Asymmetric catalysis with water: Efficient kinetic resolution of terminal epoxides by means of catalytic hydrolysis., *Science***,** 277 936 – 938.

[49] Schaus, S. E.; Branalt, J.; Jacobsen, E. N., (1998) Total synthesis of muconin by efficient assembly of chiral building blocks., *J. Org. Chem.*, 63 4876-4877.

[50] Schaus, S. E.; Brandes, B. D.; Larrow, J. F.; Tokunaga, M.; Hansen, K. B.; Gould, A. E.; Furrow, M.E.; Jacobsen, E. N., (2002) Highly selective hydrolytic kinetic resolution

of terminal epoxides catalyzed by chiral (salen)Co^{III} complexes: Practical synthesis of enantioenriched terminal epoxides and 1,2-diols., *J. Am. Chem. Soc.*, *124* 1307-1315.

[51] Bandini, M.; Cozzi, P. G.; Melchiorre, P.; Umani-Ronchi, A. (2004) Kinetic resolution of epoxides by a C-C bond-forming reaction: Highly enantioselective addition of indoles to *cis*, *trans*, and *meso* aromatic epoxides catalyzed by [Cr(salen)] complexes., *Angew. Chem. Int. Ed.*, *43* 84-87.

[52] Bartoni, G.; Bosco, M.; Carlone, A.; Locatelli, M.; Massaccesi, M.; Melchiorre, P., Sambri, L. (2004) Asymmetric aminolysis of aromatic epoxides: A facile catalytic enantioselective synthesis of *anti*-β-amino alcohols *Org. Lett.*, *6* 2173-2176.

[53] For leading references on kinetic resolution, see E. L. Elier, S. H. Wilen, L. M. Mandar, Stereochemistry of organic compounds, (Wiley-Interscience, New York, 1994, pp. 395-415.

[54] H. B. Kagan and J. C. Flaud, in Topics in Stereochemistry, N. L. Alinger and E. L. Eliel, Eds. Interscience, New York, 1987, vol. 14, p. 249.

[55] Kureshy, R. I.; Singh, S.; Khan, N. H.; Abdi, S. H. R.; Ahmad, I.; Bhatt, A.; Jasra R. V. (2005) Improved catalytic activity of homochiral dimeric cobalt salen complex in hydrolytic kinetic resolution of terminal racemic epoxides., *Chirality*, 17 590–594.

[56] Kureshy R. I.; Khan N. H.; Abdi S. H. R.; Patel S. T.; Jasra R. V. (2002) Simultaneous production of chirally enriched epoxides and 1,2 diols from racemic epoxides via hydrolytic kinetic resolution (HKR*)., J Mol Catal.,* 179 73-77.

[57] Ready, J. M.; Jacobsen, E. N. (2001) Highly active oligomeric (salen)Co catalysts for asymmetric epoxide ring opening Reaction, *J. Am. Chem. Soc.,* 123 2687-2688.

[58] Ready, J. M.; Jacobsen, E. N. (2002) A practical oligomeric{(salen)Co} catalyst for asymmetric epoxide ring opening reaction, *Angew. Chem. Int. Ed.,* 41 1374-1377.

[59] Cavazzini, M.; Quici, S.; Pozzi, G. (2002) Hydrolytic kinetic resolution of terminal epoxides catalyzed by fluorous chiral Co(salen) complexes, *Tetrahedron* 58 3943-3949.

[60] Shepperson, I.; Cavazzini, M.; Pozzi, G.; Quici, S. (2004) Fluorous biphasic hydrolytic kinetic resolution of terminal epoxides, *J. Fluor. Chem.*, *125* 175-180.

[61] Kim, G. J.; Lee, H.; Kim, S. J. (2003) Catalytic activity and recyclability of new enantioselective chiral Co–salen complexes in the hydrolytic kinetic resolution of epichlorohydrine, *Tetrahedron Lett.* 44 5005-5008.

[62] Song, Y.; Yao, X.; Chen, H.; Bai, C.; Hu, X.; Zheng, Z., (2002) Highly enantioselective resolution of terminal epoxides using polymeric catalysts., *Tetrahedron Lett.* 43 6625-6627.

[63] Song, Y.; Chen, H.; Hu, X.; Bai, C.; Zheng, Z. (2003) Highly enantioselective resolution of terminal epoxides with cross-linked polymeric salen-Co(III) complexes, *Tetrahedron Lett.*, 44 7081-7085.

[64] Zheng, X.; Jones, C. W.; Weck, M. (2006) Poly(styrene)-supported Co–salen complexes as efficient recyclable catalysts for the hydrolytic kinetic resolution of epichlorohydrin, *Chem. Eur. J.,* 12 576 – 583.

[65] Holbach, M.; Weck, M. (2006) Modular approach for the development of supported, monofunctionalized, salen catalysts., *J. Org. Chem.*, 71 1825-1836.

[66] Jain, S.; Zheng, X.; Jones, C. W.; Weck, M.; Davis, R. J. (2007) Importance of counterion reactivity on the deactivation of Co-salen catalysts in the hydrolytic kinetic resolution of epichlorohydrin, *Inorg. Chem.*, 46 8887-8896.

[67] Lee, K. Y.; Kawthekar, R. B.; Kim, G. J. (2007) Synthesis of chiral intermediates catalyzed by new chiral polymeric (salen) cobalt complexes bearing Lewis acidic metal halides., *Bull. Korean Chem. Soc.*, 28 1553-1561.

[68] Breinbauer, R.; Jacobsen, E. N. (2000) Cooperative asymmetric catalysis with dendrimeric [Co(salen)] complexes., *Angew. Chem. Int. Ed.,* 39 3604-3607.

[69] Annis, D. A. Jacobsen, E. N. (1999) Polymer supported chiral Co(salen) complexes: synthetic applications and mechanistic investigations in the hydrolytic kinetic resolution of terminal epoxides., *J. Am. Chem. Soc.*, 121 4147-4154.

[70] Welbes, L. L.; Scarrow, R. C.; Borovik, A. S. (2004) Development of porous materials for heterogeneous catalysis: Kinetic resolution of epoxides, *Chem. Commun.,* 2544-2545.

[71] Kim, G. J.; Park D. W. (2000) The catalytic activity of new chiral salen complexes immobilized on MCM-41 in the asymmetric hydrolysis of epoxides to diols., *Catalysis Today* 63 537–547.

[72] Yang, H.; Zhang, L.; Su, W. Yang, Q.; Li, C. (2007) Asymmetric ring-opening of epoxides on chiral Co(Salen) catalyst synthesized in SBA-16 through the "ship in a bottle" strategy., *J. Catal.* 248 204–212. ibid, (2007) *Chem. Commun.* 1086-1088.

[73] Choi, S. D.; Kim, G. J. (2004) Enantioselective hydrolytic kinetic resolution of epoxides catalyzed by chiral Co(III) salen complexes immobilized in the membrane reactor., *Catal. Lett.*, 92 35-40.

[74] Belser T.; Jacobsen, E. N. (2008) Cooperative catalysis in the hydrolytic kinetic resolution of epoxides by chiral [(salen)Co (III)] complexes immobilized on gold colloids., *Adv. Synth. Catal.*, DOI: 10.1002/adsc.200800028.

[75] Oh, C. R.; Choo, D. J.; Shim, W. H.; Lee, D. H.; Roh, E. J.; Lee, S. G.; Song, C. E. (2003) Chiral Co(III)(salen)-catalyzed hydrolytic kinetic resolution of racemic epoxides in ionic liquids., *Chem. Commun.,* 1100-1101.

[76] Bandini, M.; Fagioli, M.; Mclloni, A.; Ronchi, A. U-. (2004) Polymer supported indium lewis acid: Highly versatile catalyst for regio- and stereoselective ring opening of epoxides, *Adv. Synth. Catal.* 346 573-578.

[77] Jacobsen, E. N.; Wu, M. H. In Ring Opening of Epoxides and Related Reactions in Comprehensive Asymmetric Catalysis; Jacobsen, E. N., Pflaz, A.,Yamamoto, H., Eds.; Springer-Verlag: Berlin, 1999; Vol. III, p 1309.

[78] Ager, D. J.; Prakash, I.; Schaad, D. R. (1996) 1,2-Amino Alcohols and Their Heterocyclic Derivatives as Chiral Auxiliaries in Asymmetric Synthesis., *Chem. Rev.,* 96, 835-876

[79] Bergmeier, S. C. (2000) The Synthesis of Vicinal Amino Alcohols., Tetrahedron, 56 2561-2576.

[80] Angelino, M. D.; Laibinis, P. E. (1999) Polymer-supported salen complexes for heterogeneous asymmetric synthesis: stability and selectivity, *J. Polym. Sci. Part A: Polymer Chemistry*, 37 3888-3898.

[81] Gigante, B.; Corma, A.; Garcia, H.; Sabater, M. J. (2000) Assessment of the negative factors responsible for the decrease in the enantioselectivity for the ring opening of epoxides catalyzed by chiral supported Cr(III)-salen complexes *Catalysis Lett.* 68 113–119.

[82] Baleizão, C.; Gigante, B.; Sabater, M. J.; Garcia, H.; Corma, A. (2002) On the activity of chiral chromium salen complexes covalently bound to solid silicates for the enantioselective epoxide ring opening *Applied Catalysis A: General* 228 279–288.

[83] Dioos B. M. L.; Jacobs P. A. (2003) CrIII(salen) impregnated on silica for asymmetric ring opening reactions and its recovery via desorption/re-impregnation., *Tetrahedron Lett.,* 44 8815–8817.

[84] Dioos, B. M. L.; Geurtsa W. A.; Jacobs, P. A. (2004) Coordination of CrIII(salen) on functionalised silica for asymmetric ring opening reactions of epoxides., *Catalysis Lett.,* 97 125-129.

[85] Dioos, B. M. L.; Jacobs P. A. (2005) Impregnation of dimeric CrIII(salen) on silica and its application in epoxides asymmetric ring opening reactions., *Applied Catalysis A: General,* 282 181–188.

[86] Dioos, B. M. L.; Jacobs P. A. (2006) Heterogenisation of dimeric Cr(salen) with supported ionic liquids., *J. Catal.* 243 217–219.

[87] Song, C. E.; Oh, C. R.; Roh, E. J.; Choo, D. J. (2000) Cr(salen) catalysed asymmetric ring opening reactions of epoxides in room temperature ionic liquids., *Chem. Commun.,* 1743-1744.

[88] Hou, X. L.; Wu, J.; Dai, L. X.; Xia, L. J.; Tang, M. H. (1998). Desymmetric ring-opening of meso-epoxides with anilines: A simple way to chiral β-amino alcohols., *Tetrahedron: Asymmetry*, 9 1747–1752.

[89] Fu, X. L.; Wu, S. H. (1997) A regio-and stereoselective synthesis of β-amino alcohols., *Synth. Commun., 27* 1677–1683.

[90] Schneider, C.; Sreekanth, A. R.; Mai E. (2004) Scandium-bipyridine-catalyzed enantioselective addition of alcohols and amines to meso-epoxides., *Angew. Chem., Int. Ed. Engl.*, 43 5691–5694.

[91] Mai, E.; Schneider, C. (2007) Scandium-bipyridine-catalyzed enantioselective aminolysis of meso-epoxides., *Chem. Eur. J.,* 13 2729-2741.

[92] Mai, E.; Schneider, C. (2007) Indium-bipyridine catalyzed, enantioselective aminolysis of meso-epoxides., *Synlett,* 2136-2138.

[93] Ogawa, C.; Azoulay, S.; Kobayashi, S. (2005) Bismuth triflate-chiral bipyridine complex catalyzed asymmetric ring opening reactions of meso-epoxide in water., *Heterocycles***,** 66 201-206.

[94] Schneider, C. (2006) Synthesis of 1,2-difunctionalized fine chemicals through catalytic, enantioselective ring-opening reactions of epoxides., *Synthesis***,** 3919-3944.

[95] Gao, B.; Wen, Y.; Yang, Z.; Huang, X.; Liu, X.; Feng X. (2008) Asymmetric ring opening of meso-Epoxides with aromatic amines catalyzed by a new proline-based N,N'-dioxide-indium tris (triflate) complex, *Adv. Synth. Catal.,* 350 385– 390.

[96] Kureshy, R. I.; Singh, S.; Khan, N. H.; Abdi, S. H. R;. Agrawal S.; Jasra R. V. (2006) Enantioselective aminolytic kinetic resolution (AKR) of epoxides catalyzed by recyclable polymeric Cr(III) salen complexes *Tetrahedron: Asymmetry* 17 1638–1643.

[97] Kureshy, R. I.; Prathap, K. J.; Singh, S.; Agrawal, S.; Khan, N. H.; Abdi, S. H.. R.; Jasra, R. V. (2007) Chiral recyclable dimeric and polymeric Cr(III) salen complexes catalyzed aminolytic kinetic resolution of trans-aromatic epoxides under microwave irradiation., *Chirality,* 19 809-815 .

[98] Kureshy, R. I.; Prathap, K. J.; Agrawal, S.; Khan, N. H.; Abdi, S. H. R.; Jasra, R. V. (2008) Highly enantioselective syntheses of chiral β-aminoalcohols using chiral Ti(IV) Schiff base complexes as catalysts., *Eur. J. Org. Chem.* (In press)

[99] Kureshy, R. I.; Singh, S.; Khan, N. H.; Abdi, S. H. R.; Suresh, E.; Jasra. R. V. (2006) Facile enantioselective ring-opening reaction of meso-epoxides with anilines using (*S*)-(-)-BINOL-Ti complex as a catalyst., *Eur. J. Org. Chem.*, 1303–1309.

[100] Reddy L. R.; Bhanumathi, N. Rao, K. R. (2000) Dynamis kinetic asymmetric synthesis of β-aminoalcohols from racemic epoxides in cyclodextrin complexes under solid state conditions., *Chem. Commun.* 2321-2322.

In: Heterogeneous Catalysis Research Progress
Editor: Mathias B. Gunther, pp. 345-368
ISBN 978-1-60456-979-7

Chapter 9

LEAD-RUTHENATE PYROCHLORE MODIFIED NAFION® MEMBRANE FOR TUNABLE HETEROGENEOUS CATALYTIC OXIDATION REACTIONS

Jyh-Myng Zen[1a], Annamalai Senthil Kumar,[b] and Shanmugam Venketasan[a]

[a]Department of Chemistry, National Chung Hsing University, Taichung 402, Taiwan
[b]Department of Chemistry, Vellore Institute of Technology University, Vellore, India

ABSTRACT

Development of heterogeneous metal oxide catalytic system with selective redox transition and highly recyclic feature is a challenging research in synthetic organic chemistry. Yet, very limited attention has been paid in consideration with stability and reusability of the catalytic system. This article reviews our recent development on lead-ruthenate pyrochlore ($Ru_2Pb_2O_7$, Pyc) modified Nafion® membrane catalyst (designated as |NPyc|) for selective organic functional group transformation including oxidation of alcohol to aldehyde and ketone, organic sulfide to sulfoxide under ambient conditions. Note that Nafion® is a rigid perfluropolymer backbone polymer bearing ion-exchangeable sulfonic acid terminal group ($–SO_3^-H^+$) and is extensively used as a solid-state protonic conductor in fuel cell application. It is seldom used for chemical modification in synthetic organic chemistry. This article covers preparation of |NPyc|, characterization by XRD, SEM, SECM, AFM and TGA, and catalytic organic synthesis in triphasic medium with co-oxidants such as H_2O_2, NaOCl and O_2. Under optimal working conditions, the membrane catalyst showed very good selectivity and good turnover for wide range of organic compounds. The |NPyc| catalyst can be recycled over > 25 times without any catalytic degradation. As to the mechanism, a high valent-oxo ruthenium redox species, perruthenate/ruthenate redox couple ($Pyc\text{-}RuO_4^-/Pyc\text{-}RuO_4^{2-}$) exists in the |NPyc|, is believed to participate in the oxidation reaction with the co-oxidants.

[1] Corresponding author's e-mail: jmzen@dragon.nchu.edu.tw.

1. INTRODUCTION

In the last few years there has been growing interest for recyclable heterogeneous catalysts in synthetic organic chemistry.[1,2] Among the many catalysts, polymer supported reagents are considered to be relatively clean due to its easy recycling and simple handling procedures.[3,4] Nevertheless, very limited attention has been paid in consideration with stability and reusability of the catalytic system. This review focuses on the recent results on lead-ruthenate pyrochlore ($Ru_2Pb_2O_7$, Pyc) modified Nafion® membrane catalyst (designated as |NPyc|) for selective organic oxidations.[5-7] Nafion® is a rigid perfluropolymer backbone polymer bearing ion-exchangeable sulfonic acid terminal group ($–SO_3^-H^+$) (Scheme 1), extensively used as a solid-state protonic conductor in fuel cell applications.[8] Ruthenium oxide and its related composites materials are popular oxidant amongst other oxides because of tunable redox properties ranging from III/II, IV/III, VI/IV and VII/VI.[9-29] The Pyc is found to have bi-functional active sites both for oxidation and reduction purposes.[30-36] In this article, we are highlighting preparation, characterization and synthetic organic applications of the |NPyc| for selective oxidations such as selective alcohol oxidation and photocatalytic sulfoxidation reactions under heterogeneous condition.

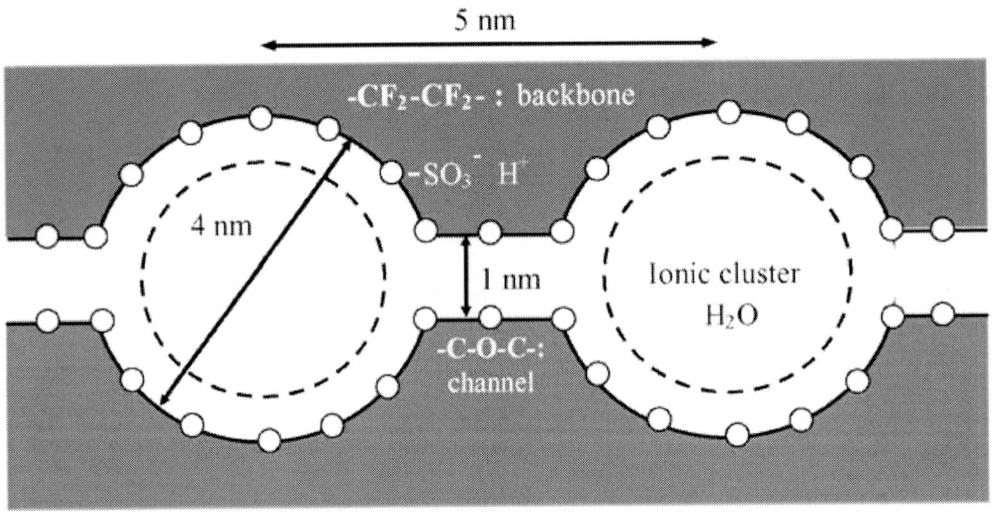

Scheme 1. Structure of the perfluorinated ionomer membrane (Nafion®).

2. PREPARATION OF |NPYC|

Scheme 2 illustrates typical preparation procedure for *in-situ* precipitation of Pyc units inside the Nafion® 417 membrane. Briefly, a piece of Nafion® 417 membrane with a geometric area of ~10 cm^2 is first ion-exchanged overnight with millimolar mixture solution of Ru^{3+} and Pb^{2+} having a ratio of 1.5, followed by thoroughly washing with deionized water to remove the non-ion exchanged cations. Then the modified membrane is submersed in 1.1 M KOH at 53 °C for 24 h with constant O_2 purging. This preparation procedure resulted in a uniform distribution of the catalytically active micro-particles throughout the Nafion® membrane matrix.

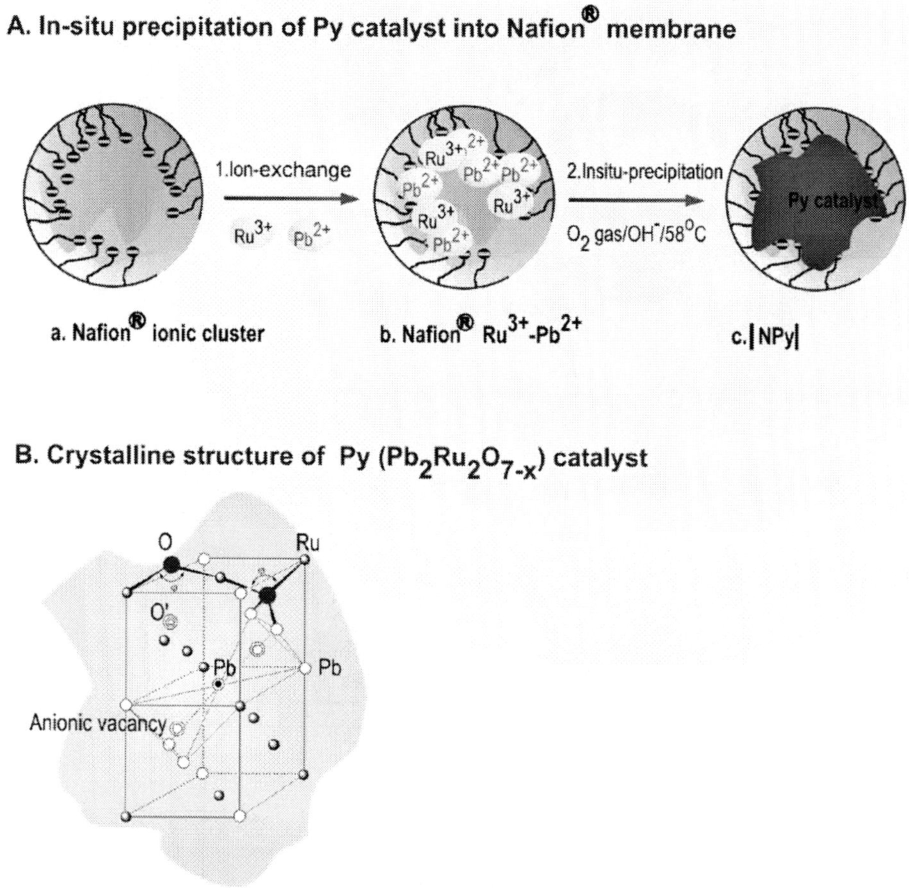

Scheme 2. (a) Preparation route of the |NPyc| (b) Crystalline structure of the active Pyc site.

Calculated weight difference of Nafion® 417 membrane before and after Pyc modification is 60 mg/g, which corresponds to 70.6 μmol of Pyc active units per gram of the membrane. Figure 1 shows the difference in the appearance of the |NPyc| and an unmodified Nafion® 417 membrane.

2.1. Physico-Chemical Characteristics of The |Npyc| Catalyst

2.1.1. X-Ray Diffraction (XRD) Analysis

X-ray diffraction (Shimadzu XRD-6000) measurements are carried out using a Cu Kα 1.54060 Å radiation source. Figure 2 shows comparative X-ray diffraction patterns of the |NPyc| along with Nafion® 417 membrane and Pyc powder control systems. As can be seen, the Pyc powder shows a broad XRD 2θ peak centered at 30° and other diffuse peaks are in the range of 40–60°.

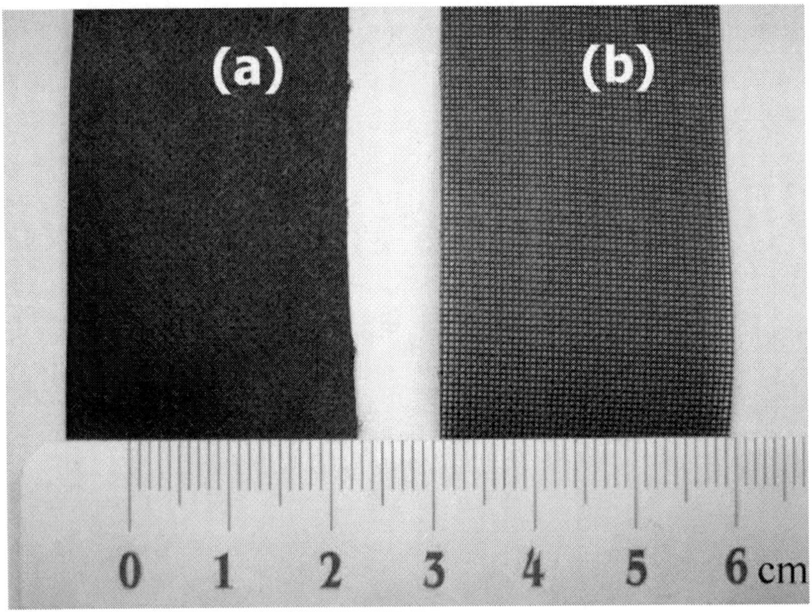

Figure 1. Typical photographs for Nafion® 417 membrane after (a) and before (b) Pyc modification.

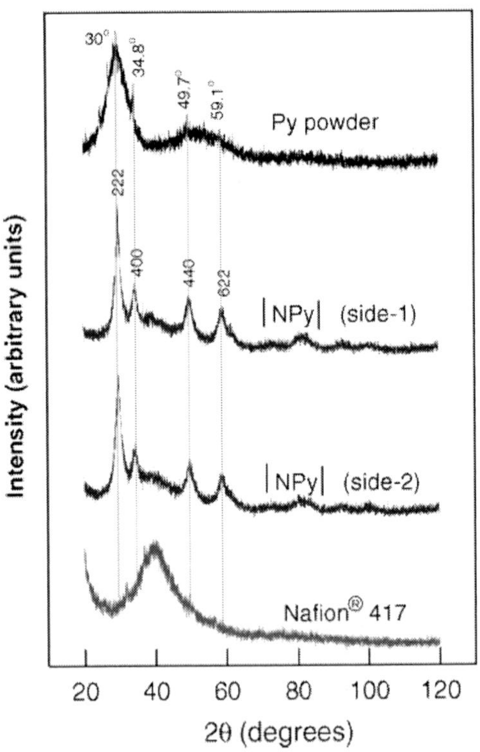

Figure 2. X-ray diffraction patterns of the |NPyc| catalyst (both sides) in comparison with unmodified Pyc powder and Nafion® 417 membrane samples.

On the other hand, the |NPyc| prepared under similar conditions shows sharp peaks at 30, 34.8, 49.7 and 59.1°, which is in agreement with the standard values of the Pyc crystalline powder. The XRD peaks of |NPyc| are relatively sharper than those of bare Pyc powder. These results indicate that enhanced crystallinity of the Pyc within the |NPyc| composite membrane. In other words, the Nafion® membrane can assist the crystallization of Pyc within the Nafion® interfacial structures. Nafion® network can be visualized as a reverse micelle-like ionic cluster (4 nm diameter) with three distinctive structural regions, i.e., perfluorinated backbone polymer network (hydrophobic zone, CF_2–CF_2), water cores (hydrophilic area, $SO_3^-H^+\cdot nH_2O$), and an interfacial domain (where water cores in neighboring clusters are interconnected through 1 nm channels, C–O–C) (Scheme 1).[8] The Pyc grain size value of 5.3 nm for the composite membrane is slightly higher over the cluster size of Nafion® membrane (4 nm). Liu et al. observed a particle size of 3.8 nm for Nafion®- composite membrane with TiO_2 [37]. In another case, Ludvigsson et al. reported a grain size of 30 nm for the composite of MnO_2 and Mn_2O_3 with Nafion® by XRD.[38] For the case of |NPyc|, presumably the grains might grow over the clusters as agglomerate like network structure. Regarding the enhanced XRD peaks with the |NPyc|, the close packing of Pb^{2+} and Ru^{3+} ions in the cluster sites of Nafion® should aid in the increased crystallinity of the Pyc. Further studies with scanning electron microscopy (SEM), atomic force microscopy (AFM), scanning electrochemical microscope (SECM), and thermo-gravimetric analysis (TGA) reveals important structural factors about the |NPyc| membrane.

2.1.2. SEM Analysis

Figure 3 shows SEM images and energy dispersive X-ray analysis (EDAX) of the |NPyc| catalyst. As can be seen, a pancake type of morphology without any marked cracks in the surface was observed. Surface morphological view is different in different sides. For example, a pancake type of morphology can be seen in the top side, while the bottom position shows a smooth and folded-wire like surface, may be due to the membrane net (cross-section SEM in Figure 3D and E). Top position is expected to be relatively rough and more hydrophilic in nature. There is no obvious difference in the XRD for those sides, further indicating similar Pyc crystalline units throughout the |NPyc| matrix (Figure 2). There is no appearance of isolated Pyc particles with the |NPyc| as per the value of 5.3 nm by XRD and unlike to the case of the TiO_2 modified Nafion® membrane.[37] Classical preparation procedures utilize Nafion® 117 as a host and the surface structures can be easily influenced by external modification procedures.[37-39] Nature of the Nafion® backbone and preparation conditions should consequently affect the critical morphological structure of the membrane composite. For example, Nafion® composite prepared at high temperatures and/or with strong chemical treatments (such as light paraffin and Pd-alloy modified films by sputtering treatment at 300 and 100 ° C, respectively[39]) leads to more surface cracks. While mild treatments (such as ion-exchange followed by hot water hydrolysis) result in uniform distribution without any surface cracks as to the case of TiO_2.[37] Considering that Nafion® 417 used in this work possesses more rigid network due to reinforced Nafion® and Decron® polymeric units, hence it is rugged enough to protect the network from surface cracks even under any harsh environments.

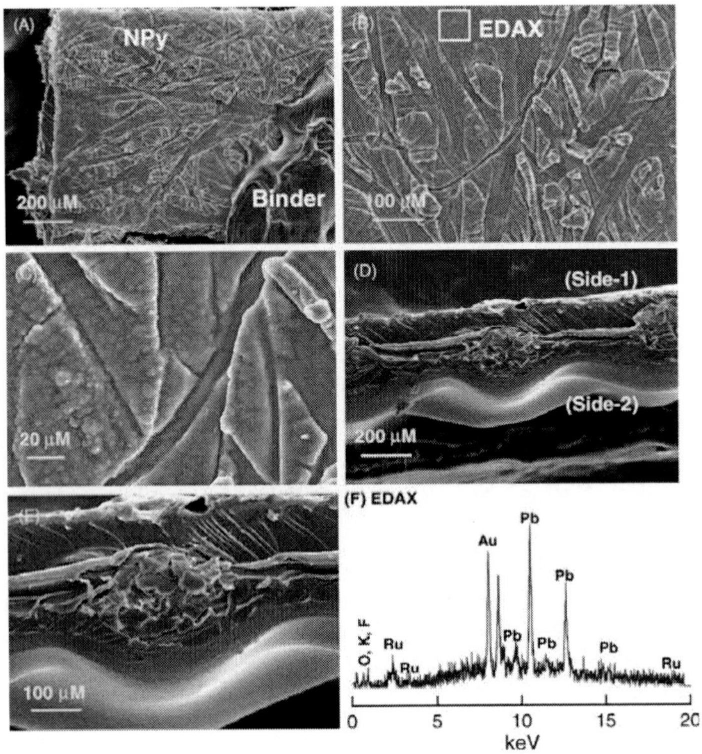

Figure 3. SEM of the |NPyc| membrane (Au coated) at different magnifications and its typical EDAX response.

2.1.3. AFM Analysis

Figure 4 shows the typical surface morphological picture of |NPyc| by AFM analysis. Even though the general morphology is similar to that of SEM, AFM reveals finer surface feature like specific contours on the networks structure. The top view AFM reveals agglomerate-like structures with different size nano-particles. Average particle size of ~30 nm is detected in remote places (Figure 4B), which is 7.5 times larger than the value obtained by previous XRD measurements. This larger particle size could be due to agglomerated units of the 5.3 nm Pyc particles. In other words, the observation provides a clue about the number of grains in the agglomerated units. For example, about six of the 5.3 nm Pyc particles were clubbed together to make the ~30 nm size agglomerate in the |NPyc| network. The inter-cluster Pyc formation dominated in |NPyc| further diminishes the individual particle characteristics.

2.1.4. SECM Analysis

SECM is a useful electrochemical technique for imaging the surface topographical structure at solid/liquid interfaces.[40,41] Briefly, the electrochemical system consists of a 10 μm Pt-ultramicroelectrode (UME) with Ag/AgCl (3 M KCl) as the reference and Pt as the counter electrode. The unmodified- and Pyc modified-Nafion® membranes (side-1) are carefully mounted on a homemade plastic plate on the bottom of the SECM cell.

(A)

(B)

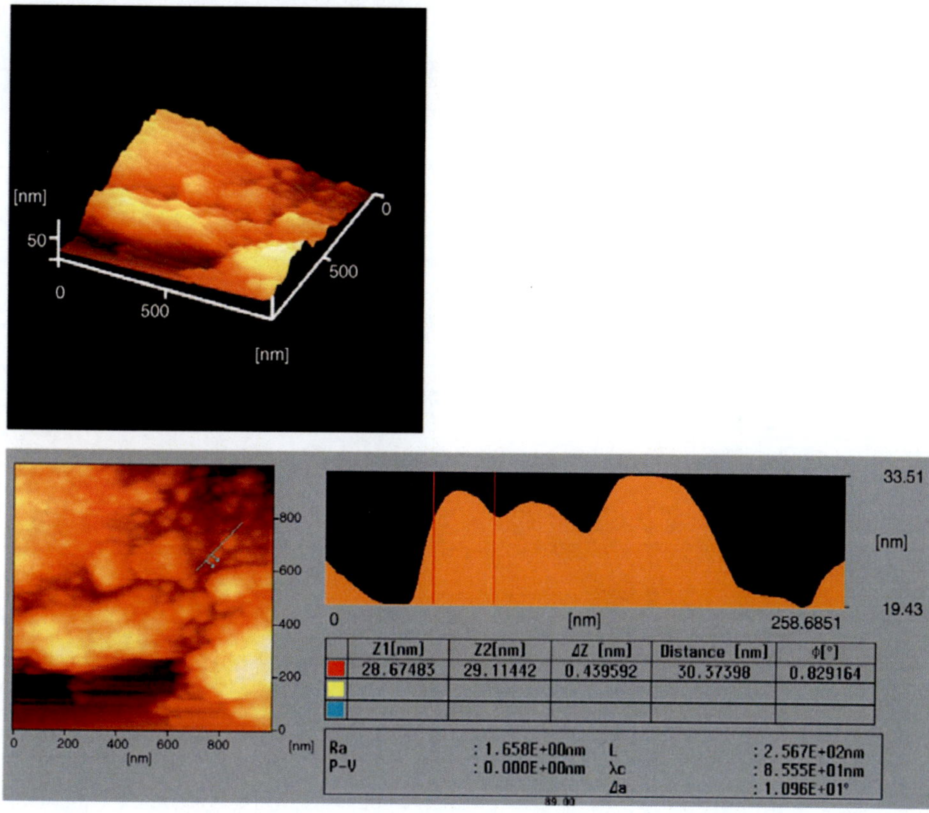

Figure 4. AFM surface morphology view of the |NPyc| catalyst (side-1) (A) and in-built program result for the calculation of particle size on the surface (B).

The top position of the cell was covered with 1 mM ferrocenemethanol (Mh-Fc, E^o = 0.31 V vs. SCE) in 0.2 M KCl. The distance of the UME is set by carefully positioning the tip (z-axis), while measuring the transport limited current (i_{tlc}) of Mh-$Fc^{+/0}$ redox system. The tip is maneuvered in the z-axis from an infinite distance to the working membrane surface at which the i_{tlc} approached 20% of the original value (usually ~90% is acceptable for smooth surfaces [40]). However, since the membrane surface is unsmooth (*via* SEM), relatively low current values and high interface distance are chosen here. Images of the transport-limited current for the one-electron oxidation of Mh-Fc ion, versus tip position, are obtained by scanning at 1.5 μm/s in a fixed plane above the working membrane surface. A negative feedback mode is adopted here, where the diffusion/transport current of the Mh-$Fc^{+/0}$ redox system hindered by the membrane contours and in turn transport-limited current, i_{tlc} at a fixed applied potential (0.3 V vs. Ag/AgCl) can be monitored as a surface topological structure (Figure 5A).

Typical three-dimensional SECM image of the bare Nafion® 417 and |NPy| using Pt-UME over a 50 μm × 50 μm area is shown in Figure 5. The |NPyc| film shows relatively higher contour levels over the bare Nafion® 417 membrane, presumably due to the Pyc modified layer. Note that the surface topology viewed by SECM represents in situ permeable regions of the underlying surface and this result further reflects that the Pyc site anchored Nafion® 417 membrane has sufficient active-layers within the surface for catalytic applications.

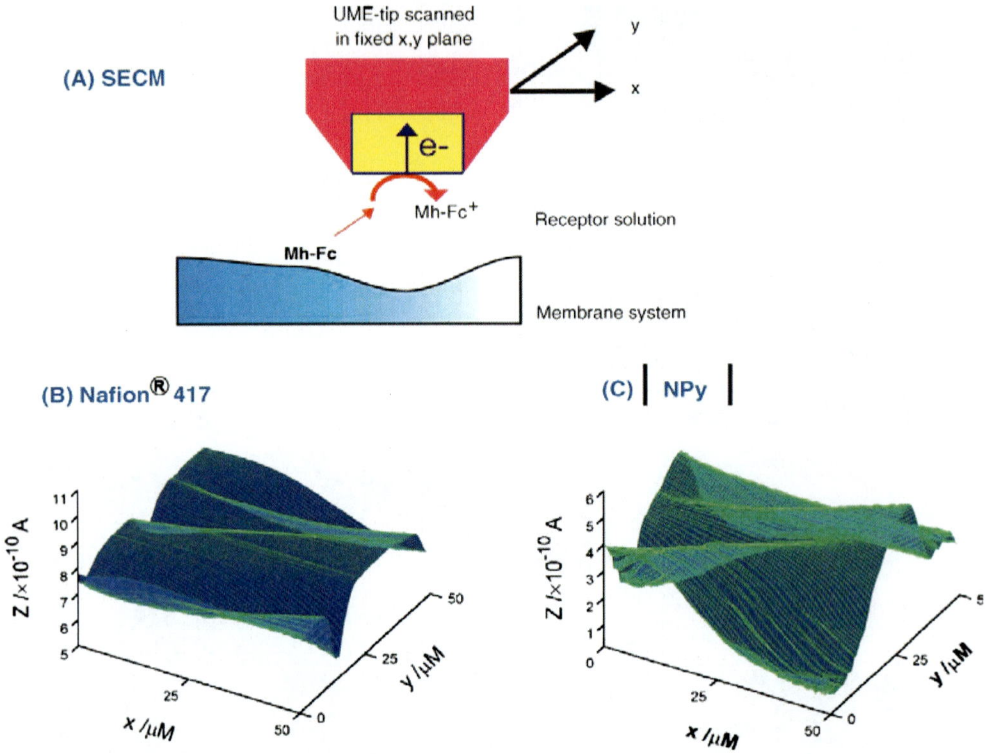

Figure 5. Schematic illustration for the SECM operation using a Pt UMP tip (A), and the topographical images for the bare Nafion® 417 (B) and |NPyc| composite membrane (C) (side-1) (the scales *x* and *y* = distance, *z* = current).

Overall, the SECM observation is in parallel with the SEM and AFM morphological views regarding the continuous distribution of Pyc catalytic sites in the |NPyc| catalyst.

2.1.5. TGA Studies

Figure 6 shows the thermal decomposition and its differential (DTGA) patterns of the Pyc powder, |NPyc| and bare Nafion® 417 membrane systems in the temperature window of 30–900 °C. The Pyc powder shows feeble weight loss peaks at ~100 and 300 °C respectively due to physisorbed, $(H_2O)_{phy}$ and chemisorbed, $(H_2O)_{chem}$ water molecules or OH, $(OH)_{chem}$ molecules, and the Pyc powder is fairly stable up to 800 °C, while the |NPyc| maintains its weight up to 400 °C without any marked degradation, but at > 400 °C it starts to lose its weights due to the decomposition of SO_3^-, C–O–C (400–500 °C) and CF_2 (500–600 °C) functional groups. This observation is qualitatively similar to the case of bare Nafion® 417 membrane except at increased temperatures (Figure 6c and Scheme 3). The reinforced nature and rigid structure of the Nafion® 417 cause the shift of the degradation temperature. Figure 6B shows the comparative TGA at extreme temperatures (500–700 °C). Interestingly the bare Nafion® 417 resulted to 100% weight loss, while the |NPyc| retained ~9% weight at 800 °C. The remaining amount in the latter case is due to the un-burn Pyc active sites and the result is comparable to that of weight loss measurement during the |NPyc| preparation procedure. Fig 6B shows possible thermal degradation pathways for the |NPyc| based on the TGA.

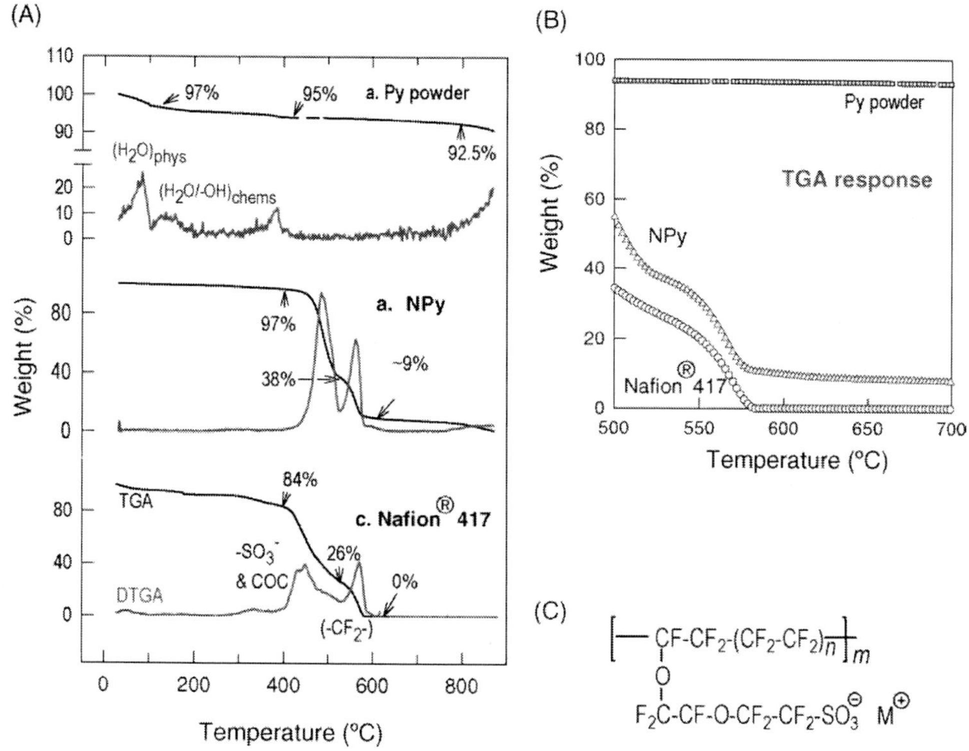

Figure 6. (A) Typical TGA and DTGA response for the |NPyc| catalytic system in comparison to the bare Pyc and Nafion® 417 samples in the temperature window of 30–900 °C. (B) 500–700 °C (C) and basic chemical structure for the Nafion®.

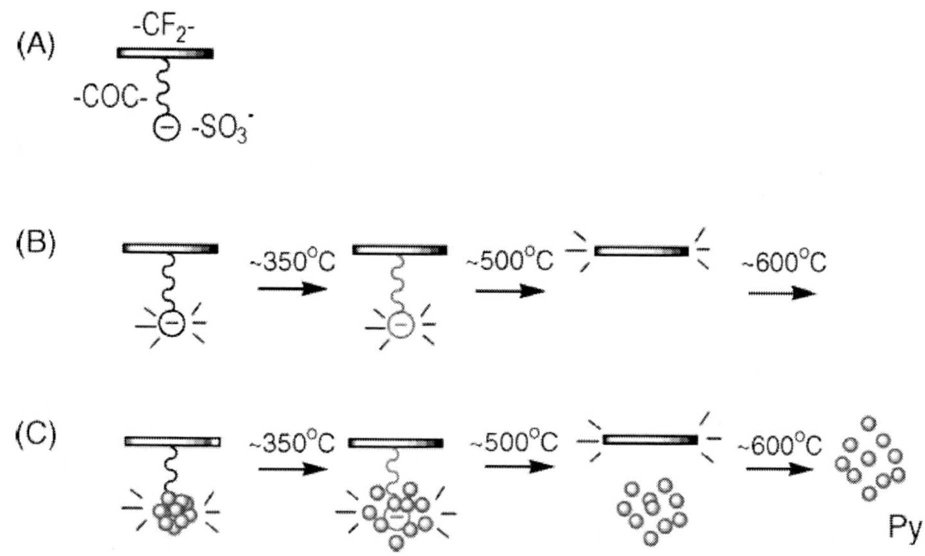

Scheme 3. Conceptional representations for the decomposition pathways of the Nafion® 417 (A and B) and |NPyc| units (C) under thermogravimetric analysis. Lighter colors denote partial decomposition of the compounds.

At extreme temperature (> 600 °C), the Nafion® units are completely burned out with the nude Pyc powder in the test system. But, at T < 400 °C, the membrane archive is rigid and stable, and is thus suitable for use in hot-bath catalytic organic synthesis.

2.1.6. Cation Exchange Property (CEP)

Since the Nafion® membrane is prone to cationic exchange behavior; the ion-exchange ability of the |NPyc| is studied. The photosensitizer, $Ru(bpy)_3^{2+}$, is taken as a probe to analyze the CEP using its specific UV-vis absorption band, λ_{max} = 456 nm. If the Pyc catalyst is completely loaded within the Nafion® membrane, then there is no free anionic sulfonic group and hence absence of the $Ru(bpy)_3^{2+}$ ion-exchange feature. On the other hand, if the Pyc is partially loaded with considerable exposure of the free anionic sulfonic groups, then specific UV-vis absorption signals could be observed corresponding to the partial ion-exchanged $Ru(bpy)_3^{2+}$. Figure 7 shows typical solid-state UV-vis absorption spectra of the bare |NPyc| and $Ru(bpy)_3^{2+}$ loaded |NPyc| composite films. It is obvious that the thick opaque |NPyc| membrane results in a highly noisy behavior, while the $Ru(bpy)_3^{2+}$ treated membrane shows a specific peak at ~450 nm corresponding to the $Ru(bpy)_3^{2+}$ absorption energy level. This observation is particularly cationic exchangeable vacancies within the |NPyc| composite system.

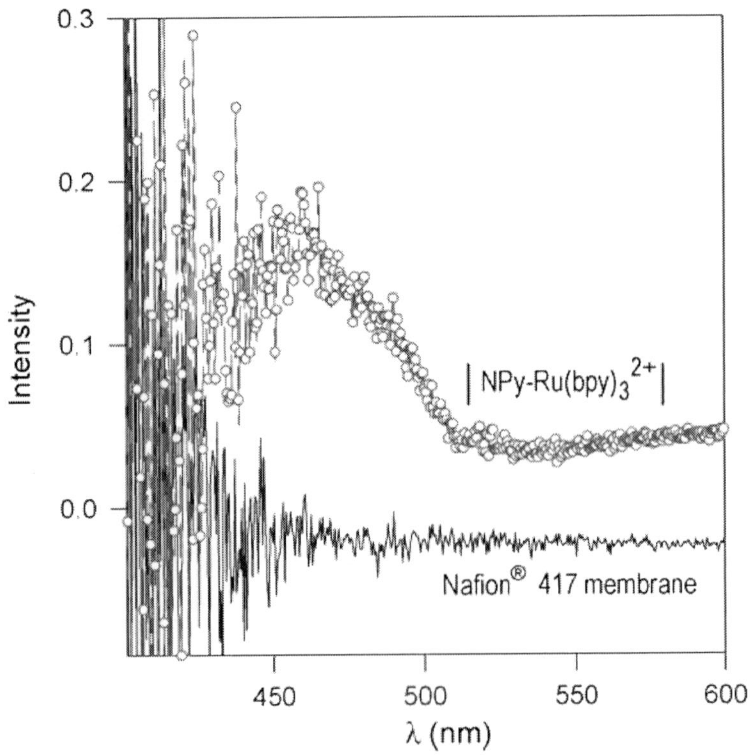

Figure 7. Solid state UV-vis absorption spectrum of Nafion® 417 and Ru(bpy)32+-modified |NPyc| catalytic system (|NPyc–Ru(bpy)32+|) in ambient conditions. In the modification procedure, 1 cm2 test membrane is first dipped in 1 mM Ru(bpy)3 2+ CH3CN:H2O (1:1) solution for ~30 min followed by water washing.

3. |NPyc| Catalytic Alcohol Oxidation Reactions

3.1. H_2O_2 co-oxidant

Catalytic reactions are carried out in a three-necked round bottom flask with 32.8 mM of alcohol (e.g., benzyl alcohol) in 15 mL CH_2Cl_2 at 80 °C. The |NPyc| catalyst (~1 g) is placed in the flask and co-oxidant, 30% H_2O_2 was slowly added (0.5 mL h^{-1}) to the reaction mixture. Progress of the reaction is continuously monitored by analyzing the non-aqueous layer during the course of reaction using gas-liquid chromatography (GLC). Figure 8 shows typical response from 1200 min of the reaction mixture having ~80% conversion of benzaldehyde. Control experiments in absence of catalyst do not show any appreciable conversion (Table 1). This result reveals the necessity of the catalytic system with the co-oxidant. Based on the GLC results, kinetic rate constant (k_{cat}, pseudo first order rate) for the benzyl alcohol oxidation is calculated as shown in Table 1. The k_{cat} values is in the range of 2–3.6 × 10^{-6} s^{-1} for the catalytic oxidations at different timings and it is about two orders higher than that of the uncatalyzed reaction. Measured turnover frequency (TOF) for the catalyst is in the range of 16–23 h^{-1}. Product analysis by GLC–MS, H^1 NMR and C^{13} NMR shows that benzaldehyde and some trace of starting material in the reaction mixture. Note that the $RuCl_3{\cdot}xH_2O$ based classical system yields unselective products of both benzaldehyde and benzoic acid.[13] Hence, the observation with the |NPyc| based catalytic system with > 99% benzaldehyde selectivity without any over-oxidation to benzoic acid is new to Nafion® based catalytic oxidation in organic synthesis.

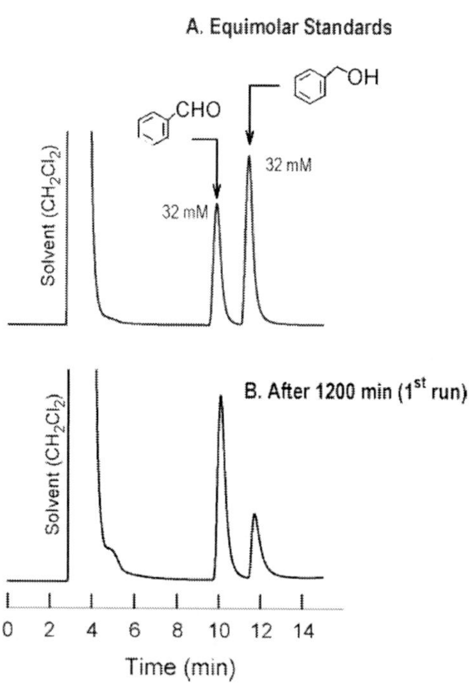

Figure 8.Typical GLC response for the equimolar standards of benzyl alcohol and benzaldehyde (A) and from the |NPyc| catalyzed benzyl alcohol reaction mixture (B) under optimized GLC conditions.

Table 1. Selective catalytic oxidation of $PhCH_2OH \rightarrow PhCHO$ at |NPyc| catalyst in CH_2Cl_2 + 30% H_2O_2 at 80 °C

Catalyst entry	$PhCH_2OH$ (mM)	|NPyc| (g) Added	|NPyc| (g) Recovered	H_2O_2 (mL)[d]	Time (h)	yield[c] (%)	k_{cat}[a] ($\times 10^{-5}$ s^{-1})	TOF (h^{-1})[b]
Blank	32	--	--	9	24	6.3	0.06	--
|NPyc| (fresh)	32	1.026	1.019	9	24	84.0	2.03	16
|NPyc| (2nd time)	32	1.019	1.019	9	20	99.0	3.59	23
|NPyc| (3rd time)	32	1.019	1.017	3	6	16.0	1.19	16
|NPyc| (4th time)	32	1.017	1.012	9	24	89.1	2.38	17

[a] First order rate constant

[b] Turn over frequency (TOF) = $[\text{Product}]_{\text{Mol}}/([\text{catalyst}]_{\text{Mol}} \times T_{\text{hour}})$; [Pyc] = 70.56 μMol./gram of NPyc

[c] by GLC

[d] Slow addition in the reaction mixture (condition not optimized).

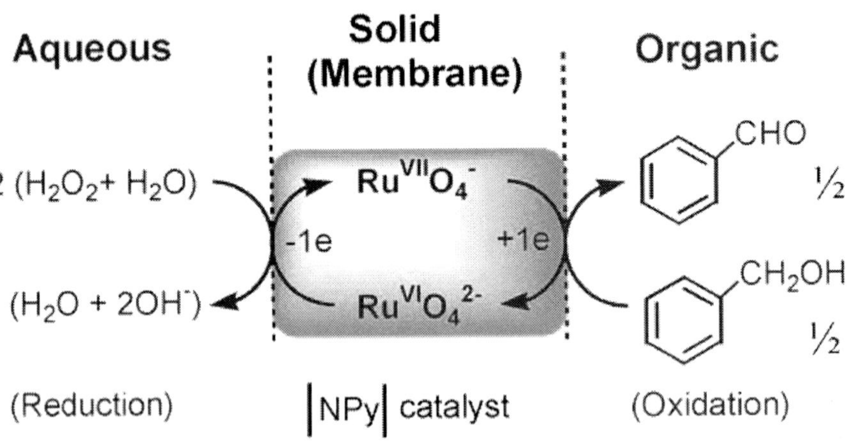

Scheme 4. Proposed triphasic benzyl alcohol oxidation mechanism on the |NPyc| catalyst with 30% H_2O_2.

As to the mechanism, a high valent perruthenate/ruthenate, $Ru^{VII}O_4^{-}/Ru^{VI}O_4^{2-}$ redox couple exists in the |NPyc| surface is believed to participate in the benzyl alcohol oxidation with the H_2O_2 as per the reaction (Scheme 4). In the first step, the co-oxidant H_2O_2 helps to convert the ruthenate (RuO_4^{2-}) to high-valent perruthenate ion (RuO_4^{-}) in aqueous solution, which in turn oxidizes the benzyl alcohol to benzaldehyde (organic phase) and return to the low-valent RuO_4^{2-} in a cyclic manner (Scheme 4). Regeneration of the perruthenate intermediate ion is the key for the catalytic alcohol oxidation, where the oxidation state is usually stabilized in alkaline condition or organic solvents.[36] In the above reaction scheme, H_2O_2 act as a sacrificial electron acceptor for the overall oxidation. It is expected that fraction of basic cationic sites exist in the |NPyc| archive helps for the easy abstraction of proton from the $PhCH_2OH$ (β-hydride elimination reaction pathway similar to Ru=O based metal complex for the alcohol oxidation) in turn to the selective benzaldehyde formation.[42] Finally, the catalytic oxidation reaction is repeated several times under similar working conditions to evaluate the multiple re-usage of the |NPyc| (Table 1). Figure 9B shows typical results for four repeated usages of the |NPyc| catalyst.

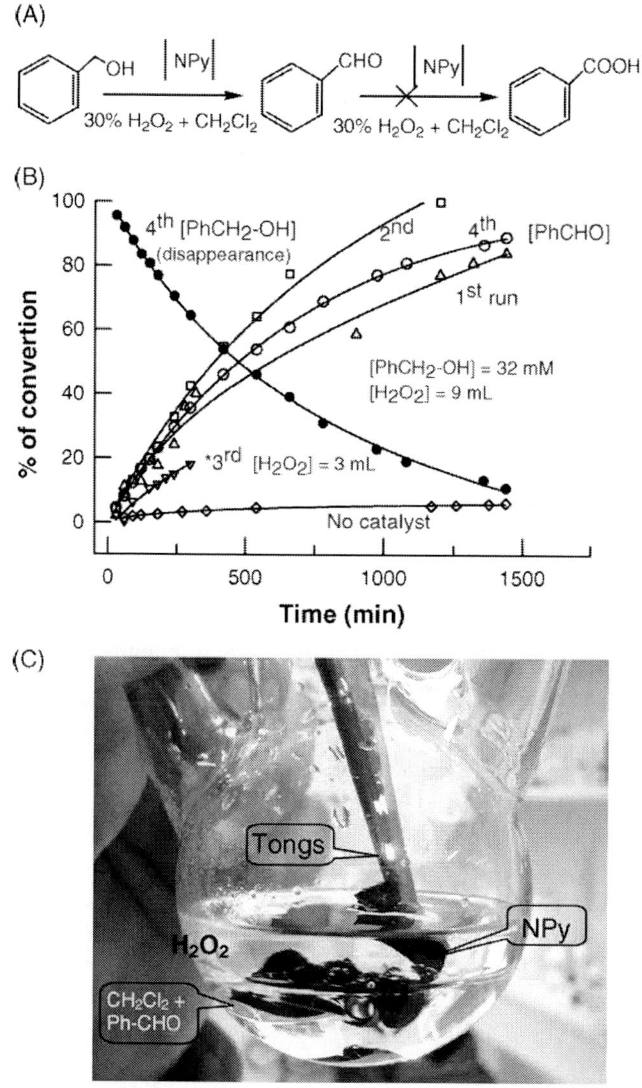

Figure 9.(A) Reaction scheme for the selective benzyl alcohol oxidation without any over-oxidation to benzoic acid. (B) Plot of % of conversion over the reaction time for the benzyl alcohol oxidation reaction catalyzed by |NPyc| at various recyclic conditions with 30%H_2O_2. (C) The reaction step up picture.

The conversion rates and yields are similar in all experiments revealing appreciable stability and reusability of the present system. Apart from this, easy separation of the |NPyc| from the reaction mixture is another clear advantage of the present approach (Figure 9C). Most importantly, > 99% recovery of the catalytic system is achieved each time. Apart from that, even at stringent working conditions like 100 h of total reaction timings (T = 80 °C) with a strong stirring, weight loss of only ~0.07% is observed with the |NPyc| catalyst. It is very difficult to expect such stability and reusability from the conventional catalytic systems, since large quantities of active site leakages are often reported in the literature.[21-24] Finally, the catalytic oxidation reaction demonstrated in this work is cleaner and environmentally nontoxic final products (H_2O), which further supports for greener organic synthetic approach.

3.2. NaOCl co-oxidant

In this study, co-oxidant NaOCl coupled |NPyc| catalytic system was demonstrated for > 99% selective oxidation of both primary and secondary alcohols in a triphasic condition of CH_2Cl_2 (org)//|NPy| (s)//NaOCl-pH 11 (aq). NaOCl is found to be considerably improving the catalytic performance of the |NPyc| membrane catalyst over the H_2O_2 co-oxidant. This is not surprised as ruthenium can easily convert into a higher oxidation state of Ru(VIII) tetraoxide in the presence of NaOCl.[12,14-17] This highest oxidation state, however, can not be attained in either H_2O_2 or molecular oxygen based co-oxidant systems. Meanwhile, the strong oxidation power of the Ru(VIII) always guides to unselective oxidation products. This is not the case in the |NPyc| system. Presumably lower oxidation states, Ru(VII) as RuO_4^- might only be produced within the membrane and in turn for the mild and selective oxidation process.

In order to understand the role of co-oxidants (i.e., 30% H_2O_2 or 12% NaOCl) and catalyst, selective oxidation of benzyl alcohol to benzaldehyde is studied as a model system uniformly. As can be seen in Table 2, |NPyc| + H_2O_2 system can yield 84% of conversion and > 99% of selectivity to benzaldehyde in 24 h at T = 80 °C with turnover number (TON) and TOF of 381 and 16 h^{-1}, respectively (Table 2, entry 2) . On the other hand, the benzyl alcohol oxidation reaction completed within 3 h with > 99% conversion and selectivity when the co-oxidant is changed from H_2O_2 into NaOCl at a lower reaction temperature of T = 40 °C (entry 4). The calculated TON and TOF are 465 and 155 h^{-1}, respectively, where the TOF improved by ~10 times over the H_2O_2-assisted reaction. Based on the above results, the role of ClO^- is believed to be bi-functional, i.e., to act as an efficient co-oxidant to push up the Ru species in the Pyc network to perruthenate ion (RuO_4^-) and as a phase transfer catalyst (PTC)[15] to enhance the TOF. Control experiments under similar experimental conditions using both RuO_2/NaOCl (pH 11) and $RuCl_3$/NaOCl (pH 11) systems fail to give such catalytic responses (entries 6 and 7). Presumably ruthenium tetraoxide might be formed as an intermediate with the RuO_2 and $RuCl_3$ systems, which further leading to over oxidation and/or decomposition of benzyl alcohol. The exact detail for the absence of catalysis is unknown for us now. Previously, using the $RuCl_3$ catalyst in the presence of H_2O_2 and a PTC system, > 99% of benzoic acid within 1.5 h at 80 °C was reported.[13] The $RuCl_3$-based catalytic system, however, can not be recycled due to irreversible formation of RuO_2 and/or ruthenium oxy-hydroxyl compounds. Meanwhile, conventional tetra-n-propyl ammonium perruthenate (TPAP) catalyst was also troublesome for recycling procedure.[19-26] The |NPyc|, on the other hand, has > 99% selective and > 99% recoverable with perfectly separated products. A simple use of a laboratory tong can fully recover the |NPyc|. This easy recovery process is an obvious choice for near perfect recyclable catalyst.

Varieties of primary and secondary alcohols are selectively oxidized to aldehyde or carbonyl compounds in moderate to excellent yields as summarized in Table 3. As can be seen, *p*-substituted benzyl alcohols (e.g., –Cl, $–CH_3$, $–OCH_3$, and $–NO_2$) yielded > 90% of product conversion in ~3–4 h of reaction time with TOF in the range of 84–155 h^{-1} (entries 2–5, Table 3), Heterocyclic alcohols with sulfur- and nitrogen-containing compounds are found to show the best catalytic yield with TOF of 1517 and 902 h^{-1} for (pyrindin-2-yl)methanol and (thiophene-2-yl) methanol, respectively (entries 9 and 10, Table 3). Some of aliphatic primary alcohols (long chain alcohols) and secondary alcohols (cyclohexanol, its methyl substituted derivatives and norbornan-2-ol) are also selectively oxidized by the membrane catalyst (entries 11–14 and 15–17, Table 3) with TOF values in the window of 8–

41 h^{-1}. Even though the TOF is relatively lower than those of the aromatic and heterocyclic compounds; there is still no over-oxidation to carboxylic acid in any of the aliphatic systems. Note that, for the case of octan-1-ol and its derivatives, it is highly difficult to control the reaction to octanal. The |NPyc|/NaOCl (at 40 °C) catalyzed secondary alcohols (cyclohexanol) performance are closer to the TPAP/NaOCl (0.7 mol) based homogenous system at room temperature (selectivity 89%).[16,17] Note that, for most of the alcohol oxidation systems, the co-oxidants are slowly fed into the reaction mixture in order to control the reaction selectivity; whereas the NaOCl in this work is added only once in the reaction mixture at the beginning in the present case.

Table 2. Benzyl alcohol oxidation reaction with different catalytic systems [a]

Entry	Catalyst (g)	Co-oxidant	Time (h)	Conversion, (%)	TOF [c] (h^{-1})	S_{CHO} [d] (%)	Recycle[e]
1[b]		H_2O_2 (80 °C)	24	6.3	--	70	--
2[b]	|NPyc| (1g)	H_2O_2 (80 °C)	24	84	16	>99	1^a(fresh)
3		NaOCl	24	20	--	80	--
4	|NPyc| (1g)	NaOCl/pH 11	3	>99	155	>99	1^b (fresh)
5		NaOCl/pH 11	3	---	---	---	
6	RuO_2 (0.06g)	NaOCl/PH 11	3	>99	--	<1	
7	$RuCl_3$ (0.06g)	NaOCl/pH 11	3	>99	--	<1	

[a] Benzyl alcohol (32 mM), CH_2Cl_2 (15 ml), NaOCl = 6 ml of 12% solution (4.1 mol), T= 40 °C.
[b] Entries 1and2 by Ref. 6
[c] Turn over frequency (TOF) = $[Product]_{Mol}/([catalyst]_{Mol}\times T_{hour})$; [Pyc] = 70.56 μMol./gram of NPyc
[d] S_{CHO} = percentage of aldehyde selectivity by GLC.
[e] The values 1^a and 1^g correspond to the |NPyc| catalyst "a" and "b", respectively, used for 1st time.

Table 3. Oxidation of various alcohols via |NPyc| / NaOCl [a]

Entry	Substrate	Product	Time (h)	Yield[b] (%)	TOF (h^{-1})	S_{CHO} [c] (%)	Recycle[d] (#)
1	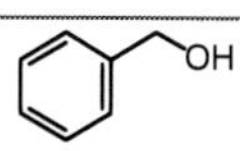	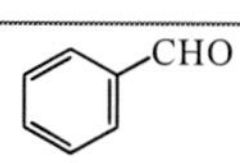	3	100	155	<99	6^e
2	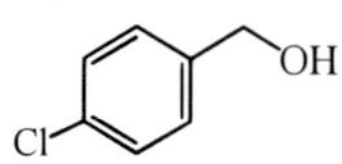	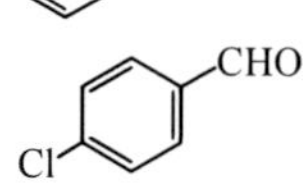	4	92	107	<99	8^e
3	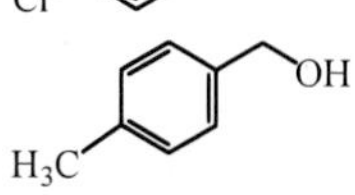	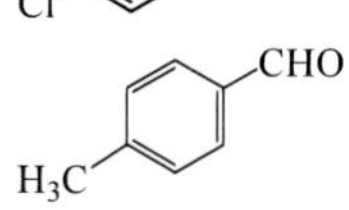	5	91	84	<99	9^e
4	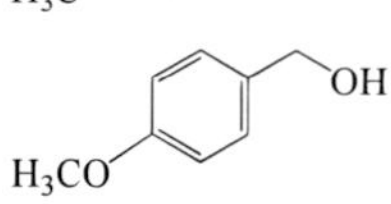	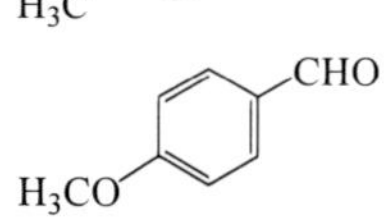	3	93	131	<99	12^e

Table 3. (Continued)

Entry	Substrate	Product	Time (h)	Yield[b] (%)	TOF (h^{-1})	S_{CHO}[c] (%)	Recycle[d] (#)
5	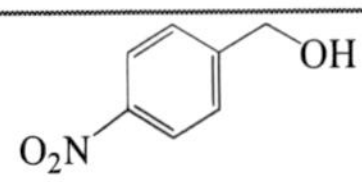	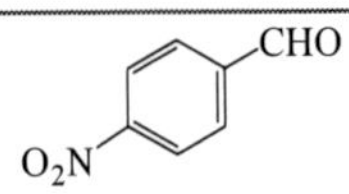	3.5	99	144	<99	13[e]
6	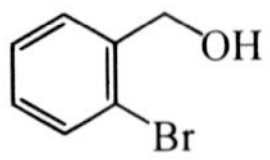	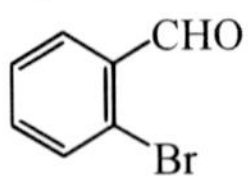	10	83	38	<99	15[e]
7	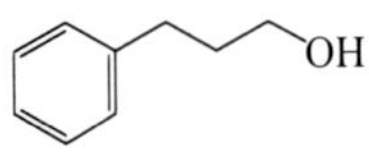	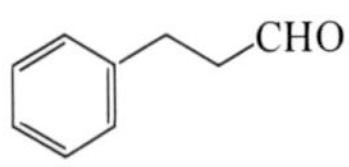	10	18	8	<99	17[e]
8	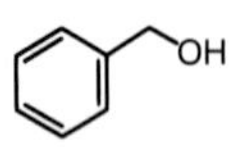	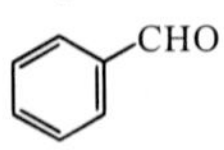	3	<99	155	<99	20[e]
9	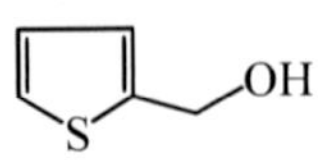	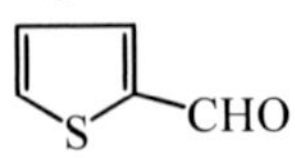	0.5	97	902	<99	1[f] (fresh)
10	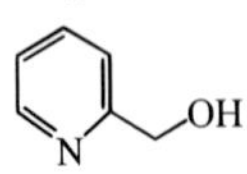	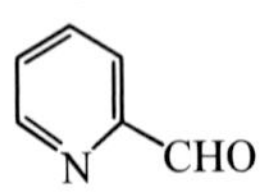	0.2	98	1517	<99	3[f]
11	CH_3-$(CH_2)_5$-CH_2OH	CH_3-$(CH_2)_5$-CHO	13	<99	36	<99	1[g] (fresh)
12	CH_3-$(CH_2)_6$-CH_2OH	CH_3-$(CH_2)_6$-CHO	11	97	41	<99	1[h] (fresh)
13	CH_3-$(CH_2)_8$-CH_2OH	CH_3-$(CH_2)_8$-CHO	15	50	15	<99	1[i] (fresh)
14	CH_3-$(CH_2)_9$-CH_2OH	CH_3-$(CH_2)_9$-CHO	15	59	18	<99	1[j] (fresh)
15	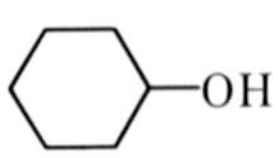	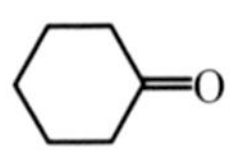	7.5	43	27	<99	11[e] (fresh)
16	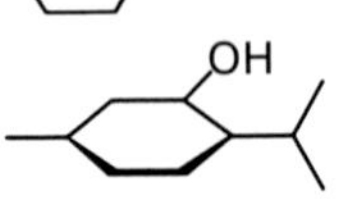	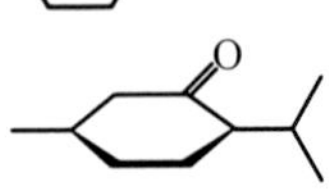	15	31	10	<99	4[f]
17	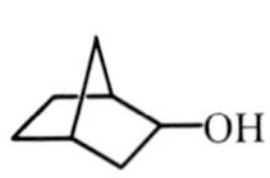	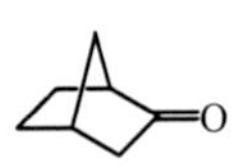	10	98	45	<99	18[e]

[a] Benzyl alcohol (32 mM), |NPyc| catalyst = 1 g, CH_2Cl_2 (15 ml), NaOCl = 6 ml of 12% solution (4.1 mol), T= 40 °C.
[b] Yield by GLC or NMR (Entry 5).
[c] S_{CHO} = percentage of aldehyde selectivity by GLC and NMR.
[d] The "e-g" were different |NPyc| catalysts used under various recyclic conditions.

3.3. Catalytic mechanism

In order to understand the reaction mechanism, the |NPyc| catalytic system is studied by electrochemical cyclic voltammetry using the Nafion®-Pyc chemically modified electrode (designated as NPycCME) under the working condition similar to that of the synthetic alcohol oxidation. The NPycCME is prepared by first spin-coating 5 μl of Nafion® (5 wt%) solution on a clean glassy carbon electrode (GCE) at 3600 rpm. The procedure can allow the coating of a thin film of Nafion® on the GCE surface and it is further subjected to in situ precipitation of Pyc units as described by our previous reports.[32,33] The NPycCME is stored in 1.1 M KOH and the film can be stable for more than 3 months. Prior to the electrochemical measurements, the NPycCME is pretreated for five continuous cycles in base electrolyte solution at a scan rate of 50 mV/s. Formation of the fine Pyc crystalline units is confirmed by X-ray diffraction technique for both |NPyc| and NPycCME systems.[6,32,33] Our previous comparative electrochemical and electrocatalytic studies of NPycCME with various redox probe molecules (glucose, ethanol, and aldehyde) reveal that, the ruthenate ion (i.e., $Ru^{VII}O_4^{2-}$) in the octahedral sites of the Pyc unit has been electro-generated as perruthenate ion, RuO_4^{2-} , at ~0.6 V vs. Ag/AgCl (i.e., 1.5 V versus RHE) selectively in alkaline solution for efficient oxidation of organic molecules. Prior to the electrochemical experiments, biphasic reaction mixture is prepared as 15 ml CH_2Cl_2 + 6 ml 12% NaOCl + 10 ml pH 11 PBS after reacted at 40 ◦C for 1 h and it is separated out. Figure 10 shows typical cyclic voltammetric responses of the NPycCME both in organic and aqueous phases discreetly without and with addition of benzyl alcohol. In aqueous solution, two cathodic peaks at 0.15 and −0.5 V versus Ag/AgCl corresponding to the OCl^- reductions and an anodic shoulder due to the electrogenerated perruthenate species are observed (Figure 10A). Upon the addition of benzyl alcohol, a profound increase in the anodic peak with a decrease in the cathodic response at a potential where the perruthenate species exist are occurred (inset of Figure 10A). This is the typical example for a redox-mediated oxidation mechanism. Parallel electrochemical experiment with organic layer (non-aqueous) is quite different, where a sharp increase in the current response starting from 0.7 V versus Ag/AgCl in the anodic side irrespective of benzyl alcohol are noticed. Net catalytic current in the anodic side, however, is not noticeable indicating relatively less catalytic behavior in the organic phase. Although the precise nature of the catalytic reaction is not exactly defined, from the available electrochemical behavior it is proposed that NaOCl acted as electron acceptor to regenerate the Pyc-RuO_4^- from Pyc-RuO_4^{2-} to mediate the alcohol oxidation reaction effectively (Scheme 5).

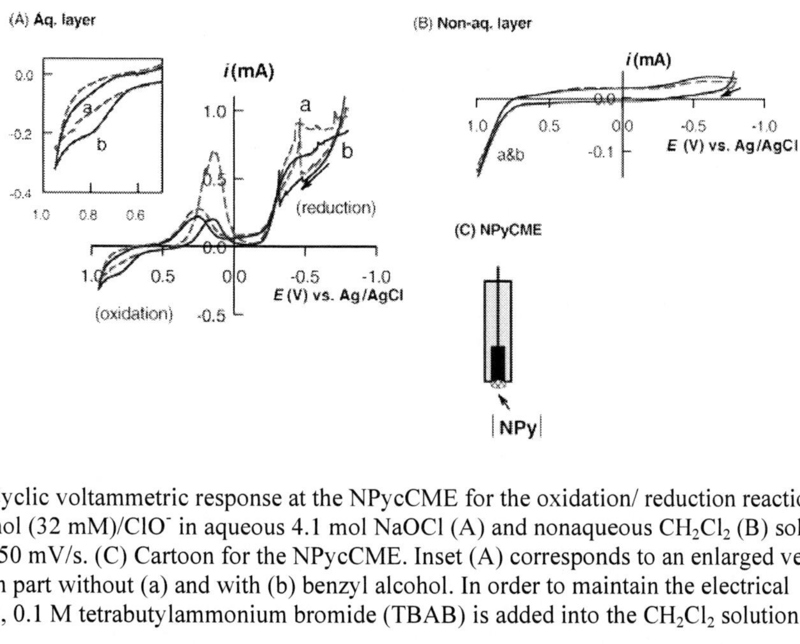

Figure 10. Cyclic voltammetric response at the NPycCME for the oxidation/ reduction reaction of benzyl alcohol (32 mM)/ClO^- in aqueous 4.1 mol NaOCl (A) and nonaqueous CH_2Cl_2 (B) solutions at a scan rate of 50 mV/s. (C) Cartoon for the NPycCME. Inset (A) corresponds to an enlarged version of the oxidation part without (a) and with (b) benzyl alcohol. In order to maintain the electrical conductivity, 0.1 M tetrabutylammonium bromide (TBAB) is added into the CH_2Cl_2 solution.

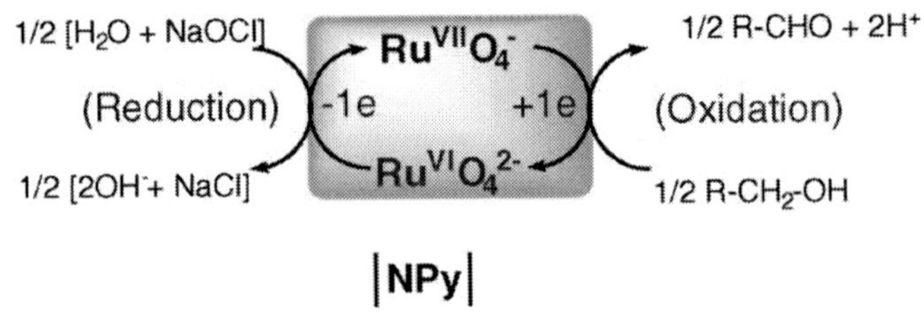

Scheme 5. Mechanism for the |NPyc| membrane catalyzed selective alcohol oxidation reaction in the presence of NaOCl co-oxidant (primary alcohol is given here as an example).

4. |NPYC-RU(BPY)$_3^{2+}$| PHOTO-CATALYTIC SULFOXIDATION REACTION

In this study, photo sensitizer, $[Ru(bpy)_3]^{2+}$ coupled |NPyc| system is used for selective oxidation of sulfides to sulfoxides under molecular oxygen at room temperature. Concern about the catalyst preparation, $[Ru(bpy)_3]^{2+}$ is doped into the |NPyc^{x-}| membrane simply by an ion-exchange process from a solution containing 1 mm $[Ru(bpy)_3]^{2+}$. This membrane (designated as |NPyc^{x-}-Ru(bpy)| is then used for the photo catalytic oxidation. Photochemical experiments are all carried out at pH 1 (adjusted with HCl) in a mixture of CH_3CN and H_2O (3:4, ca. 70 mL) in a closed round-bottomed flask sealed with a gasket-septum under constant purging of O_2 gas. Controlled experiments are carried out under various experimental conditions (Table 4) to rationalize the reaction mechanism. It is clear that the success of the system lies in the proper combination of catalyst, photosensitizer, solvent composition, pH value, O_2, and light illumination (Table 4). The products are analyzed simply by evaporation of the solution of the separated reaction product in $CHCl_3$ with a rotary vacuum system. All reactions give a single product of sulfoxide (that is, no sulfone was observed on the TLC plate and is further confirmed by NMR and mass spectroscopic studies) in > 90% yield.

In order to electrochemically characterize the system, multicomposite |NPyc^{x-}CME - Ru(bpy)| is prepared by first spin-coating 5 μl of Nafion® (5 wt%) solution on a clean glassy carbon electrode (GCE) at 3600 rpm and it is further subjected to in situ precipitation of Pyc units followed by ion exchanging with $[Ru(bpy)_3]^{2+}$. For comparison, Nafion® chemically modified electrode (NCME) without Pyc active site (NGCE) and its $Ru(bpy)_3^{2+}$ doped system (NGCE-Ru(bpy) are also studied. As shown in (Figure 12b), a well-defined redox peak at approximately 1.1 V versus Ag/AgCl corresponding to the $[Ru(bpy)_3]^{3+}/[Ru(bpy)_3]^{2+}$ couple is observed at the NPyc^{x-}CME-Ru(bpy). The anodic peak current (i_{pa}) observed is much higher than that of the NGCE-Ru(bpy). The increase in the current response clearly has something to do with the existence of active Pyc catalyst inside Nafion®. The catalyst can somehow assist the reaction of $[Ru(bpy)_3]^{3+}/[Ru(bpy)_3]^{2+}$ via $[Ru(bpy)_3]^{2+}$. The importance of the [NPyc^{x-}CME-Ru(bpy)] towards to the sulfide oxidation is further demonstrated by the following two experiments. First, an irreversible oxidation peak at the redox potential of the $[Ru(bpy)_3]^{3+}/[Ru(bpy)_3]^{2+}$couple with a much higher i_{pa} value with NPyc^{x-}CME than with NGCE is observed on addition of 5 mm $PhSCH_3$ (Figure 12b).

Table 4. The selective photochemical oxidation reaction of R-Ph-S-CH$_3$ to R-Ph-SO-CH$_3$ on the influence of reaction conditions.[a]

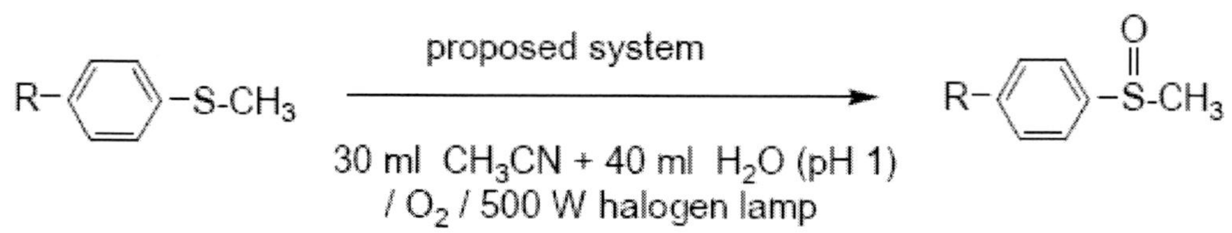

Entry	Membrane catalyst		Solution				Time /hrs	Yield /%
	Pyc	$[Ru(bpy)_3]^{2+}$	CH_3CN: H_2O	pH[b]	O_2	$h\nu$[c]		
1	√	×	3:4	1	√	√	8	×
2	×	√	3:4	1	√	√	8	×
3	√	√	3:4	1	√	×	8	×
4	√	√	1:4	×	√	√	8	×
5	√	√	3:4	1	×	√	8	23
6	√	√	14:1	1	√	√	8	37
7	×	×	3:4	1	1 ml of 30% H_2O_2 added		3	47
8	√	√	3:4	×	√	√	8	49
9	√	√	3:4	1	√	√	3	> 97

[a] √: presence, X: absence, [R-Ph-S-CH$_3$: R=H, COCH$_3$, OCH$_3$]= 17 mM, CH_3CN = 30 ml, 40 ml H_2O.

[b] adjusted/non-adjusted by dilute HCl.

[c] 500 W halogen lamp as light source. As per the case 9, the isolated yields for the substituted organic sulfides: R= COCH$_3$: 96% and R= OCH$_3$: 90%.

Secondly, virtually no change in the i_{pa} value is observed on using NGCE and NGCE-Ru(bpy); while a net increase of approximately 60 μA in the i_{pa} value is observed at the NPyc^{x-}CME-Ru(bpy) with respect to the NPyc^{x-}CME (Figure 12b) In other words, there is no intrinsic effect on the sulfoxidation reaction (SOR) by using $[Ru(bpy)_3]^{2+}$ without the assistance of the active Pyc site in Nafion®. These preliminary electrochemical results can indeed help to explain the reaction mechanism of the photochemical oxidation of RSCH$_3$ (R=Ph, PhCOCH$_3$, and PhOCH$_3$) at the |NPyc^{x-}-Ru(bpy)| membrane (Scheme 6).

A possible reaction mechanism based on these results is shown in Scheme 6, where Pyc plays a dual catalytic role both in the ORR (dark reaction) and the SOR (light reaction). It is noteworthy that, since the Pyc is opaque, only $[Ru(bpy)_3]^{3+}$ can be used to absorb the light in the membrane. The active Pyc site is reported to be an efficient catalyst for the ORR and hence, the purging O_2 is essential for the formation of H_2O_2 during the reaction. The control experiment in pure H_2O_2 gave only ~47% conversion with poor selectivity (Table 4). However, the assistance of Pyc and $[Ru(bpy)_3]^{3+}$ in the SOR is supported by the indirect electrochemical studies.

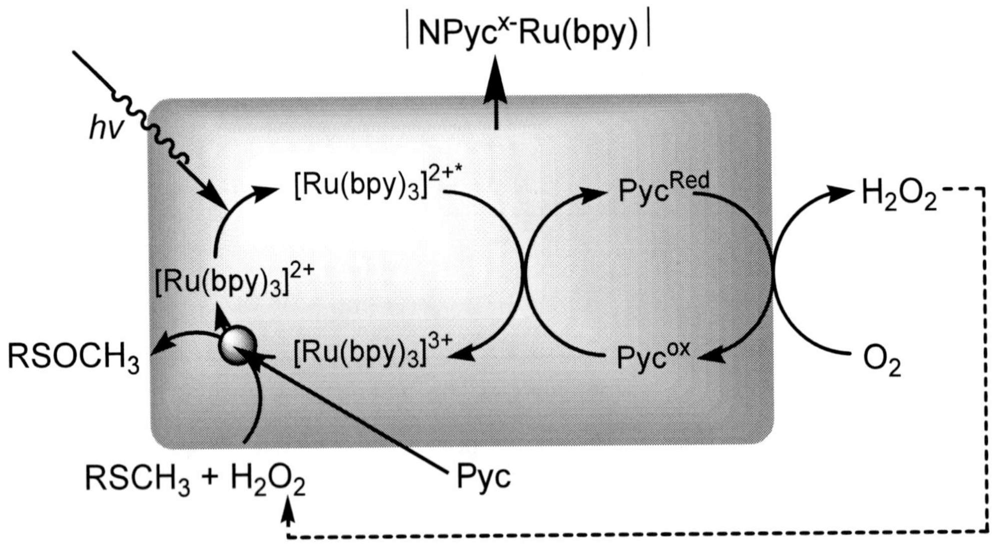

Scheme 6. Reaction mechanism for the photochemical sulfide oxidation reaction.

4.1. |NPyc| stability

For the stability check, catalytic benzyl alcohol oxidation reactions are also carried out at different temperatures (40–80 °C) with excess NaOCl under non-stiochiometric conditions to test the selectivity and stability of the catalyst. Unlike other Ru-based catalyst that aldehyde selectivity depends on mild temperature, all reactions in this study are ended with aldehyde products even under stringent conditions.[44] As confirmed by NMR, GLC, and GLC–MS, none of the primary alcohols are found to over-oxidize to carboxylic acids. Separation of reaction products is another advantage of the |NPyc| system. Benzyl alcohol oxidation reaction under the four systems of RuO_2, $RuCl_3$, Pyc, and |NPyc| is used for comparison. As can been in Figure 11, both the Ru^{3+} and RuO_4^- ions are soluble in alcohols and hence it is difficult to separate the catalyst from the reaction mixture. Only the |NPyc| system yielded clear reaction products after the filtration.

The recoverability and stability of the membrane catalyst is also excellent as the |NPyc| catalyst can be reused without any loss in the catalytic activity after used for more than 20 reactions. Similar reaction kinetics (data not enclosed) and yields in 1st run (entry 2, Table 2) and 20th run (entry 8, Table 3) indicates good recyclability of the |NPyc| system. Note that ~30% loss of activity after three recycles for $RuO4^-$ anchored sol–gel and amberlyst anion exchange resin (IR 27) containing quaternary ammonium systems were reported.[21-24] Other Ru catalyst, such as Ru-HAP, Ru/Al_2O_3, RuMnMn and zeolite confined nano-RuO_2, etc., can only be recycled for 3–9 times.[43-46] Meanwhile elemental analysis also shows no metal leaching behavior for an |NPyc|-assisted benzyl alcohol oxidation reaction as no trace of Ru and Pb ions in solution is detected. Actually the |NPyc| catalyst is still active after stored in 1.1 M KOH for ~6 month.

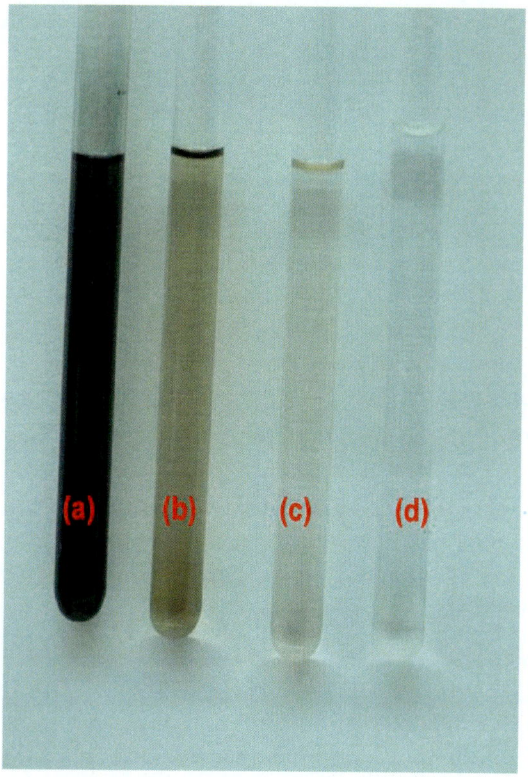

Figure 11. Photographs of the filtered organic phase for the product of benzyl alcohol oxidation reaction catalyzed by (a) RuCl3, (b) RuO2, (c) Pyc powders, and (d) |NPyc| membrane catalyst with 15 ml CH2Cl2 + 6ml NaOCl (4.1 mol)/pH 11 PBS.

5. CONCLUSION

Lead-ruthenate pyrochlore (Pyc) modified Nafion® membrane prepared by in-situ precipitation methodology is found to be a versatile model for the heterogeneous membrane based catalytic system. Nafion® membrane can unusually stabilize the Pyc-RuO_4^-/Pyc-RuO_4^{2-} redox active site within the matrix and for the mild and selective catalytic applications. Catalytic alcohol oxidation with the membrane catalyst along with co-oxidant, H_2O_2 or NaOCl under triphasic condition shows > 99% selective yields towards aldehyde or ketone products. Similarly the photosensitizer, $Ru(bpy)_3^{2+}$ modified membrane catalyst system yields > 99% selectivity towards to the organic sulfide oxidation to sulfoxide product without any over-oxidation to sulfones. The active site stabilized polymer membrane catalyst resembles to natural enzyme system, where the active site is stabilized by various amino acids and peptide units. If the membrane catalyst is properly turned with respect to the redox site and doping, it can be used for innovative applications including fuel cell and bio-mimicking enzymatic systems.

ACKNOWLEDGMENT

Authors gratefully acknowledge National Science Council of Taiwan for financial support.

REFERENCES

[1] T. Mallat, A. Baiker, Oxidation of Alcohols with Molecular Oxygen on Solid Catalysts, *Chem. Rev.* 104 (2004) 3037–3058.

[2] T. Matsumoto, M. Ueno, N. Wang, S. Kobayashi, Recent Advances in Immobilized Metal Catalysts for Environmentally Benign Oxidation of Alcohols, *Chem. Asian J.* 3 (2008) 196–214.

[3] C. A. McNamara, M. J. Dixon, M. Bradley, Recoverable Catalysts and Reagents Using Recyclable Polystyrene-based Supports, *Chem. Rev.* 102 (2002) 3275–3300.

[4] S. S. Ozdemir, M. G. Buonomenna. E. Drioli. Catalytic Polymeric Membranes: Preparation and Application, *Applied Catalysis A: General* 307 (2006) 167–183.

[5] S. Venkatesan, A. S. Kumar, J.-M. Zen, A Rugged Lead-ruthenate pyrochlore Membrane Catalyst for Highly Selective Oxidation of Alcohols, *J. Mol. Catal. A: Chem.* 250 (2006) 87–93.

[6] J.-H. Ke, A. S. Kumar, J.-W. Sue, S. Venkatesan, J.-M. Zen, Catalysis and Characterization of a Rugged Lead ruthenate pyrochlore Membrane Catalyst, *J. Mol. Catal .A: Chem.* 233 (2005) 111.

[7] J.-M. Zen, S.-L. Liou, A. S. Kumar, M.-S. Hsia, An Efficient and Selective Photocatalytic System for Oxidation of Sulfides to Sulfoxides, *Angew. Chem. Int. Ed.* 42 (2003) 577–579.

[8] M. Kenneth, R. B. Moore, State of Understanding of Nafion®, *Chem. Rev.* 104, (2004) 4535.

[9] C. Djerassi, R. Engle, Oxidations with Ruthenium Tetraoxide, *J. Am. Chem. Soc.* 75 (1953) 3838–3840.

[10] P. E. Morris, D. E. Kiely, Ruthenium Tetraoxide Phase-Transfer-Promoted Oxidation of Secondary alcohols to Ketones, *J. Org. Chem.* 52 (1987) 1149–1152.

[11] L. A. Paquette, Encyclopedia of Reagents for Organic Synthesis, *John Wiley and Sons,* New York, 1995, Volume 6. 4415–4422.

[12] H. B. Friedrich, The oxidation of Alcohols to Aldehyde or Ketones, *Platinum Metal Rev.* 43 (1999) 94–102.

[13] G. Barak, J. Dakka, Y. Sasson, Selective Oxidation Of Alcohols by a H_2O_2-$RuCl_3$ System under Phase-Transfer Conditions, *J. Org. Chem.* 53 (1988) 3553–3554.

[14] S. Wolf, S. K. Hasan, J. R. Campbel, Ruthenium Trichloride-Catalysed Hypochlorite Oxidation of Organic Compounds, *Chem. Commun.* 21 (1970) 1420–1421.

[15] I. Tabushi, N. Koga, Synergetic Combination of Catalysis of the Phase Transfer-Electron Transfer Type for the Oxidation of Alcohols or Hydrocarbons, *Tetrahedron Letters*, 2038 (1979) 3681–3684.

[15] L. Gosalvi, I. W. C. E. Arends, R. A. Sheldon, Selective Ruthenium-Catalyzed Oxidation Of 1,2:4,5–Di-*o*-isopropylidene β-D-fructopyranose and other Alcohols with NaOCl, *Org. Lett.* 4, (2002) 1659–1661.

[16] L. Gosalvi, I. W. C. E. Arends, P. Moilanen, R. A. Sheldon, The Effect of pH Control on the Selective Ruthenium-Catalyzed Oxidation of Ethers and Alcohols with Sodium Hypochlorite, *Adv. Synth. Catal.* 345 (2003) 1321–1328.

[17] L. Gosalvi, I. W. C. E. Arends, P. Moilanen, R. A. Sheldon, The Effect of pH Control on the Selective Ruthenium-Catalyzed Oxidation of Ethers and Alcohols with Sodium Hypochlorite, *Adv. Synth. Catal.* 345 (2003) 1321–1328.

[18] D. G. Lee, L. N. Congson, Kinetics and Mechanism of the Oxidation of Alcohols by Ruthenate and Perruthenate Ions, *Can. J. Chem.* 68 (1990) 1774–1779.

[19] S. V. Ley, J. Norman, W. P. Griffith, S. P. Marsden, Teterapropylammonium Perruthenate, $Pr_4N^+RuO_4^-$, TPAP: A Catalytic Oxidation for Organic Synthesis, *Synthesis* 7 (1994) 639–666.

[20] R. Lenz, S. V. Ley, Tetra-n-propyl ammonium perruthenate (TPAP)-Catalyzed Oxidations of Alcohols using Molecular Oxygen as a Co-oxidant, *J. Chem. Soc., Perkin Trans. 1*, (1997) 3291-3292.

[21] B. Hinzen, S. V. Ley, Polymer Supported Perruthenate (PSP): A New Oxidant for Clean Organic Synthesis, *J. Chem. Soc., Perkin Trans. I* (1997) 1907–1908.

[22] A. Bleloch, B. F. G. Johnsons, S. V. Ley, A. J. Price, D. S. Shepard and A. W. Thomas, Modified Mesoporous Silicate MCM-41: Immobilized Perruthenate – A New Highly Active Heterogeneos Oxidation Catalyst for Clean Organic Synthesis using Molecular Oxygen, *Chem. Commun.* 18 (1999) 1907–1908.

[23] D. L. Wu, A. P. Wight, M. E. Davis, Shape Selective Oxidation of Primary Alcohols using Perruthenate-Containing Zeolites, *Chem. Commun.* 6 (2003) 758–759.

[24] S. Campestrini, M. Carraro, R. Ciriminna, M. Pagliaro, U. Tonellato, Alcohos Oxidation with Hydrogen Peroxide Promoted by TPAP-doped ormosils, *Tetrahedron Lett.* 45 (2004) 7283–7286.

[25] M. Hasan, M. Musawir, P. N. Davey, I. V. Kozhevnikov Oxidation Of Primary alcohols to aldehydes with Oxygen Catalyzed by Tetra-n-propyl ammonium perruthenate, , *J. Mol. Catal. A: Chem.* 180 (2002) 77–84.

[26] M. Pagliaro, S. Campestrini, R.Ciriminna, Ru-based oxidation catalysis, *Chem. Soc. Rev.* 34 (2005) 837.

[27] H. B. Friedrich, N. Singh, A Study of Poly (4-vinylpyridine)-Supported Ruthenate in the Oxidation of Alcohols, *Catal. Lett.* 110 (2006) 61–70.

[28] R. Ciriminna, P. Hesemann, J. J. E. Moreau, M. Carraro, S. Campestrini, M. Pagliaro, Aerobic Oxidation of Alcohols in Carbon Dioxide with Silica- Supported Ionic Liquids Doped with Perruthenate, *Chem. Eur. J.* 12 (2006) 5220–5224.

[29] H. Hamamoto, M. Kudoh, H. Takahashi, H. Natsugari, S. Ikegami, Polyacrylamide-based Functional Polymer-Immobalized Perruthenate for Aerobic Alcohol Oxidation, *Chem. Lett.* 36 (2007) 632–633.

[30] T. R. Felthouse. P. B. Fraundorf, R. M. Friedman, C. L. Schosser, Expanded Lattice Ruthenium Pyrochlore Oxide Catalysts. I. Liquid Phase Oxidations of Vicinal Diols, Primary alcohols, and related Substrates with Molecular Oxygen, *J. Catal.* 127 (1991) 393–420.

[31] J.-M. Zen, C.-B. Wang, Oxygen Reduction on Ruthenium Oxide Pyrochlore Produced in a Proton Exchange Membrane, *J. Electrochem. Soc.* 141 (1994) 38–39.

[32] J.-M. Zen, C.-W. Wang, Determination of Dissolved Oxygen by Catalytic Reduction on a Nafion®/Ruthenium Oxide Pyrochlore Chemically Modified Electrode, *J. Electroanal. Chem.* 368 (1994) 251–256.

[33] J.-M. Zen, A. S. Kumar, A Mimicking Enzyme Analogue for Chemical Sensors, *Acc. Chem. Res.* 2001, 34, 772–780.

[34] J. Prakash, D. A. Tryk, E. B. Yeager, *J. Electrochem. Soc.* 146 (1999) 4145–4149.

[35] J.-M. Zen, A. S. Kumar and J.-C. Chen, Electrochemical Behavior of Lead-Ruthenium Oxide Pyrochlore Catalyst: Redox Characteristics in Comparison with that of Ruthenium Dioxide, *J. Mol. Catal. A: Chem.* 165 (2001) 177–188.

[36] S. Kumar, J.-M. Zen, Unusual Redox Catalysis in a Ruthenium Oxide-Prussian blue Combined Material, *Chem. Phys. Chem.* 5 (2004) 1227–1231.

[37] P. Liu, J. Bandara, Y. Lin, D. Elgin, L.F. Allard, Y.-P. Sun, Formation of Nanocrystalline Titanium dioxide in Perfluorinated Ionomer Membrane, *Langmuir* 18 (2002) 10398–10401.

[38] M. Ludvigsson, J. Lindgren, J. Tegenfeld, Incorporation and Characterization of Oxides of Manganese, Cobalt, Lithium into Nafion® 117 membranes, *J. Mater. Chem. 11* (2001) 1269–1276.

[39] Z. Q. Ma, P. Cheng, T. S. Zhao, A Palladium-Alloy Deposited Nafion membrane for Direct Methanol Fuel cells, *J. Membr. Sci.* 215 (2003) 327–336.

[40] A. J. Bard, F. F.-R. Fan, D. Pierce, P. R. Unwin, D. O. Wipf, F. Zhou, Chemical Imaging of Surfaces with the Scanning Electrochemical Microscope, *Science* 254 (1991) 68–74.

[41] J. Bard, M. V. Mirkin (Eds), Scanning Electrochemical Microscopy, Marcel Dekker, New York, 2001.

[42] B. Moyer, M. S. Thompson, T. J. Meyer, Chemically Catalyzed Net Electrochemical oxidation of Alcohols, Aldehydes, and Unsaturated Hydrocarbons using the system $(trpy)(bpy)Ru(OH_2)^{2+}/(trpy)(bpy)RuO^{2+}$, *J. Am. Chem. Soc.* 102 (1980) 2310–2312.

[43] K. Yamaguchi, K. Mori, T. Mizugaki, K. Ebitani, K. Kaneda, Creation of a Monomeric Ru Species on the Surface of Hydroxyapatite as an Efficient Heterogeneous Catalyst for Aerobic Alcohol Oxidation, *J. Am. Chem. Soc.*122 (2000) 7144–7145.

[44] K. Yamaguchi, N. Mizuno, Supported Ruthenium Catalyst for the Heterogeneous Oxidation of Alcohols. *Angew. Chem.* 41 (2002) 4538–4531.

[45] B.-Z. Zhan, M. A. White, T.-K Sham, J. A. Pincock, R. J. Doucet, K. V. R, Rao, K. N. Robertson, T. S. Cameron, Zeolite-Confined Nano-RuO_2: A Green, Selective, and Efficient Catalyst for Aerobic Alcohol Oxidation, *J. Am. Chem. Soc.* 125 (2003) 2195–2199.

[46] K. Ebitani, K. Motokura, T. Mizugaki, K. Kaneda, Heterotrimetallic RuMnMn Species on a Hydrotalcite Surface as Highly Efficient Heterogeneous Catalysts for Liquid-phase Oxidation of Alcohols with Molecular Oxygen, *Angew. Chem.* 44 (2005) 3423–3426.

In: Heterogeneous Catalysis Research Progress
Editor: Mathias B. Gunther, pp. 369-393
ISBN 978-1-60456-979-7

Chapter 10

EXTRACTING INFORMATION ABOUT SURFACE HETEROGENEITY EFFECT ON HETEROGENEOUS REACTIONS USING MULTIFRACTAL SCALING ANALYSIS

Ajay Chaudhari[1a] ***and Shyi-Long Lee***[b2]

[a]School of Physical Sciences, Swami Ramanand Teerth Marathwada University, Nanded-431 606, India

[b]Dept. of Chemistry and Biochemistry, National Chung-Cheng University, Ming-Hsiung, Chia-Yi-621, Taiwan

ABSTRACT

Solids with rough surfaces are extremely common in nature and they appear in many industrial processes. The effect of surface heterogeneity in physical, chemical and biological processes is consequently of great interest for many practical purposes. Due to its importance in both basic research and practical applications, the study of geometric heterogeneity have attracted extensive attention and propelled intensive research activities in this field. In heterogeneous catalysis, reaction performance depends strongly on the structure and geometry of the environment in which the catalytic process takes place. The structure of a real catalyst is never a perfect crystalline one. More information about the complex structure of surfaces and materials could be obtained by adopting some new approaches or theories. Role of surface heterogeneity in heterogeneous catalysis will be reviewed in this article with special attention to the surface heterogeneity effect on Eley-Rideal reaction mechanism, which is an elementary step in many complex interfacial processes. More detailed characterization of heterogeneous structures can be obtained by applying multifractal scaling analysis (MSA). Some discussion about its development and its application to the heterogeneous reaction is also given in this article. The applicability of the multifractal scaling analysis to the characterization of the heterogeneous process will be given with a focus on the heterogeneous catalytic reaction.

---

1 E-mail : AJAYCHAU5@YAHOO.COM.
2 CHESLL@CCU.EDU.TW.

The time dependence of the heterogeneous reaction over heterogeneous surface will be discussed with the help of dynamic scaling theory.

INTRODUCTION

The heterogeneity of surfaces and materials plays a key role in most natural and artificial chemical processes. In practice most surfaces are both geometrically complex and chemically non-uniform. Real solids have complex microstructures and most materials are to some extent rough. [1-16] For the adsorption system, the surface heterogeneity, which is mainly induced by the complexity of the crystallographical and geometrical structure of many solids and their complex chemical composition, plays a key role in the adsorption on crystalline solids and non-crystalline solids, nonporous and wide-porous solids. Due to its importance great efforts have been made to obtain a better understanding of heterogeneity effects on physical and chemical processes; nevertheless, the problem of how to quantify the heterogeneity is still difficult.

In heterogeneous catalysis, any heterogeneous catalytic reaction is governed by the elementary mechanism involved in the chemical kinetic scheme, by the external operating parameters but also by the geometry of the surface on which the reaction occurs. For most catalysts, the catalytic surface is usually irregular, convoluted and fractured. The problem of heterogeneous reactions is of major importance in the chemical industry. In catalytic reactions, the geometry parameter is as important as the other parameters such as nature of the chemical bond formed between the reacting molecules, the stereochemical requirements or restrictions, which govern this molecular association and the energetic profile of the reaction. The geometry parameter alone can dictate whether a reaction will take place at all. The majority of materials and their surfaces are characterized by extremely complex geometry.

One basic mechanism in catalysis is the Eley-Rideal (ER) mechanism which is a reaction of an important class in condensed phases, e.g. in solution and surface catalytic reactions and has been extensively used in analyses of catalytic reactions. [17-19] Therefore it is important to examine the surface morphology effects on this key process. It is an elementary step in many complex interfacial processes and has received much attention in the last two decade. [20-50] Numerous theoretical studies of ER reactions on fractal surfaces have appeared, primarily in the past decade and so. [32,33.37-50] In practice, one does not find real surface that are fractals due to the mechanical weakness even though aggregation processes in surface production do occur. The surfaces actually used in the laboratory and in industry are broken fractals and their fragments. Also, the catalyst surfaces are neither totally irregular (fractal) nor are perfectly regular (geometrically and energetically homogeneous). [51] Real catalyst surfaces represent an intermediate case, which can be viewed as rough surfaces. [34,35,46-51]

Avnir and coworkers [31] have reported the multifractal scaling analysis of diffusion limited reactions over two mathematician-made fractal surface, the Cantor set and Devil's staircase. They pointed out that Devil's staircase surface shows higher position sensitivity than the Cantor set surface does, even if they have the same fractal dimension. Mai et. al. [52] observed the strong influence of the lattice structure on the reaction and in case of fast particle diffusion, the effect of lattice structure became unimportant. Diffusion limited reactions over fractal surface of diffusion limited aggregation and rough surfaces were performed by Lee et. al. [32-40,43,46-50] and found a wider range of α values, the scaling exponents measuring the range of reaction probability, than Cantor set and Devil's staircase. α and its density function, $f(\alpha)$, allows a quantitative evaluation of the degree of reaction probability distribution inhomogeneity.

The aim of this article is to review the role of surface heterogeneity in heterogeneous catalysis with special attention to the surface roughness effect on ER reaction mechanism. We consider ER reaction with reaction probability 1 as well as event dependent reaction probability. The rough surfaces used in this study are generated by Random deposition model and Random Deposition with Surface diffusion model. [53,54] The former is the surface with no correlation between the neighboring columns whereas for the latter, there is correlation between the neighboring columns. Multifractal scaling analysis [55] is performed with which the complex distribution of reaction probabilities is analyzed. Dynamic scaling theory [53,54,56-60] is also applied on ER reaction mechanism over the surface with different surface roughness to study the time dependence of ER reactions over the rough surface to obtain the two scaling parameters α and β.

Rough Surface of RD and RDWD Model

We have used two models for generating rough surface, 1) Random deposition (RD) model and 2) Random deposition model with surface diffusion (RDWD) model. [53,54] In random deposition model, particles simply rain down onto a smooth surface. Particles move along straight-line trajectories until they reach the top of the column in which they are dropped, at which point they stick to the deposit and become part of the aggregate. In random deposition with surface diffusion model the particles are allowed to diffuse on the surface within a finite distance from the column where it was originally dropped, until it finds the position with the minimum height. For the former, there is no correlation between the heights of different neighboring columns whereas for the latter there is correlation between the heights of different neighboring columns. The cross sections of the two surfaces generated by RD and RDWD models are shown in Figure 1.

The surface thickness is obtained as

$$t = \left[\frac{\sum_i (h_i - H)^2}{N_s} \right]^{1/2}$$

where h_i is the height of the ith column, $H = \frac{\sum_i h_i}{N_s}$ is the mean deposition height and N_s is the number of surface sites. Here $N_s = L$. The surface roughness is defined as the difference between the maximum and minimum height among the columns. Once the rough surface is generated, the releasing particle is changed to the reacting species.

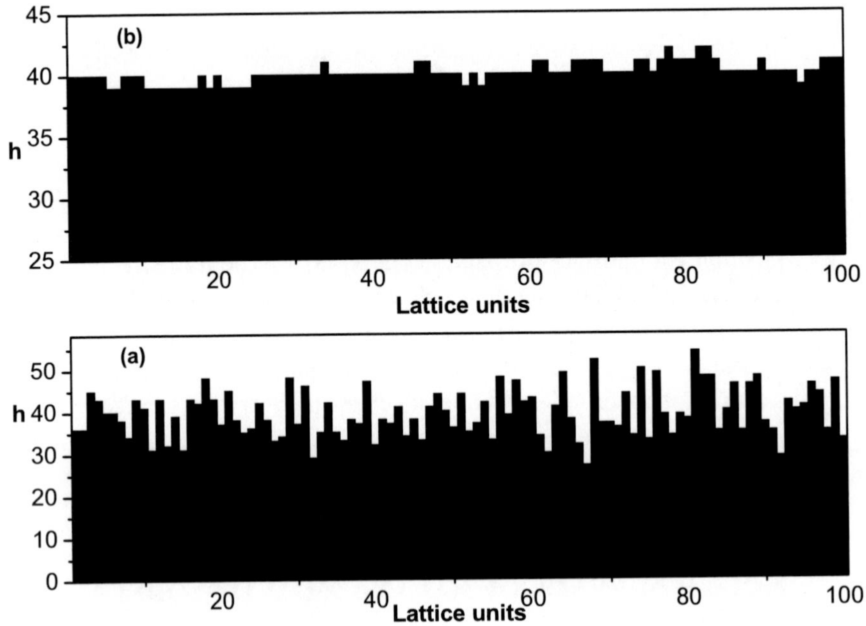

Figure 1. Rough surfaces generated by (a) RD model and (b) RDWD model Number of Particles deposited = 4 x 105 on 100x100 square lattice.

Multifractal Scaling Analysis

We have applied the multifractal scaling [55] which relates the analysis of the distribution of the reaction probabilities over the length of the rough surface. The steps in the multifractal scaling analysis are given below.

There are three basic equations in the multifractal scaling analysis.

$$M_q = \sum_i P_i^q = \sum_p n(P)\, P^q \propto L^{-\tau(q)} \qquad (1)$$

$$\tau(q) = q\,\alpha(q) - f(\alpha) \qquad (2)$$

$$\frac{d}{dq}[\tau(q)] = \alpha(q) \qquad (3)$$

where M_q is the qth order moment of reaction probability distribution, $\tau(q)$ is the scaling exponent, P_i is the reaction probability of site i, n(P) is the number of sites with reaction probability P and L is the length of the surface. The exponents α and $f(\alpha)$ are the characteristics scaling exponents for the fractal measures. Among these three equations, eq. (2) is obtained by the following scaling assumptions used in the limiting case of large L

$$P(q) \propto L^{-\alpha(q)} \qquad (4)$$

$$n[P(q)] \propto L^{f(\alpha)} \tag{5}$$

where P(q) denotes the value of P that dominates the sum in eq.(1) for the qth-order moment. Substitution of eqs. (4) and (5) into eq.(1) and taking the logarithm of the maximum elements as an approximation of the logarithm of the sum yields eq.(2). Finally, eq.(3) represents the value of α that dominates the summation term in eq.(1) and it can be transferred into a more explicit form,

$$\alpha(q) = \frac{d}{dq}[\tau(q)] = \frac{\sum_i P_i^q \ln(P_i)}{\sum_i P_i^q \ln(1/L)} \tag{6}$$

In practice, the quantities $\tau(q)$, $\alpha(q)$ and $f(\alpha)$ can be computed as follows. The probability distribution is first determined from the simulation. Then for each L, $\tau(q)$ is calculated from eq.(1). $\alpha(q)$ is then computed from eq.(6). By knowing τ and α, one can compute $f(\alpha)$ from eq.(2). The calculated values of τ, $\alpha(q)$ and $f(\alpha)$ are presented through two types of plots τ vs q and $f(\alpha)$ vs α.

ER Reactions Over Rough Surface of RD And RDWD: A Comparison

This section compares ER reactions over rough surface generated by random deposition model and random deposition with surface diffusion model. We have generated the rough surfaces by depositing 4 x 10^5 particles on three different square lattices of 50x50, 100x100 and 150x150 lattice units by using RD and RDWD models. The surface roughness is 5, 4, 3 lattice units respectively for the three surfaces generated by RDWD and that for the three surfaces of RD is 84, 46 and 32 lattice units respectively. For both types of surfaces the surface for 50x50 lattice size is the roughest among the three surfaces. After generating the rough surfaces, the releasing particles are changed to the reacting species. The number of reacting species used here is 5 x 10^5. The reaction events occurring at each site on the surface are recorded by counting the number of visits by the reacting particle and the reaction probability is calculated at each site. The plot of reaction probability as a function of active site position gives the position sensitivity of the reaction probability. The active sites are numbered from negative to positive in y direction and finally in upward direction. The reaction probability is plotted as a function of active site position for the three surfaces for RD and RDWD in Figure 2 and 3, respectively.

It can be seen from Figure 2 and 3 that the range of the reaction probability decreases with an increase in surface roughness for the three surfaces of RD as well as RDWD. Wider distribution of the reaction probability is observed for the roughest surface, indicating higher position sensitivity. The heterogeneity in the reaction probability distribution, for RD and RDWD, is due to the reacting particle reaches at the top of the column having maximum height among the neighboring columns.

The number of active sites for the three surfaces of RD are 999, 3943 and 8887 whereas that for the three surfaces of RDWD are 1453, 5790 and 12937.

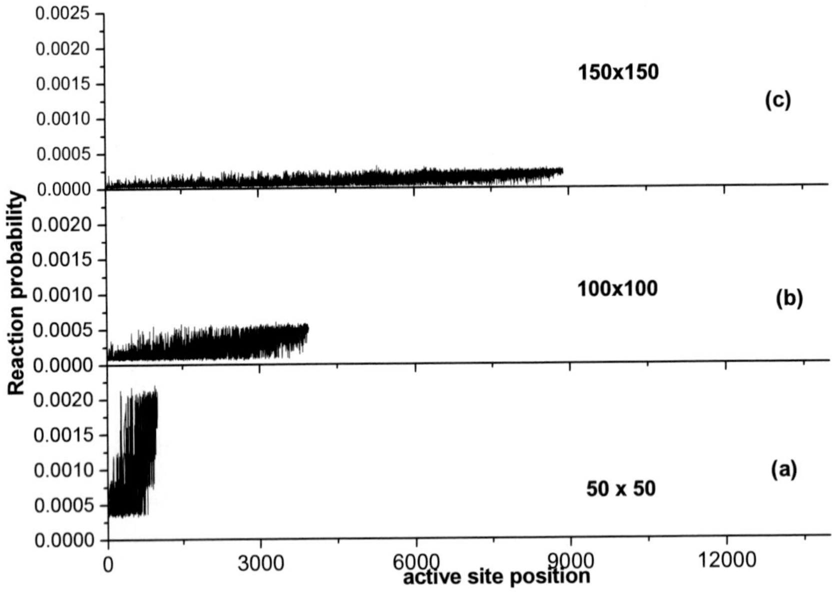

Figure 2. Reaction probability for different active sites for the three surfaces of RD model.

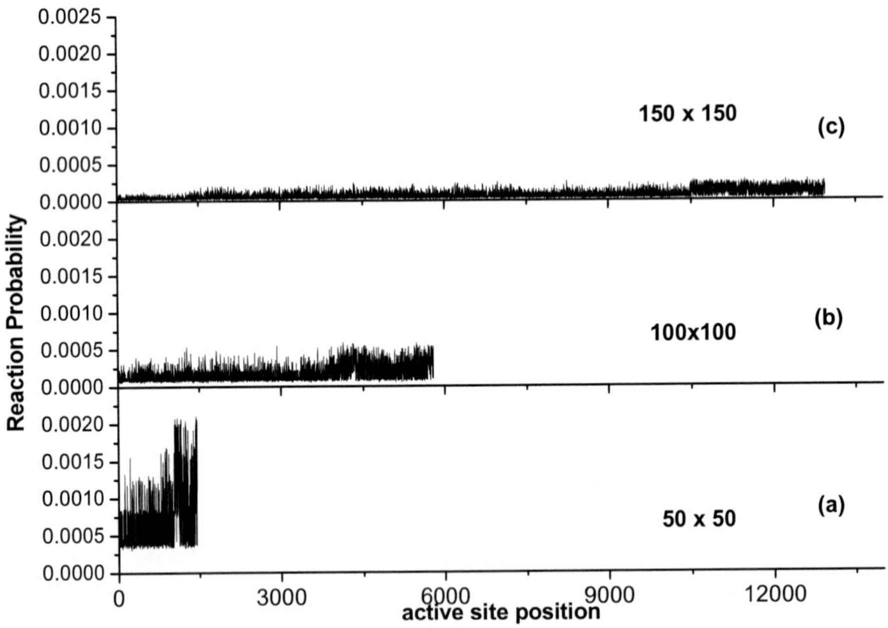

Figure 3. Reaction probability for different active sites for the three surfaces of RDWD model

It indicates that for RD as well as RDWD, the number of active sites increases with decrease in surface roughness. As the surface roughness decreases, the surface becomes smooth and the difference between the heights of the columns decreases or becomes zero. Therefore there is nearly equal probability of reaction at the top of all the columns. Hence, the number of active sites increases with decrease in surface roughness for the surfaces of RD as well as RDWD. As the surface roughness decreases, the number of active sites increases and the reaction probability distribution becomes homogeneous.

On comparing the reaction probability distribution for the surface of RD and RDWD for 50 x 50 lattice, it can be seen that the range of reaction probability for the former is wider than that for the latter. The number of active sites is higher for RDWD than RD. Since surface of RD has less number of active sites, the reaction probability at each of these active sites is greater than that for the surface of RDWD. Most of the active sites have same reaction probabilities for the surface of RDWD, and few have higher reaction probability. It is due to the correlation between the neighboring columns. For the surface with correlation between the neighboring columns, most of the columns of the surface have the similar height. Few columns, which have higher reaction probability, have greater height. Since for the surface of RD, there is no correlation between the columns, due to the screening effect the reacting particle could not reach and react at the top of the columns with smaller height than the neighboring columns. This gives wider distribution of reaction probability for RD than RDWD.

The multifractal formalism has been used to transfer the reaction probability distribution picture into a useful compact form through the $\tau(q)$ vs q and $f(\alpha)$ vs α plots. The multifractal plots for three surfaces of RD and RDWD are shown in Figure 4 and 5 respectively. For both type of surfaces the deviation from linearity for $\tau(q)$ vs q plots indicates multifractality, i.e., simple single-valued fractal scaling does not apply in this condition. From Figs. 4(a) and 5(a), rougher the surface, bigger is the deviation of the $\tau(q)$ vs q curve from linearity. The curvature also implies the degree of heterogeneity in reaction probability distribution. These results agree with the reaction probability distribution profiles for the surface with different roughness. As the surface becomes smoother, the non-linearity in the $\tau(q)$ vs q curve decreases representing a homogeneous reaction probability distribution. The effect caused by surface roughness on reaction probability distribution can also be analyzed clearly through the $f(\alpha)$ dispersion of the reaction probability distribution profiles which allows a quantitative evaluation of the degree of reaction probability distribution inhomogeneity : high non-linearity of the $\tau(q)$ vs q curves is translated into wider distribution in the $f(\alpha)$ plots.

In Figure 4(b) and 5(b), $f(\alpha)$ is plotted as a function of α to further explore the properties of reaction probability distribution for the three surface with different roughness, for RD and RDWD respectively. As can be seen from these figures, for RD as well as RDWD, $f(\alpha)$ profiles show that the reaction probability distribution is characterized by a wide range of α values, indicating the existence of multifractality. Also shown is that the range of α values decreases with decrease in surface roughness, for RD as well as RDWD. This can be connected to the reaction probability distribution shown in Figure 2 and 3, for RD and RDWD respectively. As demonstrated in Figure 2 and 3, the surface with higher roughness is found to have much higher position sensitivity than that with less roughness. For the surface with less roughness, the great number of low reaction probability sites would level out the distribution and lower the position distinction. For both type of rough surfaces RD as well as RDWD, the $f(\alpha)$ figures are asymmetric with respect to α. This is in contrast to the symmetric $f(\alpha)$ figures observed for ER reactions over the fractal surface of Cantor set and Devil's staircase. [31] On the other hand similar asymmetric $f(\alpha)$ profiles are obtained for ER reactions over the fractal surface of diffusion limited aggregation by Lee et al. [32,33]

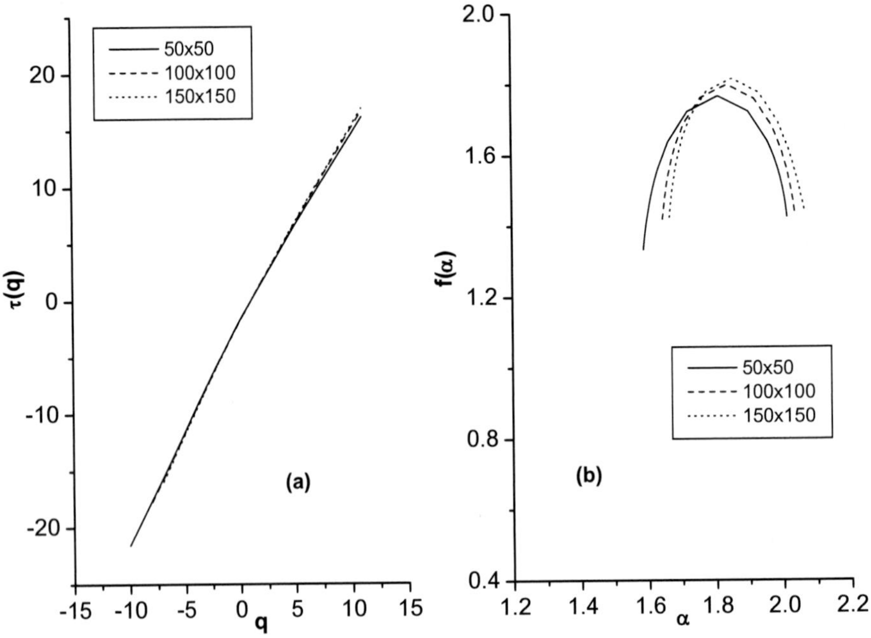

Figure 4. q-τ(q) and α-F(α) multifractal plots for the three surfaces of RD.

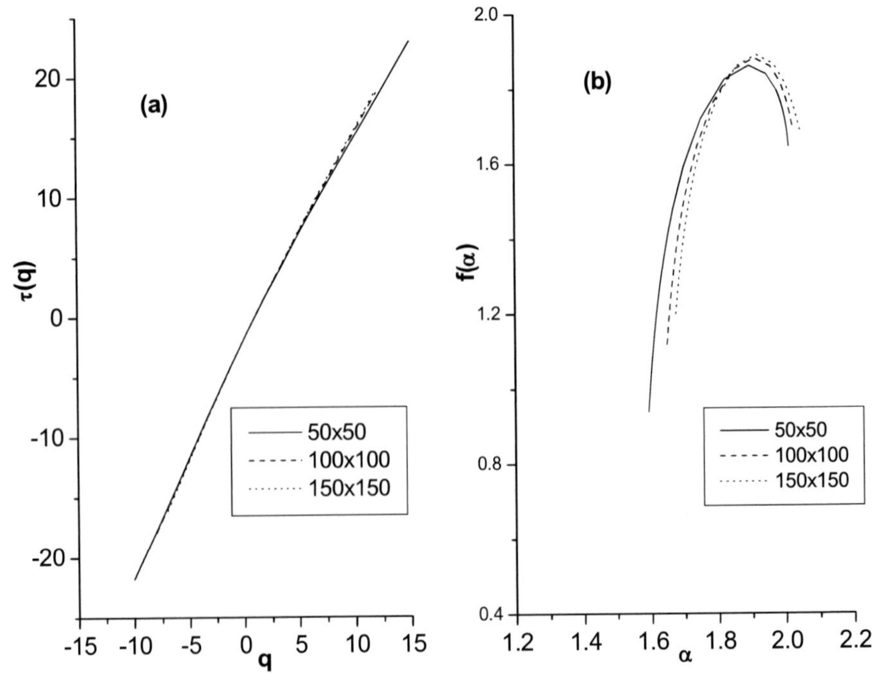

Figure 5. q-τ(q) and α-F(α) multifractal plots for the three surfaces of RDWD.

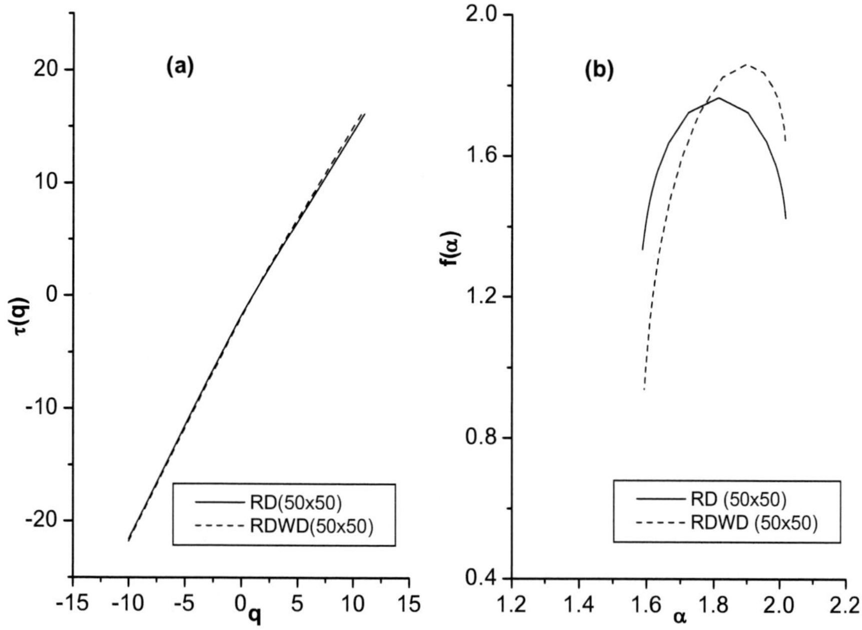

Figure 6. Comparision between the multifractal plots for the rough surface of 50x50 for RD and RDWD.

The f(α) curves here contract upwardly at high α values indicating that the number of lowest reaction probability sites and the number of large reaction probability sites are not even.

The multifractal plots for the surface of RD and RDWD are compared for 50x50 lattice in Figure 6. From Fig, 6 it is observed that the deviation of τ(q) vs q curve from linearity for RD is greater than that for RDWD in high q value region. This result can be interpreted by the fact that the number of active sites with highest reacting probability, which dominates the sum in eq. (1) in the large positive q region, is quite different for RD and RDWD. This high nonlinearity of the τ(q) vs q curve for RD is translated into wider distribution in the f(α) plots than RDWD, as shown in Figure 6(b). From Figure 6(b), the number of active sites with low reaction probability is more for the surface of RDWD than RD.

Effect of Stickling Probability on ER Reactions

The ER reactions studied in previous section are with perfect sticking case i.e. the sticking probability 1. Sticking probability =1 means zero activation energy. For the ER mechanism with sticking probability 1 each active site is assumed to have the same fixed sticking probability for each reaction event. In real catalytic systems, the reactions occurring at active sites of surface having non-zero activation energy seems to be more reasonable. Sticking probability of the reacting particles is one of the important parameters in catalytic reaction. In ER reactions mechanism, very first step is the adsorption of the reacting species A on the surface. The probability that the species A will be adsorbed generally depends on the temperature of the catalyst surface.

Adsoption (sticking) probability= f exp(-ΔE/kT)

where f is pre-exponential factor, ΔE is binding energy, k is Boltzmann's constant and T is temperature of the surface. In Monte Carlo simulation, the term f exp(-ΔE/kT) is best represented by some fixed value called sticking probability. In this section we have studied ER reaction mechanism over rough surface with different fixed sticking probability (P_{ini}) of the reacting species viz. 1, 0.7, 0.4 and 0.1. We used the rough surface generated by depositing 5 x 10^5 particles over a square lattice of size L=300 lattice units using RD model. The surface roughness of this surface is 18 lattice units.

After the rough surface is generated by depositing $5x10^5$ particles, the releasing particles are considered as reacting species. The columns are selected randomly. The reacting species are allowed to react at the top of the column having maximal height among the four nearest neighboring columns and with sticking probability P_{ini}. The probability of reaction can be achieved by generating a random number when the particle visits the surface site and checking whether it is less than or equal to the given fixed sticking probability P_{ini}. If it is less than P_{ini}, the reaction count on that surface site are added by one and if it is greater than P_{ini} then a new reacting particle is considered. After 10^6 reacting particles are launched, the reaction probabilities of different surface sites are recorded and analyzed.

The reaction events occurring at each site on the surface are recorded by counting the number of visits by the reacting particle and the reaction probability is calculated at each site. From the reaction distribution plot it was observed that the range of reaction probability decreases with decreasing sticking probability. Higher the sticking probability, wider is the distribution of the reaction probability, indicating higher position sensitivity. We also observed valleys among the spikes in reaction probability distribution profile, which arise mainly from a screening effect i.e. the inner surface sites are screened by the outer ones indicating that the inner sites are rarely visited. The inner and outer surface sites are the active sites, which are at lower height and greater height than the neighboring sites, respectively.

The multifractal plots for different sticking probability are shown in Figure 7. The q-τ(q) plots are nonlinear indicating multifractality. Higher the sticking probability, bigger is the deviation of the q-τ(q) curve from linearity. The curve curvature also implies the degree of heterogeneity in reaction probability distribution. These results agree with the reaction probability distribution profiles for the different sticking probability. As the sticking probability decreases, the non-linearity in the q-τ(q) curve decreases representing a homogeneous reaction probability distribution.

The high non-linearity of the q-τ(q) curves is translated into wider distribution in the f(α) plots. Higher the sticking probability, greater is the deviation in the q-τ(q) curve results in wider range of α values representing greater heterogeneity in the reaction probability distribution. Also, f(α) figures are asymmetric with respect to α for all the sticking probabilities. The difference in the f(α) spectrum for these four sticking probabilities can also be seen at the low and high α values. As the sticking probability decreases, the right extreme f(α) value becomes larger and left are smaller. This implies that as the sticking probability decreases, the low reaction probability part is associated with larger sets but the higher reaction probability part is connected to very small sets and as a result it produces a relatively homogeneous reaction probability distribution pattern.

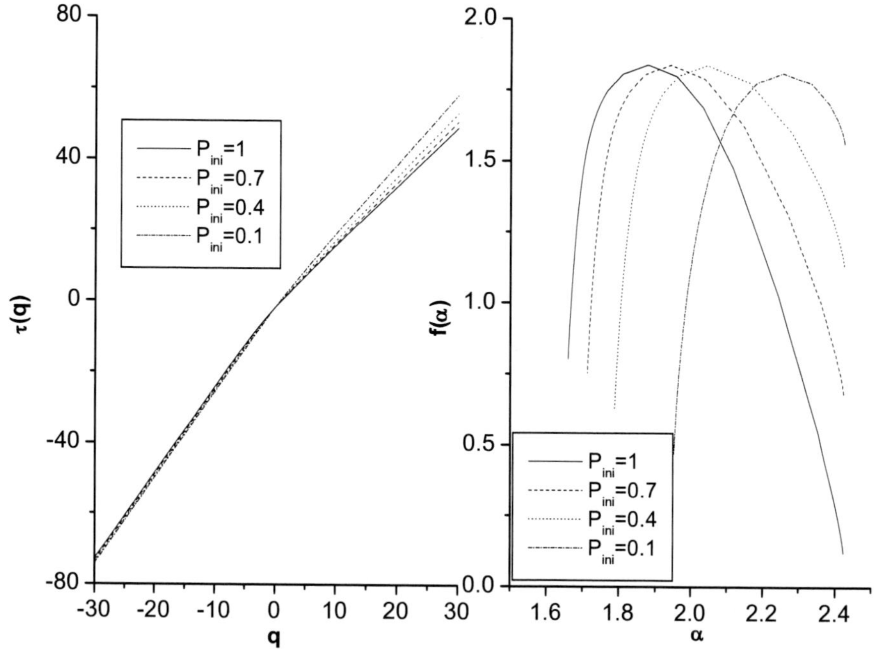

Figure 7. Effect of sticking probability on the multifractal plots.

The reaction probability is plotted as a function of height in Figure 8 to further analyze the position sensitivity, for different sticking probability. It can be seen from Figure 8 that the active region is not the topmost height of the rough surface. The reaction probability increases with increasing height, gets a maximum value and decreases thereafter and finally gets zero, for all the sticking probabilities. A maximum reaction probability (RP_{max}) is obtained for each sticking probability at a particular height. The RP_{max} increases with increasing sticking probability. The linear increase of RP_{max} with sticking probability is shown in Figure 9. The increase shows the linear behavior of the type RP_{max}=-0.00011 + 0.2216 * P_{ini}. The fits of data are also shown in Figure 9.

Decay Type Reactions Over Rough Surface

In the previous two section the ER mechanism considered with perfect sticking case i.e. reacting particle with sticking probability 1 and with some fixed sticking probability in which each active site is assumed to have the same fixed sticking probability for each reaction event. In real catalytic systems, a more general case may include an event dependent sticking probability at each active site that is a function of reaction events over that site. In this section we have studied the decay type diffusion limited reaction (DLR) over rough surface of random deposition model. In this decay type of reaction the sticking probability of every active site is defined as decay and enhancing function of reaction events. In this work, the effect of different parameters such as surface roughness, the number of reaction events (NE), the decay rate (m) and the initially sticking probability (P_{ini}) is studied on the decay type diffusion limited reactions.

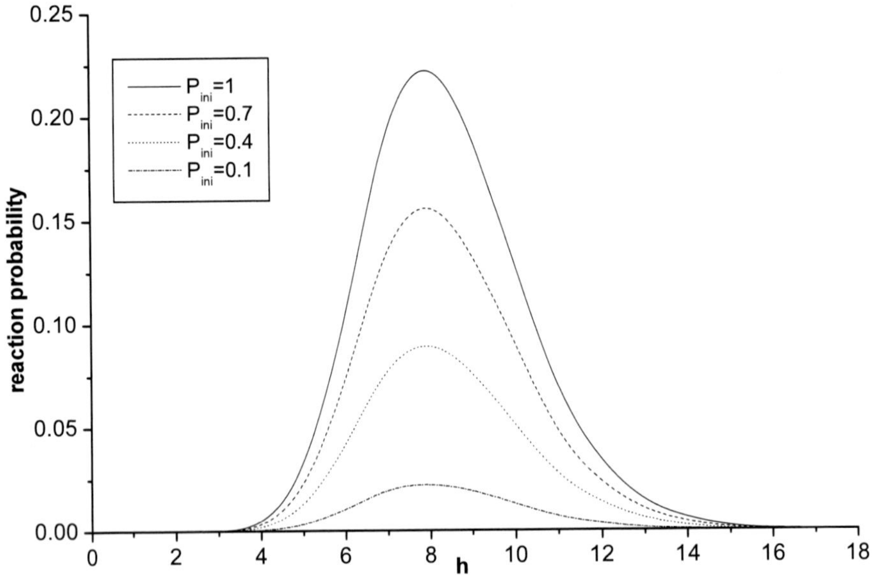

Figure 8. Reaction probability as a function of height for different sticking probability.

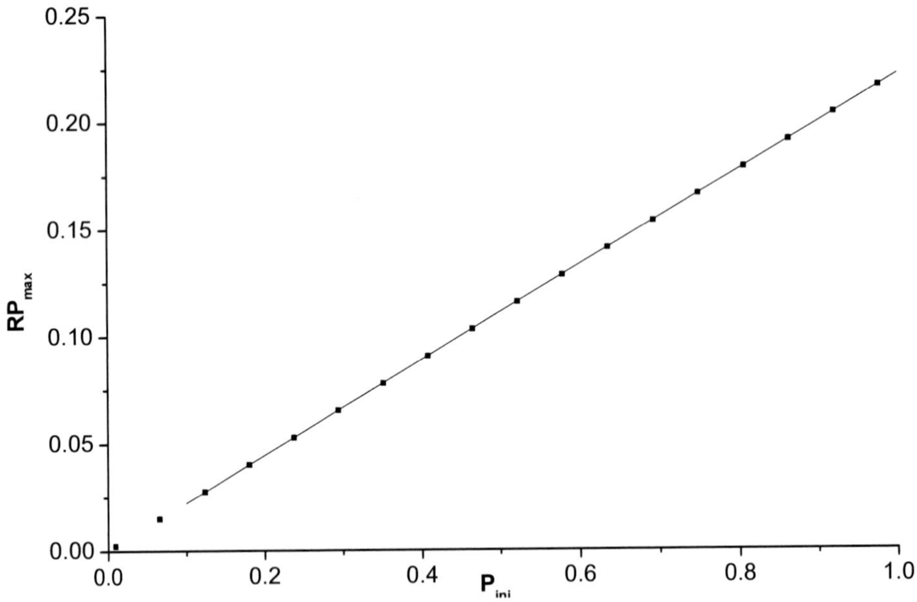

Figure 9. RPmax (Maximum reaction probability) for different sticking probability. The fits of data are also shown.

The rough surfaces for this study are generated by depositing 10^4 particles over four different lengths L viz. 200, 500, 1000 and 1500 lattice units. The values of surface thickness are found to be 6.76, 4.52, 3.23 and 2.64 lattice units, for L=200, 500, 1000 and 1500 lattice units respectively. The surface roughness is 34, 28, 20 and 17 lattice units for L=200, 500, 1000 and 1500 lattice units respectively. After generating the rough surface, the releasing particle is changed to reacting species. The columns, in which the reacting particle is to be launched, are selected randomly. The reacting particles are allowed to reach the top of the column, which is having maximum height among columns i, (i-1) and (i+1), where i is the column selected randomly. When this reacting particle comes onto the site of the rough surface, it gets absorbed with probability P. In this simulation, the event dependent sticking probabilities are introduced by a decay function defined as

$$P(n_i) = P_{ini} * \exp(-n_i/m) \tag{7}$$

where P_{ini} is the initial sticking probability for every sites, n_i is the reaction counts over active site i, and m is the tuning factor to adjust the rate of decaying. A higher and lower value of P_{ini} corresponds to the lower and higher activation energy of the active sites, respectively whereas a lower and higher value of m corresponds to the faster and slower decay rate, respectively.

We have considered different initial sticking probabilities, P_{ini}, and different tuning factor m in eq. (7) to examine the influence on reaction probability distribution. This process can be achieved by generating a random number when the reacting particle visits the rough surface site and checking whether it is less than or equal to P. When the random number generated is less than or equal to P, the reaction counts on that surface site are added by one. If it is greater than P, it is allowed to diffuse (and react) in a randomly selected direction, only if the height of the adjacent column is less than the column in which the particle is launched. Even if it does not react with the sites in the adjacent columns, it is further allowed to diffuse and react to the sites in the next column if its height is less than the height of the column under consideration, and so on. The reacting particle is not allowed to climb if the height of the adjacent column is larger than that of the present column. In such cases, a new reacting particle is considered. All other parameters are kept fixed while studying the effect of one parameter as follows :

1) Effect of m : L=500 lattice units, NE=$5x10^4$, P_{ini}=1, m=300,150,30,5.
2) Effect of P_{ini} : L=500 lattice units, NE=$5x10^4$, m=300, P_{ini}=1, 0.7, 0.4, 0.01.
3) Effect of surface roughness : NE=$5x10^4$, m=300, P_{ini}=1, L=200, 500, 1000, 1500 lattice units.
4) Effect of number of reacting events : L=500 lattice units, m=300, P_{ini}=1, NE=$2.5x10^4$, $5x10^4$, 10^5, $2x10^5$.

For each case above, the reaction probabilities of different surface sites are recorded and analyzed using multifractal analysis.

The two parameters P_{ini} and m control the properties of the decay function defined in eq. (7), and thus the reaction dynamics. The reaction events occurring at each site on the surface are recorded by counting the number of visits by the reacting particle and the reaction probability is calculated at each site. It was observed that the number of active sites increases with an increase in P_{ini}, and the number of reaction events whereas it decreases with an increase in m and the surface roughness.

To study the effect of m on the reaction probability distribution simulations were performed with P_{ini}=1.0, L=500, and m= 5, 30, 150 and 300, respectively over 5 x 10^4 reaction events.

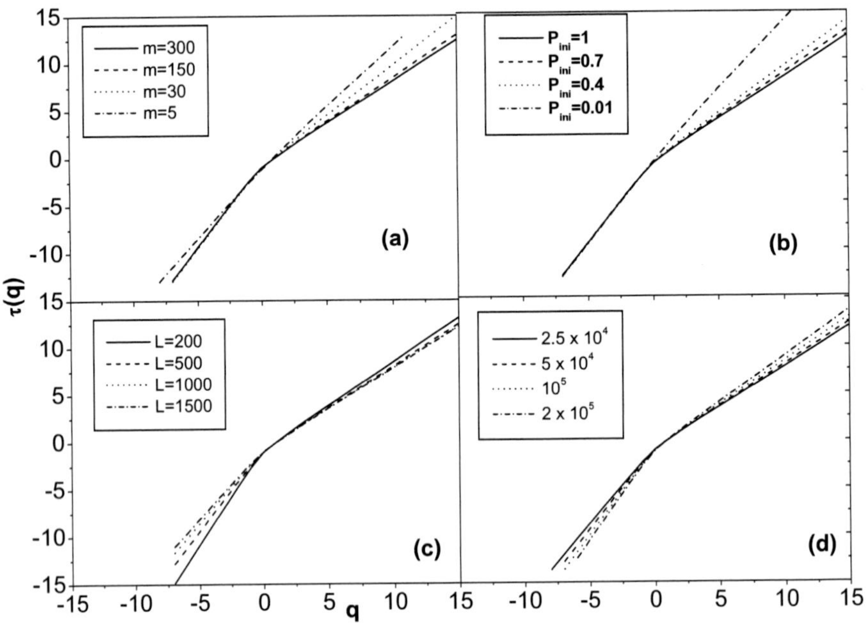

Figure 10. q-τ(q) multifractal plots for different Pini, m, surface roughness and NE.

The multifractal plots for different values of m are shown in Figure 10(a) and 11(a). The number of active sites decreases with an increase in m. The range of reaction probabilities is increased with an increase in m. The valleys between the spikes are also observed in the reaction probability distribution profile, which arise mainly from the screening effect. For different values of P_{ini} values, with m=300, L=500 and NE=5x10^4 the number of active sites increased with an increase in the P_{ini} values. The range of reaction probabilities here also is increased with an increase in P_{ini}. The surface roughness also affects the number of active sites. For different rough surface, with m=300, P_{ini}=1 and NE=5x10^4, the number of active sites decreases with an increase in surface roughness. The surface with more roughness has less number of active sites gives wider range of the reaction probability whereas for the surface with less roughness has more number of active sites and hence the range of reaction probability is narrow. Due to the increase in number of active sites, the reaction probability distribution picture becomes more homogeneous. We have also studied the effect of number of reaction events on the reaction probability distribution. To check the effect of number of reaction events, we have kept the other parameters fixed (L=500 lattice units, m=300 and P_{ini}=1). It was observed that the number of active sites increases with increase in number of reaction events. But the range of reaction probability decreases with an increase in number of reaction events. The more number of reaction events lead to more homogeneous reaction probability distribution profile because after some reaction events the sticking probability is reduced quickly to a very low value and the reaction probability distribution picture is more uniform.

The q-τ(q) curves are nonlinear for different values of m, Pini, NE and surface roughness as can be seen from Figure 10 which indicates multifractality From Figure 10(a), for q >0, as m decreases, the curvature of q-τ(q) curves also decreases gradually, indicating relatively homogeneous reaction probability distribution. This also indicates that, the number of active sites with highest reacting probability, which dominates the sum in Eq. (1) in the large q region, is quite different for different values of m. The q-τ(q) curve for m=5 exhibits a nearly linear relationship, representing a homogeneous reaction probability distribution and the slope almost equals D, implying that the homogeneity was distributed over almost every active site. The q-τ(q) curves for different m coincide in the range q <0 (except m=5) which implies that the number of active sites with a small reaction probability are nearly equal. The curve for m=5 do not coincide with the curves for higher m values, for q < 0 indicates that for m=5, the number of active sites with lowest reaction probability, which dominates the sum in Eq. (1), is quite different than those for higher values of m.

The q-τ(q) relations for different values of P_{ini}, are shown in Figure 10(b). For q >0, as P_{ini} decreases, the curvature of q-τ(q) curves also decreases indicates the homogeneous RPD. This is due to the fact that, the number of active sites with highest reacting probability, which dominates the sum in Eq. (1) in the large q region, is quite different for different values of P_{ini}. The nearly linear q- τ(q) curve for P_{ini}=0.01 representing the homogeneous RPD and the slope almost equals D implies that the homogeneity was distributed over almost every active site. The q- τ(q) curves for different P_{ini} coincide in the range q < 0 which implies that the number of active sites with a low reaction probability are nearly equal for different values of P_{ini}.

The q-τ(q) curves for the surfaces with different roughness are shown in Figure 10(c). The difference among the four different surface roughness can also be observed in the parts of low as well as high q values. The deviation of the q-τ(q) curves decreases with decrease in surface roughness. More the roughness of the surface, more is the deviation in q-τ(q) curves indicates more inhomogeneity in the RPD. As the length of the surface increases, the surface roughness decreases and the q-τ(q) curves are less deviated. The curves are well separated in the lower as well as higher q values. This indicates that the number of active sites with lowest reaction probability is quite different for the surfaces of different roughness. The number of active sites with highest reaction probability is also different for the surfaces with different roughness.

In Figure 10(d), the q-τ(q) curves are plotted to see the effect of number of reaction events. It is observed that the number of reaction events also affects this autopoisoning reactions. The curvature in q-τ(q) curves decreases with an increase in number of reaction events indicating more homogeneity in the reaction probability distribution. As these curves are well separated in lower and higher q values for different number of reaction events, we can say that the number of active sites with lowest reaction probability for different number of reaction events is not the same, also the number of active sites with highest reaction probability for different number of reaction events is not the same.

Figure 11 shows the α-f(α) plot for different values of m, Pini, NE and surface roughness. From f(α) spectra for different values of (Figure 11(a)), the range of α values increased with an increase in values of m. As m decreased, the curves contracts upwardly at high α values. The linear q- τ(q) relation for m=5 can also be seen by a narrow range of α values and nearly symmetric pattern here. The effect caused by faster decay rate on RPD can also be seen from the narrow range of α values and symmetric patter in f(α) spectrum. Lower the m, more are the number of active sites and higher is the maximum in α-f(α) curves.

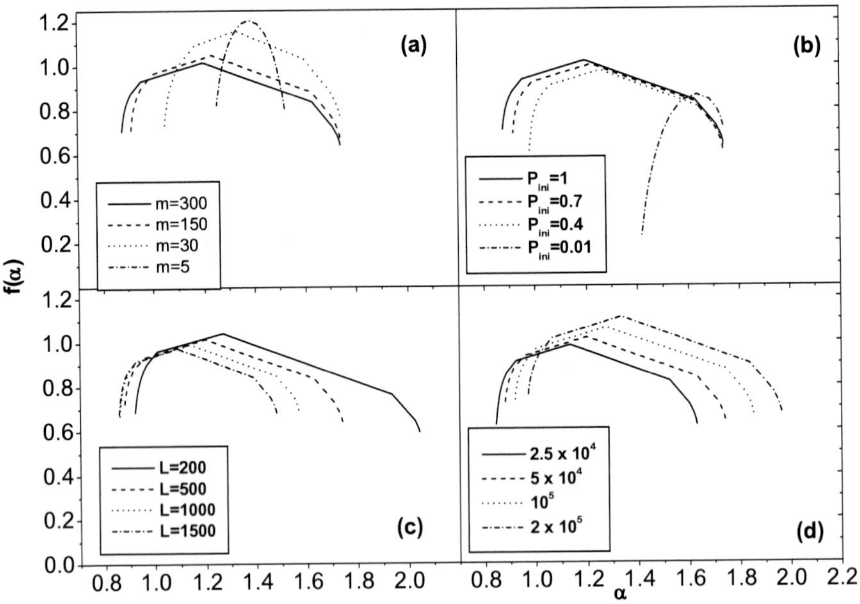

Figure 11. α-f(α) multifractal plots for different Pini, m, surface roughness and NE.

The α-f(α) curves for different P_{ini} values are shown in Figure 11(b). The range of α values increases with an increase in P_{ini}. The narrow range of α values can be seen for P_{ini}=0.01, as the q-τ(q) relation is nearly linear for this. The maxima in α-f(α) curves are not the same. Higher the P_{ini}, more is the number of active sites and higher is the maximum in α-f(α) curves.

For the surfaces with different roughness the α-f(α) curves are shown in Figure 11(c). It can be seen that the surface with more roughness has wider range of α values than that with less roughness, as the q-τ(q) is more deviated for the surface with more roughness than that with less roughness. Also the maximum of the curves are not the same but curves contract upwardly at high α value indicating the number of lowest reaction probability sites and the number of large reaction probability sites for the surfaces with different roughness is not even. More the rough surface, higher is the maximum in the α-f(α) curves. The α-f(α) curves for different number of reaction events are shown in Figure 11(d). In this case, the curves contract upwardly with increase in the number of reaction events. More the number of reaction events, higher is the maximum in the f(α) spectrum.

Figure 12 shows the plot of reaction probability as a function of height of the surface sites for different values of m and P_{ini}, L and NE. It can be seen that the most active region is not the outermost layer of the surface. From Figure 12(a), for smaller m, the number of active sites is more. It is due to the superiority of height to reaction. When reactants randomly rain down to the ith column, it will be more likely to react on the highest column among the i, i-1 and i+1 column. Much more number of active sites is found particularly in those cases with small tuning parameter m when inter-column diffusion mechanism is introduced. Smaller m means the sticking probability decrease more rapidly in auto-poisoning reaction. That causes non-reacting particles cross to lower columns by inter-column diffusion and find more active sites. Similarly, from Figure 12(b), higher the P_{ini}, more are the active sites.

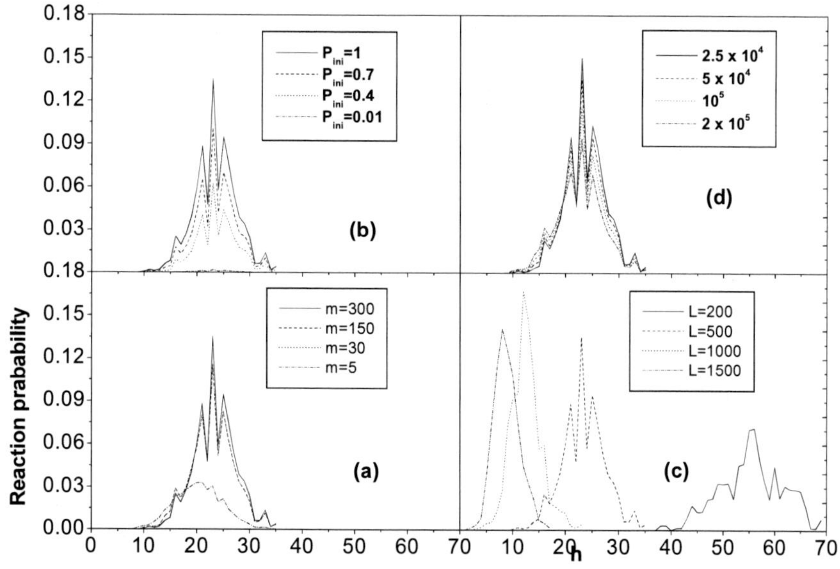

Figure 12. Reaction probability as a function of height for different m, Pini, surface roughness and number of reaction events.

Once the P_{ini} is very small, most reacting particles do not react. Even when inter-column diffusion is introduced, the reactants can reach the local minimum easily, but reactions seldom take place then and less active sites are found. From Figure 12(c), it can be seen that the total number of non-active sites decreases with decrease in surface roughness.

TIME DEPENDENCE OF ER REACTIONS OVER ROUGH SURFACE

In all the earlier studies of ER reactions emphasis was laid upon the multifractal characters involved in the fluctuation of the reaction probability distribution caused by the geometric heterogeneity. In this section dynamic scaling is introduced in exploring the time dependent effect involved in ER reactions over rough surface to obtain the scaling exponents for the time dependence of the ER mechanism. In this study, the rough surfaces are generated by deposition of N number of particles over lattice size L, using RD model. Two types of rough surfaces are considered here 1) different rough surface with same surface density and 2) different rough surface with different surface density, where surface density is defined as, Surface density = N/L. Rough surfaces with different surface density are generated by the deposition of 10^4 particles on lattice size L=100, 200, 300, 400 and 500. The surface density for these five surfaces is 100, 50, 33.3, 25 and 20 respectively. Rough surface with same surface density are generated by deposition of different number of particles on different L. The number of particles deposited on lattice size 100, 200, 300, 400 and 500 is 2000, 4000, 6000, 8000 and 10000 respectively and the surface density for these five surfaces is 20.

After generating the rough surface by depositing proper number of particles as explained above, the releasing particle is changed to reacting species. The reacting particles are allowed to react at the top of the column having maximum height among the neighboring columns. When the reacting particle reaches the top of the column the reaction count on that surface site is added by one. Total reacting species used in our simulation are 10^5. We considered total reaction event as a time parameter in our simulation.

The number of reacting particles visited at each site at time t is counted.

$$P(x, y, h; t) = \frac{r(x, y, h; t)}{r_\Sigma(t)} \tag{8}$$

where r(x,y,h;t) is the reaction event occurring at (x,y,h) at particular time t. r_Σ is the total reaction event amount all over the surface in time interval t.

The average probability $P_a(t)$ on each surface site is

$$\mathbf{P_a(t)} = \frac{\sum \mathbf{P(x,y,h;t)}}{\mathbf{N_f}} = \frac{1}{\mathbf{N_f}} \tag{9}$$

where N_f is the number of free sites available for reaction particles. Free site means the site where there is possibility of the particle to visit. The remaining sites are the inactive sites where the reacting particle can't reach due to the surrounding four nearest neighbors.

Due to the surface roughness, the reacting particle has greater probability to react at the surface sites, which are at the top causes heterogeneity in the reaction probability distribution. This heterogeneity can be estimated by standard deviation of P(x,y,h;t) as,

$$\boldsymbol{\sigma(t)} = \left[\frac{\sum (\mathbf{P(x, y, h; t) - P_a})^2}{\mathbf{N_f}} \right]^{1/2} \tag{10}$$

The scaling behavior of σ(t) i.e. correlation between σ(t) and time and also the surface roughness. The total reaction events (i.e. time) used for dynamic scaling are 10^5. A log-log plot of the fluctuation of reaction probability σ(t) on t, for different rough surfaces with different surface density and that with same surface density are shown in Figure 13 and Figure 14 respectively. It can be seen from these figures that for all the rough surfaces whether with same surface density or with different surface density, σ(t) first decreases fastly and finally becomes constant (independent of t) after experiencing a slowing down. Initially σ depends on r_Σ and the scaling relation between σ and t is

$$\sigma = t^{\beta} \tag{11}$$

The values of β obtained for different rough surfaces with same surface density and with different surface density, are given in Table 1. It should be noted that the value of β is negative and is different from the positive value obtained by Family et al. [54] For both type of surfaces, β values decrease with an increase in L and converge to –0.491± 0.001. The variation in β values for different rough surface is due to the random fluctuation in the Monte Carlo algorithm. For different rough surface with different surface density as well as for different rough surface with same surface density, the value of the scaling exponent β is the same.

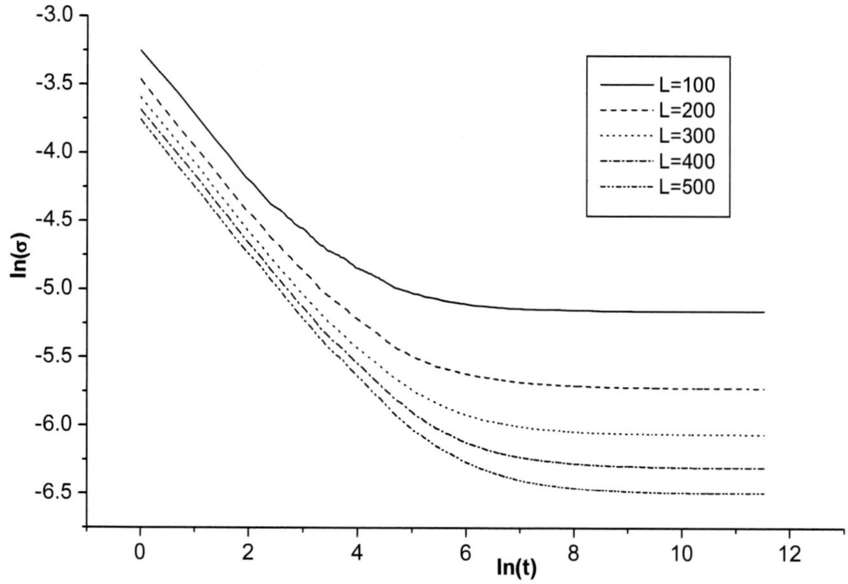

Figure 13. Dependence of σ on time t for rough surfaces with different surface density.

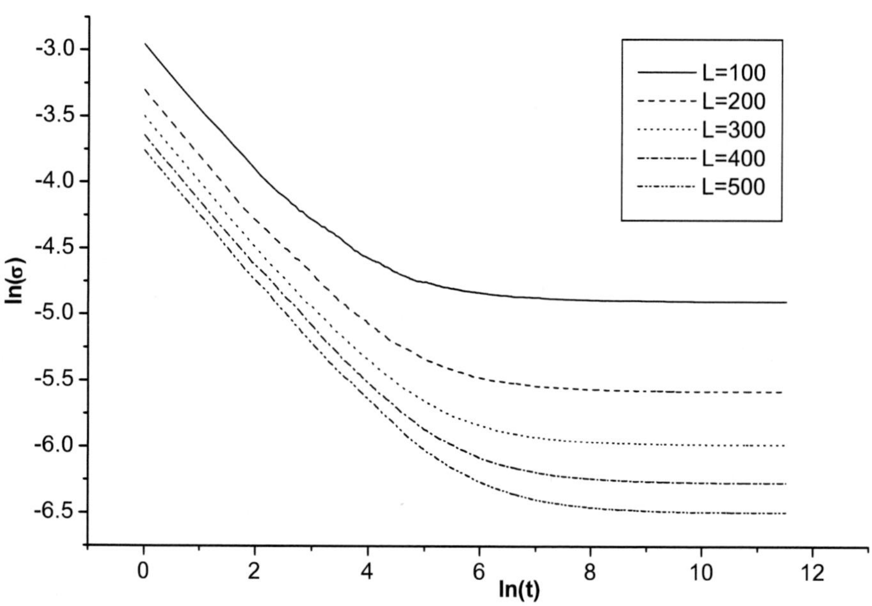

Figure 14. Dependence of σ on time t for rough surfaces with same surface density.

Therefore we can conclude that the surface roughness does not affect the scaling exponent (β) value without bothering whether we are considering the rough surfaces with same surface density or with different surface density.

Table 1. Scaling exponents in dynamic scaling theory for ER reactions over different rough surface with same surface density and different rough surface with different surface density

L (lattice Units)	Rough surfaces with same surface density		Rough surfaces with different surface density	
	β	α	β	α
100	-0.481 (±0.003)		-0.479 (±0.002)	
200	-0.485 (±0.003)		-0.485 (±0.004)	
300	-0.491 (±0.002)	-0.83 (0.002)	-0.486 (±0.003)	-1.0 (0.001)
400	-0.491 (±0.002)		-0.491 (±0.001)	
500	-0.491 (±0.001)		-0.491 (±0.002)	

At $t \rightarrow \infty$, σ becomes independent of t but depends on surface roughness, for both types of surface. It can be seen from Figures 15 and 16 that the scaling relation between σ and L is,

$$\sigma = L^{\alpha} \qquad \text{as } t \rightarrow \infty \tag{12}$$

with $\alpha = -0.83 \pm 0.002$ for different rough surface with same surface density and $\alpha = -1.0 \pm 0.001$ for different rough surface with different surface density. The dependence of σ on t and L given by eqs. (11) and (12), respectively can be combined into a single expression representing a dynamic scaling which has been argued to be universal [53,61] as

$$\sigma(t,L) = L^{\alpha} f(t/L^{\gamma}) \tag{13}$$

where $\gamma = \alpha/\beta$.

The scaling function f(x) behaves as $f(x) = x^{\beta}$ for $x << 1$ and $f(x) =$ constant for $x >> 1$. Figures 17 and 18 show plots of $\sigma(L,t)\, L^{-\alpha}$ versus t/L^{γ} for different rough surfaces with different surface density and that with same surface density, respectively, using respective values of α and γ. It can be noted that the curves for different rough surfaces, for both types of surface, all collapse into a single curve indicating excellent agreement with the scaling form given in Eq. 13.

CONCLUSION

The measures associated with physical events occurring on the heterogeneous surface may lead to spatial distribution, which possesses infinitely many singularities. The development of formalism for the description of such situations therefore needs infinitely many dimension-type exponents to characterize these distributions.

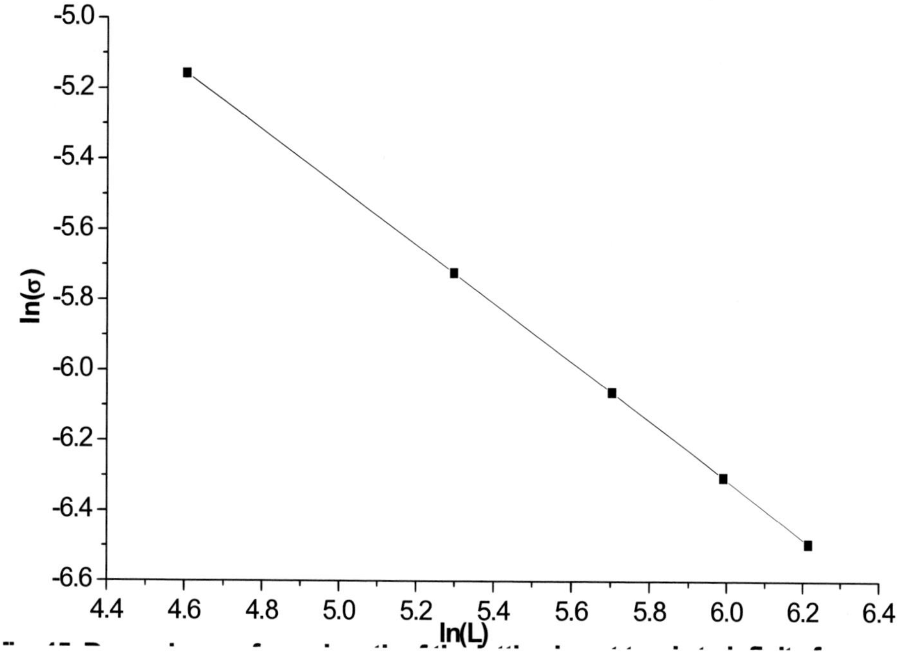

Figure 15. Dependence of σ on length of the lattice L as t tends to infinity for rough surfaces with same surface density.

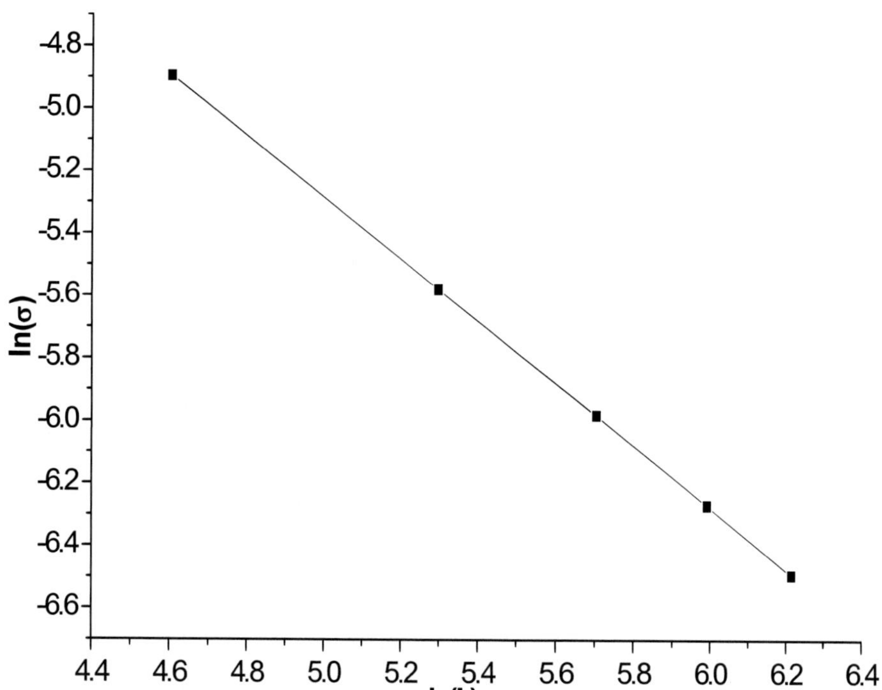

Figure 16. Dependence of σ on length of the lattice L as t tends to infinity for rough surfaces with different surface density.)

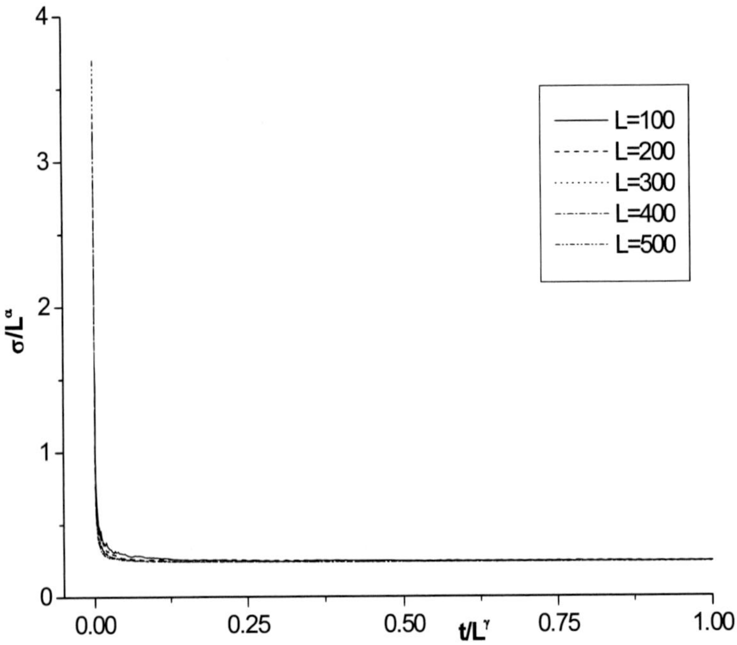

Figure 17. Plot of scaling function f(x) defined in eq.(13) shown by plotting σ(L,t)/Lα against t/Lα by selecting α=-0.491 and γ=2.04 for rough surfaces with different surface density.

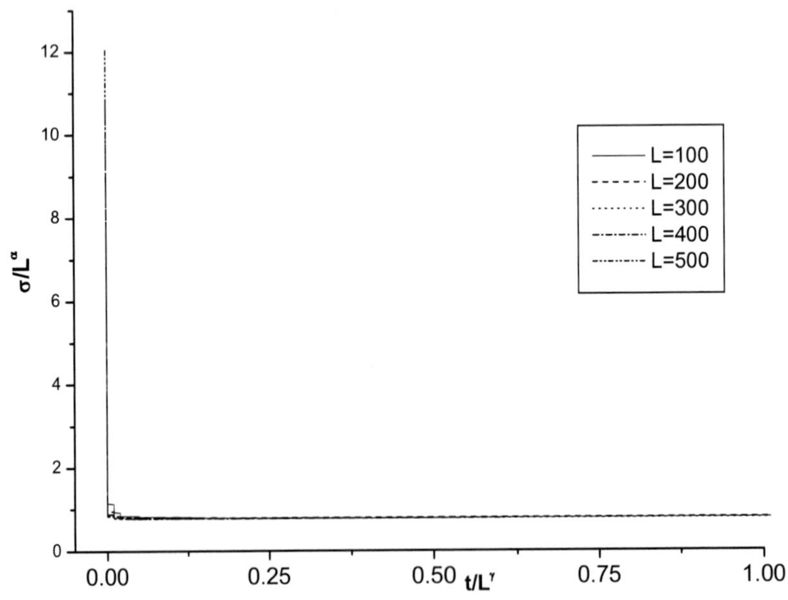

Figure 18. Plot of scaling function f(x) defined in eq.(13) shown by plotting σ(L,t)/Lα against t/Lα by selecting α=-0.491 and γ=1.69 for rough surfaces with same surface density.

The internal texture or morphology of materials is largely responsible for most of their physical properties and any correlation between internal structure or morphology and the physical properties is of considerable practical importance. It has been realized that in many cases a more detailed characterization can be obtained using multifractal formalism. In this case the structure is described by a spectrum of scaling indices α-f(α). We have applied multifractal scaling analysis to one basic mechanism in catalytic reaction namely ER reaction mechanism over the rough surface. Such characterization may be useful for the study of relationship between the microstructure and microscopic properties of the materials. Relatively little experimental work has been carried out to test the multifractal formalism. The reasons are that it is difficult to obtain well-characterized heterogeneous surface on which experiments can be carried out and to obtain accurate experimental results to compare with the multifractal formalism. However, some encouraging results are obtained here by applying multifractal analysis to the ER reaction mechanism. This will prompt more people to apply the same type of approach to number of important heterogeneous systems. Computer simulations results here indicated that it is a powerful tool in this field and indeed could qualitatively identify the important trends although it is difficult to mimic the a full real catalytic process. The multifractal scaling analysis can be applied to the high sensitivity of processes to complex geometries and chemical heterogeneity.

REFERENCES

[1] Rudzinski, W.; Everett, D. H. Adsorption of Gases on Heterogeneous Surfaces; Academic Press : London 1992.

[2] Mason, G. in Characterization of Porous Soilds II; Rodrihuez-Reinoso et al.; Ed.; Elsevier : Amsterdam, 1991.

[3] Farin, D.; Avnir, D. in Characterization of Porous Soilds; Unger, K. K. et. al.; Ed.; Elsevier : Amsterdam, 1988.

[4] Pfeifer, P.; Obert, M. in The Fractral Approach to Heterogeneous Chemistry : Surfaces, Colloids, Polymers; Avnir, D. ; Ed.; John Wiley : Chichester, 1989.

[5] Family, F.; Landau, D. P. Kinetics of Aggregation and Gelation; Elsevier/North-Holland : Amsterdam 1984.

[6] Stanley, H. E.; Ostrowski, N. On Groth and Form : Fractal and Nonuniform Patterns in Physics; Martinus Nijhoff : Dordrecht 1986.

[7] Stanley, H. E.; Ostrowski, N. Random Fluctiations and Pattern Growth : Experiments and Models; Kluwer Academic : Dordrecht 1988.

[8] Jaroniec, M.; Madey, R. Physical Adsorption on Heterogeneous Solids; Elsevier : Amsterdam 1988.

[9] Satterfield, C. N. Heterogeneous Catalysis in Practice; McGraw Hill : NY 1980.

[10] Everett, D. H. in The Solid-Gas Interface; Flood, E. A.; Ed.; Vol. 2, Edward Arnold : London 1967.

[11] Mann, R.; Thomson, G. in Adsorption Science and Technology; Rodrigues, A. E. et. al.; Eds.; Kluwer Academic : Dordrecht 1989.

[12] Farin, D.; Avnir, D. in The Fractal Approach to Heterogeneous Chemistry : Surfaces, Colloids, Polymers; Avnir, D. Ed.; John Wiley : Chichester 1989.

[13] Jullien, R.; Botet, R. Aggregation and Fractal Aggregates; World Scientific : Singapore 1987.
[14] Bunde, A.; Havlin, S. Fractals and Disordered Systems; Springer : Berlin 1991.
[15] Schmidt, P. W. in The Fractal Approach to Heterogeneous Chemistry : Surfaces, Colloids, Polymers; Avnir, D. Ed.; John Wiley : Chichester 1989.
[16] Pfeifer, P.; Kenntner, J.; Cole, M. in Fundamental of Adsorption; Mersmann, J.; Scholl, S. E.; Eds.; Engineering Foundation : NY 1989.
[17] Adamson, A. W. Physical Chemistry of Surfaces, 4th ed.; Wiley : NY 1982.
[18] Rice, S. A. Diffusion-limited reactions; Elsevier : Amsterdam 1985.
[19] Connors, K. A. Chemical Kinetics: the study of reaction rate in solution; VCH : NY 1990.
[20] Rettner, C. T. *Phys. Rev. Lett.* 1992, 69, 383.
[21] Rettner, C. T. ; Auerbach, D. *J. Surf. Sci.* 1996, 357-358, 302.
[22] Eilmsteiner, G. E.; Winkler, A. *Surf. Sci.* 1996, 366, L750.
[23] Xi, M.; Bent, B. E. *J. Phys. Chem.* 1993, 97, 4167.
[24] Park, Y. S.; Kim, J. Y.; Lee, *J. Surf. Sci.* 1996, 363, 62.
[25] Rettner, C. T.; Auerbach, D. J.; Lee, J. *J. Chem. Phys.* 1996, 105, 10115.
[26] Butin, S. A. *J. Chem. Phys.* 1996, 105, 2066.
[27] Cheng, C. C.; Lucas, S. R.; Gutleben, H.; Choyke, W. J.; Yates Jr., *J. T. Surf. Sci.* 1992, 273, L441.
[28] Su, C.; Song, K. J.; Wang, Y. L.; Lu, H. L.; Chuang, T. J.; Lin, J. C. *J. Chem. Phys.* 1997, 107, 7543.
[29] Garton, D. J.; Minton, T. K.; Alagia, M.; Balukani, N. B.; Casavecchia, P.; Volpi, G. G. *J. Chem. Phys.* 2000, 112, 5975.
[30] Jackson, B.; Persson, M.; Kay, B. D. *J. Chem. Phys.* 1992, 96, 2378.
[31] Gutfraind, R.; Sheintuch, M.; Avnir, D. *J. Chem. Phys.* 1991, 95, 6100.
[32] Lee, S. L.; Chu, C. H.; Luo, Y. L.; Ho, J. *J. Chem. Phys. Le*tt. 1993, 207, 220.
[33] Lee, C. K.; Lee, S. L. *Surf. Sci.* 1995, 325, 294.
[34] Chaudhari, A.; Yan, C. C.; Lee, S. L. *Chem. Phys. Lett.* 2002, 351, 341.
[35] Chaudhari, A.; Yan, C. C.; Lee, S. L. PCCP 2002, 4, 5330.
[36] Yan, C. C.; Chaudhari, A.; Lee, S. L. *Appl. Surf. Sci.* 2002, 196, 375.
[37] Lee, S. L.; Lee, C. K. Int. *J. Quant. Chem.* 1994, 52, 339.
[38] Lee, C. K.; Lee, S. L. Het. *Chem. Rev.* 1996, 3, 269.
[39] Lee, C. K.; Lee, S. L. *Chem. Phys. Lett.* 1994, 226, 88.
[40] Lee, C. K.; Lee, S. L. *Surf. Sci.* 1995, 339, 171.
[41] Gutfraind, R.; Sheintuch, M.; Avnir, D. *Chem. Phys. Lett.* 1990, 174, 8.
[42] Seri-Levi, A,; Avnir, D. *Surf. Sci.* 1991, 248, 258.
[43] Lee, C. K.; Lee, S. L. *Chem. Phys. Lett.* 1994, 228, 539.
[44] Seri-Levi, A,; Avnir, D. *J. Phys. Chem.* 1993, 97, 10380.
[45] Lee, C. K.; Chiang, A. S. T.; Wu, F. Y. *AICHE J.* 1992, 38, 128.
[46] Chaudhari, A.; Yan, C. C.; Lee, S. L. *Int. J. Chem. Kin.* 2005, 37, 175.
[47] Chaudhari, A.; Yan, C. C.; Lee, S. L. Int. *J. Chem. Kin.* 2004, 36, 286.
[48] Chaudhari, A.; Yan, C. C.; Lee, S. L. *J. Phys. A : Math Gen.* 2003, 36, 3757.
[49] Chaudhari, A.; Yan, C. C.; Lee, S. L. *Chem. Phys.* 2005, 309, 103.
[50] Chaudhari, A.; Rabbani, G.; Lee, S. L. J. *Chin. Chem.. Soc.* 2007, 54, 1201.
[51] Rudzinski, R.; Lee, S. L.; Yan, C. C.; Panczyk, T. *J. Phys. Chem.* B. 2001 105, 10847.

[52] Mai, J.; Casties, A.; Von Niessen, W. *Chem. Phys. Lett.* 1992, 196, 358.
[53] Family, F.; Vicsek, T. Dynamics of Fractal Surfaces; World Scientific : Singapore 1991.
[54] Family, F. *J. Phys.* A 1986, 19, L441.
[55] Halsey, T. C.; Jensen, M. H.; Kadanoff, L. P.; Procaccia, O.; Shraiman, B. *I. Phys. Rev.* A 1986, 33, 1141.
[56] Botelho, S. S.; Reis, F. D. A. *A. Phys Rev* E 2002, 65, 032101.
[57] Kotrla, M.; Slanina, F.; Predota, M. *Phys Rev* B 1998, 58, 10003.
[58] Horowitz, C. M.; Monetti, R. A.; Albano, E. V. *Phys Rev E* 2001, 63, 066132.
[59] Horowitz, C. M.; Albano, E. V. *J Phys A : Math Gen* 2001, 34, 357.
[60] Wang, W.; Cerdeira, H. A. *Phys Rev E* 1993, 47, 3357.
[61] Barabasi, A. L.; Stanley; H. E. Fractal Concepts in Surface Growth; Cambridge University Press : Cambridge 1995.

In: Heterogeneous Catalysis Research Progress
Editor: Mathias B. Gunther, pp. 395-429
ISBN 978-1-60456-979-7

Chapter 11

HETEROGENEOUS CATALYSIS UNDER CONTINUOUS FLOW CONDITIONS

Ildikó Kovács[1], Richard Jones, Zsolt Ötvös, László Ürge, György Dormán, and Ferenc Darvas
ThalesNano Inc. Záhony u. 7; H-1031 Budapest, Hungary

ABSTRACT

In the last decade increasing numbers of articles have been published focusing on the combination of heterogeneous catalysis and continuous flow methods. The publication of such an overview of the field is particularly justified by the recent practical applications involving a large number of novel instruments designed to perform continuous flow reactions in the presence of heterogeneous or immobilized homogeneous catalysts. In a typical experiment, a solution of the substrate/reactant is pumped through the heterogeneous catalyst optionally under elevated temperature and/or pressure. This application of flow methodology offers many advantages in organic chemistry. Reactions are typically performed in a continuous flow mode on a small scale, thus, reaction temperatures can easily be controlled. Products are eluted from the reactors after minutes, so information about the outcome of the reaction can be quickly obtained. Parameters, such as flow rate, temperature or pressure may be adjusted during the reaction allowing for fast reaction optimization and the results are highly reproducible. Scale up of reactions can be realized by either running the reaction over a long period of time or by performing many reactions in parallel. Automation of reactions can also be easily achieved through automated liquid handlers which inject solutions of the substrate before contacting the catalyst and collect the reaction mixture at the reactor output. In this chapter we would like to give a comprehensive overview of the field of continuous flow techniques using heterogeneous catalyst to perform reactions. The topics covered will begin with reactions under ambient conditions, continue towards reactions under supercritical conditions and conclude with the description of microwave assisted continuous flow systems.

---

[1] Ildiko.kovacs@thalesnano.com.

INTRODUCTION

The challenge to perform more efficient reactions has always stimulated the mind of chemistry practitioners. Novel synthetic methods inspire scientists to increase their knowledge of the chemical space such as combinatorial chemistry methods or using specific attribute materials such as supercritical fluids and ionic liquids. Not only have the chemicals changed and their number expanded, but also the way to produce materials developed significantly over the years. In standard laboratories reactions are performed in commercial glassware, however as the number of scientific articles describing the application of different microreactors integrated with liquid handlers is steadily growing it generates more and more interest.

Continuous-flow processes have numerous advantages and can be realized by different ways to replace the traditional batch method. In laboratory scale microreactors [1,2,3] or chip-based reactors are favored. On the other hand, other micro-constructed devices were designed for larger throughputs and they are also more suitable for mass production purposes [4]. All methods can reduce the reaction time [5] and can eliminate the adjustment of reaction parameters during scale-up. In many cases the characteristics of the reactors allow the chemists to perform reactions without solvents and catalysts [6] or reagents can be generated in-situ in the reactor. These features have made continuous flow methods particularly attractive, especially as they satisfy most of the Green Chemistry criteria [7]. These flow systems are also applicable for highly exothermic reactions without the need of external cooling, and decrease or eliminate the formation of by-products.

A number of reactors were designed covering homogeneous and heterogeneous reactions for different applications. When reactors were used for homogeneous reactions the inner diameter of the coil where the reaction takes place is a key factor in conjunction with the mixing of the reagents [8]. The dimension of the channels affects the maximum flow rate of the reactant solution, so low productivity is achieved using reduced size channels. Introduction of liquids can be realized by hydrodynamic pumps or by an external force such as electroosmotic flow. For proper mixing several solutions were designed, including T- and Y-shaped microchannels [9] or micromixers [10]. One of the major advantages of the continuous flow technique is that heat exchange can be rapidly achieved resulting in good heat dissipation and easy control of exothermic reactions.

Heterogeneous reactions are typically performed in solid and liquid phase reactors; typically fixed bed and fluidized bed reactors (Figure 1) and wall reactors (Figure 2) wherein the contact time and the ratio of solid/liquid phase are different and can be moderated.

Both fabrications allow the user to utilize on-line detections such as UV, IR or on-line analytical separation, and to produce compounds on a large scale. Heterogeneous catalysts have already proved their superiority over homogeneous catalysts under batch conditions, due to their easy handling. Such advantages underlined their application in continuous flow processes. On the other hand liquid-phase catalysts can also be used under continuous conditions in the form of immobilized reagents/catalysts and scavengers [11] anchored to solid supports such as silica or polystyrene [12]. Furthermore, in recent years encapsulation of reagents [13] and catalysts have extended the scope of routine synthetic approaches such as the use of hazardous chemicals and performing reactions under extreme conditions. The safe and simple handling of such compounds led to novel approaches such as applications with ionic liquids and supercritical fluids [14,15,16].

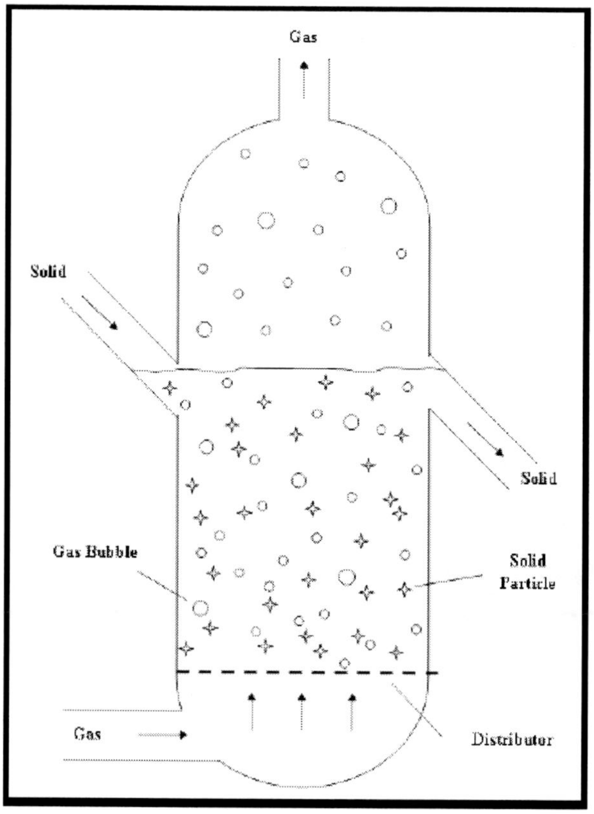

Figure 1. Fluidized bed reactor design.

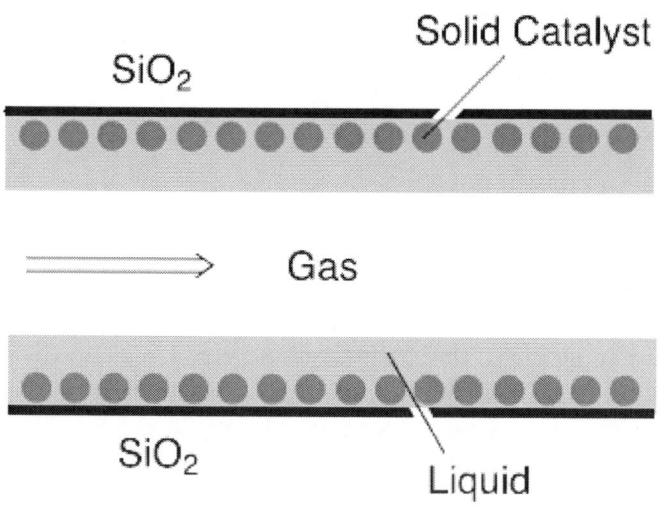

Figure 2. Wall reactor design [48].

GAS PHASE REACTIONS UNDER CONTINUOUS FLOW CONDITIONS

Reactions in gas phase are not commonly applied in laboratory scale because of the need of high temperature, other extreme conditions and difficult instruments in most cases, but three examples where gases were introduced into flow reactor will be presented here.

Gas phase synthesis of 1-halobutanes (Figure 3), namely the important starting materials of more reactions was performed by Tundo et al. [17]. where silica supported ZnCl2 and phosphonium salt were used as catalysts. Solution of 1-butanol and the haloic acid was introduced the catalyst filled column preheated to 130-170°C. According to the haloic acid the corresponding 1-chloro- and 1-bromobutene was easily synthesized with high yield (98% in case of 1-bromobutene using the onium salt). This application allows chemistry to use heterogeneous catalysis under continuous flow as an efficient and pure tool for selective synthesis of 1-halobutanes in 10 g scale.

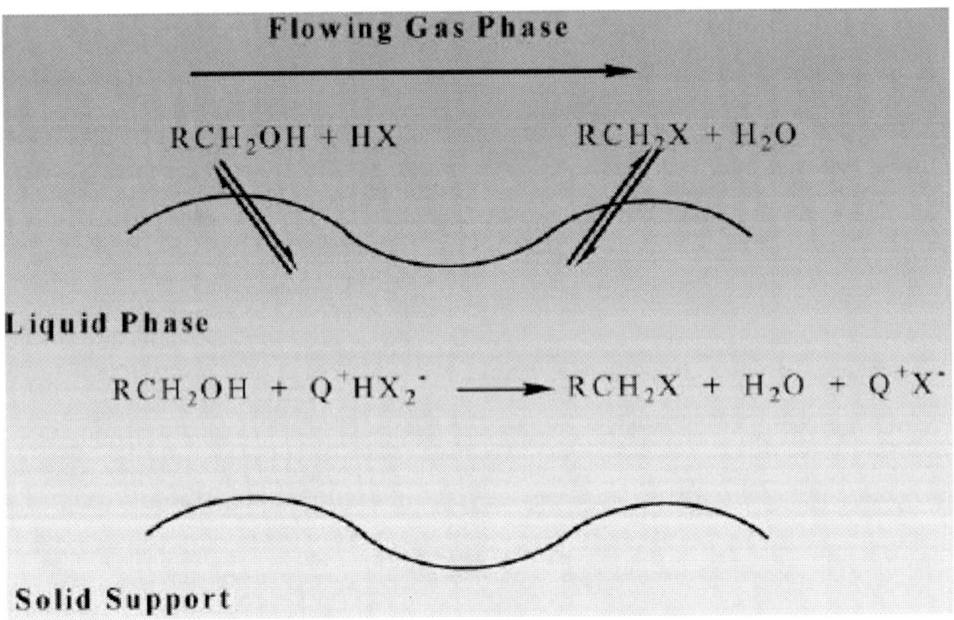

Figure 3. Synthesis of 1-halobutanes [17].

Titanium-oxide impregnated with vanadium was used as a solid support in partial oxidation of toluene [18]. Investigating the best catalyst composition for production of important intermediates, such as benzaldehyde and benzoic acid, a fixed bed reactor was used for catalyst screening reactions. Results indicate that metal addition (K, Al and Sb) and the increase of vanadium content did not result in a significant change in the catalyst's activity. While the surface area of the support was changed from 20 to 70 m^2/g, the catalytic activity was increased by one order of magnitude. According to these experiments an appropriate surface area and moderate V_2O_5 (5%) loading was selected. Since this oxidation is a highly exothermic reaction which needs to be performed in gas phase to reduce the hazards experienced with liquid phase synthesis it was carried out in a temperature-controlled microchannel reactor (total volume channel of 0.342 mL; 75 mm x 75 mm; 1 mm depth and width channels, 12 channels). Regulated oxygen gas and toluene were mixed and preheated before being introduced into the microreactor, where the reaction took place at atmospheric pressure. A cold trap linked to the microreactor enabled the isolation of the product and gas chromatography integrated into the system allowed online product analysis. Using high temperatures between 317-417°C in separate experiments the conversion and product ratio were measured and found that at around 370°C the direct combustion of toluene to CO_x increased dramatically. Furthermore, increasing the concentration of oxygen decreased the selectivity of the formation of the targeted intermediates, as expected.

Gases as reaction media are widely used in large industry based technologies both under batch and flow conditions. A mixture of CO and hydrogen gas is called 'syngas' and can be generated by gasification. Both components can be oxidized but Rossignol et al. [19] described the selective oxidation of CO in the presence of hydrogen. Three types of oxidation were performed a.) oxidation of carbon monoxide (using mixture of CO, O_2 and He) b.) selective oxidation of CO in the presence of hydrogen (2% CO, 2% O_2, 48% H_2 and 48% He) c.) oxidation of H_2 (48% H_2, 2% O_2 and 50% He). For these experiments three Au catalysts were used (Au/Al_2O_3, Au/ZrO_2 and Au/TiO_2), with the catalyst prepacked into the continuous flow fixed bed reactor with a catalyst bed of around 10 mm in length. Gas mixtures were introduced to the reactor at a typical flow rate of 50 NmL/min or can be introduced directly to a gas chromatograph. The reactor can be heated up to 450°C under atmospheric pressure and the products were analyzed on-line by GC equipped with a TCD detector. It was known that titanium is the best support for gold nanoparticles in catalytic oxidation of CO and it was confirmed when compared to the zirconia and alumina supports. The size of gold particles were also determined and when 3.7 nm gold particles were used, generated by a deposition-precipitation method, higher activity was observed than in the case of laser vaporization synthetized gold particles (size: 2.9 nm). When hydrogen was oxidized all catalysts were active and maximum hydrogen conversion was achieved at 170°C (Au/Al_2O_3), 225°C (Au/ZrO_2) and at 250°C on Au/TiO_2 therefore they observed just a less effect of the support. Finally the ability of the catalyst for selective CO oxidation was investigated in the presence of hydrogen and despite the results from CO oxidation, the same activity was found in all the catalysts.

TRIPHASIC REACTIONS

Triphase hydrogenation of 1-Phenylcyclohexene (Figure 4) was carried out in Pd-immobilized capillary column reactor by Kobayashi and co-workers [20]. The Pd catalyst was immobilized onto the inner surface of the capillaries with 200 μm inner diameters and 40 cm^2 surface by the microencapsulation and polymer-micelle-incarceration methods. The reaction parameters when the reaction proceeded quantitatively were 0.8 mol/L concentration and 0.4 mL/h liquid flow rate which resulted in 0.32 mmol/h productivity. The scale-up examination of the system, which includes the connection of nine capillaries gives decreased yield (89%) due to the inhomogeneous distribution of the gas and the liquid in each line. However, the productivity was increased to 2.8 mmol/h. Leaching of catalyst was measured and interestingly it was found to be less than 0.20 μg and reuse of catalyst showed that the immobilized catalyst can be used several times without change in activity.

Fabrication of catalyst immobilized microchannel reactors usually needs expensive, complex, and multistep methods. On the contrary, reactors with simple structures (Figure 5) can also perform efficient hydrogenation reactions. Yoswathananont et al. [21] reported an efficient hydrogenation reaction in a continuous flow system by the use of a gas-liquid-solid tube reactor.

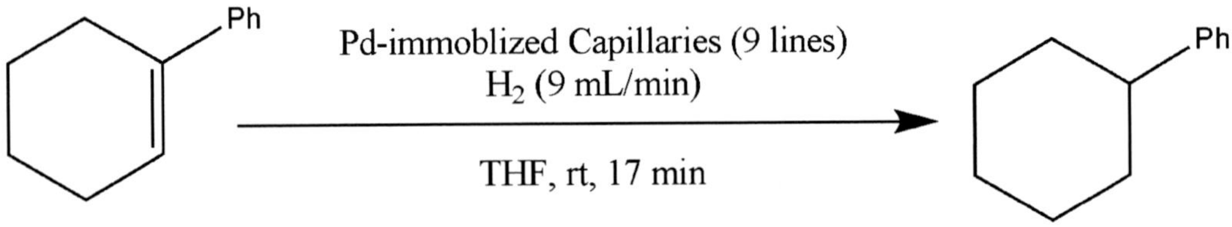

Figure 4. Hydrogenation of 1-phenylcyclohexene [20].

The applied stainless steel tube with filters at both ends was simply filled with 5% Pd/C catalyst and a mixture of substrate solution and hydrogen was pumped through the system. The mixture flowed through the narrow channels formed between the catalyst particles with approx. 20 μm i.d.. The efficiency of the flow reactor system was proved by the hydrogenation reaction of 4-cyanobenzaldehyde in methanol. Applying the same reaction parameters (temperature, pressure, residence time) the results from batch and flow experiments were compared. At each examined temperature, the efficiency of hydrogenation in flow system was found to be higher than in the corresponding batch reactions.

The production of hydrogen peroxide through the hydrogenation of 2-ethylanthraquinone in a microreactor system was examined by Halder and Lawal [22]. The reactor tube was packed with 1% Pd supported on silica and glass beads, which was added to prevent any fine catalyst particle from clogging the filters placed at the ends of the reactor tube. Investigating the effect of reaction parameters on the space-time yield (STY), increasing STY was observed with increasing reaction temperature, pressure, the reactant concentration and residence time.

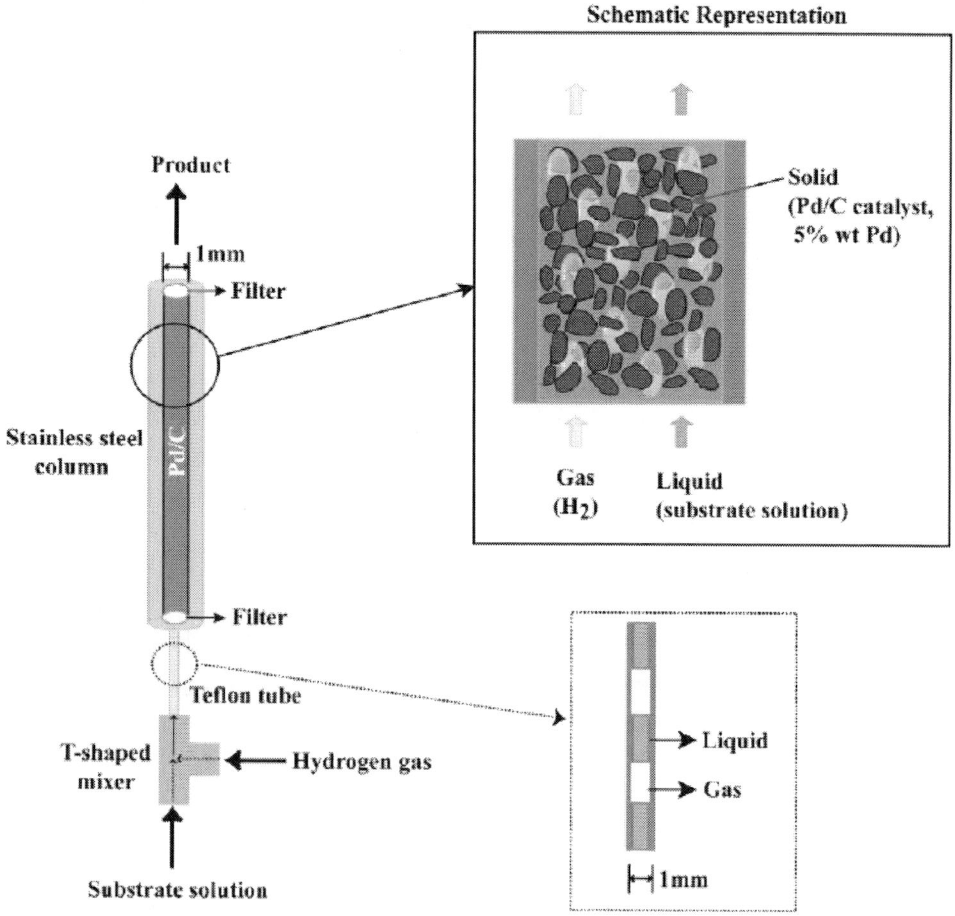

Figure 5. Reactor design in hydrogenation of 4-cyanobenzaldehyde [21].

The microreactor technique described by the authors offered 30 and 50 times higher STY values than the conventional tubular reactor and the slurry reactor, respectively, suggesting microreactors to be a successful technique in mass transfer controlled reactions.

Flow chemistry applications offer the opportunity to perform conventional multi-step reactions directly in a single step, as described in our previous work [23]. Synthesis of aryl dicarboxylic acid monoamides (Figure 6), an important series of biologically active compounds in medicinal chemistry, was successfully performed with a single step direct aminocarbonylation reaction in an X-Cube™ continuous flow reactor. The high system pressure allowed us to use tetrahydrofuran (THF) as solvent above its boiling point. Over the supported Pd catalyst the reaction of 4-Iodobenzoic acid, pyrrolidine and triethylamine, each dissolved in THF in presence of carbon monoxide, was performed simultaneously in a flow reactor and also in an autoclave and in a glass vessel with a balloon. The comparison of the results obtained from each method resulted in a remarkable difference in favor of the flow technique, namely, much higher selectivity and conversion rate. Using the optimal reaction parameters a derivatized dicarboxylic acid series was also synthesized. The results indicated that aminocarbonylation of iodobenzoic acid isomers and relatively reactive amines worked well in flow reactor.

As it is well known, enzyme catalyzed reactions can result in high enantiomer selectivity but the use of enzymes is limited by their properties and expressivity. However, enzymes have been utilized in some applications, such as the degradation of *p*-chlorophenol [24] because of the small amount of enzyme needed under continuous flow.

Performing a systematic comparison of lipase-catalyzed kinetic resolutions of several secondary alcohols in continuous flow mode (Figure 7) and shake flask batch mode using immobilized and non-mobilized lipases was reported by Csajagi and co-workers [25]. The results indicated that immobilized as well as lyophilized powder forms of liphases can be effectively used in continuous flow mode kinetic resolutions of racemic alcohols in non-aqueous solvent systems. The productivity of the lipases was higher in continuous flow reactors than in batch mode systems, whereas the enantiomer selectivities were similar.

An automated H-Cube® platform was designed, constructed and validated by Clapham et al. [26], to perform deprotection reactions and produce compound libraries. This special instrument includes 48 starting material vials for automated injection into the flow reactor. In producing the Cbz deprotected library, complete conversions and a 93% average crude yield were obtained over Pd/C catalyst at 60°C using full hydrogen mode. For the second library, 4-benzyloxy benzoic acid was coupled with a series of amides. The deprotection performed in H-Cube® resulted in complete conversions to the products and 88% average crude yield.

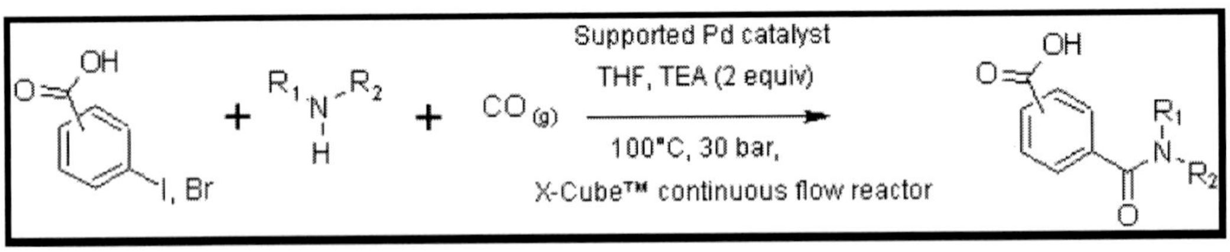

Figure 6. One-step aminocarbonylation reaction [23].

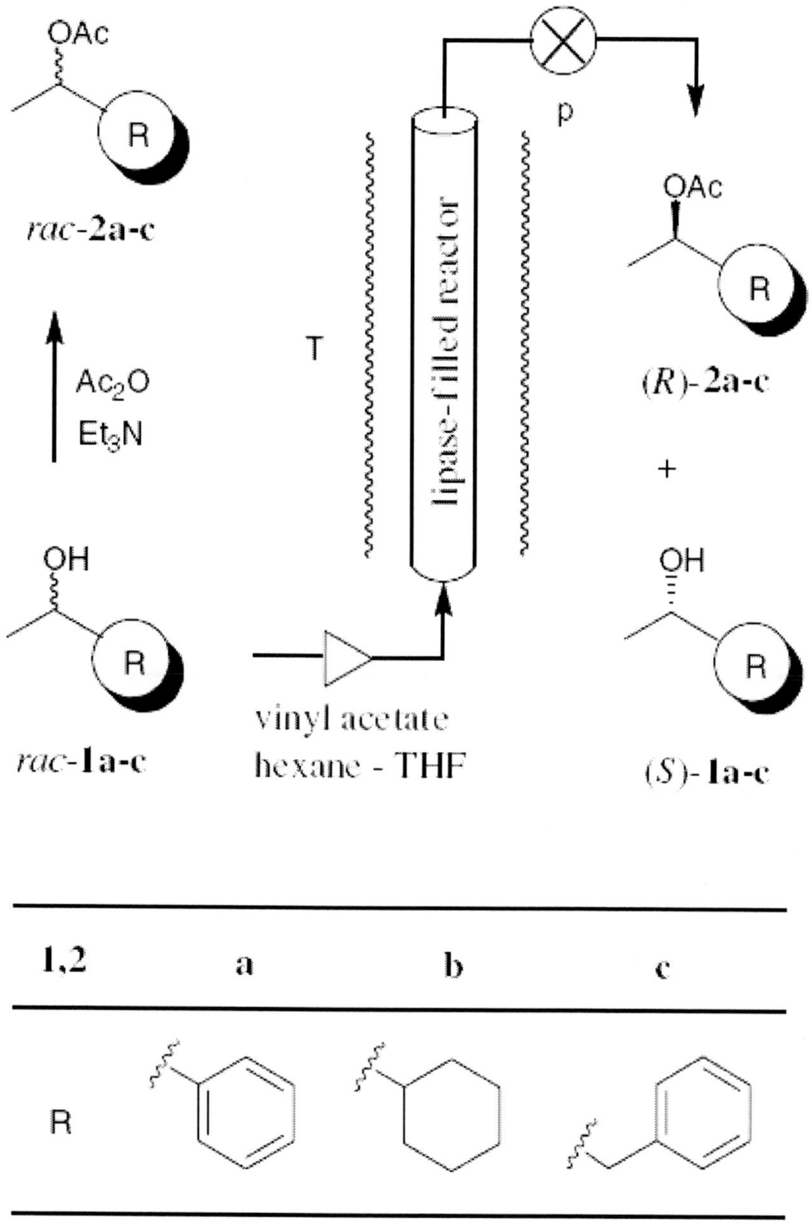

Figure 7. Acylation by enzyme filled reactor [25].

The investigation into the optimal parameters of catalytic hydrogenolysis reactions using the *O*-debenzylation of *O*-benzyl protected *N*-Boc-tyrosine and Cbz deprotection (Figure 8) in a flow reactor system was carried out by Knudsen et al. [27]. A series of experiments over 10% Pd/C catalyst was performed to probe the effects of flow rate, temperature and concentration. As a result a faster catalytic deterioration rate was observed with the increasing concentration and flow rate. However, although the increasing pressure had little effect, the increasing temperature had a significant effect in maintaining catalyst activity.

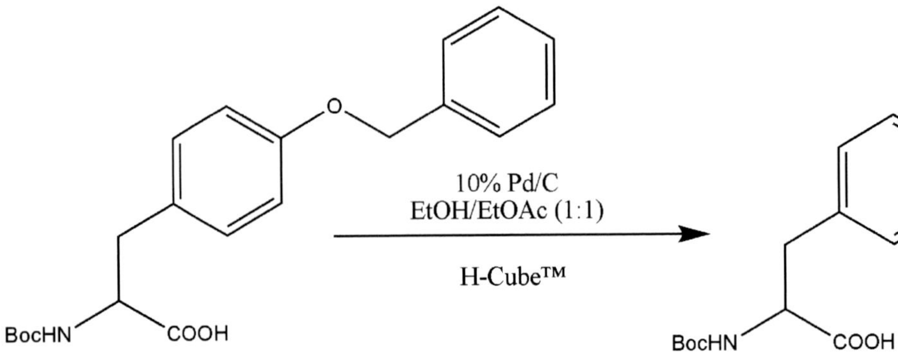

Figure 8. Deprotection under continuous flow mode [27].

In the case of *O*-debenzylation, 100% conversion was reached when the concentration of reactant solution was 0.1 M, the applied temperature was 60 °C, and the flow rate was 1 mL/min. The conditions where the Cbz deprotection was optimal were 0.05 M substrate concentration, 80 °C reaction temperature, 1 mL/ min flow rate and 1 bar hydrogen pressure over 10% Pd/C catalyst. To establish general conditions that would be suitable for the hydrogenation of a small compound library of *N*-Cbz derivatives, ε-[*N*-Cbz] lysine *tert*-butyl ester as a test substrate was selected. A small library of eight *N*-Cbz protected compounds was then compiled. In all cases the deprotected derivatives were isolated in good yields and high purities. The Cbz deprotection reactions were performed with high conversions, between 95 and 99%. In all cases the deprotection derivatives were isolated in good yields and high purities.

In these experiments palladium on carbon was used as catalyst, but other metals and supports are also used for hydrogenation such as platinum, which has been evaluated on the hydrogenation of tetralin and hydrogenolation of neopentane in flow mode by Williams and co-workers [28]. Support of the catalyst was varied, adding silica to the alumina support and testing the pure supports independently. The varying electronegativities of the supports gave diverse results.

Liquid-Solid Reactions

Comparison Between Different Types of Reactor

Yube et al. [29,30] have investigated a comparison between prepacked and catalytic wall reactors (Figure 10), two of the most widespread used reactor types, in performing selective hydroxylation reactions. To develop the best microchannel (Figure 11) and bed reactor design, a series of structures and designs were prepared and tested with the advantages/disadvantages stated in each case. In the case of the prepacked reactor, the titanium-silicate-1 catalyst has to be granulated before filled into the column, so a higher pressure drop was observed than in the catalytic wall reactors. When a packed bed microreactor was used, catalyst particles sieved to 100-150 μm were placed in a tabular tube (inner diameter = 2.0 mm; bed length = 20.0 mm). A selective oxidation reaction (Figure 9) was carried out at 60°C using low concentration of hydrogen peroxide and phenol in aqueous solvent to minimize the presence of benzoquinone. The reaction mixture emerged at the end of the used microreactors and collected product was diluted with distilled water and analyzed with several methods.

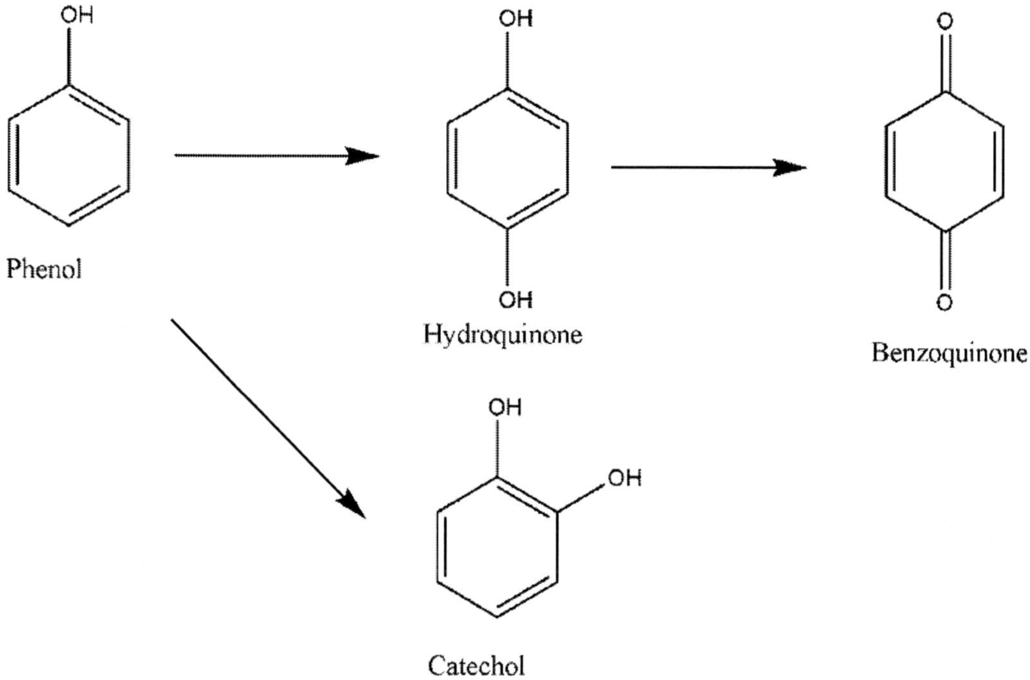

Figure 9. Reaction scheme of phenol oxidation with hydrogen peroxide [29].

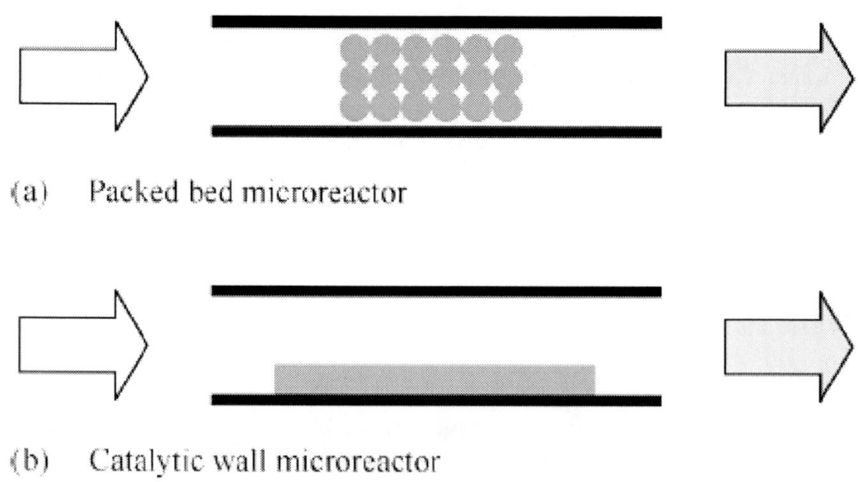

Figure 10. Reactor designs in phenol oxidation reaction [30].

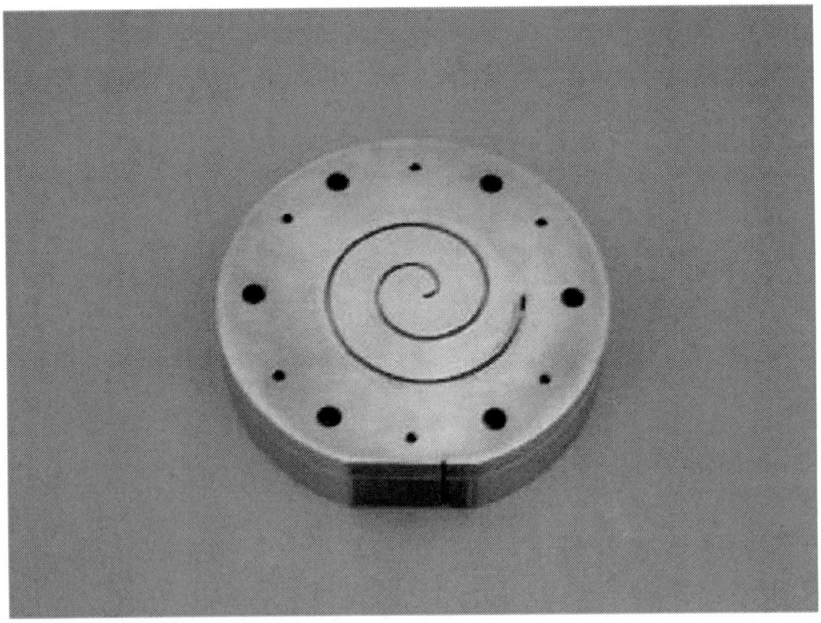

Figure 11. Best characteristic of wall reactor (channel size = 0.8 mm(height); 1.0 mm(width); 113.0 mm(length), channel geometry = vortex) [30].

Compared to the batch synthesis, higher conversion was obtained using a prepacked microreactor with a residence time of 0.5 min. However, after 2h reaction time the yield of benzodiols decreased, but catalyst could be regenerated using hydrogen peroxide.

A catalytic wall reactor, consisting of catalyst and microchannel plates, also contains catalyst without the need for granulation. As the residence time can be varied between 0.1-2.0 min, the main products using the catalytic wall reactor were the benzodiols. In summary, the catalytic wall reactor was found to be better than the prepacked reactor in terms of operability because of the easier catalyst replaceability and the low pressure loss. In terms of regioselectivity, the catalytic wall reactor gave better results; higher *para*-selectivity was achieved with larger variation. When the wall reactor was used the effect of temperature was investigated. The results showed that at higher temperatures, conversion of phenol increased but the ratio of *para/ortho* products was lower. Varying the physical parameters of the wall reactor showed that *para* selectivity was higher with increased height, however the prepacked reactor resulted in less selectivity than the smallest microchannel.

Peptide Synthesis in Flow Mode

Cbz, Boc and Fmoc protected dipeptides were synthesized (Figure 12) by Baxendale et al. [31] using immobilized reagents and scavengers under flow conditions. Boc and Cbz protected dipeptides were prepared using the same methodology by passing the solution of protected amino acid, DIPEA and PyBroP through the polymer supported HOBt filled glass Omnifit column, which has the covalently attached active ester. The solution of the HCl salt amino acid was passed though three different reagent filled columns at a flow rate of 100 μL/min. The first was a PS-DMAP filled column followed by the activated amino ester filled one. Finally for scavenging any unreacted amine, a MP-SO_3H filled column was used and the product was collected at the end. For preparation of Fmoc protected amino acids a similar method was utilized with minor modification. In all cases products can be produced with excellent purity and yield. For the third amino acid coupling to the existing protected dipeptide, the protecting groups have to be removed. It is well know that Cbz can be easily removed by utilizing another flow technology process: flow hydrogenation using H-Cube®, which was also developed for heterogeneous catalysis reactions. After deprotection of dipeptide, the tripeptide was synthesized using the previous process with 59% yield.

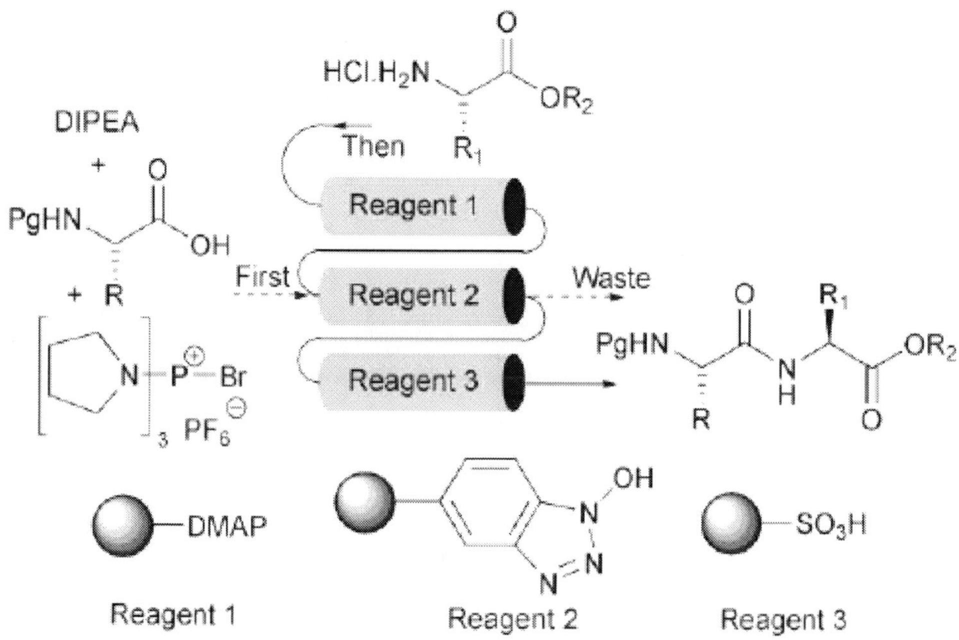

Figure 12. Peptide synthesis under continuous flow conditions [31].

Synthesis of Oxazole Derivates

Successful continuous flow synthesis of 4,5-disubstituted oxazoles (Figure 12) was demonstrated by Baumann et al. [32] where the solution of starting materials was passed through two reagent/scavenger filled cartridges. Different isocyanate derivates were reacted with substituted acyl chlorides in the presence of supported base (PS-BEMP) followed by a scavenging process (with QP-BZA) resulting in excellent purity with high yield (83-99%). Product collection and evaporation of acetonitrile solvent were fully automated. Furthermore sulfonated and phosphonated oxale derivates were also synthesized using the same methodology, resulting in excellent yield (81-94%) using the Vaportech R-4 flow reactor. To investigate the scale-up possibilities of the system, two different column sizes were used. The column for smaller-scale reactions (6 × 30 mm id.; ~ 0.3 g of PS-BEMP) is capable of producing ≤ 100 mg of desired compound and the larger column (17.5 × 45 mm id.; ~ 15 g of PS-BEMP) 5-10 g compound.

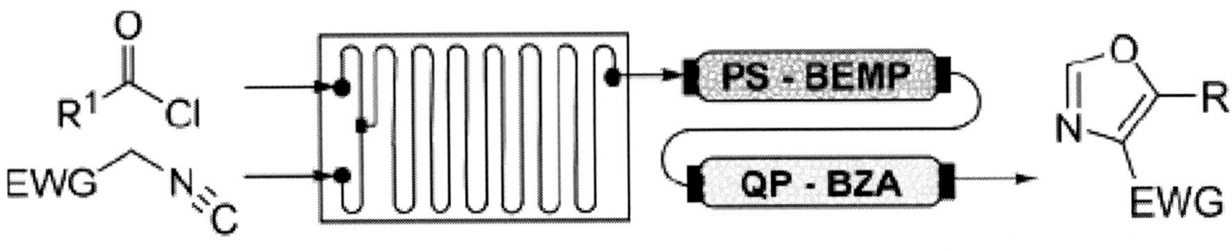

Figure 13. Synthesis of 4,5-disubstituted oxazoles [32].

CROSS-COUPLING REACTIONS

Kumada Reaction

Different cross-coupling reactions have been performed using flow, such as the Kumada reaction (Figure 14) by Phan et al. [33] using an immobilized salen-type nickel (II) catalyst on functionalized silica. The mini flow reactor (Figure 15), consisting of a syringe pump and a catalyst filled column, can be utilized only at low pressure and at room temperature, but the flow rate of the solvent can be changed between 6-33 µL/min. In this study, the reaction between 4-bromoanisole and phenyl-magnesium chloride in tetrahydrofuran was carried out without the need of nitrogen, which is necessary when performed in batch. Catalyst was filled into the column and a mixture of starting material was passed though the column, and the product collected at end of the reactor. At an optimized flow rate, namely 13 µL/min, the conversion rate reached 64% with the minimum presence of 4,4'-dimethoxy-biphenyl by-product. Unfortunately after 5 hours reaction time a conversion of only 30% is achieved.

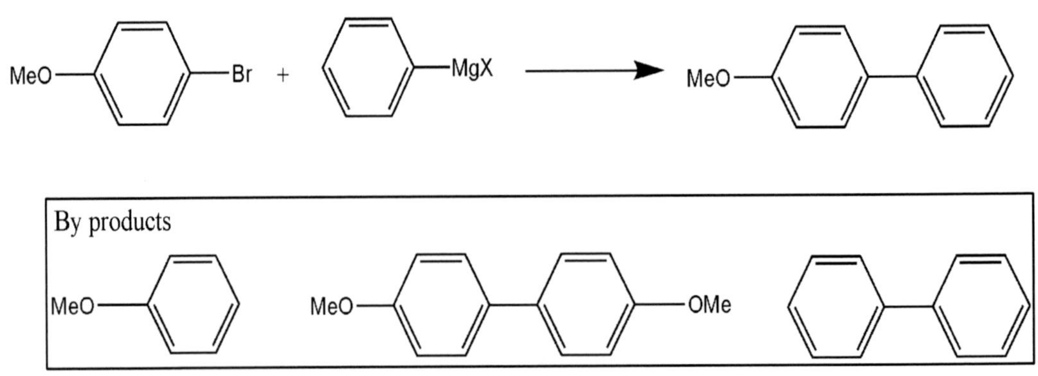

Figure 14. Reaction scheme of Kumada reaction [33].

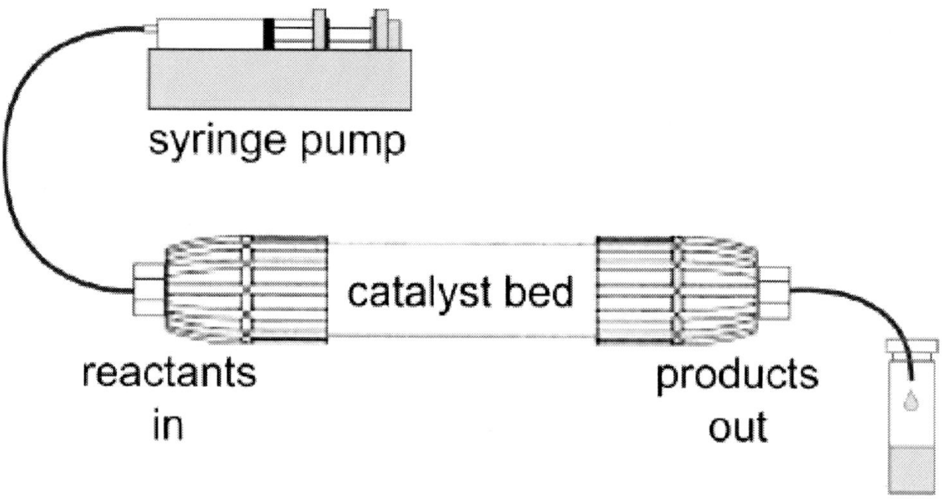

Figure 15. Reactor design in Kumada reaction [33].

Suzuki-Miyaura Reaction

Another way to produce biphenyl derivates using flow was described by Leeke et al. [34] where they performed a Pd catalyzed Suzuki-Miyaura synthesis in the presence of a base. First experiments were carried out in toluene/methanol solvent. A reaction mixture was passed through the encapsulated Pd filled column; bed length 14.5 cm (some cases 10 cm) × 25.4 mm id.; 45 g of PdEnCat. Base concentration, temperature and flow rate were optimized and at optimum parameters (0.05 M base concentration, 100°C and 9.9 mL/min) the conversion was 74%. Then the reaction was performed under supercritical conditions using supercritical CO_2 at high pressure and temperature. After optimizing the concentration of base, flow rate, pressure and temperature, the highest conversion rate (81%) was observed at 166 bar and 100°C where the reactant mixture was monophasic in the supercritical state. This system is able to produce 0.06 g/min of the desired product.

Knoevenagel Condensation

Another important chemical application, the Knoevenagel condensation reaction (Figure 17, 19) was also carried out in flow mode by Jackson [35], Zhang [36], Nikbin and Watts [37]. In the work carried out by Jackson et al, silica coating on plate (Figure 16) modified with base was used and the solventless reaction mixture of ethyl cyannoacetate and benzaldehyde was passed through the reaction zone heated to 90°C at a flow rate of 6.6 μL/min. Conversion was high and remained nearly constant for 8 hours reaction time.

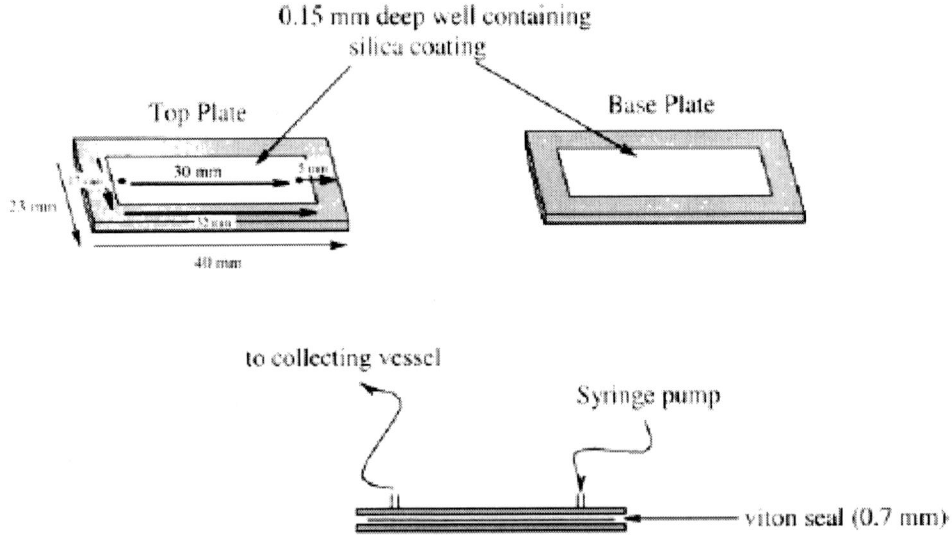

Figure 16. Flowcell reactor constructed from silica coated aluminium plates sealed with a viton spacer [35].

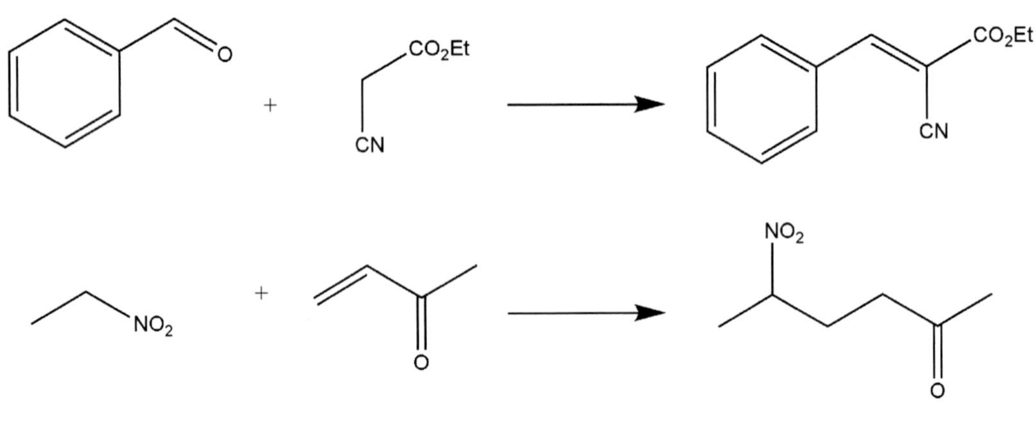

Figure 17. Test reactions of Knoevenagel and Michael reactions [35].

Zhang et al. performed the same reaction but they used different zeolite catalysts and manually filled a microreactor consisting of 35 multichannels (dimension of channel: 25 mm × 25 mm, 0.2 μm pores; Figure 18) and a stainless steel housing. Aminopropylated CsNaX catalyst was found to be the most efficient for the Knoevenagel condensation. Using a catalyst coated multi-channel membrane plate, the activity of the catalyst was unchanged for 60 h and yield of 2-cyano-3-phenylacrylate was 19.8% at 100°C. The zeolite catalyst was also used in a packed bed reactor and a comparative study between the two types of microreactors showed that the packed bed reactor as a function of residence time resulted in higher yield but less productivity compared to the multi-channel reactor.

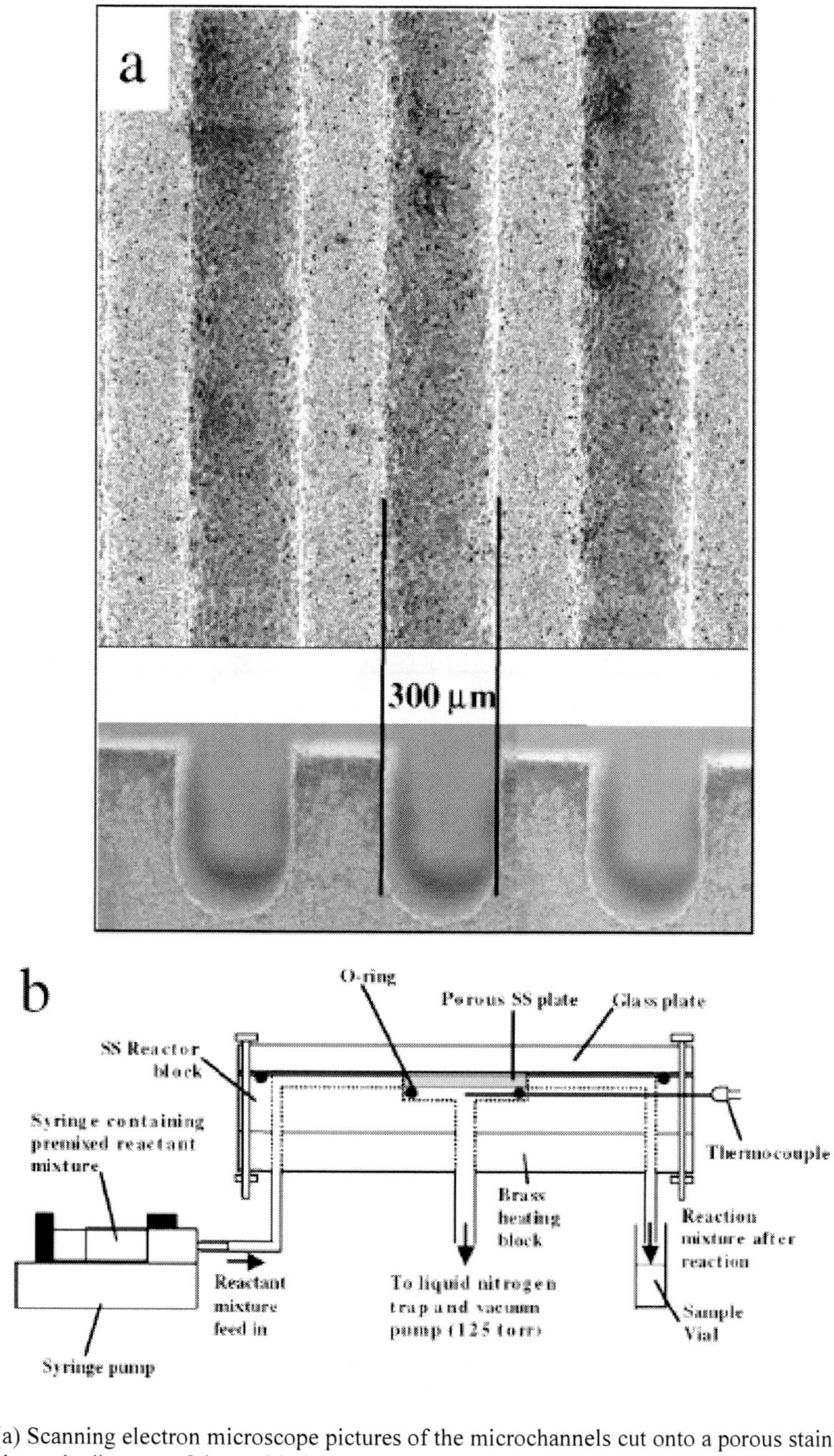

Figure 18. (a) Scanning electron microscope pictures of the microchannels cut onto a porous stainless steel. (b) Schematic diagram of the multi-channel microreactor set-up [36].

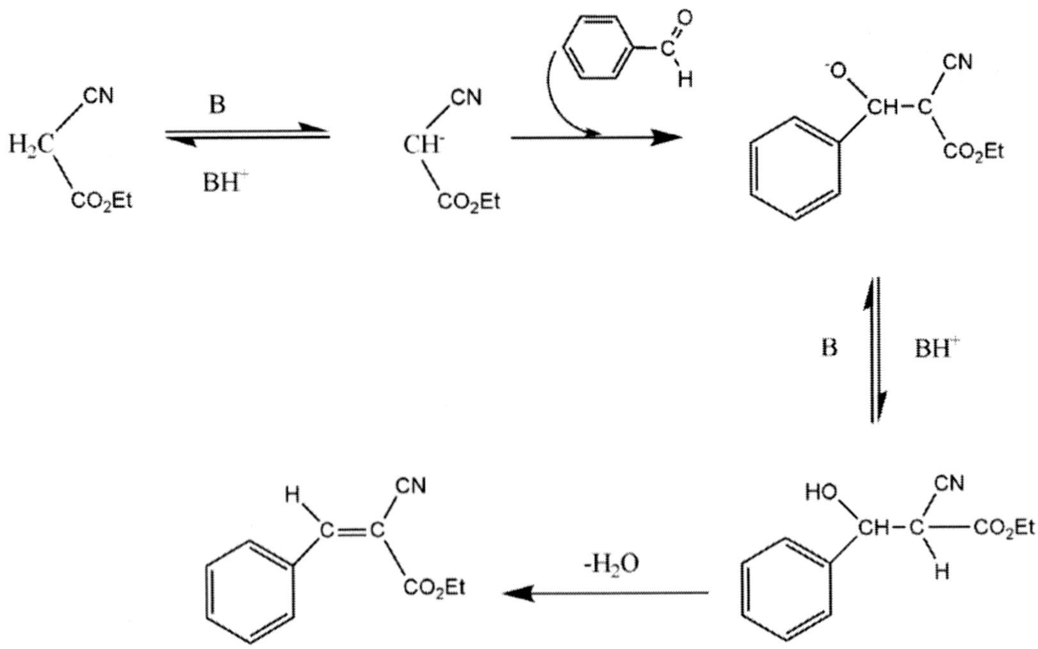

Figure 19. Reaction scheme for solid-base catalysed Knoevenagel condensation reaction [36].

Finally this reaction type was also performed by Watts's group in a prepacked bed reactor (Figure 20) where the reaction solution was controlled using electroozmotic flow. This method has been used previously by others [38]. The reaction took only 5 minutes to complete at room temperature. The reproduced and optimized reaction was successfully carried out with a conversion rate of 82% using immobilized piperazine onto benzyl chloride functionalized silica gel. After optimization of reaction conditions, a series of other Knoevenagel condensation reactions were performed using ethyl cyanoacetate and more aldehydes. To investigate the scale-up possibility of the reactor system, reaction time was increased to 20 minutes without effect on the conversion.

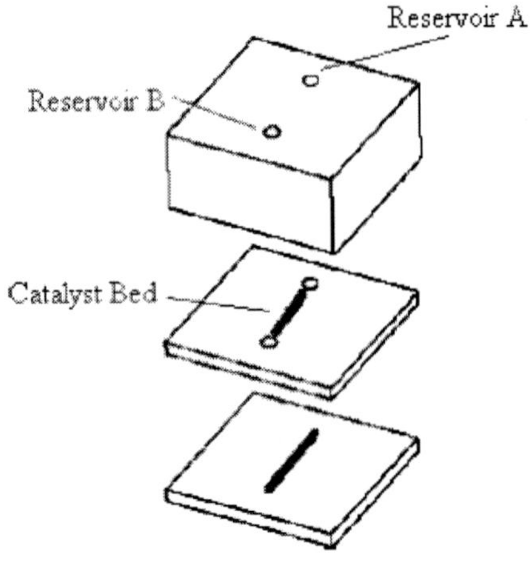

Figure 20. Schematic design of microreactor [37].

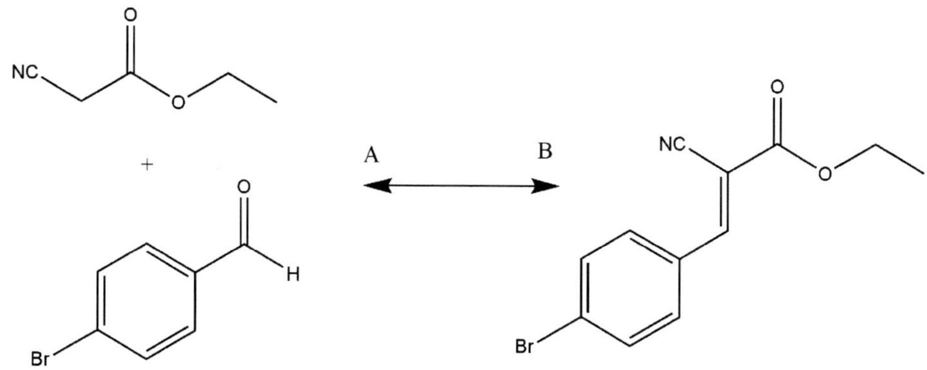

Figure 21. Microreactor manifold for Knoevenagel reactions [37].

Michael Addition

In conjunction with the Knoevenegal reaction, a Michael addition (Figure 22) reaction was also described by Jackson et al. using the same silica supported tertiary amines for the catalysis of conjugate addition of nitroalkenes to *α,β*-unsaturated carbonyl compounds. At a flow rate of 6.6 μL/min the conversion rate was constant with a high yield for 7 hours reaction time.

In an inhouse prepared reaction tube (Figure 23), a Michael reaction was carried out by Bonfils et al. [39] and enantioselective synthesis was also discussed. A test reaction between 1-oxoindan-2-carboxylate and methyl vinyl ketone was performed in batch and flow mode and adopted for asymmetric synthesis. Gel type beads were filled into a 55 mL volume of reactor tube (glass tube 36 cm long and 14 mm wide). A solution of starting materials in toluene were introduced separately to the bottom of the tube and then passed up through the fluid bed of beads and the product collected at the top of the bed using syringe. In the first experiment Amberlyst A21, an ion exchange resin, was filled into a bath heated tube. At 7.0 mL/min the average yield was 98% after 72 hours reaction time. Finally, the reaction was adopted for enantioselective synthesis using cinchona alkaloids supported to polystyrene beads. This chiral inductor was also used in the enantioselective continuous-flow triphasic hydrogenation of ethylpyruvate [40] over Pt/Al_2O_3. Reactions with supported cinchonidine were detailed resulting in the corresponding (*S*)-enantiomer. Both in batch and flow mode the immobilized catalyst was active for long time. In flow mode the catalyst was active for 72 hours at a flow rate of 5 mL/min giving a high yield (96-98%) and enantiomer excess (47-52%). Interestingly, the immobilization had only a slight effect on the enantiomer excess.

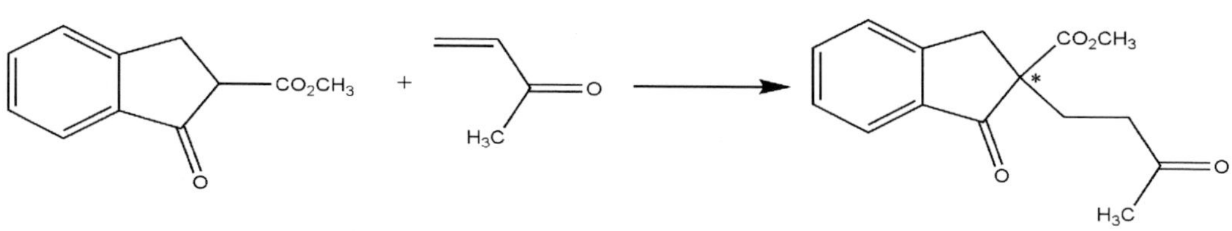

Figure 22. Michael reaction [39].

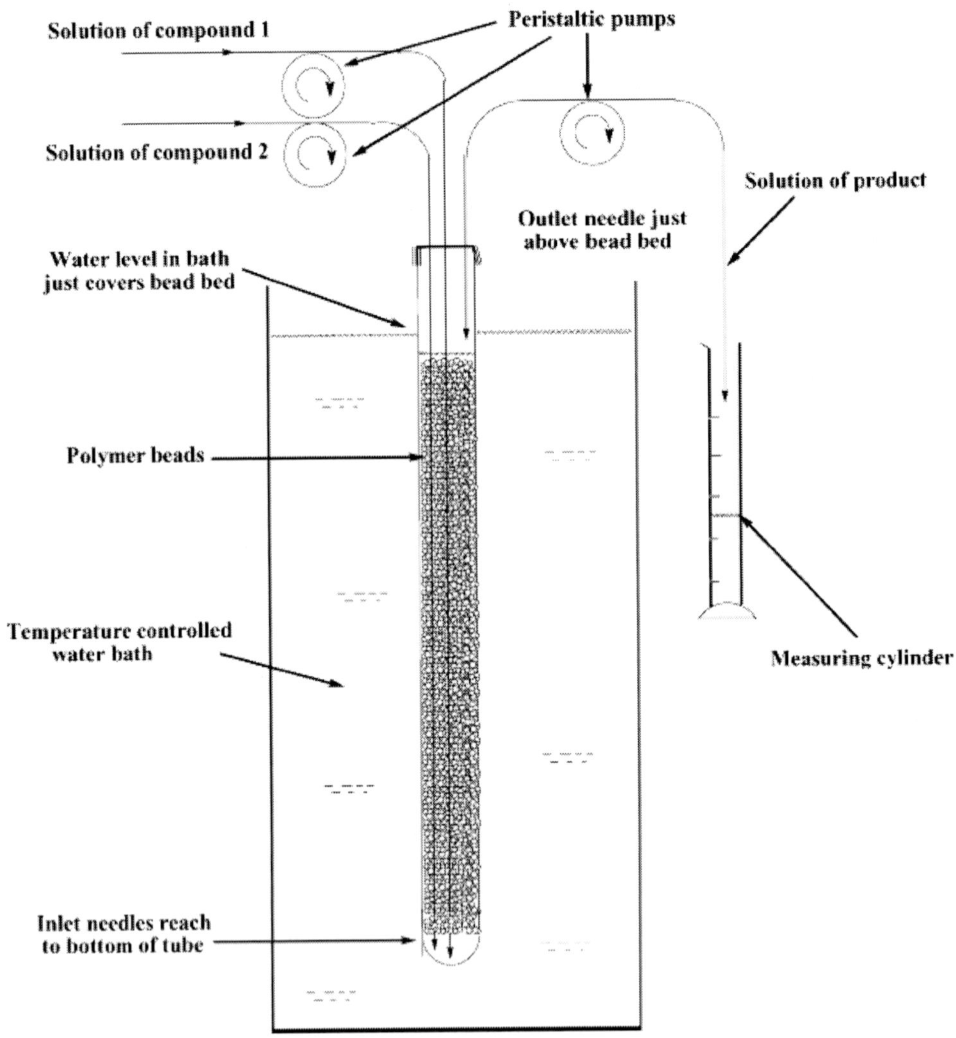

Figure 23. General arrangement of the flow reactor [39].

Protection of Ketones and Aldehydes

Ketones and aldehydes can be protected as dithioacetals and –ketals (Figure 25). Moreover, the corresponding products are used in many further reactions such as in C-C coupling reactions. Wiles and co-workers at the University of Hull [41] presented a solution for selective aldehyde protection in the presence of a ketone group under continuous flow conditions. Firstly, the protection of aldehyde was optimized using a prepacked reactor with Amberlyst A-15 catalyst. Reactor design allowed the group to utilize a wide range of flow rates (0.1 – 500.0 μL/min). The pumping mechanism utilized was electroosmotic flow, which generates minimal back-pressure when compared to pressure driven flow. To perform the reactions the borosilicate glass capillary was packed with the catalyst and the packed bed was filled with anhydrous MeCN (Figure 24). Reservoir A was filled with the mixture of starting materials and Reservoir B with an aliquot of solvent where the product will be collected. Using 2-phenyl-1,3-dithiane synthesis as a model reaction, reaction parameters were optimized (optimum flow rate: 63.7 μL/min) to reach 100% conversion. Once optimization was successful, more aldehydes were reacted with 2 dithiols (1,2-ethanedithiol and 1,3-propanedithiol) to produce the corresponding dithiolanes and dithianes. The library was synthesized with excellent yields. Furthermore, although the protection of highly substituted ketones is kinetically less favorable, the reactions resulted in excellent yields. Comparative batch reaction was performed resulting in lower yields after 24 h reaction time compared to a residence time of 76 s in flow mode. In summary, 128.5 mmol of thioacetals and ketals were synthesized using 0.231 mmol of catalyst. Finally, selective aldehyde protection of 4-acetylbenzaldehyde in the presence of a ketone was achieved, under reaction conditions previously optimized for the protection of aldehydes, resulting in the desired product with quantitative yield.

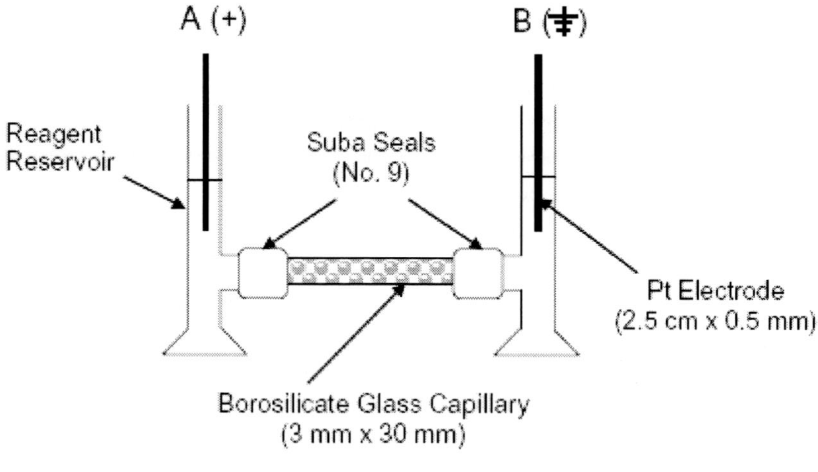

Figure 24. Schematic illustrating the reaction set-up [41].

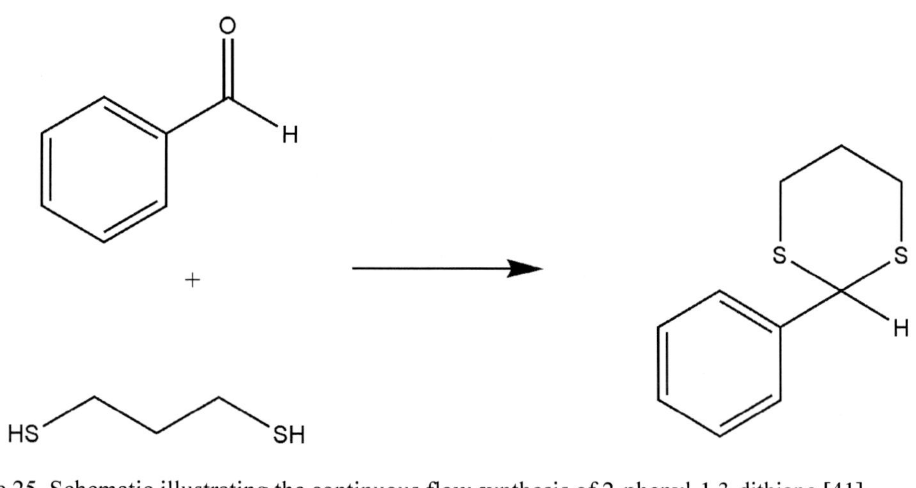

Figure 25. Schematic illustrating the continuous flow synthesis of 2-phenyl-1,3-dithiane [41].

ENANTIOSELECTIVE REACTIONS

Enantioselective Synthesis of Grossamide

The first enantioselective synthesis of grossamide was implemented by Baxendale et al. [42] under continuous flow conditions by coupling tyramide and ferrulic acid followed by oxidative dimerization and intramolecular cyclization. The first synthesis step was carried out in an automated fashion where a solution of ferrulic acid, PyBrOP and DIPEA was passed though a PS-HOBt filled column to prepare the activated ester. Then the solution of the tyramide was pumped through this column and the product was analyzed with inline UV/VIS indicating that 90% of conversion could be reached. After a scavenging process, using a sulfonic acid resin filled column, the emerging solution was mixed with hydrogen peroxide in a buffer and passed towards a peroxidase filled column where the oxidation and cyclization steps were performed. At the end of the process, online LCMS measurement showed that this process is capable of synthesizing gram quantities of compound without manual purification steps between each step.

Enantioselective Glyoxylate-Ene Reaction

Mandoli et al. [43] described how flow heterogeneous catalytic reactions using chiral catalysts can result in enantiomerically pure products using glyoxylate-ene reactions (Figure 27) as an example. Evans et al. had developed a mild homogeneous protocol previously[44,45]. First the Mandoli group prepared the homogeneous and the heterogeneous catalysts, containing copper and a chiral ligand (box class of ligands, Figure 26), from a starting material of modified malonic acid.. The insoluble polymer-bound bis(oxazoline) ligand was successfully prepared with yield of 87% and a ligand loading of 0.22 mmol/g. After suspending the ligand in $Cu(OTf)_2$ solution, a metal uptake of 0.18 mmol/g was estimated. Once the catalysts were prepared, reactions between 1,1-disubstituted alkenes, 1-methylcyclohexene and ethyl glyoxylate were performed and compared to Evans's results. Results from the batch reactions showed that immobilization of ligand had no effect on the asymmetric induction with a reaction time between 6-48 h. A pre-packed column (25 cm x 4.6 mm) filled with catalyst contains 12 ligand and 0.19 mmol of $Cu(OTf)_2$ per gram of ligand was utilized for the flow experiments. A solution of isopropenylbenzene and ethyl glioxylate was passed through the column at a reduced flow rate (0.015 – 0.025 mL/min) at 0°C. Five successive batches of starting materials were fed to the column. Overall, 23 mmol of isopropenylbenzene was converted in 80 h without change in enantiomeric excess. Yield of the desired product was 78% with an enantiomeric excess of 88%.

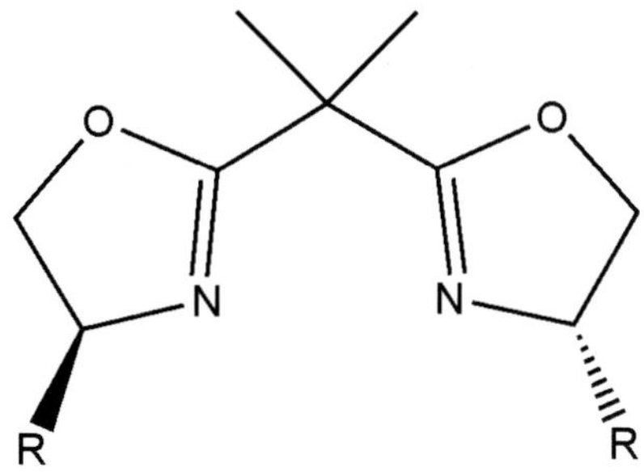

Figure 26. Structure of ligand [43].

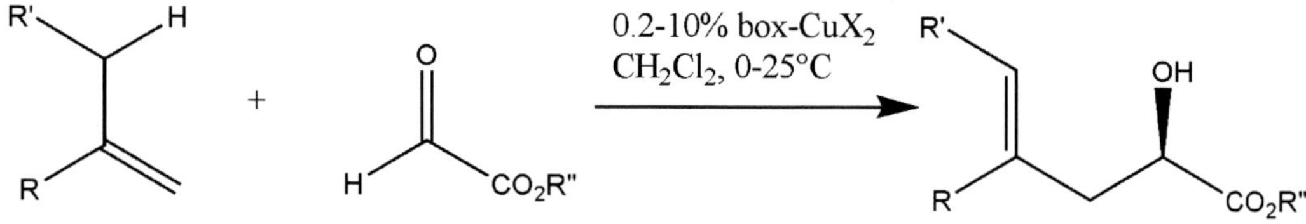

Figure 27. Glyoxylate-ene reaction [42].

HETEROGENEOUS CATALYSIS UNDER SUPERCRITICAL FLOW CONDITIONS

Supercritical solvents have a number of advantages which make them excellent reaction media, such as the low cost, non-toxicity, and low viscosity. These advantages have meant that they are increasingly utilized in reactions. Supercritical solvents can be described as fluids with attributes of both liquids and gases. Solubility of the solute in the fluid depends on the vapour pressure of the solute however, addition of different polar/nonpolar compounds can change the solubility. What makes the supercritical solvent to be so unique is that properties, such as solubility, can be tuned by varying the pressure, so the solvent becomes more gas-like or liquid-like [46].

One the most common supercritical solvents is supercritical carbon-dioxide, because of its relatively low critical parameters ($T_{c,}$= 31.1 °C and p_c = 72.9 bar) and easy and low-cost usability. However, nitrogen gas ($T_{c,}$= -147.0 °C and p_c = 33.5 bar) and ethane ($T_{c,}$= 32.2°C and p_c = 48.2 bar) are also used.

HYDROGENATION UNDER SUPERCRITICAL FLOW CONDITIONS

Kobayashi [47] et al. performed catalytic hydrogenation reactions using a microchannel reactor developed by the group [48]. Supercritical carbon-dioxide dissolved hydrogen was introduced with the substrate solution into the microencapsulated Pd-immobilized glass microchannel reactor (200 μm width, 100 μm depth, 40 cm length) at a constant flow rate of CO_2 (1 mL/min). Using this apparatus, a series of double and triple bond containing compounds were successfully hydrogenated in quantitative yields. In these experiments, the calculated residence time was less than 1 second and the hydrogenation of 0.20 mmol starting material can be performed over 2 hours after immobilizing only 10 μg of palladium contains catalyst to the reactor wall.

HYDROFORMYLATION UNDER SUPERCRITICAL FLOW CONDITIONS

Hydroformylation of 1-octene (Figure 28) was carried out using a homogeneous catalyst [49] , and the same catalyst adsorbed into a microporous solid, in an ionic liquid phase [50] with $scCO_2$, CO, and H_2. The reaction results in the corresponding linear and branched aldehydes. The supercritical carbon dioxide was used to help diffuse the substrates and reagent gases into the ionic liquids. Reaction conditions of 100°C and high pressures (80-150 bar) were used. The turnover frequency (TOF), conversion and linear:branch ratio were were all recorded and a comparison made between the heterogeneous catalyst approach and the homogeneous. In homogeneous flow conditions, the maximum TOF value was 239 (mol catalyst h)$^{-1}$ with a conversion rate of 57%. However, in heterogeneous flow mode, the TOF reached 800 (mol catalyst h)$^{-1}$ with a conversion rate of 100% with a similar ratio of linear to branched aldehydes (3:1) in both experiments. It is important to note that the leaching of the rhodium metal was also measured and was found to be under 1 ppm concentration and that the reaction time can be dramatically reduced using the supported ionic liquid catalyst.

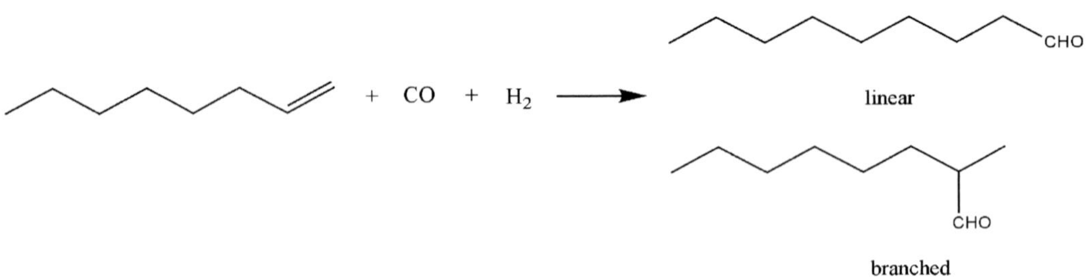

Figure 28. Hydroformylation of 1-octene to aldehyde [50].

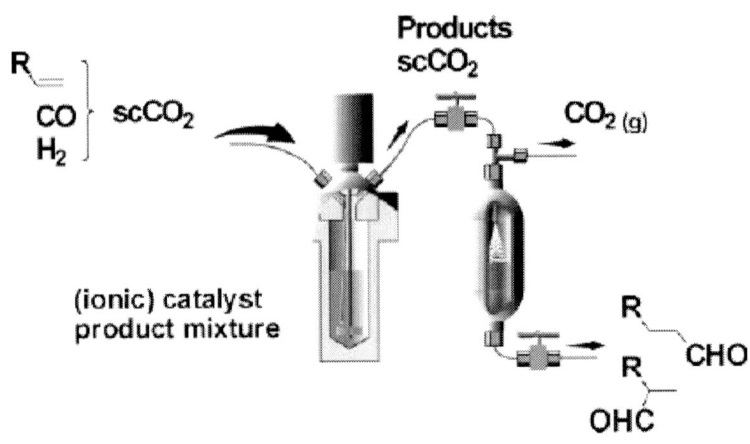

Figure 29. Concept for continuous flow hydroformylation [49].

ENZYMATIC REACTIONS IN SUPERCRITICAL FLOW MODE

A novel continuous-flow $scCO_2$ process for the kinetic resolution of 1-phenyethanol enantiomers (Figure 30) using Novozym 435 immobilized enzyme from *Candida antarctica* was described by Matsuda et al. [51]. The lipase enzyme, selectively acetylated the (*R*)-alcohol component. A mixture of starting material and vinyl acetate was passed through the enzyme with supercritical carbon-dioxide (Figure 31). The reaction zone was pressurized and heated, so the reaction could be performed under supercritical conditions, synthesizing the desired (*R*)-acetate with 99.7% ee. and 47% yield.

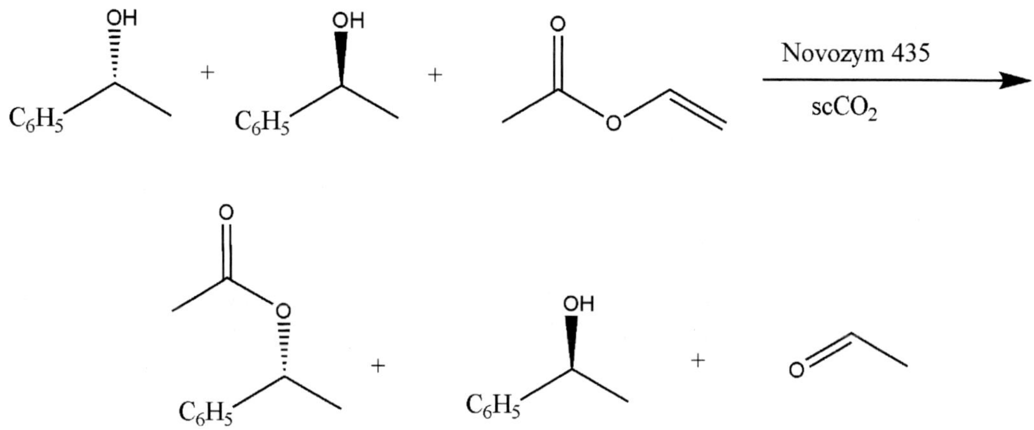

Figure 30. Kinetic resolution by enzyme [51].

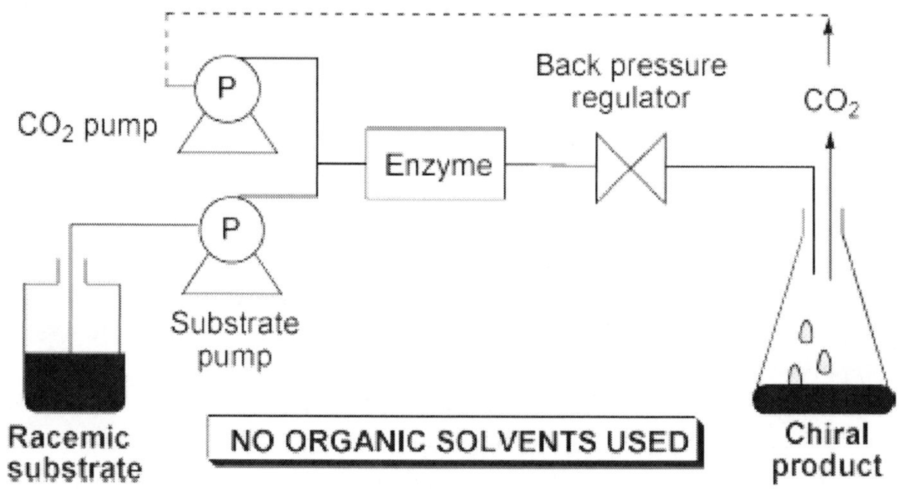

Figure 31. Design of enzyme filled reactor [51].

Microwave Assisted Synthesisunder Continuous-Flow Conditions

In the past decade, microwave assisted synthesis has become a routine technique in many laboratories. Microwave-assisted flow applications for reactions such as acylation [52], esterification [53], biodiesel production [54], coupling reaction [55], Fisher indole synthesis [56], and also for aseptic processes [57, 58] have been developed. The efficiency and advantages of this technique, such as reduced reaction time, controllable and uniform heating, made the use of microwave heating popular, but the technique is limited when it comes to scaling up reactions. To solve this issue two solutions can be proposed: 1.) increased volume of reactant in conjunction with novel instrument development 2.) flow technology combined with microwave heating from the modification of existing instruments. When the volume of the vessel was increased to react more compound at any one time, complications developed with the uniform heating. At the typical operating frequency of most microwave reactors of 2.45 GHz, the penetration depth of microwave radiation is typically only a few centimeters. Therefore, solvent or reagents in the centre of a vessel of a size >1 L are heated by convection and not by direct microwave dielectric heating. Also, as the reaction volume increases more microwave power is required to reach identical conditions. Therefore, most microwave experts consider the solution to microwave scale-up to be continuous-flow.

In recent years, two excellent reviews by the Kappe group [59] and by Bowman et al. [60] gave details about flow scale-up possibilities using microwave heating. Two types of reactor were developed for this purpose: the continuous- and the stop flow reactors. Domínguez et al. [61] used a prepacked flow quartz reactor (length: 45 cm x 2.2 cm i.d.) to utilize microwave irradiation for the decomposition of methane to produce hydrogen. Using 36-65 mL/min flow rate of methane at different temperatures, it was found that at 900°C higher conversion of methane was measured, filling the reactor with activated carbon. When compared to the conventional (electric) heating it was found that microwave heating gave higher conversion at a lower temperature. Although after 2 h at 800°C, a decrease in catalyst activation was noted because of the catalyst deposition.

Cross-Coupling Reactions in Microwave-Flow Mode

Suzuki type coupling reactions (Figure 34) were performed by He et al. [62,63] in three different reactor types (Figure 32,33). Two consisted of plates and the third one was a U-type reactor prepacked with catalyst.

Catalyst with a particle size of 45-63 μm was introduced into the plate reactors resulting in a monolayer catalyst bed over the catalyst channel area. During the reactions, a mixture of reagents were passed through the catalyst region irradiated with 2.45 GHz (0-300 W) using a CEM Discover instrument and product was collected at the other side of the reactor. Retention time, catalyst and starting materials were varied to find the optimum conditions, but authors suggested that this system was far from optimal. Reactor design B was preferred and the group found that Pd catalyst with a gold coating was the best at performing Suzuki reactions using low microwave irradiation to reach an average temperature of 95°C at a higher flow rate (5 μL/min) when compared to design A.

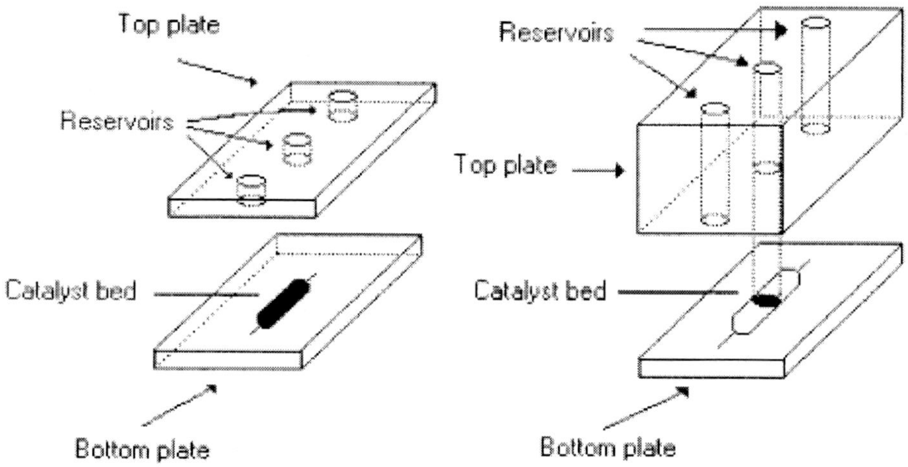

Figure 32. Reactor design A and B. (dimensions of 1.5 mm wide, 80 μm deep and 15 mm long, catalyst: 130 μm wide, 50 μm deep, 15 mm long) [62].

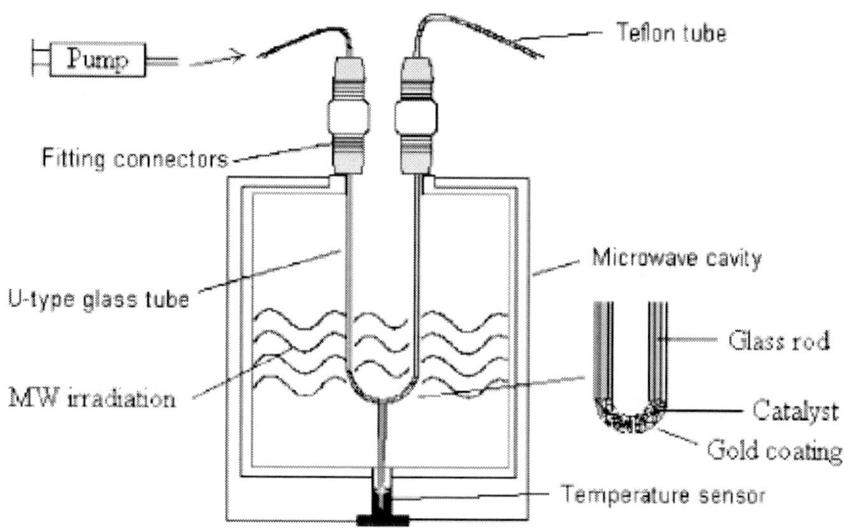

Figure 33. Design of U type reactor (inner diameter 800 μm, total length 138 mm) [63].

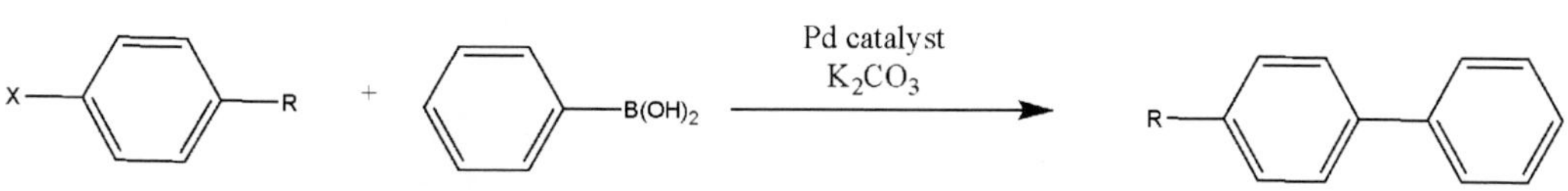

Figure 34. Scheme of Suzuki reaction [62].

It was also discovered that when the group used Pd-polystyrene, which contained 20% less Pd than the Pd-alumina catalyst, higher conversion was achieved.

The third glass U-tube type reactor was then tested using Suzuki reactions. To investigate the best conditions, a model reaction between aryl halide and phenylboronic acid was chosen. The reaction was carried out in the presence of base in DMF/H2O (75:25) using a Pd containing catalyst filled into a U-tube where gold coating was also used. After optimization of parameters, the group found that when using K_2CO_3 as base, Pd/Al_2O_3 as catalyst, low microwave heating to 80°C, and a 0.04 mL/min flow rate resulted in the highest conversion and compared well to the Pd-silica catalyst. Comparison reactions were performed, which proved the necessity for the presence of gold coating and that conventional heating resulted in lower yields than. When the optimum parameters were established, more aryl halides were reacted with phenyl boronic acid giving yields varying between 30-74%.

Shore and coworkers [64] used a capillary reactor with a Pd thin film and microwave-assisted continuous-flow conditions for Suzuki-Miyara and Heck coupling reactions. The Pd film was prepared by passing $Pd(OAc)_2$ solution into the 1150 μm capillary at 150°C resulting in a highly porous catalyst composed of nanometer-size grains.

When the capillary was ready for reaction, a mixture of aryl boronic acid, aryl bromide and base in solvent was flowed through the capillary and heated using microwave irradiation from a Biotage Smith Creator Synthesizer (Figure 36) to perform the Suzuki-Miyara reaction (Figure 35). Control experiments proved that the microwave irradiation was necessary for the reactions to progress in good yields most cases.

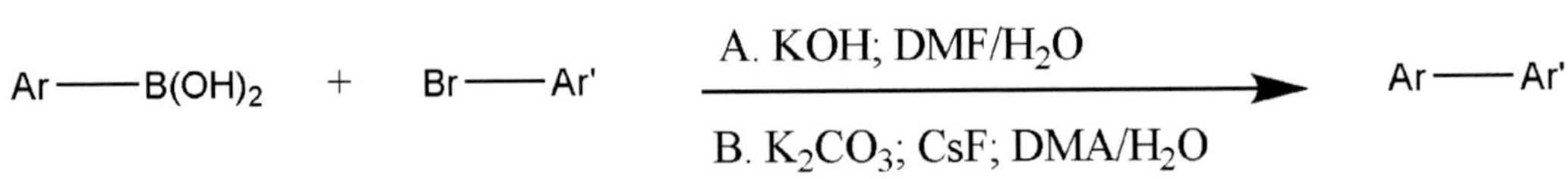

Figure 35. Scheme of Suzuki-Miyara reaction [64].

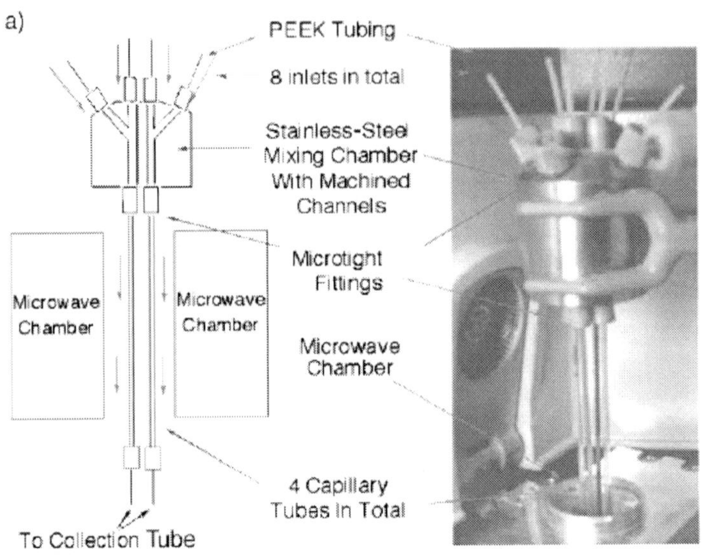

Figure 36. Design of continuous flow microwave reactor [64].

After these experiments, Heck reactions between aryl halides and acrylates were investigated in the presence of triethylamine as base in dimethylacetamide at 200°C using the same techniques and Pd-film capillary reactor described previously. The reactions typically resulted in high conversions. This process was capable of producing 10 g of desired product in minutes and scale-up of the reaction can be achieved by flowing large volumes of starting materials through the reactor without modification of optimized conditions.

Oxidation Reactions Under Microwave-Flow Conditions

Bo and co-workers [65] investigated a methodology where phenol derivates can be oxidized using microwave-assisted continuous-flow as a technique for removing phenolic pollutants in aqueous solution. Oxidation was carried out by mixing air with a phenolic solution and passing it into a granular activated carbon-support palladium filled column. Catalyst preparation was well detailed and resulted in a 1% Pt content surface with a particle size range of 0.1-0.8 μm. Catalyst characterization showed that the carbon surface had the same microwave irradiation absorbance as the Pt dispersed one. Before the reaction, catalyst was pretreated with phenolic pollutant to abate the absorption effect of the granulated activated carbon. In the experiments, degradation of two starting materials, *p*-nitrophenol and pentachlorophenol, were achieved using 400 W of microwave irradiation and an influent flow of 6.4 mL/min (hydraulic retention time of 16 min) at atmospheric pressure. When *p*-nitrophenol was introduced to the system, after 5 h reaction time a removal efficiency of 85-86% was observed using the platinum containing catalyst compared to the pretreated activated carbon where just 63.5-69% was observed.

Batch Versus Flow Conditions

One of the major advantages of flow reactors is the short residence time of reactions in the reactor. This allows selective reactions to pass through the system and out again before any side reaction can take place. This is very well illustrated in the synthesis of dithioketal and –acetals, where the selective reaction resulted in superior conversion using continuous flow when compared to batch synthesis. A dramatic increase of yield was noted in the hydrogenation reaction performed by Kobayashi et al. when a residence time of less than 1 min was used. The yield increased from 1% to 97% using the continuous-flow reactor.

The use of continuous-flow for rapid optimization followed by scale up was demonstrated by Steven Ley's group [66]. Imine reductions were performed in the presence of Pd/C as catalyst on the H-Cube flow hydrogenation system. After optimization of conditions on small quantities, 1.0 g of desired product was synthesized within 70 min with quantitative yield and excellent purity. Worth noting is that the reactions progressed quantitatively without the need for further purification.

When Novozyme 435 enzyme was filled into a column to achieve kinetic resolution of enantiomers, comparative batch reactions were also carried using supercritical carbon dioxide. The results showed that both the enantiomeric excess (90.6% in batch; 99.5% in flow for (S)-alcohol) and the conversion (48% in batch and 50% in flow) were higher when performing the reaction in continuous-flow

Hydrogenation is widely used for deprotection of benzyl and benzyloxycarbonyl groups. Kappe's group [67,68] deprotected benzyl esters using continuous flow and then performed comparative experiments using conventional heating and microwave assisted transfer hydrogenation, which resulted in lower yields (53-65%) than when compared to flow (80-85%).

CONCLUSION

These above examples show the widespread utilization of heterogeneous catalysts under flow conditions including its advantages. In some of the examples, employing a heterogeneous catalyst eliminated time-consuming purification steps and reduced leaching of the active catalyst. Comparisons to reactions performed with homogeneous catalysts showed that higher conversions and more selective reactions can be performed using heterogeneous catalysts. Moreover distribution of reaction products can be manipulated by changing the support or other properties of the catalyst.

Combining advantages of heterogeneous catalysts and flow conditions gives us a new synthetic solution and opens the door to more efficient and greener chemistry with reduction in the time spent performing and optimizing reactions. Low quantities of compounds can be used, allowing the user to gain maximum information from using minimal amounts of chemicals. Also, when performing reactions in batch, reaction parameters are more limited and less closely controlled, whereas in continuous flow, reaction parameters can be performed more precisely, resulting in better reproducibility. This is particularly important when performing highly exothermic reactions. Furthermore, combination of flow reactors with automated technologies allows the chemist to automate reactions, which weren't possible before when performed in batch.

The number of examples utilizing heterogeneous catalysts with flow from the literature has increased substantially in the last several years. The different applications with which this technique has been applied to from frequently used lab reactions to exploring the efficient, easy and practical use of supercritical media, shows this area to be very adaptable and progressive. Either way, flow catalysis will continue to have a large amount of focus for years to come.

REFERENCES

[1] Yoshida, J. and Nagaki, A. and Iwasaki, T. and Suga, S. (2005). Enhancement of Chemical Selectivity by Microreactors. *Chem. Eng. Technol., 28 (3),* 259-266.

[2] Haswell, S.J. and Middleton, R.J. and O'Sullivan, B. and Skelton, V. and Watts, P. and Styring, P. (2001). The application of micro reactors to synthetic chemistry. *Chem. Commun., 5*. 391-398.

[3] Wirth, T. (Ed.), *Microreactors in Organic Synthesis and Catalysis*. Weinheim, Germany: Wiley-VCH)

[4] Pennemann, H. and Hessel, V. and Löwe, H. (2004). Chemical microprocess technology – from laboratory-scale to production. *Chemical Engineering Science, 59*, 4789-4794.

[5] De Bellefon, C. and Tanchoux, N. and Caravieilhes, S. and Grenouillet, P. and Hessel, V. (2000). Microreactors for Dynamic, High Throughput Screening of Fluid/Liquid Molecular Catalysis. *Angew. Chem. Int. Ed., 39 (19),* 3442-3445.

[6] Kirschning, A. and Jas, G. (2004). Applications of Immobilized Catalysts in Continuous Flow Processes. *Topics in Current Chemistry, 242,* 209-239.

[7] Haswell, S.J. and Watts, P. (2003). Green chemistry: synthesis in micro reactors. *Green Chemistry, 5,* 240-249.

[8] Mae, K. (2007). Advanced chemical processing using microspace. *Chemical Engineering Science, 62*, 4842-4851.

[9] Engler, M. and Kockmann, N. and Kiefer, T. and Woias, P. (2004). Numerical and experimental investigation on liquid mixing in static micromixer. *Chemical Engineering Journal, 101*, 315-322.

[10] Nagasawa, H. and Mae, K. (2006). Development of a New Microreactor Based on Annular Microsegments for Fine Particle Production. *Ind. Eng. Chem. Res., 45,* 2179-2186.

[11] Griffiths-Jones, C.M. and Hopkin, M.D. and Jönsson, D. and Ley, S.V. and Tapolczay, D.J. and Vickerstaffe, E. and Ladlow, M. (2007). Fully Automated Flow-Through Synthesis of Secondary Sulfonamides in a Binary Reactor System. *J. Comb. Chem., 9,* 422-430.

[12] Hodge, P. (2005). Synthesis of Organic Compounds Using Polymer-Supported Reagents, Catalysts, and/or Scavengers in Benchtop Flow Systems. *Ind. Eng. Chem. Res., 44*, 8542-8553.

[13] Ley, S.V. and Ramarao, C. and Gordon, R.S. and Holmes, A. B. and Morrison, A.J. and McConvey, I.F. and Shirley, I.M. and Smith, S.C. and Smith, M.D. (2002). Polyurea-encapsulated palladium(II) acetate: a robust and recyclable catalyst for use in conventional and supercritical media. *Chem. Commun., 10*, 1134-1135.

[14] Webb, P.B.and Sellin, M.F. and Kunene, T.E. and Williamson, S. and Slawin, A.M.Z. and Cole-Hamilton, D.J. (2003). Continuous Flow Hydroformylation of Alkenes in Supercritical Fluid-Ionic Liquid Biphasic System. *J. Am. Chem. Soc., 125,* 15577-15588.

[15] Oku, T. and Arita, Y. and Tsuneki, H. and Ikariya, T. (2004). Continuous Chemoselective Methylation of Functionalized Amines and Diols with Supercritical Methanol over Solid Acid and Acid-Base Bifunctional Catalysts. *J. Am. Chem. Soc., 126*, 7368-7377.

[16] Keskin, S. and Kayrak-Talay, D. and Akman, U. and Hortacsu, Ö. (2007). A review of ionic liquids towards supercritical fluid applications. *J. of Supercritical Fluids, 43,* 150-180.

[17] Tundo, P. and Selva, M. (2005). Continuous-flow, gas phase synthesis of 1-chlorobutane (1-bromobutane) from 1-butanol and aqueous HCl (HBr) over silica-supported quaternary phosphonium salt. *Green Chem., 7*, 464-467.

[18] Ge, H. and Chen, G. and Yuan, Q. and Li, H. (2005). Gas phase catalytic partial oxidation of toluene in a microchannel reactor, *Catalysis Today, 110*, 171-178.

[19] Rossignol, C. and Arrii, S. and Morfin, F. and Piccolo, L. and Caps, V. and Rousset, J. (2005). Selective oxidation of CO over model gold-based catalysts in the presence of H_2. *Journal of Catalysis, 230*, 476-483.

[20] Kobayashi, J. and Mori, Y. and Kobayashi, S. (2005). Triphase Hydrogenation Reactions Utilizing Palladium-Immobilized Capillary Column Reactors and a Demonstration of Suitability for Large Scale Synthesis. *Adv. Synth. Catal.*, 347, 1889-1892.

[21] Yoswathananont, N. and Nitta, K. and Nishiuchi, Y. and Sato, M. (2005). Continuous hydrogenation reactions in a tube reactor packed with Pd/C. *Chem. Commun., 1,* 40-42.

[22] Halder, R. and Lawal, A. (2007). Experimental studies on hydrogenation of antraquinone derivative in a microreactor. *Catalysis Today,125,* 48-55.

[23] Csajagi, Cs. and Borcsek, B. and Niesz, K. and Kovacs, I. and Szekelyhidi, Zs. and Bajko, Z. and Urge, L. and Darvas, F. (2008). High-Efficiency Aminocarbonylation by Introducing CO to a pressurized Continuous Flow Reactor. *Org. Lett., 10 (8),* 1589-1592.

[24] Maruyama, T. and Uchida, J. and Ohkawa, T. and Futami, T. and Katayama, K. and Nishizawa, K. and Sotowa, K. and Kubota, F. and Kamiya, N. and Goto, M. (2003). Enzymatic degradation of *p*-chlorophenol in a two-phase flow microchannel system. *Lab Chip, 3,* 308-312.

[25] Csajagi, Cs. and Szatzker, G. and Toke, E.R. and Urge, L. and Darvas, F. and Poppe, L. (2008). Enantiomer selective acylation of racemic alcohols by lipases in contunous-flow bioreactors. *Tetrahedron: Asymmetry, 19,* 237-246.

[26] Clapham, B. and Wilson, N.S. and Michmerhuizen, M.J. and Blanchard, D.P. and Dingle, D.M. and Nemcek, T.A. and Pan, J.Y. and Sauer, D.R. (2008). Construction and Validation of an Automated Flow Hydrogenation Instrument for Application in High-Throughput Organic Chemistry. *J. Comb. Chem., 10,* 88-93.

[27] Knudsen, K.R. and Holden, J. and Ley, S.V. and Ladlow, M. (2007). Optimization of Conditions for *O*-Benzyl and *N*-Benzyloxycarbonyl Protecting Group Removal using an Automated Flow Hydrogenator. *Adv. Synth. Catal., 349,* 535-538.

[28] Williams, M.F. and Fonfé, B. and Woltz, C.and Jentys, A. and van Veen, J.A.R. and Lercher, J.A. (2007). Hydrogenation of tetralin on silica-alumina-supported Pt catalysts II. Influence of the support on catalytic activity, *Journal of Catalysis, 251,* 497-506.

[29] Yube, K. and Furuta, M. and Aoki, N. and Mae, K. (2007). Control of selectivity in phenol hydroxylation using microstructured catalytic wall reactors. *Applied Catalysis A: General, 327*, 278-286

[30] Yube, K. and Furuta, M. and Mae, K. (2007). Selective oxidation of phenol with hydrogen peroxide using two types of catalytic microreactor, *Catalysis Today, 125*, 56-63.

[31] Baxendale, I.R. and Ley, S.V. and Smith, C.D. and Tranmer, G.K. (2006). A flow reactor process for the synthesis of peptides utilizing immobilized reagents, scavengers and catch and release protocol. *Chem. Comm., 46*, 4835-4837.

[32] Baumann, M. and Baxendale, I.R. and Ley, S.V. and Smith, C.D. and Tranmer, G.K. (2006). Fully Automated Continuous Flow Synthesis of 4,5-Disubstituted Oxazoles. *Org. Lett., 8(23),* 5231-5234.

[33] Phan, N.T.S. and Brown, D.H. and Styring, P. (2004). A facile method for catalyst immobilisation on silica: nickel catalysed Kumada reactions in mini-continuous flow and batch reactors, *Green Chemistry, 6,* 526-532.

[34] Leeke, G.A. and Santos, R.C.D. and Al-Duri, B. and Seville, J.P.K. and Smith, C.J. and Lee, C.K.Y. and Holmes, A.B and McConvey, I.F. (2007). Continuous-Flow Suzuki-Miyaura Reaction in Supercritical Carbon Dioxide. *OPRand D, 11*, 144-148.

[35] Jackson, T. and Clark, J.H. and Macquarrie, D.J. and Brophy, J. H. (2004). Base catalysts immobilised on silica coated reactor walls for use in continuous flow systems. *Green Chem., 6,* 193-195.

[36] Zhang, X. and Man Lai, E.S. and Martin-Aranda, R. and Yeung, K.L. (2004). An investigation of Knoevenagel condensation reaction in microreactors using zeolite catalyst, *Applied Catalysis A: General, 261*, 109-118.

[37] Nikbin, N. and Watts, P. (2004). Solid-Supported Continuous Flow Synthesis in Microreactors Using Electroosmotic Flow. *OPRand D, 8*, 942-944.

[38] Greenway, G.M. and Haswell, S.J. and Morgan, D.O. and Skelton, V. and Styring, P. (2000). The use of novel microreactor for high throughput continuous flow organic synthesis. *Sensors and Actuators B, 63,* 153-158.

[39] Bonfils, F. and Cazaux, I. and Hodge, P. and Caze, C. (2006). Michael reaction carried out using a bench-top flow system, *Org. Biomol. Chem., 4,* 493-497.

[40] Abdallah, R. and Fumey, B. and Meille, V. and de Bellefon, C. (2007). Micro-structured reactors as a tool for chiral modifier screening in gas-liquid-solid asymmetric hydrogenation. *Catalysis Today, 125,* 34-39.

[41] Wiles C. and Watts, P. and Haswell, S. J. (2007). An efficient, continuous flow technique for the chemoselective synthesis of thioacetals. *Tetrahedron Letters, 48,* 7362-7365.

[42] Baxendale, I.R. and Griffiths-Jones, C.M. and Ley, S.V. and Tranmer, G.K. (2006). Preparation of the Neolignan Natural Product Grossamide by a Continuous-Flow Process. *Synlett, 3,* 427-430.

[43] Mandoli, A. and Orlandi, S. and Pini, D. and Salvadori, P. (2004). Insoluble polystyrene-bound bis(oxazoline): batch and continuous-flow heterogeneous enantioselective glyoxylate-ene reaction. *Tetrahedron: Asymmetry, 15,* 3233-3244.

[44] Evans, D.A. and Burgey, C.S. and Paras, N.A. and Vojkovsky, T. and Tregay, S.W. (1998). C_2-Symmetric Copper(II) Complexes as Chiral Lewis Acids. Enantioselective Catalysis of the Glyoxylate-Ene Reaction, *J. Am. Chem. Soc., 120,* 5824-5825.

[45] Evans, D.A. and Tregay, S. W. and Burgey, C.S. and Paras, N.A. and Vojkovsky, T. (2000). C_2-Symmetric Copper(II) Complexes as Chiral Lewis Acids. Enantioselective Glyoxylate-Ene Reaction with Glyoxylate and Pyruvate Esters. *J. Am. Chem. Soc., 122,* 7936-7943.

[46] Hyde, J.R. and Licence, P. and Carter, D. and Poliakoff, M. (2001). Continuous catalytic reactions in supercritical fluids. *Applied Catalysis A: General. 222.* 119-131.

[47] Kobayashi, J. and Mori, Y. and Kobayashi, S (2005). Hydrogenation reactions using $scCO_2$ as a solvent in microchannel reactors. *Chem. Comm., 20*, 2567-2568.

[48] Kobayashi, Y. and Mori, Y. and Okamoto, K. and Akiyama, R. and Ueno, M. and Kitamori, T. and Kobayashi, S. (2004). A Microfluidic Device for Conducting Gas-Liquid-Solid Hydrogenation Reactions. *Science, 304*, 1305-1306.

[49] Frisch, A.C. and Webb, P.B. and Zhao, G. and Muldoon, M. J. and Pogorzelec, P.J. and Cole-Hamilton, D.J. (2007). "Solventless" continuous flow homogeneous hydroformylationof 1-octene. *Dalton Trans., 47*, 5531-5538.

[50] Hintermair, U. and Zhao, G. and Santini, C.C. and Muldoon, M.J. and Cole-Hamilton, D.J. (2007). Supported ionic liquid phase catalysis with supercritical flow. *Chem. Commun., 14*, 1462-1464.

[51] Matsuda, T. and Watanabe, K. and Harada, T. and Nakamura, K. and Arita, Y. and Misumi, Y. and Ichikawa, S. and Ikariya, T. (2004). High-efficiency and minimum-waste continuous kinetic resolution of racemic alcohols by using lipase in supercritical carbon dioxide. *Chem. Commun., 20*, 2286-2287.

[52] Marquié, J. and Salmoria, G. and Poux, M. and Laporterie, A. and Dubac, J. and Roques, N. (2001). Acylation and Related Reactions under Microwaves. 5. Development to Large Laboratory Scale with a Continuous-Flow Process. *Ind. Eng. Chem. Res., 40*, 4485-4490.

[53] Shieh, W. and Dell, S. and Repic, O. (2002). Large scale microwave-accelerated esterification of carboxylic acids with dimethyl carbonate. *Tetrahedron Letters, 43*, 5607-5609.

[54] Barnard, T.M. and Leadbeater, N.E. and Boucher, M.B. and Stencel, L.M. and Wilhite, B.A. (2007). Continuous-Flow Preparation of Biodiesel Using Microwave Heating. *Energy and Fuels, 21*, 1777-1781.

[55] Comer, E. and Organ, M.G. (2005). A Microreactor for Microwave-Assisted Capillary (Continuous Flow) Organic Synthesis. *J. Am. Chem. Soc., 127,* 8160-8167.

[56] Bagley, M.C. and Jenkins, R.L. and Lubinu, M.C. and Mason, C. and Wood, R. (2005). A Simple Continuous Flow Microwave Reactor. *J. Org. Chem., 70*, 7003-7006.

[57] Kumar, P. and Coronel, P. and Truong, V.D. and Simunovic, J. and Swartzel, K.R. and Sandeep, K.P. and Cartwright, G. (2007). Overcoming Issues Associated with the Scale-up of a Continuous Flow Microwave System for Aseptic Processing of Vegetable Purees. *Food Research International, Available online 23 November 2007.*

[58] Coronel, P. and Simunovic, J. and Sandeep, K.P. and Cartwright, G.D. and Kumar, P. (2008). Sterilization solutions for aseptic processing using a continuous flow microwave system. *Journal of Food Engineering, 85*, 528-536.

[59] Glasnov, T. N. and Kappe, C.O. (2007). Microwave-Assisted Synthesis under Continuous-Flow Conditions. *Macromol. Rapid Commun., 28,* 395-410.

[60] Bowman, M.D. and Holcomb, J.L. and Kormos, C.M. and Leadbeater, N.E. and Williams, V.A. (2008). Approaches for Scale-Up of Microwave-Promoted Reactions, *OPRand D, 12,* 41-57.

[61] Domínguez, A. and Fidalgo, B. and Fernández, Y.and Pis, J.J.and Menéndez, J.A. (2007) Microwave-assisted catalytic decomposition of methane over activated carbon for CO_2-free hydrogen production. *International Journal of Hydrogen Energy, 32*, 4792-4799.

[62] He, P. and Heswell, S.J. and Fletcher, P.D.I. (2004). Microwave heating of heterogeneously catalysed Suzuki reactions in a micro reactor. *Lab Chip, 4*, 38-41.

[63] He, P. and Heswell, S.J. and Fletcher, P.D.I. (2004). Microwave-assisted Suzuki reactions in a continuous flow capillary reactor. *Applied Catalysis A: General, 274*, 111-114.

[64] Shore, G. and Morin, S. and Organ, M.G. (2006). Catalysis in Capillaries by Pd Thin Films Using Microwave-Assisted Continuous-Flow Organic Synthesis (MACOS). *Angew. Chem. Int. Ed., 45,* 2761-2766.

[65] Bo, L. and Quan, X. and Wang, X. and Chen, S. (2008). Preparation and characteristics of carbon-supported platinum catalyst and its application in the removal of phenolic pollutants in aqueous solution by microwave-assisted catalytic oxidation. *Journal of Hazardous Materials, Available online 12 January 2008*

[66] Saaby, S. and Knudsen, K.R. and Ladlow, M. and Ley, S.V. (2005). The use of a continuous flow-reactor employing a mixed hydrogen-liquid stream for the efficient reduction of imines to amines. *Chem. Commun., 23*, 2909-2911.

[67] Desai, B. and Kappe, C.O. (2005). Heterogeneous Hydrogenation Reactions Using a Continuous Flow High Pressure Device, *J. Comb. Chem., 7*, 641-643.

[68] Desai, B. and Dallinger, D. and Kappe, C.O. (2006). Microwave-assisted solution phase synthesis of dihydropyrimidine C5 amides and esters, *Tetrahedron, 62,* 4651-4664.

INDEX

#

A

B

C

F

G

H

I

M

N

O

P

Q

R

T

U

V

W

X

Y

Z